AF478592

ENERGY AND ENVIRONMENTAL PROGRESS - I

Volume B

SOLAR ENERGY APPLICATIONS, BIOCONVERSION AND SYNFUELS

ENERGY AND ENVIRONMENTAL PROGRESS - I

Volume A - Direct Solar Energy

Volume B - Solar Energy Applications, Bioconversion and Synfuels

Volume C - Indirect Solar, Geothermal and Nuclear Energy

Volume D - Hydrogen Energy and Power Generation

Volume E - Conservation and Heat Transfer

Volume F - Environment and Energy

Volume G - Societal Issues and Energy/Environment Economics

ENERGY AND ENVIRONMENTAL PROGRESS - I

Volume B

SOLAR ENERGY APPLICATIONS, BIOCONVERSION AND SYNFUELS

Edited by T. Nejat Veziroğlu

Nova Science Publishers
New York

Nova Science Publishers, Inc.
283 Commack Road
Suite 300
Commack, New York 11725

Library of Congress Cataloging-in-Publication Data available upon request

ISBN 0-941743-97-7

Graphic Design by Elenor Kallberg and Peggy Harvey

Printed in the United States of America

STAFF

Conference Coordinator	Lucille Walter
Executive Secretaries	Donna Pressley Aymara Schmidt
Graduate Assistants	Mehmet Akcin Frano Barbir Sezgin Gürsu Labeed Kazi Nabil Lutfi Tamay Özgökmen Manoj Padki Muhammad T. Syed
Undergraduate Assistants	Charlene Blaisure Kristin Serrati

Contents

HEATING AND COOLING

ENERGY TRANSMISSION

PHOTOVOLTAICS

SOLAR POWER

INDUSTRIAL APPLICATIONS

BIOCONVERSION

SYNFUELS

HEATING AND COOLING

NEW SOLAR HEAT INSTALLATIONS AND THE UTILIZATION THEREOF

Z. A. Kabilov, M. I. Umarova, V. P. Tsoi, G. Sh. Khamedov, V. R. Mangasaryan, and R.Sh. Burnashev
Physical - Technical Institute of Tadjik SSR
Academy of Sciences, 734063 Aini Street, 299/1, Dushanbe, USSR

The energy and the environment protection problems in most cases are confined to those, associated with the equipment impact on the environment. This impact can be simulated in terms of thermodynamics and it can be treated on the engineering level providing some principles which are basic for the creation of new latest generation technique.

In treating the installations for heat supply from this aspect all of them despite its variety can be divided into two classes:

- those provided for conversion and redistribution of energy flows already existing in the environment, and
- those provided for the formation of new energy flows.

The device of the first class is presented, for example, by the solar collector or the solar collector-based heat supply system.

The second class presents chemical and nuclear energy sources.

From the viewpoint of the environmental safety best installations are those relating to the first class; their development should be given a major attention.

The present paper deals with the specific example of the practical utilization of solar energy heat supply systems based on such a phenomenon as the liquid-vapour phase transition as well as some engineering solutions for the design of convection open-circuit heat supply installations with air heat carrier.

THE SOLAR WATER PUMP

Fig.1 shows a diagram of the solar water pump provided,mainly,for local agricultural irrigation in foothill and mountainous regions. Apart from elevating water this installation heats the pumped water to some optimal temperature necessary for irrigation. The specific feature of the installation compared to conventional solar heaters with flat-plate collector systems is that its absorption panel has two groups of channels so arranged as to provide the heat exchange between heat carriers,contained in the channels through the panel without intermixing thereof.

The solar water pump comprises a flat-plate collector with the said absorption panel,the suction and expelling chambers,each comprising an elastic piston,for example,the rubber bellows. The input and the output of the pump is provided with back valves. The inside of the elastic piston of the expelling chamber is provided with the easily boiling liquid. The condenser is introduced within the same cavity and presents a portion of the pressure pipe with the extended surface.

The solar water pump operates as follows. The absorption of the solar energy in the flat-plate collector causes the heating of the pumped water in one of the collector groups,shown in the drawings by the solid line,as well as of the auxilliary heat carrier in the closed loop,shown by the dashed line.

The pre-degassed water can be utilized as the auxilliary heat carrier. In the closed loop the circulation of the auxilliary heat carrier and the warm-up of the expelling chamber to the boiling temperature of the easily boiling liquid occur. This results in the expansion in volume of the expelling chamber bellows while the heated auxilliary heat carrier is displaced into the bellows of the suction chamber which in its extension extends the cold water from suction chambers through corresponding channels of the absorption panel and the condenser into the pressure pipe to the user. The easily boiling liquid keeps boiling during the time the water heated in the flat-plate collector to the same temperature as the auxilliary heat carrier passes through the condenser. However as soon as the temperature of the water in the condenser falls below that of the phase transition the condensation initiates. The bellows is compressed and a new portion of water sucks in through input and back valves, correspondingly, and a new cycle takes place.

SOLAR WATER HEATERS

The natural circulation systems are generally not used to heat up a large volume of water, e. g. in solar premises heating systems; pumps, providing a forced circulation are used instead. However an additional source of energy, e. g. electric energy as well as energy utilizing automation equipment is needed in this case. However the demand for additional energy sources in some cases can not be satisfied. For this reason the utilization of the portion of the heat energy, obtained in flat-plate solar collectors for its conversion into the mechanical energy and the stimulation of the forced circulation mode is preferrable.

In fig. 2 and 3 two diagrams of solar water heaters designed by the authors within the framework of the regional program for the protection of the environment are shown. The "Gera" type installation contains a conventional solar collector, connected to the intermediate storage, the inner portion of which contains a closed elastic chamber (bellows) with the easily boiling liquid. The volume of the intermediate storage approximately equals that of the heat carrier, contained in the channels of the flat-plate collector. Above the intermediate storage is the air chamber connected to it and the auxilliary tank with the swimmer lever-driven valve of the air chamber disposed at the same level as the air chamber. The auxilliary tank is connected through the oppositely connected back valves to the main and the auxilliary storages.

The installation operates as follows. In the initial state the installation is filled up to the level of the air chamber. In the natural circulation mode the water contained in the intermediate storage is heated up to the boiling temperature of the easily boiling liquid. The volume of the bellows in the process of heating increases and a portion of hot water is displaced to the main storage, while the cold water from the main storage is passed to the auxilliary tank. The swimmer gliding along the rod is raised up to the upper stop and by this actuates the valve drive lever of the air chamber. The valve opens, the air chamber is communicated to the environment. In accordance with the communicating vessels the cold water thus begins to overflow from the auxilliary tank to the air chamber through the back valve and the cavity of the intermediate storage. With that the bellows begins to cool, the easily boiling liquid is condensed and the system is turned to the initial state with the only difference being that the cold water appears in the intermediate storage.

Fig. 3 shows another modified version of the solar water heater. This installation comprises the same parts as those of the above solar heater. The intermediate storage comprises a boiler, containing the easily boiling liquid. In the initial state this system is filled up with water up to the level of the delaying capacitance. As soon as the temperature of the intermediate storage reaches that of the easily boiling liquid the latter boils up and its vapours begin to press the elastic diaphragm, while the air from the air chamber begins to displace the heated up portion of water into the main storage. The cold water from the main storage first fills the delaying capacitance, then passes to the condenser and through it to the auxilliary tank. The time the delaying capacitance is filled determines the duration of the complete cycle. The passage of cold water through the condenser provides the condensation of vapours of the easily boiling liquid, the elastic diaphragm returns to its initial state, while the cold water from the auxilliary tank, the condenser and the delaying capacitance drains off to the intermediate storage and the cycle is renewed.

The described solar water heaters are provided, mainly, for solar premises with seasonal heat accumulation. They do not demand additional energy sources and operate in the automated mode.

SOLAR DRIERS

The mode of natural circulation can not be obtained in practice for stationary driers, so that fans are usually used, which demand for the additional energy source. To exclude the demand for such sources the authors design the structure of driers in which the generation of the convection flow is obtained also by the heat energy.

Fig. 4 shows the air circulation in a stationary drier, equipped with transverse circulation air flow heat exchanger. The dry air heated up in a flat-plate collector passes through the longitudinal channels of the heat exchanger while transferring a portion of its heat to the humid air, passing from the drying chamber, by which the air draught in the exhaust tube is provided. The dry air is further submitted to the underground gravelly accumulator and still further to the drying chamber which is to be of the sealed type. In the night time the drier operates due to the heat stored up in the gravelly accumulator. As the calculations show the air rate and its thermal parameters, obtained in the above drier are sufficient for a high quality drying of fruits, herbs or wood in summer.

Fig. 5 shows a diagram of the solar drier, utilizing a combined air collector. The surface of the exhaust air conduit, located inside the drier chamber, serves a portion of the absorption panel. This permits the utilization of a portion of the absorbed solar energy to warm up the air in longitudinal exhaust air conduit and in the transversal exhaust tube which are interconnected. As the drying chamber is sealed in a manner similar to that described in a previous section the draught in the air exhaust tube actuates the air along the complete solar collector -North wall - underground accumulator - drying chamber - exhaust air conduit - exhaust tube circuit. By their desighn the above driers can be referred to self - regulation systems as the air rate and the temperature are inversely related.

SOLAR PREMISES

The additional expenses for the provision of solar heat will be soon compensated if the solar premise collector operates for the whole year running,thereby providing cooling of the premise in summer and its heating in winter. With that the arrangements,which do not comprise fans,demanding for additional energy sources and operating in the automated mode are more preferrable. To realize the above requirements a premises air conditioning system of the "Alfa" type is designed by the authors.

Fig.6 shows simplified diagrams of air flow circulation in the heating mode (fig.6a) and that in the cooling mode (fig.6b). The set-up of "Alfa" type utilizes an air collector with two chambers for air circulation above the absorption panel and that below it. In the heating mode the environmental air or the waste air from the ventillation system is heated up within the interior cavity of the lower portion of the collector,passes through the channels of the ceiling in the North wall of the premise,below the floor of the South wall to the interior space of the upper portion of the collector,whereby it is additionally heated to provide the draught. The heat accumulation is provided within the volume of walls and ceiling. In the cooling mode the air from the ventillation system passes over the wet heat conduction surface of the cooling chamber to the outer portion of the collector in which the draught is provided by the heating and therefrom is exhausted to the environment. Meanwhile the heating of air in the interior cavity of the upper portion of the collector provides the draught in the circuit: cooling chamber - floor channels - channels in the North wall - channels in the ceiling - the interior upper cavity of the collector. As the result the cooled air,passing through the channels provided in the walls and ceilings cools them,thereby creating comfort within the premise.

The combination of heating and cooling functions,the provision of the premise parts with additional heating engineering functions,according to the authors,provides a construction of solar energy premises of the new type.

HEAT PUMPS

The widely known constructions of heat pumps designed for heat supply comprise freon as a working medium which deteriorates the environment. Their utilization in the solar energy heat supply systems demands for the utilization of large - sized heat exchangers of sophisticated design. Hence,the search for alternative versions has been undertaken in past years as to the utilization of harmless working fluids and the desighn of heat pumps of the new type. According to the authors,the offered diagram permits a combination of simplicity and reliability of solar energy heat supply systems and the advantages of heat pumps within one assembly with the simultaneous safety for the environment.

Fig. 7 shows the diagram of the solar energy heat supply system operating according to the principle of the heat pump. The specific feature of the offered construction is the utilization of the ejector as the compressor. It is actuated by the conventional water pump, e. g. equipped with the electric motor. The ejector, in which the air is compressed and condensed, is placed inside the storage tank providing a minimum of heat losses. The water can be used as a working fluid.

The installation operates as follows. On switching on the water pump the ejector provides a rarefaction of the air in the solar collector where the water from the heat storage is directed through the reducing valve. The water effected by solar irradiation is heated up to the boiling temperature, determined by the extent of air rarefaction, which in its turn is controlled by the thermally actuated valve. The steam is pumped from the collector and is compressed within a jet in the ejector critical cross-section to the condensation state to transfer the condensation heat to the water, contained in the heat accumulator. As the processes of water boiling in the collector and water condensation in the ejector occur at different pressures (the condensation occurs at atmospheric pressure), the condensation temperature turns to be higher than that of the boiling. Thereby a principle of the heat pump is realized as the accumulator collects in terms of the thermal energy both the heat obtained by the absorption of the solar energy and the dissipated kinetic energy of the ejector jet. The installations of this type, which do not contain substances deleterious for the environment, according to the authors, can be widely practised.

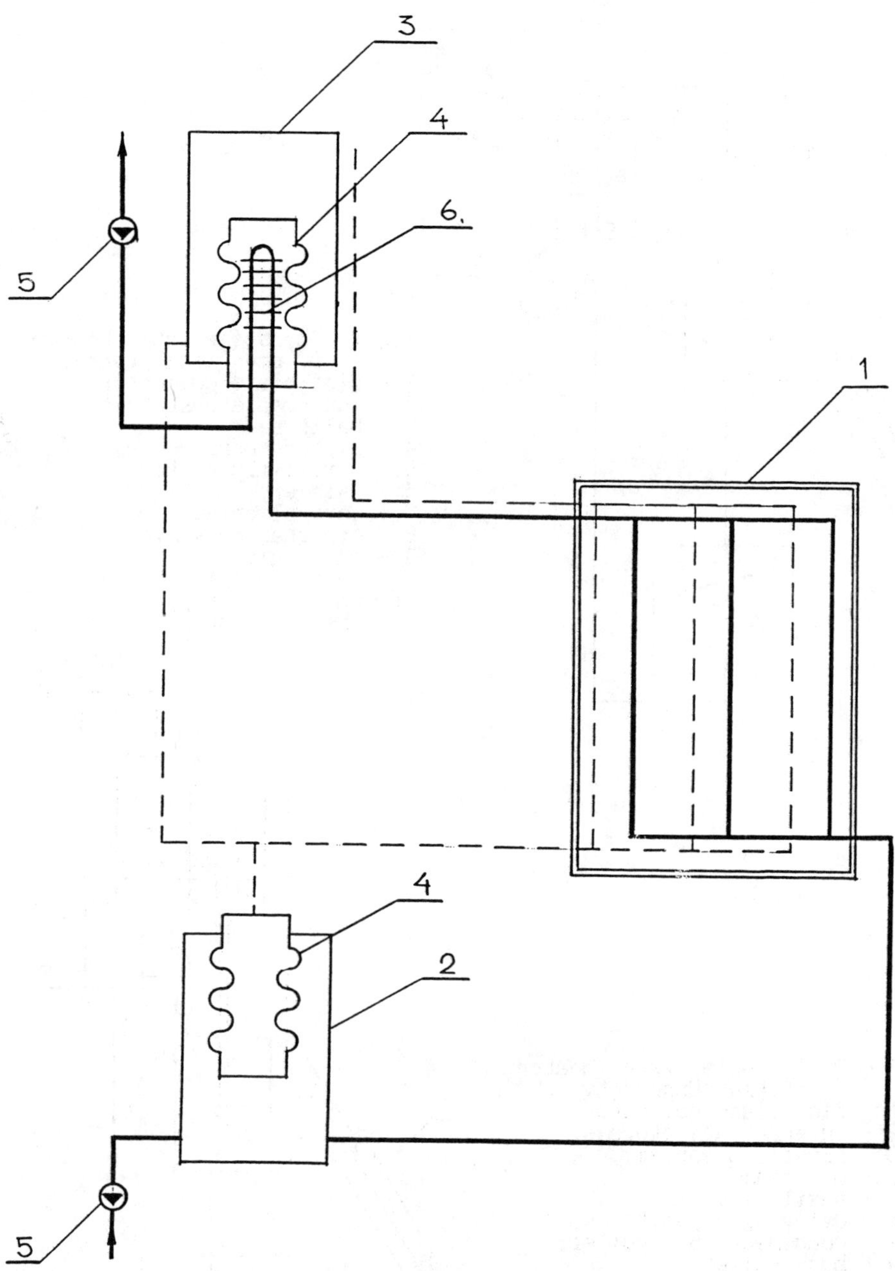

Fig. 1 The solar water pump.

1 - flat-plate collector, 2 - suckling chamber,
3 - expelling chamber, 4 - elastic pistons,
5 - back valves, 6 - condenser.

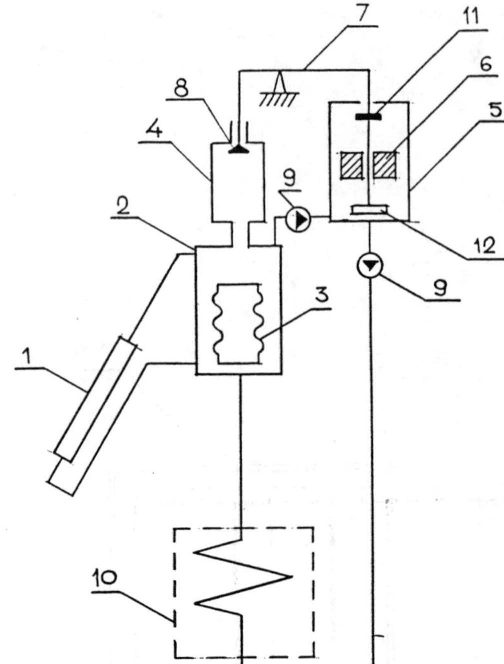

Fig. 2 The solar water heater of the "Gera" type.
1 - flat-plate collector,
2 - intermediate storage,
3 - bellows,
4 - air chamber,
5 - auxilliary tank,
6 - swimmer,
7 - leverel drive,
8 - air chamber valve,
9 - back valves,
10- main storage,
11- the upper stop,
12- the lover stop.

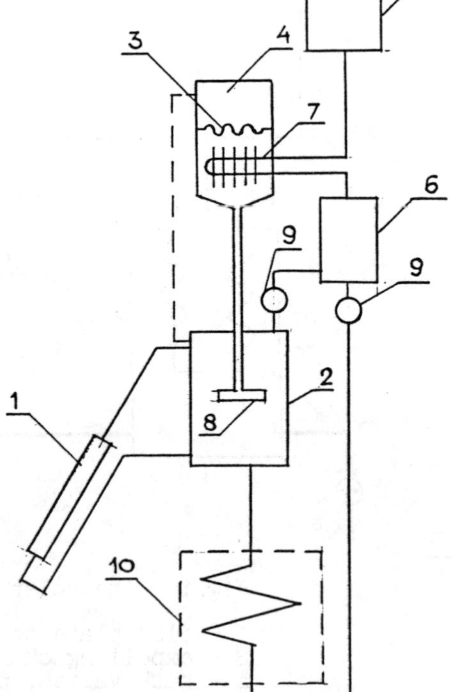

Fig. 3 The solar water heater of the "Unona" type.
1 - flat-plate collector,
2 - intermediate storage,
3 - elastic diaphragm,
4 - air chamber,
5 - auxilliary tank,
6 - delaying capacitance,
7 - condenser, 8 - boiler,
9 - back valves,
10- mane storage.

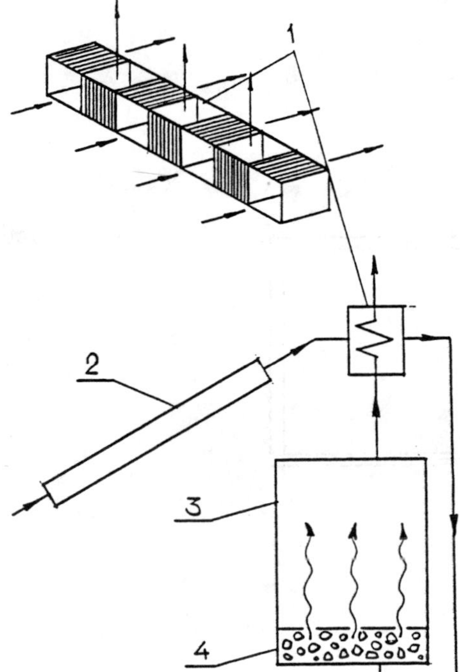

Fig. 4 The diagram of the solar drier with the heat exchanger.

1 - heat exchanger,
2 - solar collector,
3 - drying chamber,
4 - heat storage.

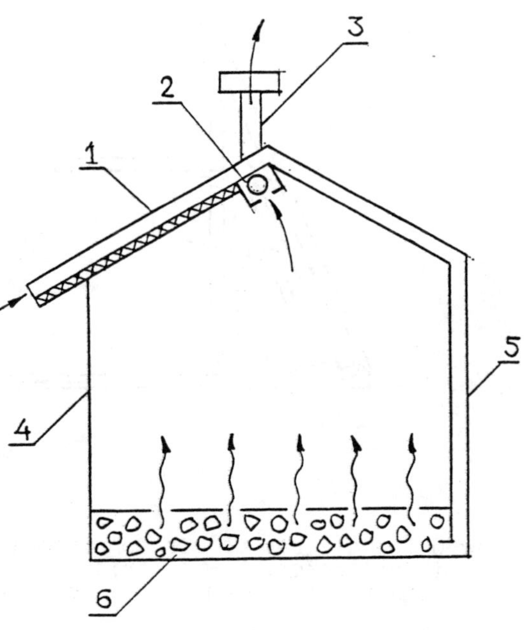

Fig. 5 The solar drier with the combined air collector.

1 - air heater,
2 - exhaust air conduit
3 - exhaust pipe,
4 - drying chamber,
5 - air channels,
6 - heat storage.

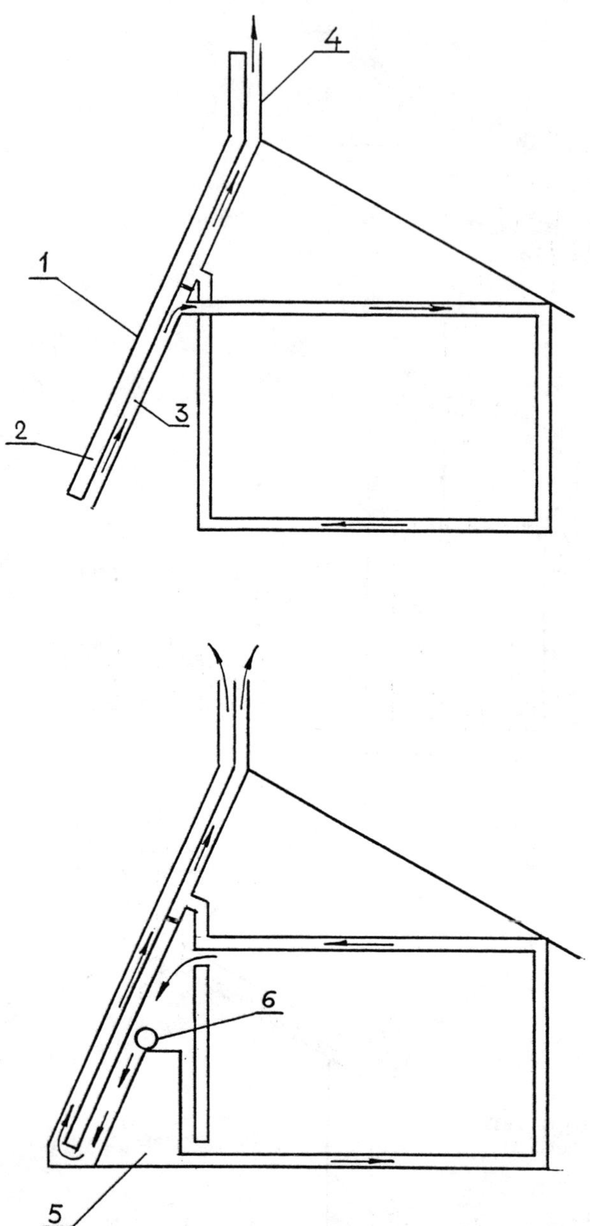

Fig. 6 The diagram of air conditioning of solar premise.
1- air collector, 2 - exterior cavity, 3 - interior cavity,
4 - exhaust air conduit, 5 - cooling chamber, 6 - sprinkler

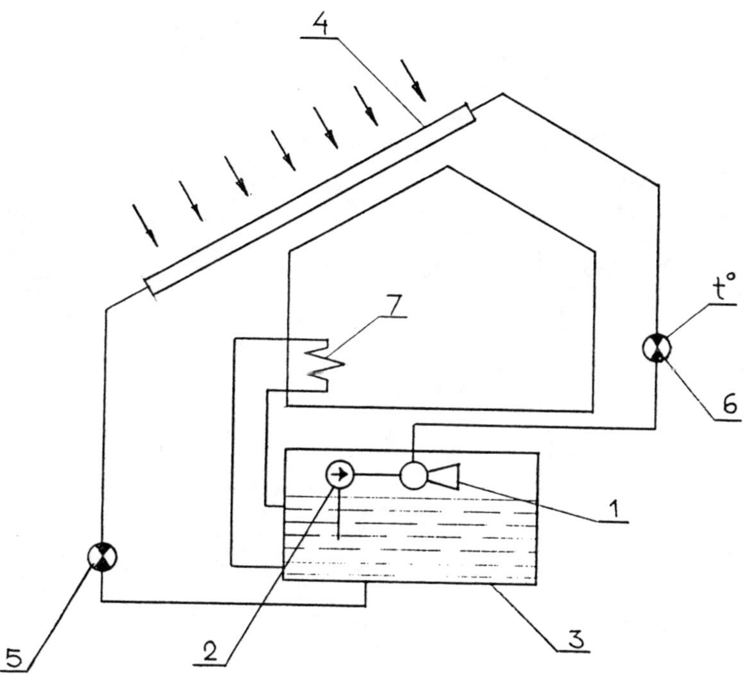

Fig. 7 Solar heat pump system with the ejector-type compressor.
1 - ejector, 2 - water pump, 3 - heat storage, 4 - solar collector, 5 - reducing valve, 6 - thermovalve, 7 - heating radiator.

COMPARATIVE PREDICTIONS OF A COMPUTER SIMULATION TO EXPERIMENTAL DATA OF A THERMOSYPHON DRIVEN DOMESTIC SOLAR HEATER

Samir F. Moujaes, Charles Wilhems
Department of Civil and Mechanical Engineering
Howard R. Hughes College of Engineering
University of Nevada Las Vegas, 4505 S. Maryland Pkwy
Las Vegas, NV 89154, U.S.A.

Abstract

An experimental study was performed to compare the predictions obtained from a computer simulation with respect to actual experimental data of a thermosyphon driven domestic solar water heater. The model uses a Runge–Kutta scheme to solve nodal equations which simulate the transient behavior of the glass cover, receiver surface , average water temperature inside collector and the average water temperature inside the reservoir. An estimate of the instantaneous water flow rate is also made by considering an instantaneous momentum balance around the loop. The model was used to simulate a rectangular cross–section type of solar collector rather than the more usual tube and fin type where the first was expected to reduce the overall losses from the system due to lower overall average temperatures of the solar collector. The experimental data was digitized through the use of a computer aided data acquisition system which monitored the system for several days. The predicted and measured temperatures showed good qualitative trends when performing in–situ experiments rather than controlled ones. No attempt was made to measure the flow rates as this needed expensive non–intrusive instrumentation which was not available. The model predicted properly the physical trends and could be improved on by having more accurate input data used for its simulation and a differential equation rather than an algebraic one to predict flow rates. For a future study the use of an LDA system to measure the flow rate and compare with the predicted value is recommended.

1. INTRODUCTION

The use of domestic solar water heaters is widely used in several countries. A fair majority of those systems employ the principle of thermosyphon to transport the useful heat accumulated in the solar collector panels to the storage reservoir. The thermosyphon is especially used in Third World countries where the availability and repair of pumps and major hardware is expensive and in some cases non–existing. The chosen system for the study relies mainly on the relative position of the collector and reservoir to ensure the proper flow of liquid through the system. The usual practice is to place the collector's top at a lower point than the lowest point of the reservoir as seen by Figure 1. Although this creates a favorable condition for flow during solar irradiation on the collector's surface it might create a "flow reversal" during the night where the reservoir is expected to be warmer than the collector. This undesirable condition can be aided also by the cooling due to the "nocturnal radiation" effect on the surface of the collector.

A voluminous number of studies is found in the literature both theoretical as well as experimental to try to understand this phenomenon[1–43]. The studies include working fluids such as air and liquid water as well as applications in two–phase flows with Freon and some with heat pipes. This noncomprehensive survey only highlights the facts that the field is of intense interest to many researchers and also that some basic phenomena are still under study because of the complex flow patterns which develop due to the nature of the heat transfer and fluid flow interactions to achieve a natural circulation pattern through a maze of pipes, collector and reservoir cross–sections which are considered highly developing flows both thermally and velocitywise.

The solar water heater used has parallel flat–plates which comprise the collector cross–section as opposed to perhaps the more common fin and tube type geometry. With this geometry it is expected that the time averaged heat losses are smaller in general than other configurations because the average temperature of the surface of the absorber surface will usually be lower as the surface of the collector is in complete contact with water. A higher temperature gradient seems to be needed to drive the heat transfer through the metal and from there to the liquid than is needed in a flat plate collector. Of course this effect is a strong function of how far the parallel tubes are apart from each other in the tube–fin geometry. The other reason for studying this geometry is simply the fact that it is an alternative collector construction and merits its own study because it presents its own set of simulation problems. The closest work which came to the present study is by Huang (6) where the same geometry for the collector was studied. His study did not include a detailed analysis of the transient behavior of the different materials used in the collector while the present one did i.e. glass plate, black metal absorber and water separately. It is felt that a treatment of those parameters individually should be done because they have completely different thermal inertia responses to solar insolation which can give rise to marked differences in temperature differences between glass plate, absorber surface and water within. The second difference is that the previous study did not handle turbulent flow conditions while this model is written in such a way that in case turbulent conditions exist anywhere in the system they can be handled by providing the necessary correlations. The third difference is that the heat lost from the reservoir ,in the present model is accounted for while in Huang's study it is not. The fourth is that water draw–off anytime in the day was allowed in this study as well as a more detailed consideration of the heat transfer mechanisms to determine eventually how much solar energy was absorbed into the collector. The previous study assumes that the solar energy absorbed is a sine function approximation which depends only on the time of day, total daily sunshine duration and solar radiation intensity at noon. Huang's paper however through non–dimensional analysis identified ten non–dimensional groups which could describe the characteristics of the parallel flat–plate solar collector uniquely using his assumptions. A more recent work by Yeh and Chen (30) have studied this system experimentally and have compared their results with a previously developed theory. An interesting result of their experimental work is that for reservoirs ,used for residential hot water, there is a very small vertical temperature gradient in them as time proceeds during the day. For example at around 3:00 p.m. there exists an average gradient of $4^{o}C/m$. The study by Yeh and Chen showed that keeping the total daily flow in the system the same the thermal efficiency is higher for the natural convection mode than the forced one in the same geometry. It is thought that the losses from the tube–in–sheet will be higher for the same daily average flow rates because of the higher average absorber temperature would be reached using the same insolation as input. It is interesting to mention that the present model supports this conclusion as it is found that the instantaneous flow for the thermosyphon system is usually higher than that of the forced circulation case hence improving the removal of heat from the collector.

The purpose of this paper is to present an attempt to model the thermal performance of a simple parallel flat plate residential size solar collector and compare the model with in–situ data obtained in the field. Difficulties and future improvements will be discussed.

2. MODEL DEVELOPMENT

The computer model describing the behavior of the system in Figure 1 was developed using a nodal analysis approach. The assumption is that the thermal transient behavior of the different components of the system are approximated by concentrated thermal mass nodes i.e. hence no temperature gradients are allowed in any node. Four nodes were used as follows:

1– Temperature of glass cover of collector represented by T_g.

2– Temperature of black metal absorber represented by T_s.

3– Temperature of water in collector represented by T_w.

4– Temperature of water in reservoir represented by T_r.

Having made these assumptions the next step is to set up the heat balance equations for each of them.

Nodal Heat Balance Equations

Glass Cover (T_g)

The glass cover is thermally interacting with the wind, the sky, the sun and the black metal absorber simultaneously. The heat balance equation is:

$$M_g C_g \frac{DT_g}{D\tau} = G_1 + G_2 + G_3 + G_4 + G_5 \qquad (1)$$

$$G_1 = A_g Q_1 \alpha$$

$$G_2 = A_g \epsilon_{sg} (T_s^4 - T_g^4)$$

$$\epsilon_{sg} = \frac{1}{\frac{1}{\epsilon_g} + \frac{1}{\epsilon_s} - 1}$$

$$G_3 = A_g H_{sg}(T_s - T_g)$$

$$G_4 = A_g \sigma \epsilon_g (T_s^4 - T_{sk}^4)$$

$$G_5 = A_g H_{ga}(T_g - T_a)$$

The determination for the effective sky temperature, for radiation purposes, is offered by Idso and Jackson (44):

$$T_{sk}^4 = T_a^4 * (1-0.261*\exp(-7.77\times10^4(0.555T_a-273.)^2)$$

The coefficient H_{sg} was determined using information from McAdams (45) and for H_{ga} from Chapman (46).

Collector Surface (T_s)

The equation expressing the heat balance for the absorber metal surface is as follows:

$$M_s C_s \frac{DT_s}{D\tau} = S_1 - G_2 - G_3 - S_2 \qquad (2)$$

$$S_1 = A_g Q_1 \gamma$$

$$S_2 = A_g H_{sw} (T_s - T_w)$$

$$\text{where } H_{sw} = 2.4 \frac{K_w}{D_h}$$

This heat transfer coefficient is considered as the asymptotic value of a developing flow situation in a rectangular channel (in laminar flow) with one major side of the channel considered under adiabatic conditions. In our case this side of course is the bottom side of the collector.

Water in Collector (T_w)

The equation expressing the transient response of the water temperature in the collector is as follows:

$$M_w C_w \frac{DT_w}{D\tau} = S_2 - H_{wa} A_g (T_w - T_a) - mC_w (T_o - T_i) \qquad (3)$$

Here the heat losses from the back of the collector ,although it is assumed insulated, were not neglected. The heat transfer coefficient between the water and the backside of the collector was approximated due the influence of the insulation as well as the average water heat transfer coefficient between insulation and ambient air as:

$$H_{wa} = \frac{1}{\frac{t_{ins}}{K_{ins}} + \frac{1}{H_{wob}}}$$

Water in Reservoir (T_r)

Finally the equation governing the heat balance in the reservoir is as follows:

$$M_r C_r \frac{DT_r}{D\tau} = \dot{m} C_w (T_o - T_i) - R_1 - R_2 \qquad (4)$$

$$R_1 = A_r H_{ra}(T_r - T_a)$$

$$R_2 = m_u C_w(T_r - T_u)$$

The H_{ra} is calculated by considering the reservoir as a cylinder and obtaining an estimate of the coefficient according to Chapman (46).

3. METHOD OF SOLUTION

Beforé discussing the method of solution one needs to recount how many unknowns and equations does one have to make sure no problem of closure arises here. The equations so far presented for solution are four heat balance equations while the number of unknowns are seven and the rest are either given or need only information which is a function of the time of day such as the insolation Q_1. The unknowns are $T_g, T_g, T_w, T_r, T_o, T_i$ and m. T_o and T_i are respectively the outlet and inlet temperatures to the solar water collector. The physical situation here is of water flowing along a finite length rectangular channel which means that technically speaking the temperature of the water coming out from the collector should be lower than that of the collector surface at that location. The temperature of the outlet water was assumed approximately equal to that of the surface T_s at any instant. This is motivated by the fact that the temperature of the water to the inlet of the collector is usually less than that of the average temperature of the reservoir, but in this study T_i was assumed equal to T_r. What is assumed here of course is the fact that the water in the reservoir is well mixed with no appreciable axial temperature gradients and that the heat lost from the connecting pipes between reservoir and collector is negligible. What is assumed here are two temperature values which are higher than what they physically are expected to be i.e. $T_o < T_s$ and $T_i < T_r$. The justification for the above assumptions for T_o and T_i being equal to T_s and T_r respectively is the fact that the important variable for this system is not so much the values of T_o and T_i as it is the difference between $\Delta T = T_o - T_i$ and that of $T_s - T_r$. This fact has been verified by numerical results of Huang (6). The latter term as will be seen later is the contributing force to the flow in the thermosyphon cycle, since this flow changes with time. By making the parallel flat plate collector more slender i.e. by increasing the ratio of axial length to the hydraulic diameter , the approximation of $T_o - T_i \simeq T_s - T_r$ is achieved. This will be the case for the exiting water temperature due to the fact that the variation of the water density is *linear* in the range of temperatures of interest is linear. Hence with this assumption the forcing function for a thermosyphon loop will be the ΔT and not just the absolute value of the temperatures around the collector.

With these two assumptions one has four equations for 5 unknowns and we still have the instantaneous mass flow rate,m, to be dealt with. This has been quantified by looking at the thermosyphon loop and recognizing that certain sections of it have different but uniform temperatures. An instantaneous momentum balance around this loop assuming inertial effects are negligible will give the following equation:

$$\oint H_i \rho_i dz = \oint \frac{dp_i}{dl} dl \qquad (5)$$

Equation 5 states that the cyclic integral of the unbalanced force due to buoyancy calculated as a function of the temperatures T_i around the loop should be balanced by the cyclic integral of the frictional pressure drop around the loop. It is important to note that the "z" variable denotes the *elevations* of the different components *only* as any horizontal runs of piping even if there are temperature differences will not contribute to buoyancy. On the other hand the variable "l" denotes the actual linear run of each of the components be it collector, reservoir or connecting piping. Hence it is seen that qualitatively to optimize the flow in the system one needs to shorten unnecessary horizontal runs of piping because they will increase the friction pressure drop at the expense of the buoyancy force. The pressure drop term ,$\frac{dp_i}{dl}$, will have in it second and first order terms of m and some constants. The first order terms are estimates of pressure drops in the collectors and piping and the second order terms result from pipe fittings. The eventual result is a quadratic equation in m with one of its roots being unacceptable due to established physical facts which leaves the other root as the only choice. In mathematical form:

$$A\dot{m}^2 + B\dot{m} + C = 0 \qquad (6)$$

Where A,B and C are non zero constants. Also A and B are positive while C is a function of temperature distribution around the loop and hence could be positive or negative. The two roots are:

$$\dot{m}(\pm) = \frac{-B \pm \sqrt{B^2 - 4AC}}{2A}$$

The (+) root is the sought after root because it is argued that when there is no buoyancy force i.e. the temperature around the loop is uniform then the "C" term is zero and one should not expect any measurable flow around the loop. This can only be achieved by the aforementioned root.

Hence we have a way to determine algebraically the flow rate in the thermosyphon loop and so have five equations for five unknowns.

<u>Numerical Solution</u>

Equations 1–4 are four ordinary first order non–linear differential equations in time. The problem is an initial value problem with some uniform given temperature ascribed to the four variables T_g,T_s,T_w and T_r. The fourth order Runge–Kutta technique is employed to solve these equations. Due to the differences of thermal inertia for the four nodes under consideration potential numerical divergence problems resulted if the chosen time step was too large. Hence by trial and error it was determined as to what would be the maximum allowable time step to start up the program. As time progressed the time duration was relaxed and larger time steps were considered to speed up the simulation. The criterion for

a "good choice" of a time step was that if by integration using half this time step the absolute difference in the temperature predictions over a 24 hrs period is about 1–2% then it was deemed acceptable. The program was run using a cyclic forcing function for the solar insolation and other time dependent input variables and was determined that the time duration after which the system response would be insensitive to the actual initial conditions is 24 hrs. This was mainly due to the fact that the reservoir had the highest thermal inertia and took the longest to adjust.

The program was written as a series of subroutines whereby any geometrical or thermophysical property could be varied at will. Also an option for running the program in a forced circulation or natural one can be made.

The intent of this paper is to present an idea about how the program was developed rather than going in detail about numerical testing results for the particular scheme.

4. EXPERIMENTAL SETUP AND INSTRUMENTATION

The experimental solar water heater used is shown in Figures 1 and 2. The solar collector is approximately 84×183 cm long and is about 3.8 cm deep. Stiffeners were used around the collector to hold the relatively thin sheet metal in place. The surface of the collector was placed at about 30° with the horizontal which allows an optimized operation of such a collector year round. The inlet piping into the collector is achieved by breaking up the flow into three equally sized PVC schedule 40 pipes. A similar arrangement was provided for the outlet flow. The reservoir tank was made from stainless steel and is about 222.0 liters. The bottom of the collector was about 50 cm above the top of the collector to achieve a reasonable thermosyphon circulation. The back of the collector was insulated with an R–13 fiberglass batt insulation to minimize heat losses. The supply and discharge piping to the collector and reservoir were well insulated so as to improve natural circulation in the system.

Instrumentation

Twelve temperatures were sampled around the solar collector, inlet and outlet piping and reservoir using type "T"(copper/constantan) junction. The sampling was done through the use of an electromechanical switcher which fed one sample at a time to the thermocouple linearizer and hence to the A/D. Each 15 seconds one channel was sampled several times and averaged during the 15 seconds to obtain a more accurate reading over that interval. See Figures 1 and 2 for details of thermocouple locations. An attempt was made to use a floater type flow meter to measure the instantaneous flow through the thermosyphon but was eventually discarded as a viable possibility because of the rather large friction pressure drop it offers throttling down the capacity of the collector. A portable solar meter was placed very close to the solar collector and at the same inclination to measure the instantaneous solar intensity impinging on the glass cover. The output of the devise was read manually rather than being sampled automatically by the computer system. The wind speed was approximated from the local national weather bureau in the Las Vegas area.

5. RESULTS CONCLUSIONS AND RECOMMENDATIONS

The experimental performance of the solar collector was compared with one day's worth of simulated data. Although experimental data were obtained for about four consecutive days it was felt that the best comparison would be made with the fourth day where conditions inside the collector would have stabilized and the effects of initial conditions of water temperature would have already been forgotten. This was verified by the model to be the case and showed that for a representative collector system of that size the effects of initial conditions are forgotten in about a maximum of two days. The model of course on the other hand was run for four consecutive days in real time assuming the same cyclic conditions exist as far as solar insolation and wind velocity. The numerical values of the fourth day were used for purposes of comparison.

It is interesting before discussing the comparative results to show how the effect of the difference in height between the bottom of the reservoir and that of the top of the collector would affect the flow. This is presented in Figure 3 because as mentioned before the instantaneous average flow was not measured. As one would expect initially there would be an increase in the flow rate as the height is increased because the unbalanced buoyancy force is increasing and is able to drive a higher flow through the system. Note of course that there is no domestic water withdrawal at this time. Also it is interesting to see that at zero difference in height between the bottom of reservoir and top of collector there is a non–zero flow. The final remark about this numerical experiment is the result that as the height increases the rate of increase of the maximum flow rate decreases with this height. This can be explained by the fact that the friction drag force is "catching up" with the increased buoyancy force as the total length of the closed circuit is increased. It is expected of course that with an additional height increase there will be a point where the maximum flow will start decreasing and will eventually reach to zero. The reason the flow abruptly goes to zero at 18:00 hrs is because the system is assumed to have a check valve that will not allow the relatively warmer water to circulate in reverse at night and cool off as it flows from the reservoir to the collector. Figure 4 shows a plot of the solar insolation for that particular day which was measured on an hourly based and is represented here with a best curve fit. This insolation was used as input for the theoretical model also. The data collected was for an August day in the Las Vegas area in 1988.

Figure 5 shows the plots of the experimental versus the model simulation data. Both model and experiment show the black surface temperature (solid triangle) to have the highest value during the day because of its high emissivity. The second in line is the glass instantaneous temperature which seems to have very similar trends as the actual experimental values although it seems to drop faster at night in the model simulation than the experimental case. Although the thermocouple was in very good contact with the glass it was realized the perhaps a small portion of it might have been exposed to the direct sunlight. This would have made experimental readings for glass slightly higher as the emissivity and absorptivity of the junction material for the thermocouple and that for the glass are quite different for short wave radiation. This might explain also why at night the shift between the two temperature predictions. The third set of data to compare is the internal water temperature in the collector at a point midway between the inlet and outlet manifolds (hollow triangle). The maximum temperature values are slightly lower (by about $5^{O}C$) for the model than the experiment. This might be caused by perhaps microcirculation patterns that might have developed along the collector depth i.e. normal to the black surface which might push the temperature of the localized temperature somewhat higher. This is not inconceivable as the water flow for this type of arrangement will be relatively low making the assumption of a fully developed laminar flow somewhat suspect. The fourth temperature to be discussed is the average reservoir temperature (solid square). Again this shows very similar trends between experiment and theory although the values are almost shifted by about $6–7^{O}C$ with the experimental ones being higher. It is not clear at this point why this behavior is so.

It is interesting however to point out that the time of day at which each material reaches its temperature peak compared well between theory and experiment. The materials that possess the smaller thermal inertia as expected peaked earlier in the afternoon such as the black surface and the glass. While those with more massive thermal inertias peak later in the afternoon such as the reservoir and the average water temperature inside the collector.

The last two temperatures to be compared are the temperature into the collector (hollow circle) and the temperature out of the collector during normal operation (solid circle). Here the comparison will only be made between the hours 6:00 A.M. – 6:00 P.M. because a "check valve" concept is built in the model which shuts off the flow through the system to prevent any flow reversal later on at night. This is not the case with the experimental setup because that will entail some time activated solenoid operated valve which would be rather expensive. Both the experiment and the model show a realistic positive difference in temperature between the outlet and inlet temperatures to the solar collector. Of course the model does not solve explicitly for the inlet and outlet temperatures but rather assumes (as discussed previously) that in the limit the water temperature at the outlet attains the average black surface temperature and the inlet flow to the collector has a temperature close to the reservoir temperature. The difference between the maximum temperatures for the inlet conditions is about $8^{o}C$ and that for the outlet temperatures is $6^{o}C$. However the times at which these two temperatures reach their maxima during the day closely agree between theory and experiment.

Conclusions and Recommendations

A theoretical model has been developed and compared with in–situ experimental data to show the validity of the model. It was found that the model predicts the trends successfully and at this time has qualitative agreement with the experiment. On the other hand it needs to be mentioned that the experiment was not a laboratory controlled one where solar or simulated insolation on the unit can be controlled to a good degree of accuracy. As was mentioned earlier the insolation input information was obtained on an hourly basis while the model required more detailed information as small time steps were used for its stable solution. The wind velocity could have been a factor in the model prediction because the assumption was made that a constant wind speed is blowing parallel to the surface of the glass cover which is not usually the case where gusts of wind could affect the local heat transfer mechanisms around the collector. No attempt was made experimentally at determining the actual values for the emissivities of glass and the black surface paint which was used on the collector. The values used were available in the literature.

It is felt that the depth of the collector should be reduced to make sure that the assumptions used in the model for predicting the heat transfer coefficient between the black surface and the water be realistically approximately as a fully developed Nusselt number for fully developed laminar flow. In other words an attempt should be made to minimize the hydraulic diameter of the rectangular cross–section to minimize developing effects. More attention should be made to shielding the thermocouples which measure temperatures of surfaces exposed to direct radiation. Also improve on the quality of the transfer piping between the different components of the system to make sure no additional heat losses occur. An effort should be made to investigate means to measure the average flow rate instantaneously due to the thermosyphon effect such as a Laser Doppler Velocimeter (LDV). This is preferable to any intrusive techniques which might generate unacceptable pressure drop levels as the total friction drop for these systems is estimated to be a few mm of water. Lastly the model should include the effect of flow inertia to incorporate the transient nature of fluid flow in the system.

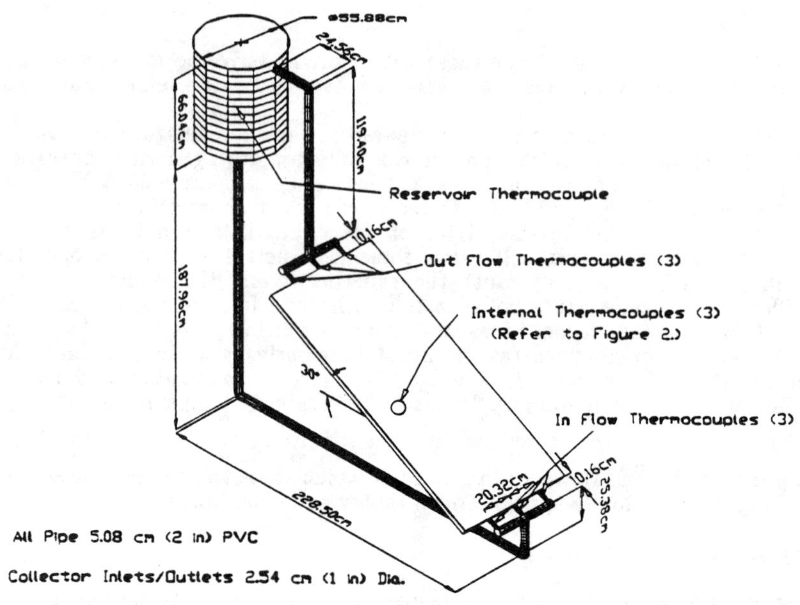

Figure 1. Scematic Diagram of Solar Water Heater

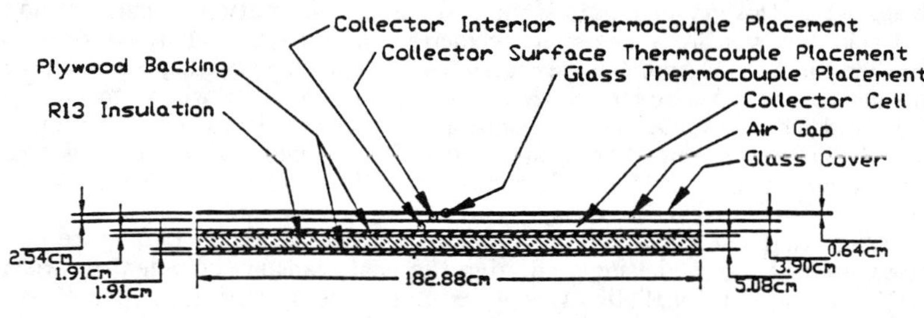

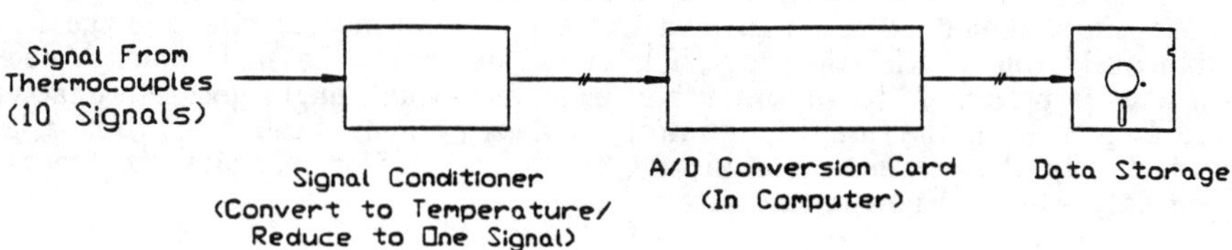

Figure 2. Cross-Section of Collector and Instrumentation Chart

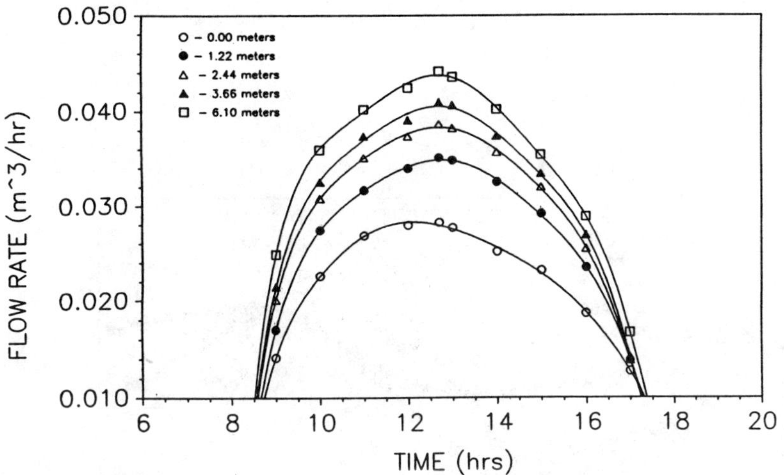

Figure 3. Flow Rate vs. Time for Various Reservoir Elevations

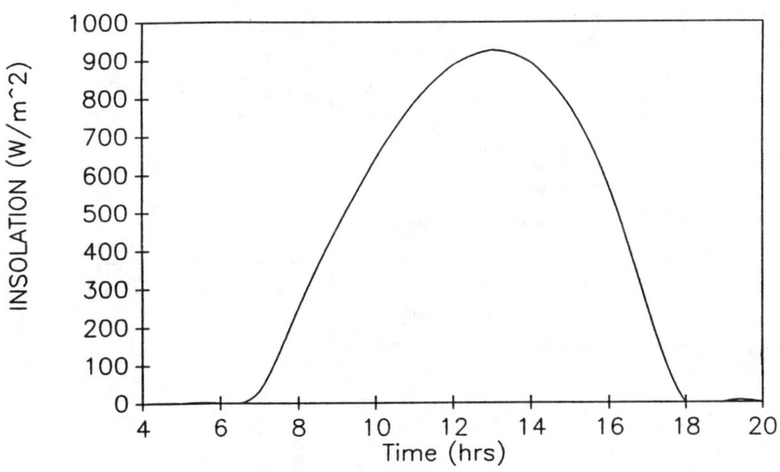

Figure 4. Experimental Variation of Solar Insolation with Time

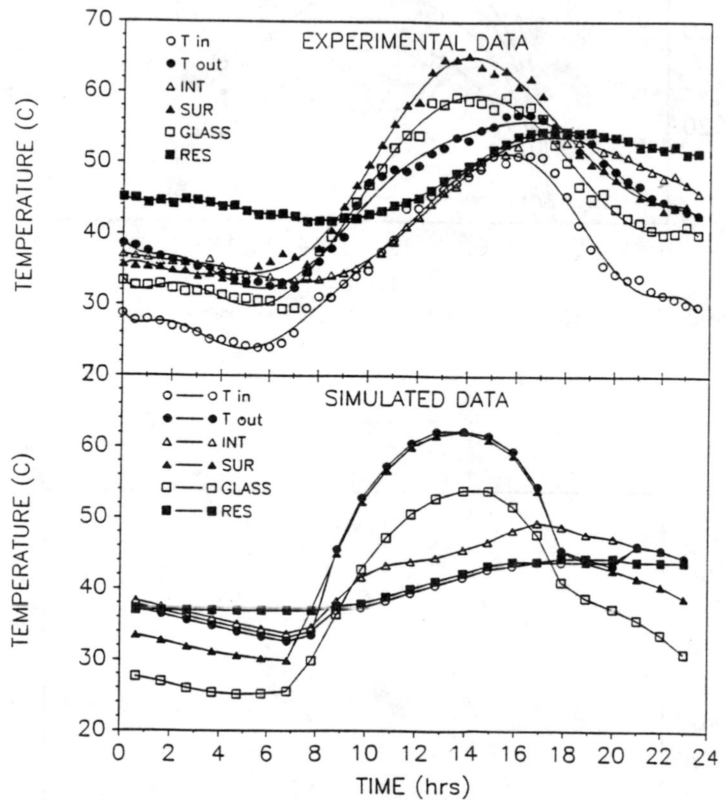

Figure 5. Experimental vs. Simulated data

REFERENCES

1. D.J. Close, The Performance of Solar Water Heaters with Natural Circulation, Solar Energy **6**, 33–40(1962).

2. S. Moujaes, A. Sfeir and M. Menguy, Computer Simulation Of a Solar Water Heater, International Conference on Helio–Techniques and Developments, (UPM)Dahran, Saudi Arabia, Sousou Publishers Boston MA, USA,February(1976).

3. G.L. Morrison and D.B. J. Ranatunga, Thermosyphon Circulation in Solar Collectors, Solar Energy **24**, 191–198(1980).

4. J.M. Gordon and Y. Zarmi, Thermosyphon Systems: Single vs Multi–pass, Solar Energy **27**, 441–442(1981).

5. G.L. Morrison and D.B. J. Ranatunga, Transient Response of Thermosyphon Solar Collectors, Solar Energy **24**, 55–61(1980).

6. B.J. Huang, Similarity Theory of Solar Water Heater With Natural Circulation, Solar Energy **25**, 105–116(1980).

7. A. Mertol, W. Place, T. Webster and R. Grief, Detailed Loop Model(DLM) Analysis of Liquid Solar Thermosiphons With Heat Exchangers, Solar Energy **27**, 367–386(1981).

8. P. Hobson and B. Norton, verified Accurate Performance Simulation Model of Direct Thermosyphon, J. of Solar Energy Eng'g, ASME Trans. **110**, 282–292(1988).

9. M. Sokolov and M. Vaxman, On the Dependence of Thermosyphonic Solar System performance on Collectors' Azimuthal Orientation, Energy Conv. and Management **28**, 257–263(1988).

10. M.J. Carvalho, M. Collares–Pereira, F.M. Cunha and C. Vitorino, Experimental Comparison of Operating Strategies for Solar Water Systems, Solae Energy **41**, 33–39(1988).

11. K.R.D. Braven, Heat Transfer in a Tilted Two–Phase Closed Thermosyphon with Application to Evacuated Tube Solar Collectors, Proceedings of the 10th Annual ASME Solar Energy Conference, 539–545(1988).

12. R.L. Sawhney, N.K. Bansal and M.S. Sodha, Solar Passive Heating Through a Thermosyphon Air Panel, Int'l J. of Energy research **12**, 217–225(1988).

13. R.E. Collins, D. Mackey and G.L. Morrison, Comparative Performance of Evacuated Tubular Collectors and Flate Plate Collectors in Thermosyphoning Systems, Proceedings of the 9th Biennial Congress of the International Solar Energy Society, Published by Pergamon Press **2**, New York, 1284–1288(1986).

14. G.L. Morrison , Performance of Evacuated Tubular and Flat Plate Solar Water Heaters. Proceedings of the 9th Biennial Congress of the International Solar Energy Society, Published by Pergamon Press **2**, New York, 1184–1188(1986).

15. S.S. Mathur and N.K. bansal, Domestic Thermosyphon Water Heating Systems, Proceedings of the Workshop of Solar Water Heating Systems, New Delhi, Published by D. Reidel Co., Dordrecht, Neth and Boston, MA, 299–326(1986).

16. B. Norton, S.D. Probert and J.T. Gidney, Diurnal Performance of Thermosyphonic Solar Water heaters–An Empirical Prediction Method, Solar Energy **39**, 257–265(1987).

17. M.p. Malkin, S.A. Klein, J.A. Duffie and A.B. Copsey, Design Method for Thermosyphon Solar Domestic Hot Water Systems, J. of Solar Energy Eng'g, Trans. ASME **109**, 150–155(1987).

18. W.Y. Saman, A. Nassir and A.A. Mohamed, Development of an Indirect Thermosyphon Solar Water and Comparison With The Direct System, Solar Energy prospect in the Arab World, Second Arab International Solar Energy Conference, Publisher Pergamon Press, Oxford England, 137–141(1986).

19. W.A. Kamal, Naturally Circulating Solar Water heater With Hot water Withdrawal, Solar Energy prospect in the Arab World, Second Arab International Solar Energy Conference, Publisher Pergamon Press, Oxford England, 130–136(1986).

20. M. Behnia, G.L. Morrison and S. Paramasivam, Heat Transfer and Flow in Inclined Open Thermosyphons, ASME–JSME Thermal Engineering Joint Conference, Published by ASME **4**, 7–14(1987).

21. K.C. Cheng, K. Fukuda and C.A. Lee, Some Operating Characteristics of a Closed–Loop Two–Phase Thermosyphon Flat–Plate Solar Collector System Using Freons R–11 and R–113 by Indoor Test, ASME–JSME Thermal Engineering Joint Conference, Published by ASME **2**, 1035–1043(1987).

22. H.H. Huang, R.J. Shyu and L.J. Fang, Experimental Study of Solar Thermosyphon Hot Water Systems With Horizontal Storage Tank, ASME–JSME Thermal Engineering Joint Conference, Published by ASME **2**, 1000–1005(1987).

23. P.V. Pedersen, 2nd Generation Thermosyphon Solar Water heater, Proceedings of The First E.C. Conference on Solar Heating, Published by D. Reidel Co., Neth and Boston MA, 658–661(1984).

24. B. Norton and S.D. Probert, Open–Loop Thermosyphon Solar–Energy Space Heating, Proceedings of The First E.C. Conference on Solar Heating, Published by D. Reidel Co., Neth and Boston MA, 341–345(1984).

25. E. Vazeos, P. Canellopoulos and C.D. Rakopoulos, Experience in Modelling and Experimenting of Thermosyphonic Solar Water Heaters With Heat Exchanger, Proceedings of The First E.C. Conference on Solar Heating, Published by D. Reidel Co., Neth and Boston MA, 203–207(1984).

26. T.L. Webster, J.P. Coutier, J.W. Place and M. Tavena, Experimental Evaluation of Solar Thermosyphons with Heat Exchangers, Solar Energy **38**, 219–231(1987).

27. M. Vaxman and M. Sokolov, Effects of Connecting Pipes in Thermosyphonic Solar Systems, Solar Energy **37**, 323–330(1986).

28. S.C. Carpenter, Design of Thermosyphon Solar D.H.W. Systems For The Canadian Climate, Proceedings– SESCI 84 Calgary, Alberta, Published by Solar Energy Soc. of

Canada Inc. Winnipeg, Manit., 116–121(1984).

29. B. Norton and S.D. Probert, Thermosyphon Solar Energy Water Heaters, Adv. in Solar Energy An Annu. Rev. of R&D **3**, Published by American Solar Energy Soc. Inc. Boulder, Co. and Plenum Press, New York, 125–170(1986).

30. H.M. Yeh and L.C. Chen, Study on Thermosyphon Solar Water Heater With Parallel Flat–Plate Collector, Energy **11**, 579–588(1986).

31. G.L. Morrison, Reverse Circulation in Thermosyphon Solar Water Heaters, Solar Energy **36**, 377–379(1986).

32. V.F. Gershkovich and A.R. Fert, Calculation of a Two–Loop Solar System With Thermosyphon Circulation, Applied Solar Energy(English Translation of Geliotekhnika) **21**, 71–73(1985).

33. G.N. Tiwari, Transient Performance of a Thermosyphon Water Heating System With N–Units Connected in Series, Solar Energy **35**, 371–375(1985).

34. W.A. Kamal, Performance Evaluation of a Refrigerant–Charged Solar Collector, Proceedings of ENERGEX 84', Regina, Sask. Canada, Published by Pergamon Press, Oxford England, 193–197(1984).

35. W.A. Kamal, On The Optimization of The Naturally Circulating Flat Plate Collector System Dimensions and Operating Conditions, 1st Arab International Solar Energy Conference, Kuwait, Published by Pergamon Press, Oxford England, 129–135(1983).

36. G.N. Tiwari, S.N. Shukla and M.S. Sodha, Performance of Large Solar Water heating System: Thermosyphon Mode, Energy Conv. and Management **25**, 29–38(1985).

37. B.J. Huang and C.T. Hsieh, Simulation Method for Solar Thermosyphon Collector, Solar Energy **35**, 31–43(1985).

38. G.L. Morrison and J.E. Braun, System Modeling and Operation Characteristics of Thermosyphon Solar Water Heaters, Solar Energy **34**, 389–405(1985).

39. G.L. Morrison and H.N. Tran, Simulation of the Long Term Performance of Thermosyphon Solar Water Heaters, Solar Energy **33**, 515–526(1984).

40. H.A.A. Abd–Al Zahra and K.A. Joudi, Experimental Investigation Into The Performance of a Domestic Thermosyphon Solar Water Heater Under Varying Operating Conditions, Energy Conv. and Management **24**, 205–214(1984).

41. A. Litvak and G.L. Morrison, Study of Flow Characteristics in Thermosyphon Solar System, Proceedings of the 8th Australian Fluid Mechanics Conference, Newcastle NSW, Aust., Published by Uni. of newcastle NSW **2**, 11c.5–11c.8(1983).

42. A.H. Fanney, Experimental Technique for Testing Thermosyphon Solar Hot Water Systems, Proceedings of the ASME Solar Energy Division 6th Annual Conference, Las Vegas, Published by ASME New York, 154–161(1984).

43. H.T. Whitehouse, Drain_Down Freeze Protection For Thermosyphon Solar Water Heaters, Proceedings of the Passive and Hybrid Solar Energy Update, Published by DOE, Washington D.C., 346–360(1982).

44. S.B. Idso and R.D. Jackson, Thermal Radiation from the Atmosphere, J. of Geophys. Res., Vol. 74, p. 5397 (1969).

45. W.H. Mcadams, Heat Transmission, 3rd edition, McGraw Hill, New York (1954).

46. A.M. Chapman, Heat Transfer, 2nd edition, McMillan Co., New York (1969).

PERFORMANCE INVESTIGATION OF INDIRECT EVAPORATIVE COOLING

G . D . Mathur, Ph . D .
Research & Design Engineer
QDT Ltd., Dallas, TX 75212, U.S.A.

Abstract

In this paper, the principle of indirect evaporative cooling is used with an air-to-air heat pipe heat exchanger system to increase the precooling effectiveness of the heat exchanger. Different types of evaporative cooling systems which may be used with heat pipe heat exchangers have been discussed. The performance of the system with and without evaporative cooling for hot climate is investigated and the results have been presented in this paper. Calculated results show a significant reduction in the cooling operation and equipment costs, if indirect evaporative cooling principle is applied to a heat pipe heat exchanger system.

1. INTRODUCTION

The evaporative cooling process is man's oldest method of cooling air and water. In ancient times, in Egypt and India, wetted grass mats and porous jars utilized the evaporative process to cool air and water. Today direct evaporative coolers are still being used in which air is passed through wetted media pads or water sprays to provide lower dry bulb temperature air. The energy for the evaporation process comes from the air stream, so that it is cooled, resulting in a lower dry bulb temperature of the air. Throughout this process, air and water are in direct contact and both mass and heat transfer takes place. In direct evaporative cooling process, a reduction in sensible energy is obtained with a corresponding increase of the latent energy due to addition of the moisture. A direct evaporative cooler does not reduce the total load on the airstream; it merely exchanges one form of the load to another. Direct evaporative coolers are used almost exclusively in hot and arid climates (low wet bulb temperature), because the introduction of moisture to the supply airstream can be tolerated.

The indirect evaporative coolers are devices that use the evaporating cooling effect in such a manner that no moisture is added to the supply air streams. This is achieved with the use of an air-to-air heat exchanger using evaporatively cooled air on the exhaust (sink) of the heat exchanger, with outside air to the space passing through the supply (source) of the heat exchanger. This process uses two separate air streams that never mix or come in direct contact. The cooling of the exhaust air allows a larger overall temperature difference across the heat exchanger resulting in a greater cooling of the supply air. The indirect evaporative cooling process requires no energy input besides that required to overcome fan and water pumping power. Coefficient of performance for these systems, therefore, tend to be high. Energy recovery is further enhanced by the improvement in heat transfer coefficient due to the wetted surface of the exhaust side of the heat exchanger. The additional summer savings with the use of indirect evaporative cooling means that the cooling equipment can be further down sized to reduce total project costs. Since less cooling is required, energy consumption and peak demand load are both reduced, yielding lower energy bills. The principle of indirect evaporative cooling is effective in air conditioned buildings, regardless of their geographical location, because the evaporative cooling process is applied to the conditioned exhaust air rather than to the outside air. Indirect evaporative cooling has been shown to be very effective in any application requiring ventilation air, which encompasses almost every HVAC system that is designed.

There are several types of heat exchangers available to recover energy from building exhaust: plate type heat exchanger, rotary wheel, two-phase thermosiphon loop heat exchanger, run around coil, and heat pipe heat exchanger. The versatile design features of a heat pipe heat exchanger makes it attractive for energy recovery applications. The recovery performance of a heat pipe heat exchanger is simply a measure of its ability to recover the available sensible thermal energy from the hotter air stream and transfer it to a cooler air stream. Although it is basically a sensible heat transfer device, condensation on the fins of a heat pipe heat exchanger does allow latent heat transfer, resulting in improved recovery performance. In a heat pipe heat exchanger, hot gas is passed through one of the heat exchanger and cold air is passed through the other side of the heat exchanger.

Heat pipe heat exchangers offer a number of significant advantages over other types of heat exchangers. One of the main advantage is that in comparison to other types of heat exchangers with the exception of two-phase thermosiphon loop heat exchangers, they require no external power to circulate the working fluid, and the heat can be transferred in either direction.

This paper deals with indirect evaporative cooling process using such an air-to-air heat pipe heat exchanger system to increase the precooling effectiveness of the heat exchanger. The performance of the system with and without evaporative cooling for hot climate is calculated and the results have been presented in this paper.

2. PRINCIPLE OF OPERATION WITH INDIRECT EVAPORATIVE COOLING

A heat pipe heat exchanger system is shown schematically in Figure 1. In its simplest form, a heat pipe is a tube that has been evacuated, partially filled with a working fluid, and permanently sealed at both ends. It is used to transfer heat from a hot fluid stream to a cooler fluid stream. The driving force for heat transfer is a hot fluid stream at one end of the heat pipe, known as the evaporator; and a cooler fluid stream at the other end of the heat pipe, known as the condenser. The heat transfer fluid inside the heat pipe will absorb heat from the hotter fluid and release the heat to the cooler fluid. Typical heat transfer fluids utilized are Freon for low temperature applications. In practice, a heat pipe heat exchanger consists of a bank of tubes (heat pipes) mounted in a casing, with inlet and outlet connections for both evaporator and condenser sections. A divider plate separates the evaporator section from the condenser section.

Hot fluid flowing over the evaporator end of the heat pipe heat exchanger causes the working fluid to vaporize. The vapor travel to the condenser due to the difference in the saturation pressures in the evaporator and condenser. The vapor condenses in the condenser end of the heat exchanger and the latent energy of the working fluid is released. The condensed liquid inside the heat pipe flows back to the evaporator due to the capillary forces, where it is revaporaized, thus completing the cycle. The cycle will continue indefinitely, as long as the hot fluid temperature passing over the evaporator exceeds the cold fluid temperature passing over the condenser end.

The operation of an air-to-air heat pipe heat exchanger for winter and summer seasons is shown in Figure 2. The section of the heat pipe heat exchanger that works as an evaporator in the winter will work as a condenser when the outside temperature exceeds the exhaust air temperature, thus reversing the roles of the evaporator and the condenser. Similarly, the section of the heat pipe heat exchanger that works as a condenser in the winter will work as an evaporator when the outside air temperature exceeds the exhaust air temperature. Hence, the existing ductwork used for the winter season will not have to be modified in any way for the system to operate in the summer season without a change in the air flow direction. Figure 3 shows a heat pipe heat exchanger system with evaporative cooling. Indirect evaporative cooling process is the evaporative cooling process in which the exhaust air stream is cooled by passing the air stream either through a water spray section, or through a filter bed of some type through which water is allowed to trickle. A part of the water evaporates and the energy for evaporation process comes from the exhaust air stream, so that it is cooled. This process is adiabatic which follows a constant wet-bulb line on a psychrometric chart. The cooled exhaust air then enters the heat exchanger resulting in a greater overall temperature difference between the supply and exhaust air streams. Hence more heat is transferred and the supply air is cooled further by a reduction in sensible load under a constant latent load, i.e. no moisture is added to the supply.

3. TYPES OF EVAPORATIVE COOLING SYSTEMS

The following three types of the evaporative cooling systems may be used with an air-to-air heat pipe heat exchanger.

(a) Media Pads

This type of system utilizes a wetted media material in which the exhaust air passes through, producing evaporative cooling. This system is shown in Figure 4. At low coil face velocities, 500-600 fpm, a relatively short chamber length can be designed into the system. Media pads generally on average have a life span of 10 years and require relatively low maintenance. With these systems, special coating on the condenser coil is not needed. Algaecide water treatment may be needed with this system. The added thickness contributes for increased pressure drop across the media pads. Two types of media pads are suited for use with heat recovery.

(i) Spray type: This unit is a hybrid air washer/cooler, which utilizes a spray system to wet the media elements. These media elements usually are glass fibers, synthetic fabric, or polyurethane foam. The thickness of the media cells is relatively large because it has to trap and retain water droplets.

(ii) Non spray type: The media in this system usually consists of a wood product fabricated into a honeycomb of cells arranged in vertical columns. The water is introduced from the top through a header or a trough. The water flows down by gravity to the sump at the bottom on the unit.

(b) Direct Spray on the Condenser Coil

The exhaust air is conditioned by contact with atomized water sprayed directly on the condenser coil. Some of the water is evaporated directly into the air producing cooling of the exhaust air. This arrangement is shown in Figure 5. The air and the surface of the coil is also cleaned to a certain degree by removing some of the airborne contaminants. At low coil face velocities, 500-600 fpm, the water trickles down the coil into the drain pan and sump. This type of system has a better performance than the media pads and requires least amount of space. Water treatment, nozzle maintenance, mineral deposits on the coil, coating on the condenser coil, are some of the disadvantages of this type of system.

<u>(c) Air Washer</u>

The exhaust air to be conditioned is passed through a chamber where an atomized water spray is directed into the oncoming air stream. This system is shown in Figure 6. Some of the water is evaporated, producing cooling of the exhaust air. The air is also cleaned to a certain degree. At the outlet end of the evaporative chamber, a bank of mist eliminator blades remove water droplets from the exhaust air stream. This system has the least pressure drop amongst the three system and works over a larger range of design conditions. Water treatment and nozzle maintenance are some of the disadvantages of this type of system.

Techniques described above will achieve saturation efficiency between 80% and 90% depending on the method of cooling the air stream.

4. PRESENT INVESTIGATION

The summer design data for Dallas, Texas, has been selected for the investigation. A heat pipe heat exchanger with 6 rows and 14 fins/inch (550 fins/m) has been selected. A typical performance curve [1] of a heat pipe heat exchanger, arranged in a counterflow configuration, for a coil face velocity of 500 fpm (2.5 m/s), and with equal mass flow rates is shown in Figure 7. The length of the heat pipe heat exchanger is taken as 10 feet with equal exhaust and supply lengths of 5 feet each. The height of the heat exchanger is taken as 4 feet. Media pad evaporative cooling method is used in this investigation.

5. RESULTS

In summer time operation, a heat pipe heat exchanger can be used to precool the supply air. As can be seen from the following data , application of indirect evaporative cooling principle to the heat pipe heat exchanger system will add more cooling of the supply air because of the greater temperature difference between the two air streams. Indirect evaporative cooling process is the evaporative cooling process in which the exhaust air stream is cooled by passing the air stream either through a water spray section, or through a filter bed of some type through which water is allowed to trickle. Calculations were done for the following summer design conditions for Dallas, Texas.

Supply entering dry bulb temperature	101 F
Supply entering wet bulb temperature	79 F
Exhaust entering dry bulb temperature	75 F
Exhaust entering wet bulb temperature	62.5 F
Saturation efficiency of 8" media pads	80%

Supply and exhaust air flow rates 10000 scfm

The calculated performance data is shown in Table I.

In order to clearly understand the mechanics of the principle of indirect evaporative cooling applied to a heat pipe heat exchanger system, the details of the calculation of Table I are given in the following section. The block diagram in Figure 8 shows the two systems with and without evaporative cooling. The plan view of the systems are shown.

The dry bulb temperature of the air leaving the evaporative cooler may be determined by

$$
\begin{aligned}
Tml &= Tme - SE\,(Tme - Tme,wbt) \\
&= 75 - 0.8\,(75 - 62.5) \\
&= 65\ F
\end{aligned}
$$

The temperature of the supply air leaving the evaporator can now be calculated for each case. Note that the exhaust entering temperature to the condenser for the evaporative cooling system has now been lowered to 65 degrees F.

Without Evaporative Cooling

$$
\begin{aligned}
Tsl &= Tse - E\,(Tse - Tee) \\
&= 101 - 0.60\,(101 - 75) \\
&= 85.4\ F
\end{aligned}
$$

With Evaporative Cooling

$$
\begin{aligned}
Tsl &= Tse - E\,(Tse - Tee) \\
&= 101 - 0.60\,(101 - 65) \\
&= 79.4\ F
\end{aligned}
$$

It may be seen from Figure 7 that the effectiveness of the heat pipe heat exchanger for 6 rows is 0.6. Figure 9 is the system block diagram with the above calculated values included.

6. SAVINGS

In order to calculate the cooling operation and equipment savings, a cooling equipment efficiency of 1.4 kW/Ton, and the cooling equipment cost of $1000/Ton is used. The reduction in cooling equipment cost may be determined by the following equation.

$$Tons\ saved = 1.08\ CFMs\ E\ (Tse - Tee)/12000$$

Assuming fan and motor efficiencies of 75 and 90%, the fan power requirements can be calculated from the following equation.

$$Power = SCFM\ Ha/(6346 * 0.75 * 0.90)$$

From the manufacturer's catalog of media pad evaporative cooler [2], a water flow rate of 5 gpm is needed for evaporative cooling with a pad which is 8 inches thick and 5 feet long (The manufacturer recommends 1.5 gpm per square foot of horizontal area of the media pad). For 5 gpm of water flow rate inside a 1.5 inch tube, the total head loss (frictional, hydrostatic, and minor losses) is 5.5 feet of water column. The power requirements for the recirculating water pump with an efficiency of 90 % may be calculated from the following equation.

Power = GPM Hw/(3965 * 0.9)

Without Evaporative Cooling	**With Evaporative Cooling**
Tons saved = 1.08 * 10000 * 0.60 * (101 - 75)/12000 = 14.0 Tons	= 1.08 * 10000 * 0.60 * (101 - 65)/12000 = 19.4 Tons
Equipment cost saved = 14.0 * 1000 = $ 14000	= 19.4 * 1000 = $ 19400
Power savings = 14.0 * 1.4 = 19.6 kW	= 19.4 * 1.4 = 27.2 kW
Power required by fans =10000 * 0.66 * 2/(6346 * 0.9*0.75) * 0.746 = 2.30 kW	= same as without evap. cooling = 2.30 kW
Power required for Media pad = 0	= 10000 * 0.14/(6346 * 0.9 * 0.75) * 0.746 = 0.24 kW
Power required by the recirculating pump = 0	= 5 * 5.5/(3965 * 0.9) * 0.746 = 0.006 kW
Net Power Savings = 17.3 kW	**= 24.7 kW**

As can be seen from the above calculations, savings of 38% and 43% is achieved in cooling tonnage and power. Again, this is due to increasing the initial temperature difference by using evaporative cooling.

7. CONCLUSIONS

The performance of an air-to-air heat pipe heat exchanger with media pad evaporative cooling has been investigated. The investigation reveals that by using evaporative cooling, the precooling of the air is increased resulting in a lower cooling operation and equipment costs.

8. NOMENCLATURE

CFM - Air volumetric flowrate, SCFM

E	-	Effectiveness of the system, %
GPM	-	Water flow rate, gallons per minute
Ha	-	Air pressure drop, inches W.C.
Hw	-	Water pressure drop, feet W.C.
SE	-	Saturation efficiency, %
Tee	-	Exhaust air entering temperature, F
Tel	-	Exhaust air leaving temperature, F
Tme	-	Media pad entering temperature, F
Tml	-	Media pad leaving temperature, F
Tse	-	Supply air entering temperature, F
Tsl	-	Supply air leaving temperature, F

<u>Subscripts</u>

dbt	-	dry bulb
wbt	-	wet bulb

REFERENCES

1. ASHRAE Equipment Handbook, Chapters 4 & 34, 1988.

2. Munters Catalog, "Evaporative Cooling Media," Fort Myers, FL, Jan. 1986.

TABLE I

Evap. Cool.	Sup. leaving (F) DBT	WBT	Exh. leaving (F) DBT	WBT	Press drop (" W.C) Supp	Exh	Med Pad
No	85.4	75.0	90.6	67.5	0.66	0.66	---
Yes	79.4	73.2	86.6	69.4	0.66	0.66	0.14

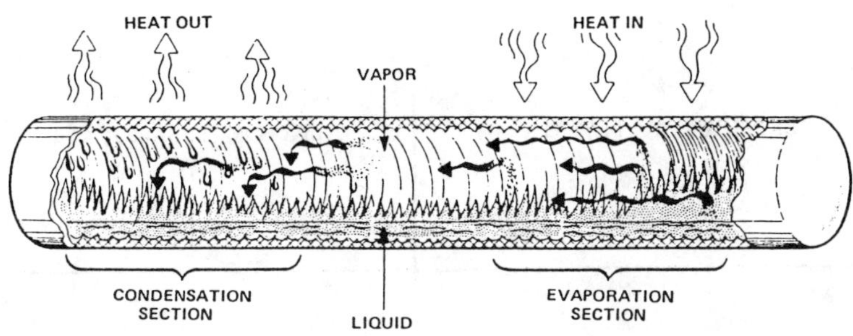

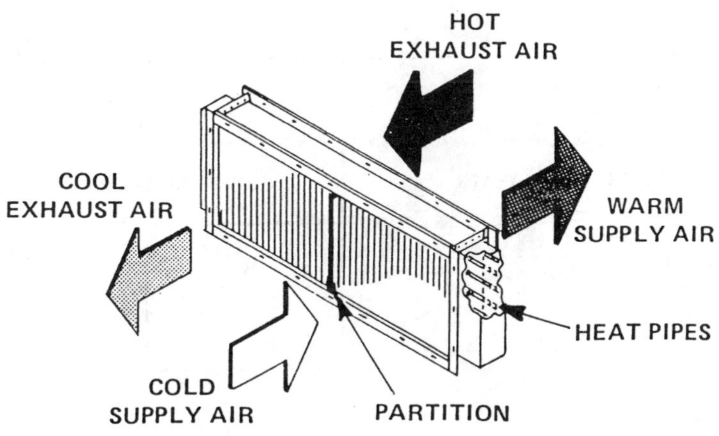

Figure 1. A Heat Pipe and a Heat Pipe Heat Exchanger

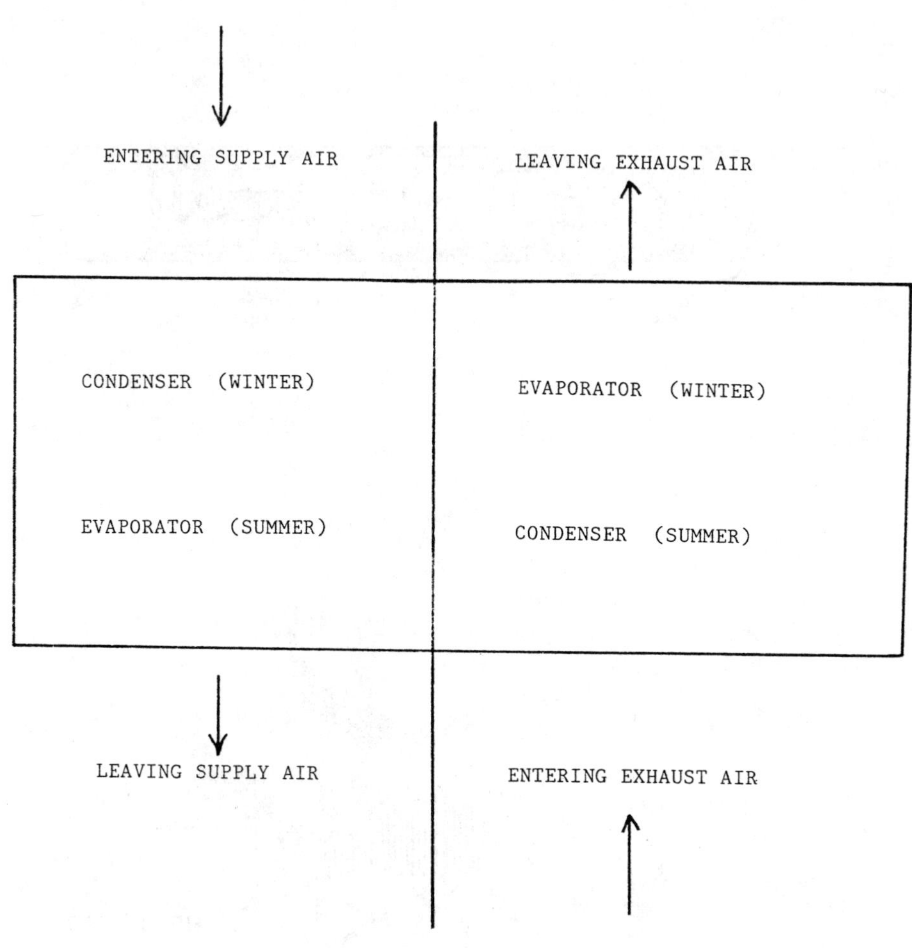

Figure 2. Operation of a Heat Pipe Heat Exchanger in Winter and Summer Seasons

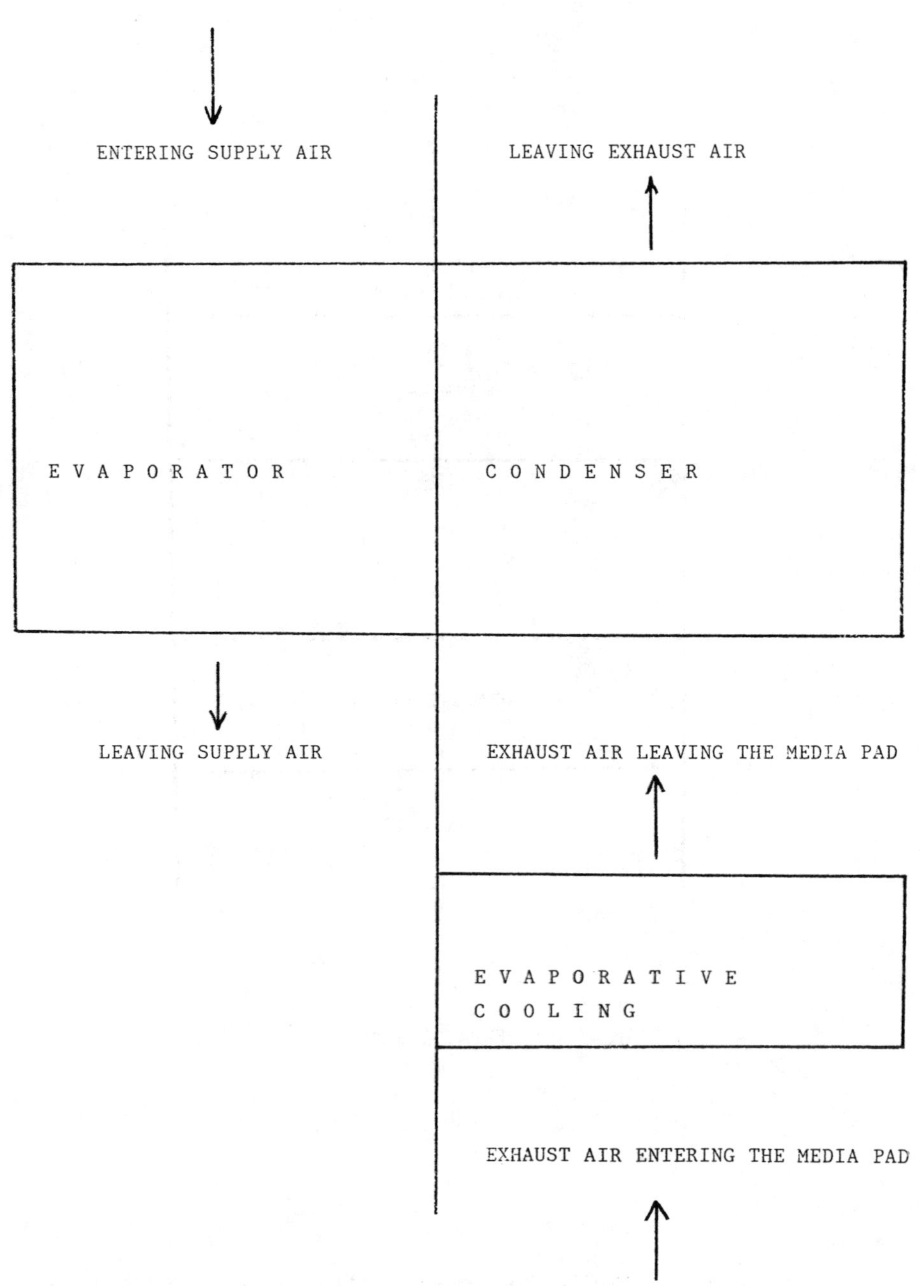

Figure 3. A Heat Pipe Heat Exchanger with Evaporative Cooling

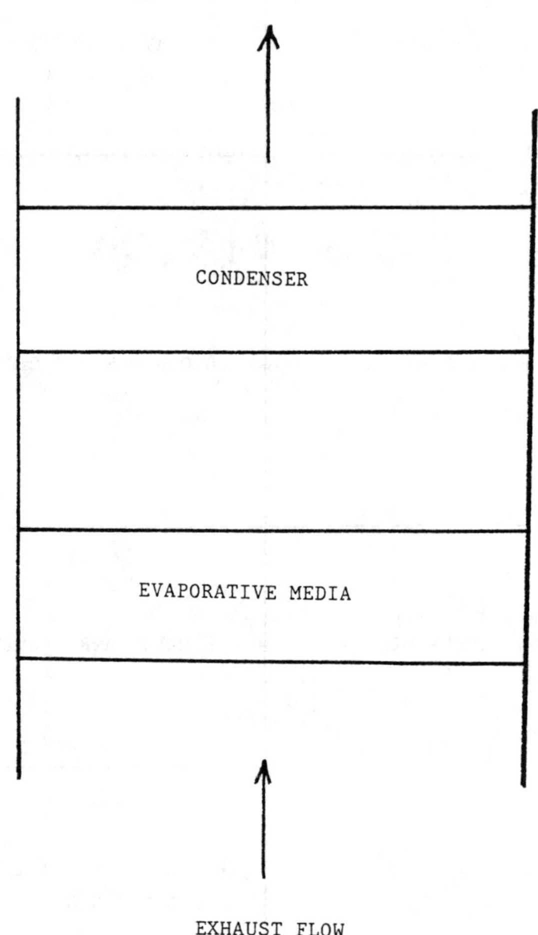

Figure 4. Media Pad Evaporative Cooling System

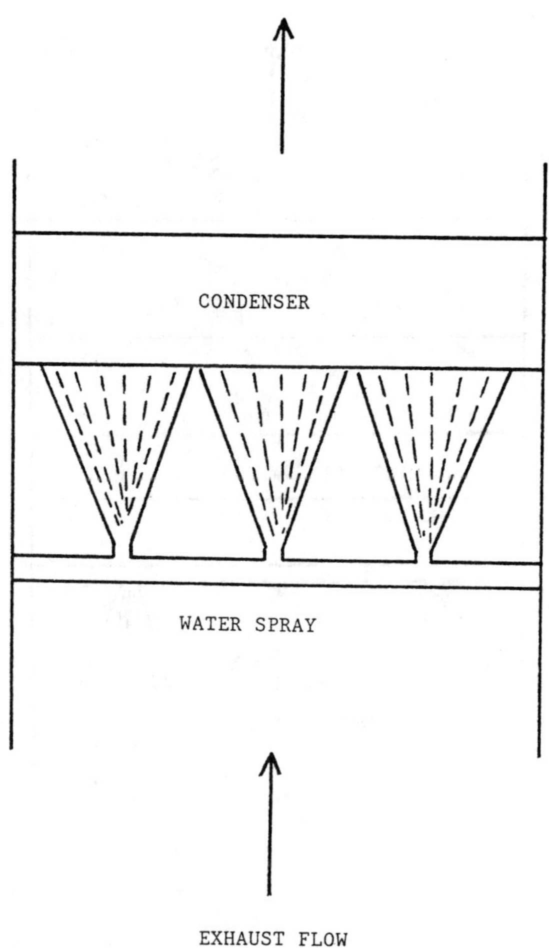

Figure 5. Direct Spray Evaporative Cooling System

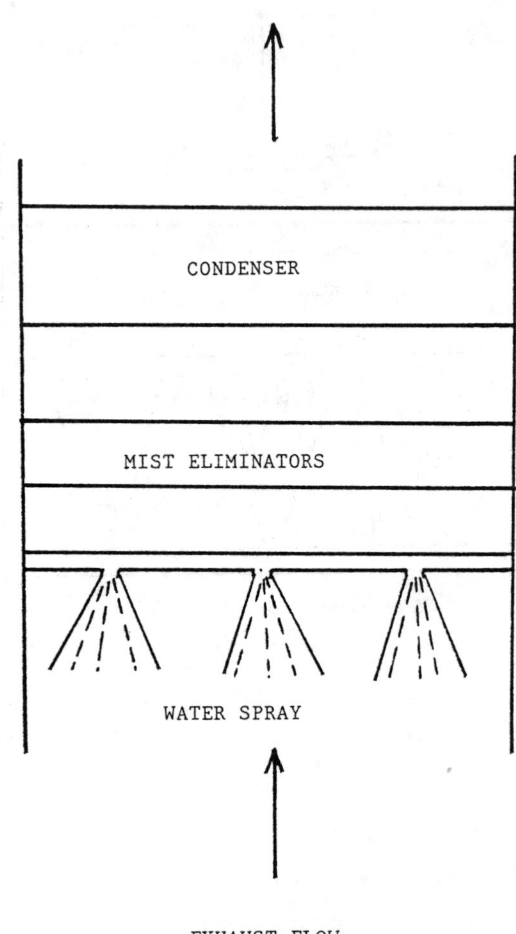

Figure 6. Air Washer Evaporative Cooling System

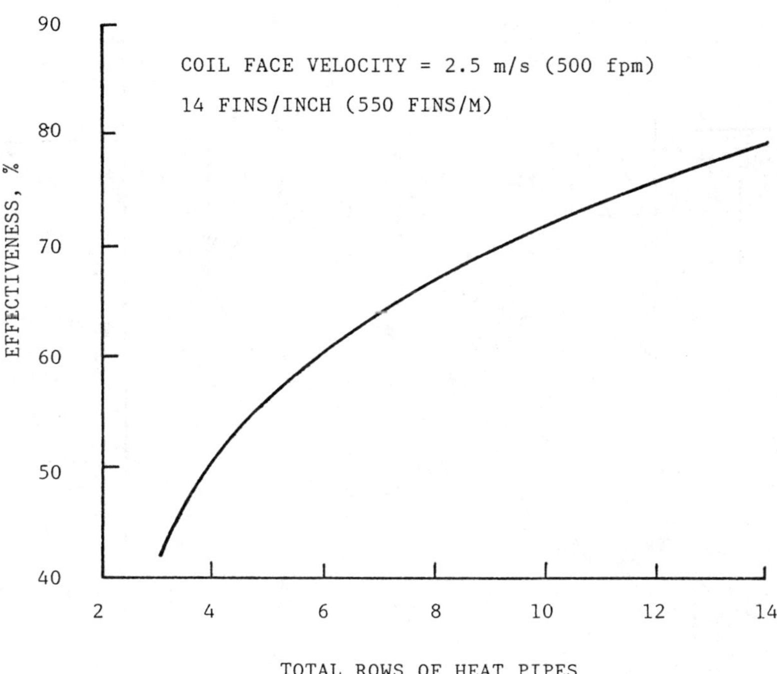

Figure 7. Effectiveness of a Heat Pipe Heat Exchanger versus Number of Rows [1]

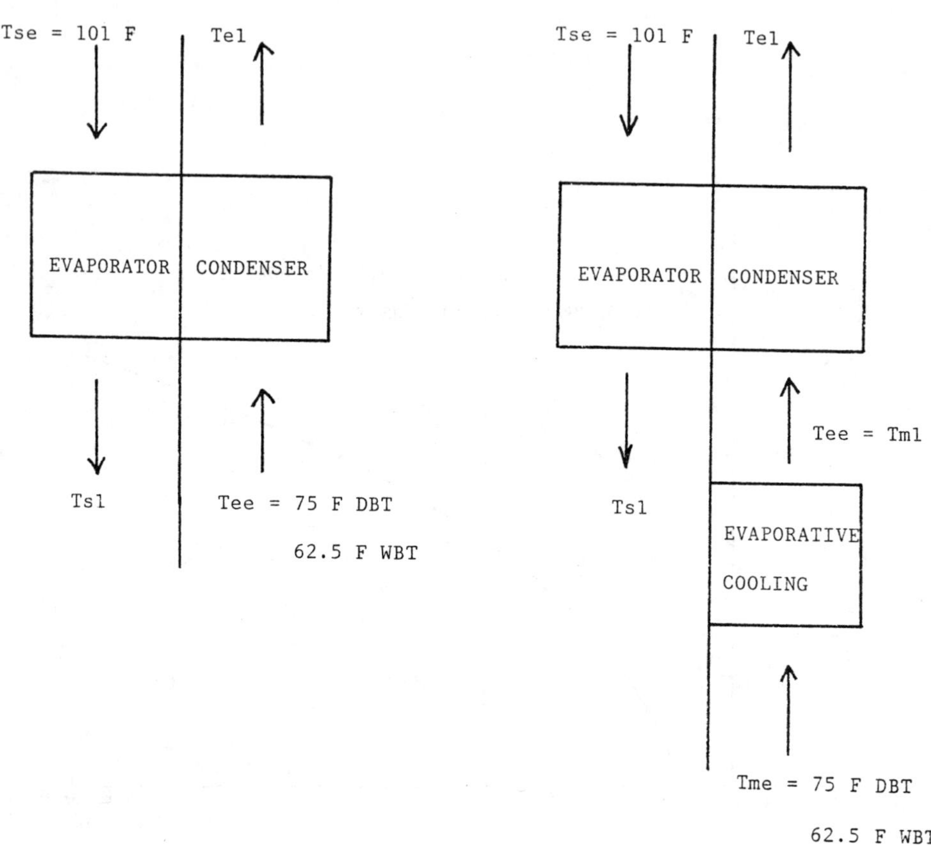

Figure 8. Block Diagram of the System

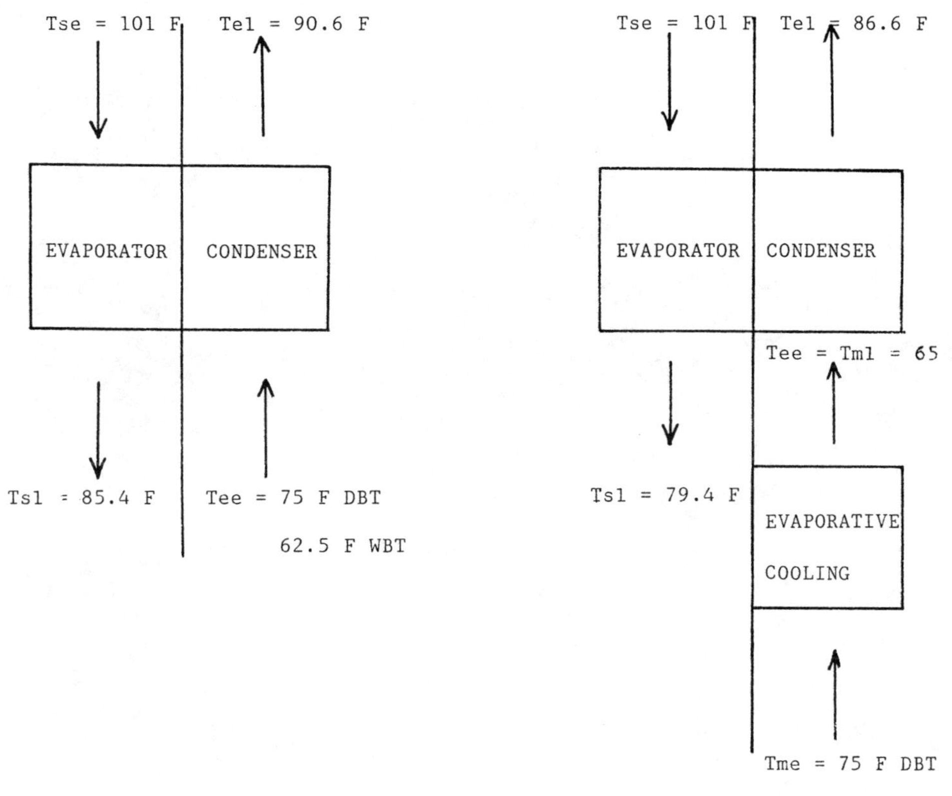

Figure 9. Block Diagram of the System with Computed Values

ENHANCED AND EXTENDED SOLAR EVAPORATION

Aghareed M. Tayeb
Faculty of Engineering , El-Minia University, El-Minia, Egypt

ABSTRACT :

The conventional solar distillation apparatus (commonly known as the solar still) , was first designed and fabricated in 1872 in Northern Chile . However no method has been indicated in the literature , up till now , for extending the operating time of the still as well as enhancing its productivity .

The present work deals with using different materials in the still for storing solar energy so as to increase and extend its productivity . Among the materials used were iron slag , white and colored wax and coal .

Comparing the productivity of stills containg these materials with a reference still showed that all these materials increased and extended the productivity with different ratios . The highest percentage increase in the productivity was due to coal followed by slag (29.4% and 23% , respectively) in the morning period , while the highest increase in productivity after sunset was due to slag followed by coal (62.8% and 39.9% , respectively) . Concerning colored wax , it caused an increase of 44% in the morning and 165.6% after sunset when compared to white wax . The increase in productivity of these stills compared with the reference still becomes more clear in the afternoon period , especially after sunset .

1. INTRODUCTION :

Water is a basic necessity for man along with food and air , thus the importance of supplying hygienic potable/fresh water can hardly be overstressed . Thus the impact of many diseases afficating mankind is drastically reduced if fresh hygienic water is provided for drinking . Further , the rapid industrial growth and population explosion all over the world has resulted in a large escalation of demand for fresh water , this in variably leads to acute fresh water shortages since the natural sources of water can meet the demands to a very limited extend .Added to this is the problem of pollution of the rivers and lakes by the industrial wastes and the large amounts of sewage , thus there is scarcity of fresh water even in cities , towns and villages near lakes and rivers .

The only inexhaustible sources of water are the oceans, lakes , seas and underground natural reservoir containing salt/brackish water . The chief drawback obviously is the very high salinity of such water . One of the attractive schemes to tackle the problem of water shortage is the distillation of such water , resulting in desalinated water .

The rapid escalation in the cost of fuels has made the solar alternative more attractive . The least that can be said in favor of solar distillation is that it is an available option for providing hygenic potable water for a single house or a small community in most places of the earth (1) .

The new concept of solar energy storage is used for the first time in this article for the purpose of enhancing and extending the productivity of a solar still . The materials used for this purpose are iron slag , coal , white and colored paraffin wax .

2. THEORETICAL

Basic Priciple of Solar Stills :

A conventional solar still is simply an air-tight basin , usually made out of galvanized iron sheet in rectangular shape. It has a top cover of any transparent material , e.g. glass , and the interior surface of its base is blackened to enable absorption of solar energy to the maximum possible extent .Brackish or saline water is poured into the still to fill it partially , and is then exposed to the sun . The glass cover permits solar radiation to get into the still , which is absorbed predominantly by the blackened basin . Consequently , the water gets heated up and hence the moisture content of the air trapped between the water surface and the glass cover increases . The base also radiates energy in the infra-red region which is reflected back into the still by the glass cover . Glass is not transparent in the long wave length region , thus the glass cover traps the solar energy inside the still , it also reduces the convective heat losses . The glass cover is usually sloped on one side to enable the water vapor , which condenses on the interior surface , to trickle into a collector .

Parameters Affecting the Productivity of A Solar Still :

The most important parameter affecting the output of a solar still is obviously the intensity of the solar radiation incident on the still and the transparent cover (2-4) , wind speed (5-15) , ambient air temperature (11-15) , and the operating parameters such as absorptance-transmittance properties of the still (17) and initial temperature of the water , brine depth (4,11,16) , basin insulation and heat losses (4,11,16,17) , still shape and orientation (4) , and cover slope (8,18) . Other operating parameters of importance are adding coal or dye to the saline water (4,19,21) , cooling cover (20) , stirring and channeling (32) and vapor tightness (19) .

3. EXPERIMENTAL APPARATUS AND PROCEDURE

The experimental set up consists of four separate single-sloped stills . These stills are numbered as follows :

Still No.	Characteristic Features
1	Containg slag as a heat absorbing and storage material
2	Containg only water , used as a refernce .
3	Containg white paraffin wax with the water
4	containing colored paraffin wax .
5	Containg coal as a heat absorbing and storage material .

The stills are supported on stands of 100 cm height and are covered with 3 mm thick window glass . The stills are constructed of 1 mm thick galvanized iron and are insulated at the bottom with asbestos. Each still is fitted with a small copper pipe for feeding water to be distilled , a thermometer for measuring the water temperature and a small copper pipe connected to a rubber tube for collecting the distilled water in a graduated cylinder .

The materials used to enhance and extend the productivity of the stills are :

Paraffin wax , colored paraffin wax , by adding 4% of its weight of a red color , crushed iron slag , used as a heat absorbing and storage material and put on the still bottom and crushed coal also used as a heat absorbing and storage material .

The parameters which were periodically recorded during test runs are : ambient air temperature (wet and dry) , intensity of solar radiation , wind speed , water temperature in the stills and glass cover temperature .

Stills are supplied with the same amount of saline water and left exposed to sun rays for evaporation to take place . The glass cover of the stills is swept with a wet fabric periodically to facilitate condensation of the evaporated water . The condensed water produced from each still is collected in a graduated cylinder .

The previously-mentioned parameters are recorded at one-hour intervals together with the volume of condensed water collected .

4. RESULTS AND DISCUSSION

Experiments were run for the object of enhancing and extending the productivity of stills through using different heat absorbing and storage materials for this purpose .

Effect of Using Slag on Productivity:

The results of this test are given in Table 1 and are plotted in Fig. 1 . The results could be summarized as :
a. A relatively small increase in water temperature results from the presence of slag in the still .
b. The percentage increase in productivity of the unit containing slag is comparatively much larger than the percentage increase in its temperature compared with the refernce still .
c. The percentage increse in productivity of unit 1 (containing slag) increases as time proceeds along the day and the maximum increase in productivity is noticed after sunset.

Effect of Using Paraffin Wax (White and Coloured) on Productivity :

The results of productivity of still 3 (containing white wax) and 4 (containg colored wax) compared with those of the reference still are shown in Table 2 and Fig. 2 .

The results indicate that adding a pigment to the paraffin wax increases its absorptivity for solar radiation , thus increasing its productivity compared with the still containing white wax and with the reference still .

In both cases , paraffin wax was placed in test tubes and the tubes are closed tightly and the corks are glued .

Effect of Coal on Productivity :

The results obtained for this test are given in Table 3 and Fig. 3.

It is cler that adding coal at the bottom of the still increases the productivity highly when compared with the reference still . The increase in productivity is becoming larger in the afternoon period which indicates the suitability of coal as a heat absorbing and storage material .

5. CONCLUSION

It is finally concluded that some additives when used as heat absorbing and storage materials increase the still productivity highly and extend its time even after sunset . The increase in productivity always becoming larger in the afternoon period .

The materials which were tried arranged according to the increase in productivity it cause were as follows : slag , coal , colored wax and , white wax .

The effect of these materials on productivity is given in Table 4 and Fig. 4 .

The percentage increase in productivity due to these materials ranges from 23% to 165.6% based on the productivity of the reference still .

This percentage increase varies from the morning period to the afternoon period or after sunset period depending on the absorptance and storage characteristics of the material used .

Coal led to the highest increase in productivity in the morning period due to its black color and its high absorption for solar energy , while slag gave the highest increase after sunset due to its high storage capacity . Thus a mixture of coal and slag is expected to give the highest productivity all along the day as well as after sun set .

REFERENCES

(1) Malik,M.S. and Tiwari , G.N. " Solar Distillation , Construction and Performance ", Pergamon press , (1982).

(2) Nasr , M.A. " Factors Affecting the Improvement of Solar Still Performance " M.Sc.Thesis , El-Minia University , El-Minia , Egypt (1982).

(3) Riera , J., et. al. " Efficiency of a Solar Still as A Function of its Main Parameters , Production Estimation for Valencia Spair " Proceedings , Cairo Symposium on Energy Resources , Vol. 2 , p.1474-95 , (1987) .

(4) Earg , H.P. and Mann , H.S. " Effect of Climatic Operational and Design Parameters on the Year Round Performance of Single-Sloped and Double-Sloped Solar Stills Under Indian and Zone Conditions ", J. Solar Energy , vol. 18 , p. 159-64 (1947) .

(5) Telkes , M. , Development Program Report " No. 13 , Dec. 1956 .

(6) Lof,G.O. and Eibling , J.A. " Energy Balance in Solar Distillation " AIChE. J. No. 4 (1961).

(7) Holard , K.G. " The Generation of Lithium Chloride Brine in Solar Still for Use in Solar Conditioning " J. Solar Energy 7,39 , (1963) .

(8) Cooper , P.F. " Digital Simulation of Transient Solar Still Problems " J. Solar Energy , 12, 313 (1969).

(9) Morse , P.N. " A Rational Basis for Engineering Development of a Solar Still " J. Solar Energy 7,39, (1963) .

(10)"Final Three Years Progress on Study and Field Test of Solar Sea Water Stills " O.S.W. Research and Development Progress , Report No. 190 (1966) .

(11)Elbling , J.A. et. al. " Solar Still Country Use Digest of Technology " J. Solar Energy Vol. 13 , p. 263-76, (1972) .

(12)Choraba , M.M., " Effect of Solar Still Internal Pressure on Yield " M.Sc. Thesis , Cairo University , Cairo , Egypt (1981) .

(13)Soliman , S.H., " Effect of Wind on Solar Distillation " J. Solar Energy , Vol. 13 , pp. 403-415 (1972) .

(14)Achilov , B.M. , et. al. , " Results of Year Long Tests of Tilted-Stepped Solar Still " J. Applied Solar Eneregy , vol. 8 , No.3 , pp.78-83 (1972) .

(15)Mohamed , M.A. , " Investigation of Engineering and Economical Evaluation of Solar Distillation " M.Sc. Thesis , Cairo University , Cairo , Egypt (1974).

(16) Bloemer , J.W., et.al. , " A Practical Basin- Type Solar Still " , J. solar Energy , vol. 9, No. 4, pp. 197-200 (1965).

(17)Bahadori , M.N. , and ElDin , F.E. " Improvement of Solar Still by the Surface Treatment of the Glass " J. Solar Energy , Vol. 23 , p. 17 (1979) .

(18) Soliman , S.H. and Kobayashi , M. , " Water Desalination by Solar Energy " , Paper presented at 1970 International Solar Energy Society Conference , Melbourne , Australlia (1970) .

(19)Akinsete , M. " A Cheap Method of Improving the Performance of the Roof Type Solar Stills " , J. Solar Energy , Vol. 23 , pp. 271-72 (1979) .

(20)Clueck , A.R. , " The Solar Desalination Process " , Third International Symposium on Fresh Water from the Sea " , Vol. 1 . pp. 633-54 (1970) .

(21)Tayeb , A.M. and El-Bassuoni , A.A. " A Study on the Effect of Different Colors on Enhancing Solar Distillation " Proceedings Great Plains Forum on Energy Management and Policy Analysis ,North Dacota , U.S.A. (1986) .

TABLE I: EFFECT OF USING SLAG AS A HEAT ABSORBING AND STORAGE MATERIAL ON PRODUCTIVITY(JUNE 13,1988)

Time No	Productivity , Cm^3		Temp. of Water, C	
	Still No. 1	Still No. 2	Still No. 1	Still No 2
9:30	0	0	34	34
10:30	16	13	45	45
11:30	28	19	54	55
12:30	90	60	58	60
13:30	145	95	60	60
14:30	215	135	61	61
15:30	280	175	61	62
16:30	340	212	58	59
17:30	340	240	53	53.5
18:30	410	250	46	48
19:30	420	258	40	40

TABLE II : EFFECT OF USING PARAFFIN WAX (WHITE AND COLORED) ON PRODUCTIVITY AND TEMP. OF WATER IN THE STILL (June 23,1988)

Time	Productivity, Cm^3			Temp. of water , C^o		
	Still No.4	Still No. 2	Still No.3	Still No. 4	Still No. 2	Still No.3
11:30	0.0	0.0	0.0	48	44	46
12:30	34.0	12.5	18	54	51	53
13:30	80	28.5	49	57	54	57
14:30	140	44	87	56	56	59
15:30	199	56	128	56	55	57.5
16:30	236	64.5	151	52	52	49
17:30	268	57.5	170	47.5	47	45.5
18:30	290	68.5	180	43	45	43
19:30	306	70	182	37	38	37
20:30	312	70	186	33	33	32

TABLE II: EFFECT OF ADDING COAL ON PRODUCTIVITY OF THE STILL (JUNE 22,1988)

Time	Productivity, Cm^3		Temp. of water, C^o	
	Still No. 5	Still No. 2	Still No. 2	Stll No. 5
13:30	0	0	56	60
14:30	72.5	56	56	61
15:30	140	102	55	59
16:30	197	133	49.5	54
17:30	230	155	44.5	49
18:30	250	173	4040	43.5
19:30	258	183	35	40
20:30	263	188	30	33

TABLE IV: ACCUMULATED RESULTS FOR THE EFFECT OF DIFFERENT ADDITIONS ON PRODUCTIVITY (JUNE 22,1988)

Time	Solar Intensity W/m^2	Ambient Temp. C	Still productivity, Cm^3				
			1	2	3	4	5
13:30	851.3	36.7	0	0	0	0	0
14:30	851.3	35.3	87.5	56	19.5	37	72
15:30	912.8	35.3	145	102	35	73	140
16:30	882.1	34.4	190	133	45	100	197
17:30	707.7	33.3	238	155	50.5	120	230
18:30	297.4	33.3	260	173	55	130	250
19:30	123.1	41.1	262	183	58	137	258
20:30	0.0	28.9	269	188	59	141	263

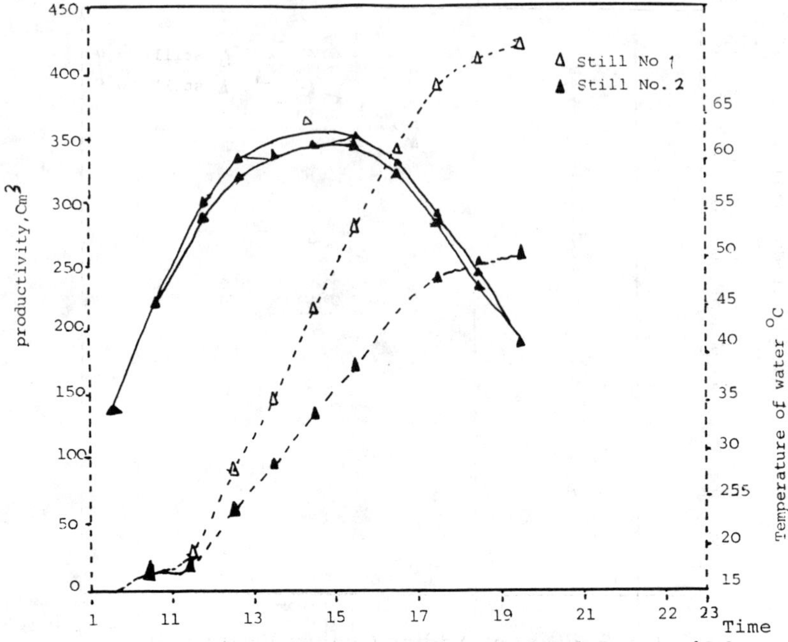

Fig. 1- Effect of Using Slag on Productivity and Temp. of Water (June 13, 1988)

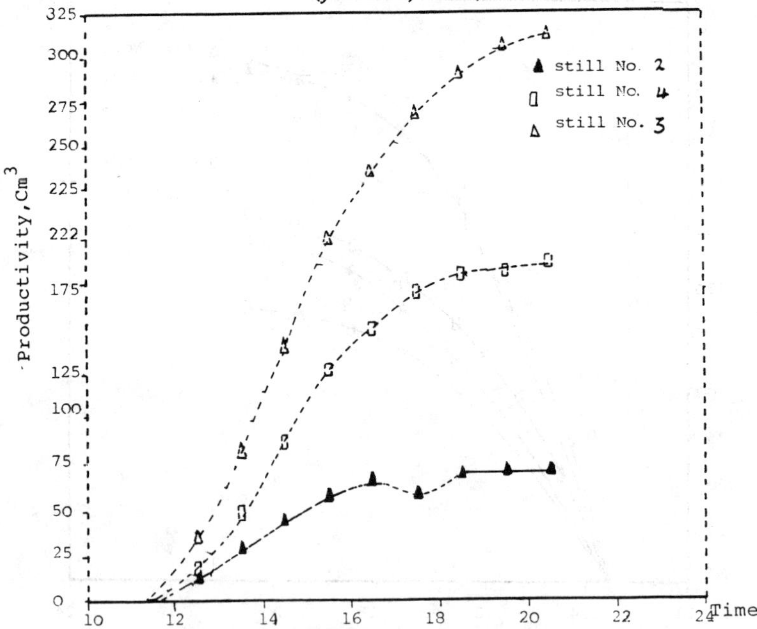

Fig. 2- Effect of Using Paraffin Wax (White and Colored) on Productivity (June 23, 1988)

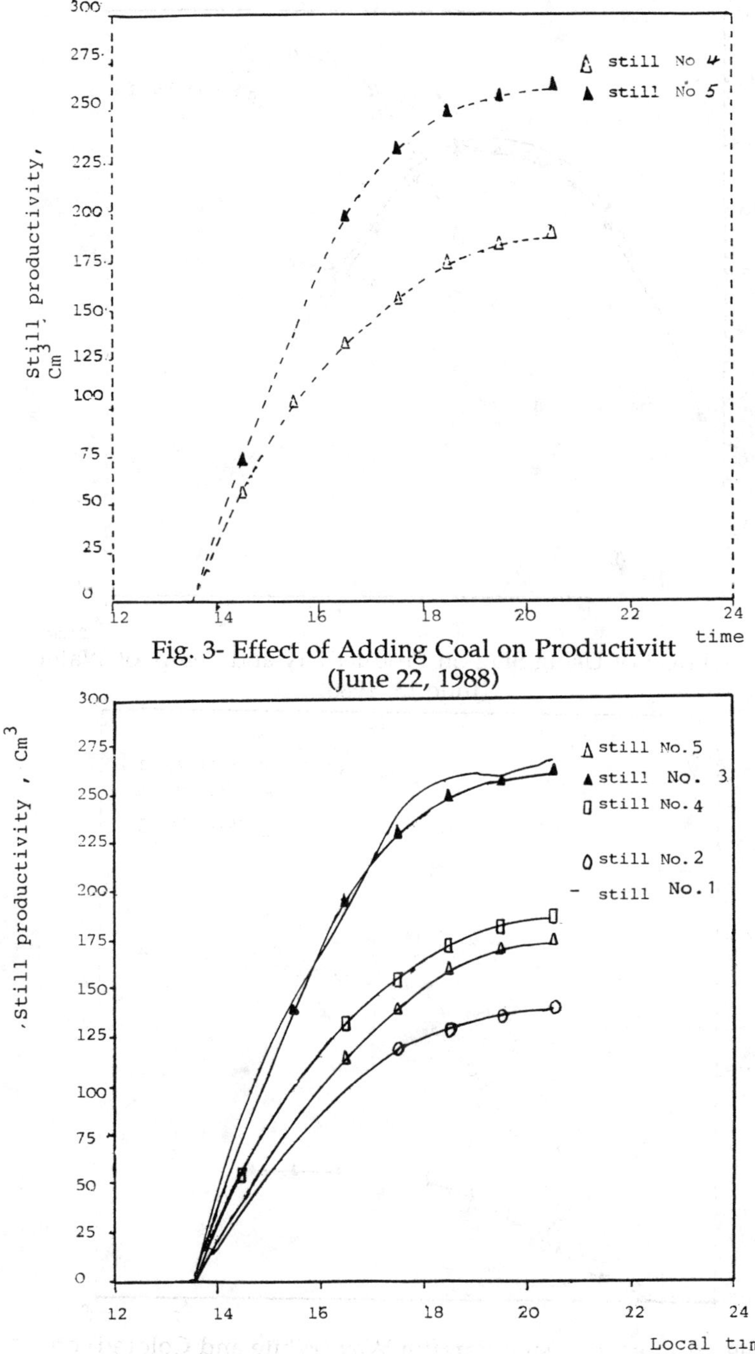

Fig. 3- Effect of Adding Coal on Productivitt (June 22, 1988)

Fig. 4- Effect of Different Additives on Still Productivity

PERFORMANCE EVALUATION OF HEAT SENSING TEMPERATURE CONTROLLERS FOR SOLAR DHW SYSTEM

C.S. Kochhar
Department of Mechanical Engineering
The University of the West Indies, St. Augustine, Trinidad, W.I.

ABSTRACT

This paper presents [illegible] of two heat sensing temperature control devices. One is mechanical [illegible] and the other an electro-mechanical device. The performance characteristics of both devices are analysed and their effects on [illegible] characteristics of Solar DHW is compared with the [illegible] Solar DHW system. To compare the performance, two identical solar DHW systems [illegible] were fabricated. One [illegible] conventional one and the other was installed with the control devices. From [illegible] results, the performance evaluation indicate a small improvement in the performance [illegible] device in the lower [illegible] range.

1 INTRODUCTION

The most commonly used solar energy device today is solar DHW system. Two basic types of systems are in general use. The first one is the thermosyphon type with either built-in collector hot water storage system or with separate collector and hot water storage systems [1, 2, 3]. The second type is the forced circulation type. The thermosyphon type of solar DHW system has found the most use because of its simplicity of design, construction and maintenance.

Solar DHW system can make significant contributions to national energy budget of small developing countries. Because of the increasing demand for solar DHW, small scale manufacturing plants have been established in Jamaica and Barbados in the Caribbean.

Conventional domestic water heaters ranges from 1 to 4 m² collector coupled to a 180-360 L (40 to 95 gal U.S.) storage tank which may be horizontal or vertical cylinder. Most of these domestic units are designed on the thermosyphon principle to transfer hot water from the collector to the storage tank.

The conventional Solar DHW systems have the inherent problem of temperature reduction in the hot water tank upon the withdrawal of hot water for domestic use. As hot water is drawn it is accompanied by cold water replacement at the bottom of the tank thus resulting in a constant dropping of the temperature of the water being used. Elimination of this problem could to a large extent determine the suitability of conventional solar DHW

PERFORMANCE EVALUATION OF HEAT SENSING TEMPERATURE CONTROLLERS FOR SOLAR DHW SYSTEM

G.S. Kochhar
Department of Mechanical Engineering
The University of the West Indies, St. Augustine, Trinidad, W I.

Abstract

This paper presents designs of two heat sensing temperature control devices. One a mechanical device (see fig.3,4) and the other an electro-mechanical device. The performance characteristics of both devices is analysed and their effects on performance characteristics of Solar DHW is compared with the conventional Solar DHW system. To compare the performance, two identical Solar DHW systems, using 1.75m^2 of collector surface, were fabricated. One was the conventional unit and the other was installed with the control device. Preliminary results of the performance evaluation indicate a small improvement in the performance of the system using control device in the lower collection temperature (45 to 55°C) ranges.

1. INTRODUCTION

The most commonly used solar energy device today is solar DHW system. Two basic types of systems are in general use. The first one is the thermosyphon type with either built-in collector hot water storage system or with separate collector and hot water storage systems [1],[2],[3]. The second type is the forced circulation type. Thermosyphon type of solar DHW system has found the largest use because of its simplicity of design, construction, operation and maintenance.

Solar DHW system can make significant contributions to national energy budgets of small dleveloping countries. Because of the increasing demand for solar DHW small size manufacturing plants have been established in Jamaica and Barbados in the Caribbean.

Common size for domestic water heaters ranges from 3 to 4m of collector coupled to a 180-360 ℓ (48 to 95 gal U.S.) storage tank which may be horizontal or vertical cylinder. Most of these domestic units are designed on the thermosyphon principle to transfer hot water from the collector to the storage tank.

The conventional Solar DHW systems have the inherent problem of temperature reduction in the hot water tank upon the withdrawal of hot water for domestic use. As hot water is drawn it is accompanied by cold water replacement at the bottom of the tank thus resulting in a constant dropping of the temperature of the water being used. Elimination of this problem could to a large extent, determine the viability of conventional solar DHW systems. Therefore there is need to develop a system, at reasonable cost and collector efficiency to overcome this problem.

2. EXPERIMENTAL INVESTIGATION

Description of the System

A schematic diagram of conventional thermosyphon solar DHW is shown in figure 1. Figure 2 gives a schematic of Solar DHW system with heat sensing temperature controller installed; this system is called control system. Table 1 gives some details of the two systems.

In the control system the flat plate solar collector is connected to the control storage tank (C.T.) about one eight the capacity of the main storage tank. Both these tanks are insulated, the main storage tank is not in the thermosyphon circuit.

The control tank functions as a storage tank in the thermosyphon circuit and delivers a batch of hot water (18 litres) at pre set temperature to the storage tank through the outlet-pipe. The control tank carries a float valve that regulates the supply from the domestic mains, the temperature control device, the cold water inlet, the bypass line and outlet control valve. Description here uses the mechanical temperature control device however, it is applicable to the electro-mechanical device as well.

The mechanical temperature control device is preset to the required temperature, by adjusting the caliberated barrel see figure 3 for detailed drawing of the device. When water in C.T. reaches the preset value the sensor sensing this valve activates the control device, causing the valve linkage to open the outlet valve. A batch of hot water then is released through the control valve into the main storage tank. This transfer lowers the float and the domestic main water valve opens replacing the exact quantity of hot water drawn out of the C.T. The domestic mains water bypasses the C.T. and is introduced into the thermosyphon circuit at a point just outside the C.T. as shown in figure 2. This bypass prevents premature dilution of hot water entering the main storage tank, and thus allows the control device to empty the C.T. preventing its premature closure. The inlet water temperature to the collector is also reduced.

The domestic main water travels the path a-b-c-d finally reaching the C.T. where temperature begins to drop. The control sensor sensing the change in temperature causes the control device to close the valve and prevent hot water below required temperature from entering the storage tank.

This cycle is repeated several times during the day at the preset temperature. Therefore the temperature of water in a properly insulated main storage tank should remain fairly constant without the dilution problem encountered in conventional systems. The stored hot water can be utilized for domestic or industrial applications.

The Mechanical Control Device

Details of design are given in figure 3. The main components of this device are:-

(i) Heat sensing cylinder

(ii) Pressure transmitting tube

(iii) Piston and cylinder

(iv) Control valve and connecting rod

(v) The presetting temperature barrel and spring

The heat sensing cylinder containing a very volatile fluid is placed in C.T. Vapour pressure increases with increasing temperature. This pressure is transmitted via the pressure transmission tube to act on the piston moving inside the cylinder. The movement of the piston is controlled by a preset spring (k_s = 1.65 kg/mm). The pressure exerts a force on the piston which moves when this force is sufficient to overcome both the preset spring force and friction in the piston cylinder system.

A valve connecting rod is attached to the piston. As the piston rises the connecting rod transmits this movement via the link to open the outlet valve, empting the C.T. into the main storage tank. Cold water replaces the hot water being drawn from C.T. thereby lowering the temperature in the C.T. This causes lowering of vapour pressure of the fluid in the heat sensing cylinder. When the force exerted on the piston reduces to a value lower than that exerted by the spring, outlet valve is returned to its closed position, stopping hot water flow.

The Electrical Control Device

Figures 4 & 5 show a schematic of this device. It consists of a gas thermometer with mercury contact switches, two batteries, a relay and a solenoid valve unit.

The gas-thermometer bulb is immersed in the control tank containing hot water. The principle of operation basically depends upon the expansion of mercury in a column or increase in temperature. As the mercury rises in the column it closes a different preselected switch, placed at various heights along the column. Each position corresponds to a particular temperature, determined by calibration. The switch in turn closes a circuit that activates the solenoid valve and allows the release of hot water at a preset temperature. The cold water from main supply causes the temperature in C.T. to fall and consequently the mercury column to drop opening the switch and closing the solenoid valve.

Measurements

The two systems were mounted side by side and tests were carried out simultaneously on both units. Solar insolation was measured by a pyranometer inclined at the same angle as the two solar collectors. Temperature measurements were made at various points in the system using copper constantan thermocouples. Efficiencies were determined for the two systems using recorded values of temperatures, insolation, stored water etc. using calorimetric principle [3].

3. DISCUSSION OF RESULTS

Table II gives sample set of data over an eight hour operation of the two systems. From table II it can be seen that for a control setting of 45°C for the Control unit and normal operation of conventional unit the efficiency of control unit is about 2% higher than that of the conventional unit. The advantage of the control unit is that the storage tank temperature has a higher constant value.

4. CONCLUSION

Preliminary results indicate a small improvement in efficiency of the system using control device in the lower collection temperature (45 to 55°C) range.

There still remains at this time need for more detailed experimentation to established conclusively the effects of control devices on performance characteristics of domestic DHW. However, it has been established that the problem of temperature reduction in the hot water storage tank upon withdrawal of hot water can be overcome using one of the proposed control device. with some improvement of the design.

ACKNOWLEDGEMENT

The author wishes to acknowledge the assistance of H.A. Brown, C. Cawley, W. Johnson and O. Lawrence in setting up of the experimental rig.

REFERENCES

1. Muneer, T. and Hawes, M.M., Experimental Study of the thermosyphonic and built in storage type solar water heaters. Proceedings ENERGEX 84. The global energy forum, Regina, Saskatchewan 14-19 May 1984, p 497-500.

2. Tet, Yin , Bong; Performance comparison of two types of solar collectors, Solar Energy Applications in the Tropics, D. Reidel Publishing Company 1983, p 129-137.

3. Duffie John A., and Beckman William A., Solar Engineering of Thermal processes. New York: Wiley C 1980.

4. Jordon Richard, C. and Benjamin Y.H., Editors "Applications of Solar Energy for Heating and Cooling of Buildings" ASHRAE GRP 170 (1977).

5. Miles Victor C., "Thermostatic Control Principles and Practices" 2nd Ed. London: Newnes Butterworths 1975.

6. Benham Peter P. and Wacnock F.V. "Mechanics of Solids and Structures" London: Pitman 1973 , 1976 Rev. ed.

7. Healey Martin, "Principles of Automatic Controls", London: Hodder and Stoughton 1975 3rd ed.

8. Sorour M.M., Development in solar water heaters, Energy Conservation Management Vol.25 No. 3 1985, p. 365-372.

9. Yellot, J.I. and Sobatka, R., "An Investigation of Solar Water Heater Performance" Trans. ASHRAE 70, 425 (1964).

10. Cohen, A.B., "Thermal Optimization of Compact Solar Water Heaters" Solar Energy 20, 193-196 (1978).

11. Chauhan, R.S. and Kadambi, V., "Performance of a collector - cum - storage type of solar water heater" Solar Energy 18, 327-335 (1976).

12. Garg, H.P. and Rani, U., "Theoretical and experimental studies on collector/storage type solar water heater" Solar Energy 29, 467-478 (1982).

TABLE 1: CONSTRUCTION DETAILS OF THE TWO SOLAR DHW SYSTEMS

Item	Conventional Thermosyphon Unit	Control tank Unit
Flat plate Solar collector	Tube in strip absorber area 1.75m^2	Tube in strip absorber area 1.75m^2
Storage tank volume	231.6 litres	18 litres control tank
Glass covers	two 3mm	two 3mm
Collector tilt angle	10° due south	10° due south
Collector insulation	5cm styrofoam plus reflective surface on casing	5cm styrofoam plus reflective surface on casting

TABLE 2: SAMPLE EXPERIMENTAL RESULTS

	TEST NO 1		TEST NO 2	
	CONVENTIAL UNIT	CONTROL UNIT	CONVENTIAL UNIT	CONTROL UNIT
Average Ambient Temp $\bar{T}_A$ ^oC	28	28	29	29
Average (HR) W/m^2	382.74 W/m^2	382.74 W/m^2	516	516
Average Initial Temp $\bar{T}_1$ ^oC	28.5	28.5	28.5	28.5
Average Finial Temp $\bar{T}_2$ ^oC	38.54	44	35.5	45
Experimental Duration $\Delta\tau$ Sec.	27000	27000	234006.5	23400
Volume of Heated Water ℓ	231.638	153.6	231.638	106.1
Average Energy Collected Q_u	$9.767 \times 10^6 J$	$9.999 \times 10^6 J$	$681 \times 10^6 J$	$7.353 \times 10^6 J$
Energy Rate Per m^2 q_u	212.8 W/m^2	217.9 W/m^2	171.19W/m^2	184.84
η = Efficiency q_u/(HR) %	55.6	57	33	36
Average Absorber Plate Temp T_p ^oC	63.0	65.0	76	75.0
Average Inlet Water Temp $(T_I)^oC$	38.0	39.0	29	40
Average Water Outlet Temp $(T_o)^oC$	44.0	45.0	49	48
Average Tube Temp Tr	55.5	55.0	62.2	61

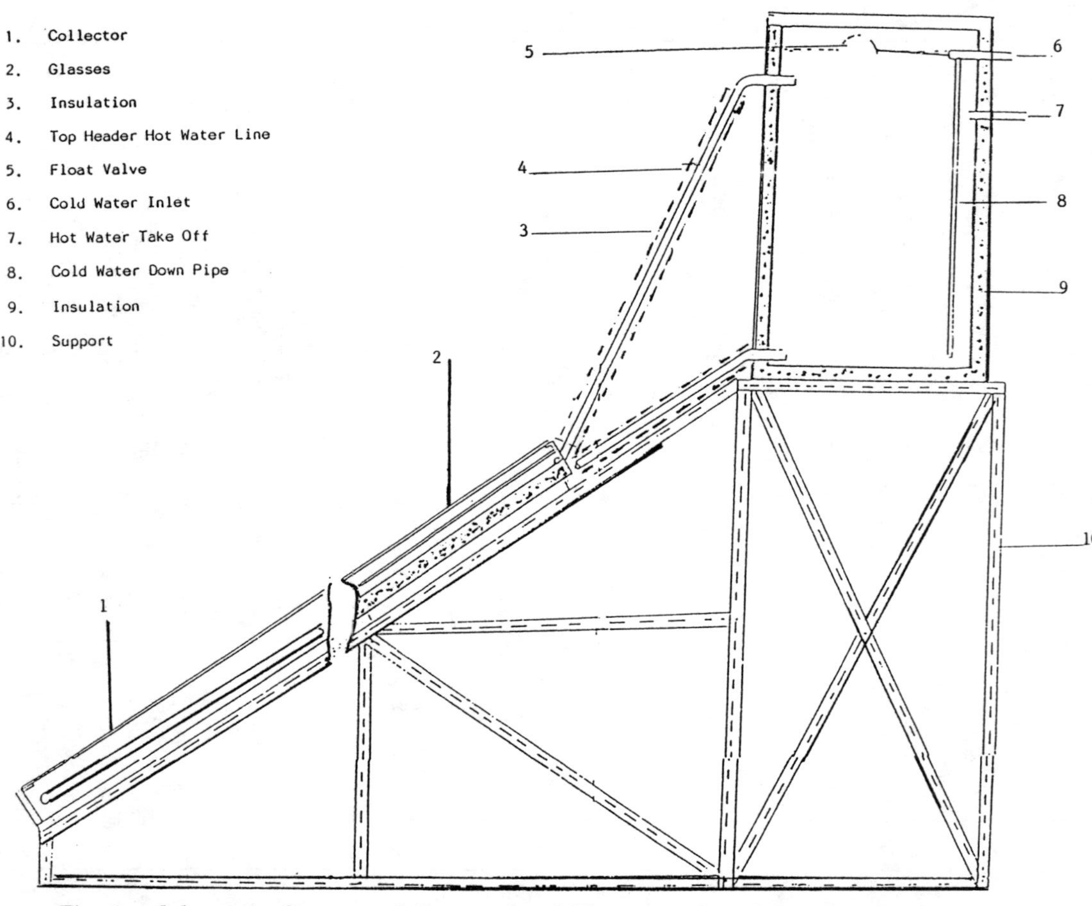

Fig. 1. - Schematic diagram of Conventional Thermosyphon Solar DHW System

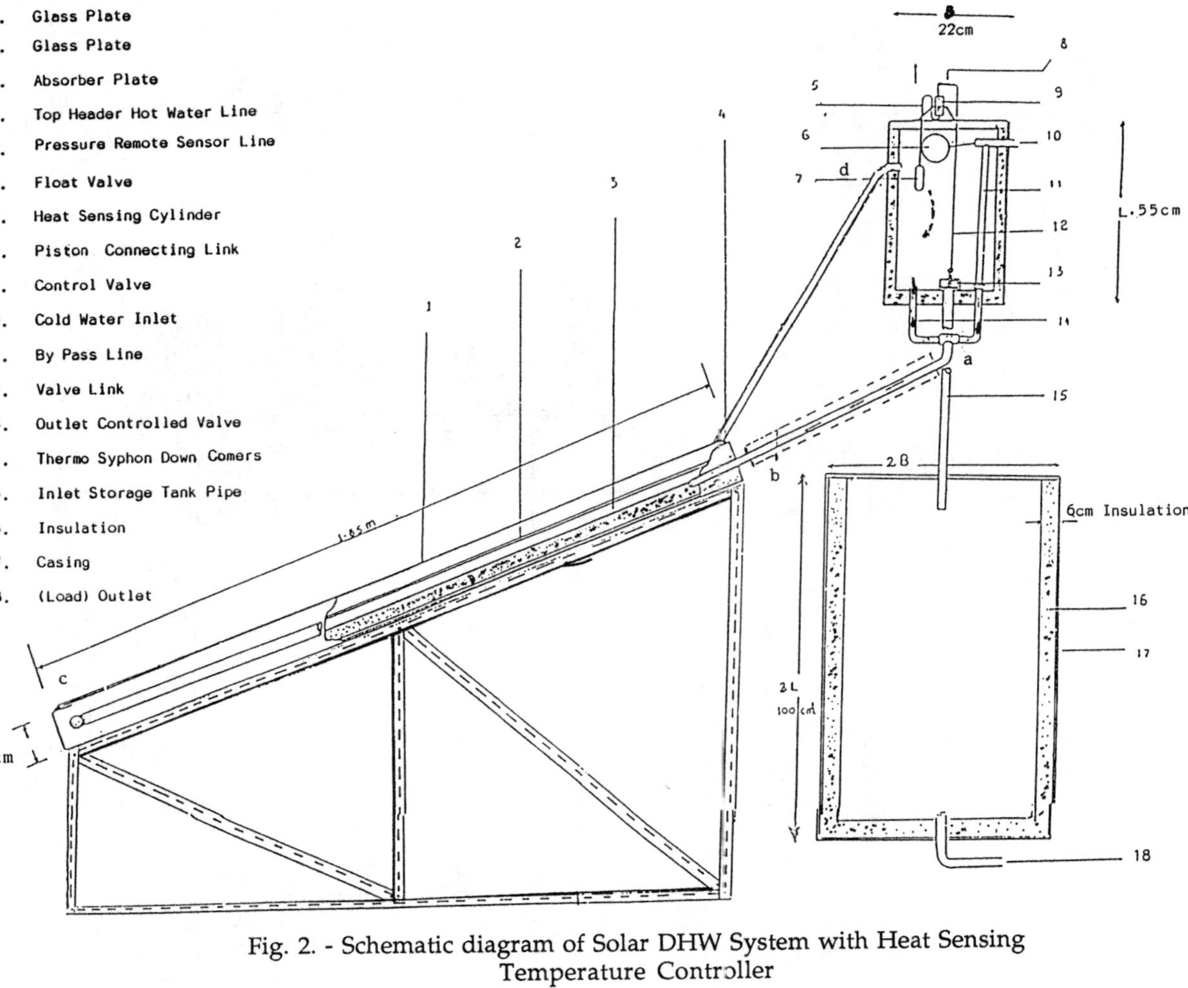

Fig. 2. - Schematic diagram of Solar DHW System with Heat Sensing Temperature Controller

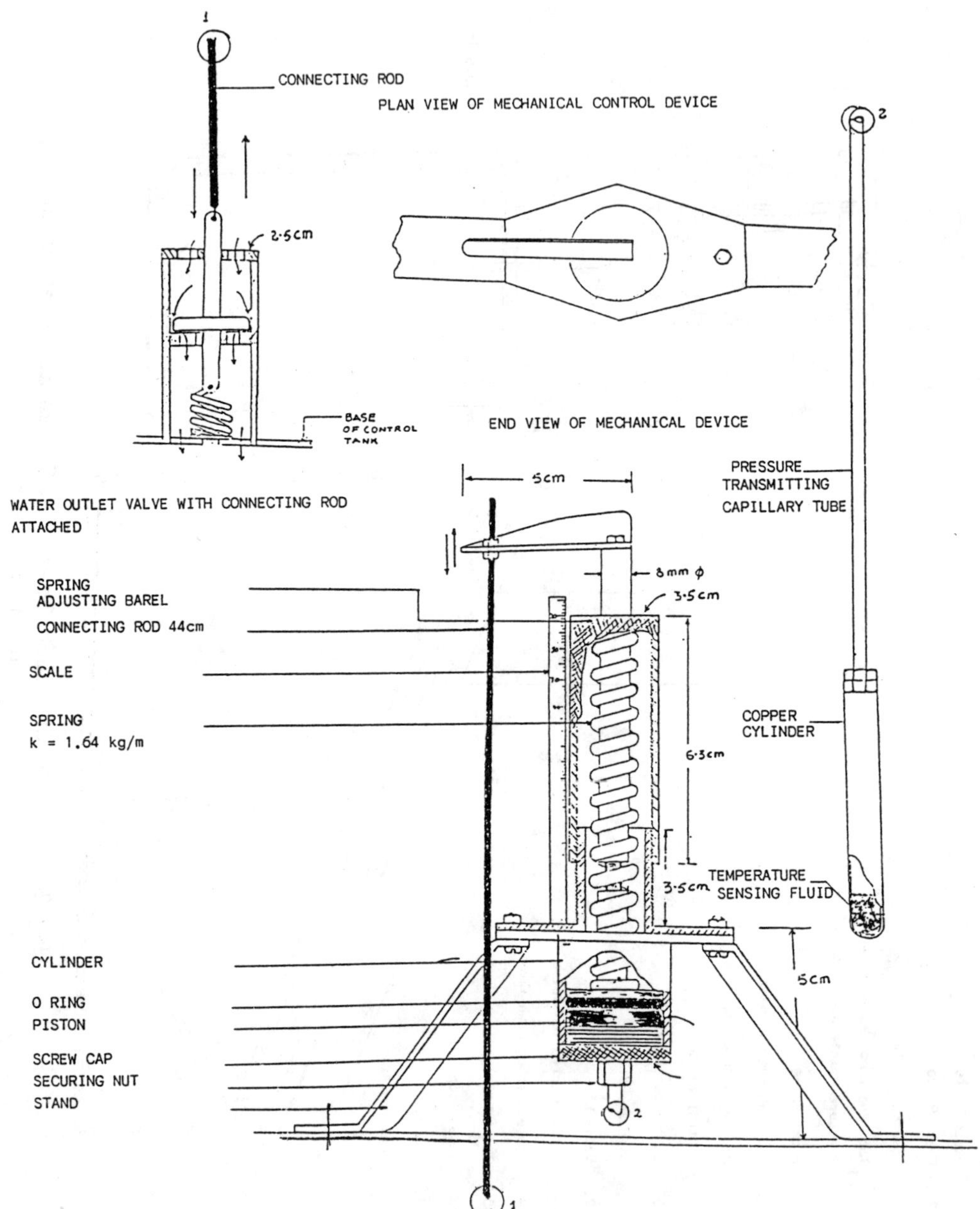

Fig. 3 - Mechanical Heat Sensing Temperature Control Device

REMOTE CONTROL TEMPERATURE SENSOR

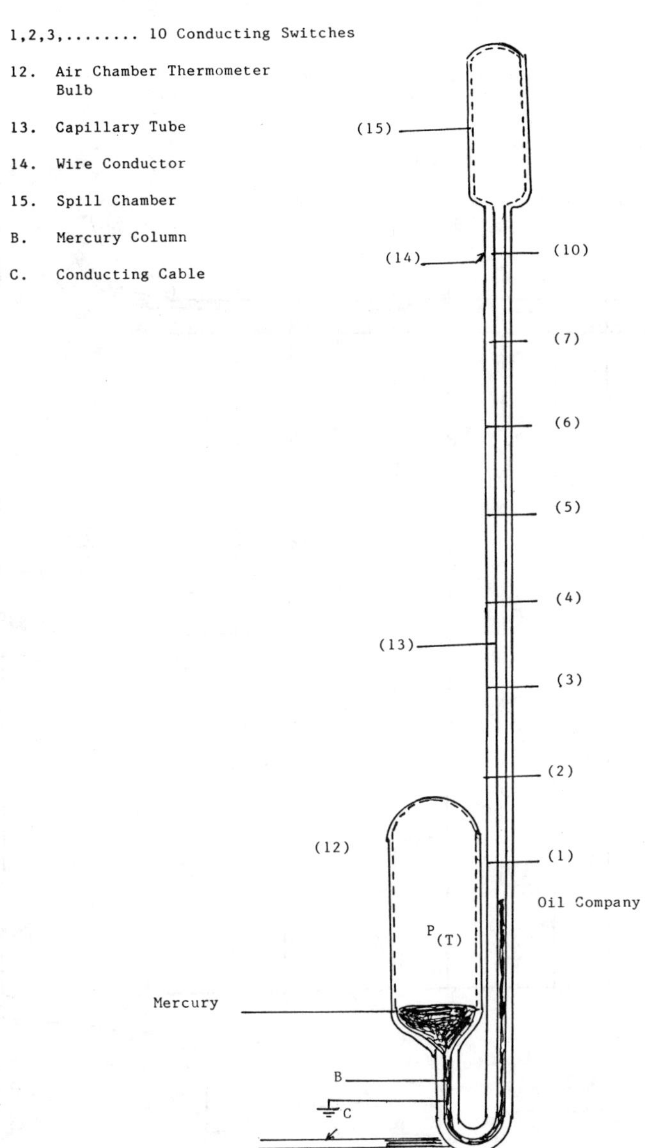

Fig. 4. - Schematic of Electro-Mechanical Heat Sensing Temperature Control Device

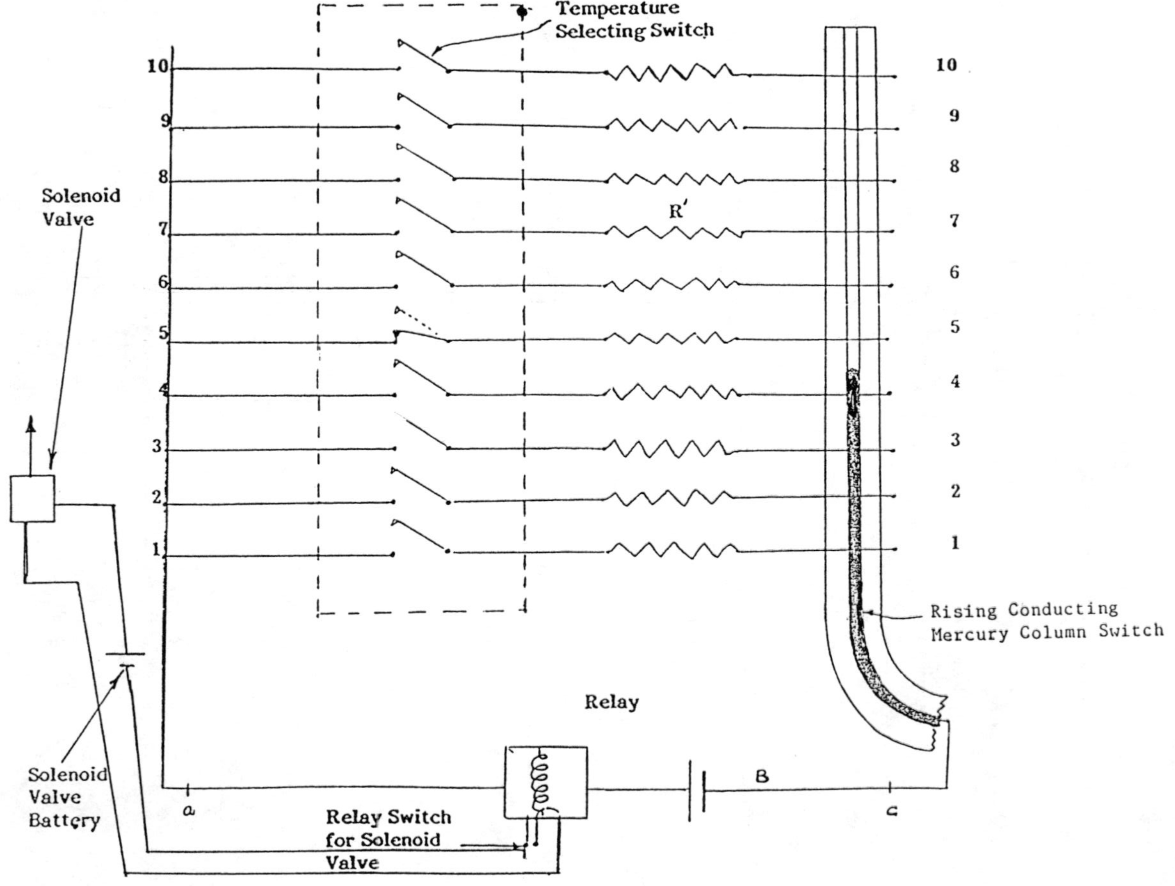

Fig. 5. - Schematic of Electro-Mechanical Control Device Switching Mechanism

PRESENT STATE AND PROSPECTS OF SOLAR HEAT-SUPPLY IN THE SOUTH OF THE EUROPEAN PART OF THE USSR

G.G. Gamkrelidze, N.V.Meladze, and G.V.Mindeli
PPO "Spetsgelioteplomontazh" Gruzglavmontazhstroi
USSR Ministry of Assembly and Special Building Works, 380002,
Tbilisi, Merkviladze st., 4-b, USSR

Abstract

Industrial utilization of solar energy in the USSR started in the first half of the 80's. By the middle of 1989 about 60000 m^2 of solar collectors were put into operation in the European part of the country, including more than 40000 m^2 in Georgia. The value of solar collectors area per capita in the Georgian SSR is about 0.01 m^2. Yearly thermal output from each m^2 of solar collectors reaches 2.9-4.4 GJ. Introduction of solar systems in Georgia and some other regions of the South of the European part of the country has been accomplished by the firm "Spetsgelioteplomontazh", established in 1984. By 2000 only in the European part of the USSR building of solar heat-supply systems with total collectors area about 1.5 million m^2 is expected.

The beginning of solar heat-supply studies in the USSR dates from the 30's. They consisted in development, construction, and test of experimental and pilot-plants solar systems of various types and purposes in order to determine their heat characteristics. According to experts' estimations, during this period several hundred solar systems have been built with the total collectors area about 3.000-4.000 m^2. However, being intended for solving research tasks, the great majority of these installations did not represent long-term and efficient solar systems which could have permanent heat consumers. Sometimes the materials used for their construction were deficient, while methods of production and assembling of elements of these systems were poorly adaptible to streamline production. Service life of such solar heat-supply (SHS) systems depended on duration of experiments and, as a rule, did not exceed one or two seasons. After winter pause these systems deprived of proper technical maintenance (since the users had no specialists in heliotechnology) not always could regain the level necessary for high quality work. As a result pretty soon they stopped working as ineffective or their working condition was maintained by interested research groups by their own efforts and at their own expenses. Nowadays such plants are almost in every Southern Republic of the USSR. However, even a large number of experimental and pilot SHS systems does not mean that solar heat-supply has reached needed range of application as a new practical alternative source of energy replacing traditional plants working on organic fuel.

The beginning of serial production of solar collectors was marked by putting into use a plant in the town of Bratsk (Siberia). As practice has shown, transportation of fully prefabricated collectors from Bratsk to southern parts of the country for a distance of several thousand kilometres with the following transfer to local transport leads to loss of marketable appearance, break-down of glass, deformation of the frame, etc. Then even after a thorough rejection these collectors are often installed by organizations for which the construction of SHS systems is beyond the scope of their bazic activity. In the absence of specialists in heliotechnology in these organizations the quality of installation work as well as adjustment and alignment is rather low. Thus installed SHS systems are transferred to users who, in their turn, have no operating experience as far as these systems are concerned. Hence, the cost of SHS systems is high, and their reliability is low. This results in discredit of solar heat-supply and hinders successful development of energy saving ecologically pure SHS systems.

In Georgia the problems of practical implementation of solar energy have been studying since the beginning of the 80's when under the Ministry of Assembly and Special Building Works (Glavupromontazhspetsstroi) of the Georgian SSR a small holding which produced solar collectors was established. The first solar system of hot water supply was put into use at one of the plants in Tbilisi in 1982. During 1982-83 still more projects of the Republic were supplied with such systems. Operating experience of SHS systems has shown their reliability and efficiency.

In April, 1984 the firm "Spetsgelioteplomontazh" engaged in production, assembly and adjustment of solar systems was founded. Since that time a new type of organization of SHS systems introduction has been forming. Concentration of production of solar collectors, assembly and adjustment of SHS systems in one specialized organization with subsequent handing them over to a customer on a turnkey basis has increased quality of the works connected with utilization of solar energy and as a consequence made the beginning of mass introduction of SHS systems into various branches of national economy.

However, the first three years have shown that such organization can be efficient only in the case of relatively small outputs and introduction of solar systems. Lack of a design department at the firm "Spetsgelioteplomontazh" has made the customers draw in their own or other nonprofiling design organizations for designing SHS systems. Naturally, this influenced the quality of design solutions as well as design work rates.

To increase introduction of the SHS systems and, hence, production of solar collectors, it became necessary to markedly expand the scope of activity and structure of the firm "Spetsgelioteplomontazh" which was accomplished in February, 1988, when the firm acquired its present-day character.

The firm "Spetsgelioteplomontazh", which is to service Transcaucasia and Southern regions of the European part of the USSR, is in charge of all kinds of works connected with introduction of the SHS systems, including design, manufacturing of solar collectors and mounting preforms; assembling and start and adjustment works; maintenance and repair of the SHS systems provided by the firm as well as design and technological works related to further improvement of the SHS systems constructions and lowering their specific consumption of materials. Therefore, the firm "Spetsgelioteplomontazh" incorporates all the stages of the heliosystem introduction process within a unified technological line which has led to an increase in economy and reliability of solar heat-supply and to awakening of interest in the SHS systems in potential customers and users.

The organizational structure of the firm consists of the managerial board, technical council, experimental research and design laboratory, special design office, a plant of prefabricated heliosystems and assembly exploitation sectors. The main task of the experimental research and design laboratory is to develop new solar systems:

- of industrial application which operate in the temperature range from 20 to 70°C with special regimes of daily, weekly, monthly and seasonal work in combination with boilers, heat pumps, cool-supply systems with peculiar resolutions over specifications, etc.;
- of individual application mainly intended for satisfaction of the demand of the people for hot water and heating;
- for drying various agricultural production, building materials and concrete, wood, died surfaces, etc.;

Tests of new construction materials for radiators, frames, thermal isolation of solar collectors, storage tanks, pipelines and connectors as well as tests for translucence and "vitality" of glass and polymeric films, nontoxic antifreezing agents, paints, selective coatings, etc., play an important role in activities of the laboratory.

All the materials for building the SHS systems are checked for correspondence of ratings to actual data. This allows us to work out concrete recommendations for engineers, designers, and installers of the SHS systems concerning utilization of each material.

For example, as a result of efforts made at the laboratory, a solar-assisted heat pump system for energy supply of health resort "Gumista" (Sukhumi) has been constructed. Besides, the laboratory in cooperation with research organization has developed new types of selective coatings. Deposition technology of these selective coatings demands less than 3 rates of electricity expense to compare with the world practice and, moreover, this technology provides complete ecological purity. The absorption coefficient of the above mentioned selective coating is 0.96, while the degree of blackness - 0.18.

The special design office performs design works in accordance with concrete applications and conditions of functioning of customers' heliosystems.

Today in connection with considerable increase in the volume of design works for the SHS systems introduction, the firm "Spetsgelioteplomontazh" with the Georgian Research Institute of Energetics and Hydrotechnical Constructions and other interested organizations has developed an automated system of design of solar heat-supply. The package of application programs is intended for optimal synthesis and design of the SHS systems, representing a complex of programs realizing the methods of circuit and structural and circuit and parametric optimization; hydraulic and heat engineering estimations of the SHS systems, formation and correction of data bases. The package is organized according to the modular principle, unified over all the initial, operating and resulting data massifs and is open, i.e. after unification and matching "input" and "output", any new module extending the package capacities can be connected to it.

The basic function of the plant producing mounting preforms, is manufacturing solar collectors, metalworks, pipe-lines, storage tanks and other auxiliary equipment for the heliosystems. At its disposal the plant has autonomous assembly and preform sectors, each of them services the adjacent territory within the radius up to 200 ÷ 300 m.

The main preform for producing solar collectors (a frame, an absorbing element, etc.) are manufactured at the plant. Assembly of the simplest metalworks and other auxiliary equipment is carried out at autonomous assembly and preform sectors for the projects situated within their activity. Large metalworks, individual storage tanks of large capacity, fixtures, pumps, and other basic complete materials are delivered to the autonomous sectors by the plant. This organization of production and delivery as a complete set allows us to reduce transportation expenses, to bring production of elements of the SHS systems closer to customers and to shorten the dates of these systems construction.

Along with production of solar collectors and other equipment for the SHS systems, the plant and autonomous sectors produce small individual solar systems to orders of population. As a rule, these works are performed on a self-sufficiency basis, since the sectors are supplied with all the necessary equipment for fulfilment of such orders in full extent.

Installation, adjustment, operation and maintenance of the SHS systems is carried out by mounting and operating sectors within the zone of their activity. Today our firm includes such sectors established in Tbilisi, Kutaisi and Rustavi. As practice has shown, each sector can service a territory within the range of 100-150 km from the place of its location. This choice is dictated by necessity of providing service to a customer during a working day so that the worker could return to the base of the sector at reasonable time.

As the operating experience of our firm (1984-1987) has revealed, sometimes having withstood adjustment tests, a SHS system did not attain required characteristics in service (especially one or two seasons later) and even stopped working. After examining several such systems we have come to a conclusion that ineffective work was caused by the following reasons. Having put the SHS system into operation, the customers not always kept to the service instructions. Also, due to lack of experience in treatment of solar systems, a small fault which would have been easily cleared by a specialist, could result in shut-down.

The above mentioned factors as well as growing introduction of the SHS systems and hence an increase of the number of projects where such systems are employed, made it possible to conclude that the most reasonable solution of the problem of high-quality operation of the SHS systems is establishment of a special service for technical maintenance and repair of heliosystems. As a result, within the mounting and operating sectors brigades were formed providing technical maintenance and repair of installed SHS systems which made it possible to improve real operating characteristics and reliability of active heliosystems as well as of those under construction.

The results of the activity of the firm "Spetsgelioteplomontazh" in the sphere of introduction of the SHS systems (the data for 01.01.1989) are summed up in Table I which permits us to trace the dynamics of solar systems erection. During the first six months of 1989 additional 13 thousand m^2 of the SHS systems were introduced.

The practice has shown that the values of the operating characteristics of the SHS systems (including, apart from efficiency, a number of factors relating to peculiarities and errors of design, mounting, adjustment, operation) have reduced from 0.45 ÷ 0.55 for the systems with the collector area of $10 m^2$ and less down to 0.3 ÷ 0.4 for the systems with the collector area of 500 m^2 and more.

The comparison of specific investments in the SHS systems over the whole range of constructed systems evidences of slight differences in extreme values: from 122 roubles/m^2 for the systems with area of $\leq$ 10 m^2 down to 117 roubles/m^2 for the systems with area of 1000 m^2. Presently in connection with rise in price of the materials used for production of solar collectors and other elements of the SHS systems, specific cost of the solar systems rose on the average up to 140-160 roubles/m^2.

Information on structure of solar systems introduced by the firm "Spetsgelioteplomontazh" concerning the area of solar collectors and composition of consumers is presented in Tables II and III, respectively. By the beginning of 1989 fuel savings attained during the time of the SHS systems utilization amounted to $>$ 7 thousand t.c.f.

The experience gained by the firm "Spetsgelioteplomontazh" as a result of introduction of the SHS systems has revealed characteristic problems and prospects of solar energy utilization.

Nowadays one of the most typical cases of solar energy utilization in the USSR is employment of the SHS systems as a source of hot water supply during non-heating periods. On the basis of our experience it can be concluded that to use solar water systems for heating is not very efficient. It seems more preferential to construct passive air solar systems for this purpose.

It is most of all expedient to utilize the SHS systems in the Southern parts of the USSR, in the regions of decentralized heat supply where the sources of heat energy are represented by small installations on liquid and solid (coal, wood) fuel. Of special interest is utilization of the SHS systems for hot water supply instead of electrical installations in recreation zones where other finds of energy are not desirable because of ecological requirements and during hot summer periods in remote regions where solar energy can replace very expensive fuel and electric power received from diesel power plants.

Solar energy for hot water supply should be considered as seasonal energy of various duration of practical application (Table IV). At projects of twelve months' operation (a dwelling house for one family of a farmstead type, a block of flats standing alone, sanatoriums, holiday homes, amenity rooms of enterprises, etc.) which have their own boiler rooms for heating and hot water supply all year round, application of the SHS systems during non-heating period (in the South it lasts for 8 months and more) leads to savings of energy sources of boiler rooms, to preventing from noxious emissions into the atmosphere, to partial dismissal of personnel of boiler rooms and to reduction of operating expenditures in this period.

For the SHS systems to operate all year round it is necessary to have an unfreezing heat-transfer agent in the first circuit of the system, which results in rising its price, or automatic equipment permitting to empty this circuit before coming of subzero temperatures which after repeated use will adversely affect reliability, ease of operating and loss of water.

For the SHS systems to go to twelve months' operation it is necessary to create a inexpensive nontoxic antifreezing agent corresponding to hygiene standards which will not freeze down to -20 ... -30°C.

Another sphere of large-scale utilization of the SHS systems should be projects of seasonal operation of various operational duration: long-term ones functioning during all the main recreation periods - 6 months (tourist camps, campings, etc.) and short-term ones functioning for 3 months, e.g. pioneer camps during school holidays, or summer cottages with permanent dwelling in them in summer and periodic visits (week-ends) during the rest of warm season which makes one more month.

The number of short-term operation projects is especially large in recreation zones where during mass summer holidays the number of public utility establishments and trade enterprises increases drastically (shower-bath installations and pavilions, temporary dining-rooms and cafes, etc.). Here in many cases as hot water supplies various electric water-heaters of direct action were used as well as small furnaces with a small utilization factor (UF) and low level of servicing and reliability of supply, in which fuel compbustion results in noticeable pollunation of environment. Certainly, electric heating of water with heat accumulation should be considered as one of efficient means of hot water supply at a comparative economic estimation of the SHS systems. However, it should be born in mind that massive utilization of accumulating electrical water heaters is possible only in the case of joint energetic systems with inexpensive summer electric energy in the night gap of the load curve schedule with free capacities of power networks and transformer capacities in the zone of decentralized power supply. Otherwise, expenses on electric power of franco-electroaccumulating installation considerably increase and approach the expenses on direct electric water-heater.

The characteristic feature of thermal load of hot water supply is its differentiation by temperatures needed for heating (see Table IV). Thus, water temperature of 40°C is sufficient for hygiene needs; water temperature of 60°C and higher is necessary for washing dishes and linen, which requires additional heating of water intended for this purpose. The cited temperature levels are peculiar to hot water supply. Also, it should be noted that a potential large-scale consumer of solar heat are swimming pools, for which water temperature should be kept at 28°C and 10% of the water capacity should be replaced daily.

The temperature for heating water in the SHS systems should be kept at 40°C. Temperature rise at the exit of the SHS system up to higher values requires economic substatiation, in which a solar system with 40°C at the exit and hot water supply on conventional energy suppliers should be considered as a basic version. Under these conditions the conventional heater can also be used as a duplicate source for the most important group of consumers or for all consumers of a seasonal project. Differentiation of the demand for heat according to heat temperatures makes it possible to attain flexibility of fitting temperature characteristics of the SHS systems to real conditions of a concrete user. This, on the one hand, rises consumer efficiency of the SHS systems, while, on the other hand, reduces expenses on such systems by 5-15%.

In summary, we should like to emphasize that the described structure of the firm "Spetsgelioteplomontazh" allows us to provide complete quality control of the produced solar collectors and other elements of the SHS systems, to carry out construction and installation work connected with erection of these systems in due time to provide their long-term operation with estimated performance, to quickly respond to consumers' demands in order to expand the sphers of effective utilization of solar energy in national economy.

It is expected that by 2000 only on the territory of the European part of the USSR the SHS systems of various applications will have been introduced with total area of 1.5 million m^2.

TABLE I: DYNAMICS OF INTRODUCTION OF THE SOLAR HEAT-SUPPLY SYSTEMS BY THE FIRM "SPTSGELIOTEPLOMONTAZH"

Years	Total for a year, m^2	including				
		industry, construction, transport	agri-culture	sanatoriums, rest-homes, holiday-hotels, hotels	sports and health es-tablish-ments	individual dwelling sector
Till 1984	50	50	-	-	-	-
1984	335	145	124	-	66	-
1985	5284	1997	235	2801	202	49
1986	10018	3776	1142	5019	-	81
1987	10426	3806	1904	3946	711	59
1988	20790	8620	3087	8251	682	150
TOTAL:	46903	18394	6492	20017	1661	339

TABLE II: STRUCTURE OF SOLAR HEAT-SUPPLY SYSTEMS BY THE COLLECTORS AREA

Collector area of one SHS system m^2	The number of SHS systems pcs	Total area of collectors m^2	Total fuel saving per annum t.c.f.	Specific fuel saving $\frac{t.c.f.}{1000\ m^2}$
Till 10	90	364	49	135
11-50	56	1754	210	120
51-100	49	3804	418	110
101-200	76	10819	1104	102
201-500	59	18155	1763	97
more than 500	16	12007	1140	95
TOTAL:	346	46903	4684	-

TABLE III: STRUCTURE OF SOLAR HEAT-SUPPLY SYSTEMS BY TYPES OF CONSUMERS

Consumer	Number of SHS systems pcs	Unit area of SHS system m^2	Total area m^2
Industry, construction, transport	140	2 - 1500	18394
Agriculture	43	3 - 648	6492
Sanatoriums, rest-homes, holiday-hotels, hotels	68	24 - 1036	20017
Sports and health establishments	10	12 - 569	1661
Individual dwelling sector	85	less than 10	339
TOTAL:	346	-	46903

TABLE IV: TYPICAL CONSUMERS OF HOT WATER SUPPLY AND TRADITIONAL HEAT GENERATING PLANTS IN THE ZONE OF DECENTRALIZED HEAT SUPPLY

Typical consumers			Replaced traditional heat generating plants			
Period (season) of operation, months	Type of consumer	Character of hot water(HW) consumption	Type of traditional heat-generating plant	Replaced power resources	Operating utility factor(UF) %	Specific reduced expenses on heat generation*) roubles/GS
I	2	3	4	5	6	7
	Individual dwelling house	75% $40^{\circ}C$ 25% $65^{\circ}C$	In-house heat generator	Condensed gas, coal, wood, electric power	45-50	10-15
12 months	Block of flats, hostel, hotel, sanatorium, rest-home, holiday hotel, sport structures, car-washing installation, etc. enterprises of industry, construction, transport, agriculture	Hygiene needs - $40^{\circ}C$, Public catering establishments - $70^{\circ}C$, swimming-pool - $28^{\circ}C$	In-house boiler rooms, small industrial boiler rooms, electric boiler rooms	coal, residual oil, electric power	50-60	7-13

1	2	3	4	5	6	7
5-6 months, middle of IV-middle of X	Recreation centres, trampers' bases, campings, temporary projects of public and seasonal utilities in the period of massive holidays	Hygiene needs 40°C public catering establishment 70°C, swimming-pool 28°C	Small hot-water boiler rooms, water heaters	Coal, residual oil, electric power	50-60	10-15
3-4 months, week-ends in autumn and spring months	Country-cottage, garden-plot	Shower-bath system 40°C	Hot-water column, electric water heater	Condensed gas, wood, electric power	45-50	13-15
3 months, VI-VIII	Pioneer camps, summer mountaneering and tourist camps	Hygiene needs 40°C public catering establishments 70°C	Small hot-water boiler rooms, water heaters	Coal, residual oil, electric power	50-55	12-15

*) without considering ecological damage

POSSIBILITIES OF UTILIZING ALTERNATIVE ENERGY SOURCES FOR COMBINED HEAT SUPPLY SYSTEMS IN THE BALTIC

Dr. Eng. P. Shipkovs, Dr. Eng. V. Grislis, and Dr. Sc. V. Zebergs
Physico-Energetical Institute
Academy of Science of Latvia Riga, Aizkraulkes Street 21, USSR

Abstract

The problem of alternative energy sources is an issue of major importance for the Baltic republics because of the limited supply of conventional energy resources. One of the ways to solve this problem could be the introduction of combined heat supply systems (CHSS). The combined heat supply systems are such systems where various energy sources in different regimes are made use of to ensure the optimum temperature on residential and industrial premises. The influence of climatic conditions on the selection of heat supply systems has been studied at large. In the present paper the use of alternative energy sources (AES) in combined heat supply systems (CHSS) is described.

Keywords:

Alternative energy sources, combined heat supply systems, climatic factors.

The basic issue is the choice of capacities for various heat supply installations involved in CHSS. The minimum total costs for CHSS as a whole constitute the chief economical criterion here;

$$C_{\Sigma} = C_1 + C_2 + \dots + C_i = min \qquad (1)$$

where C_Σ - total costs for CHSS
C_1, C_2, C_i - costs for the respective heat installations involved in CHSS.
These requirements could well be satisfied in case the costs for a unit of heat energy produced by the installation involved in CHSS were reduced to the minimum, i.e.;

$$C_\Sigma = Q_1 c_{1s} + Q_2 c_{2s} + \ldots + Q_i c_{is} = \min \quad (2)$$

where c_{1s}, c_{2s}, c_{is} - costs for a unit of heat energy produced by the installations from 1 to i ;
Q_1, Q_2, Q_i - amount of heat energy generated by the installations from 1 to i ;

$$Q_1 + Q_2 + Q_3 + \cdots + Q_i = Q_\Sigma \quad (3)$$

where Q_Σ - total heat energy supplied by CHSS.
For a majority of heat installations based on AES the unit costs for heat energy produced are exclusively dependent on the costs of the installation itself;

$$c_{is} = f(K) \quad (4)$$

where K - investment costs of the installation per unit of heat energy produced.
For those heat installations involved in CHSS and operating on conventional energy sources the unit costs are calculated in terms of their consumption rates;

$$c_{is} = aK + c_w \quad (5)$$

where c_w - are the costs for fuel and electric energy necessary to produce a unit of heat energy;
a - coefficient related to K
The costs per unit of heat energy produced, being directly related to K , are influenced by climatic factors, i. e. by

the utilization time (in hours) of the respective installation over a year;

$$c_{is} = \frac{aK_i}{Q_i} = \frac{aK_i}{N_i T_i} \qquad (6)$$

where K_i - investment costs of the heat installation i;
N_i and T_i - capacity and utilization time (in hours) of the respective installation over a year.
Fig.I. demonstrates the utilization time (in hours) of two heat installations involved in CHSS, but Fig.2. shows the determination of Q_i and T_i allowing for local climatic conditions.

For those CHSS which comprise a heat installation utilizing electricity it is expedient to use electricity during off-peak hours of night-time minimum load or when redundant capacity is available in the distribution network. However, the power capacity of the network is restricted and its increase brings about considerable expenses. As a matter of fact, it is possible to plot a characteristic curve for every region;

$$\Delta C_e = F(\Delta N_e) \qquad (7)$$

where ΔC_e - extra expenses for power capacity increase in the electric network.
ΔN_e - additionally connected electrical load .

There is a vast system of electrical network established throughout the Baltic republics, including rural areas as well. Taking into account that a large number of night-time electricity consumers are being connected to the network, the capacity of electroheat installations, forming part of CHSS, should be 0.4 - 0.6 of its total calculated capacity. To ensure a more efficient operation of power plants and electrical network it is necessary for the electroheat installation to operate in the base-load part of heat consumption chart for the combined heat supply systems. In case the capacity of the electrical installation exceeds that of CHSS by 0.4 - 0.6 due to the sharp drop of T_i (see Fig.1), ΔC_e increases to such an extent that the application of heat installations utilizing fuel becomes more profitable (Fig.3). The electroheat installation with a capacity of 0.4 - 0.6 from the total CHSS capacity accounts for 0.75 - 0.8 of the total energy output.

Provided alternative energy sources, such as, solar heat and wind energy are used, they should be made redundant to the full of their capacity, in view of specific climatic conditions of the Baltic region. In this climatic zone windless periods occur (also in winter, especially at low temperatures), besides, there are separate small areas in Latvia which stand out markedly as regards their peculiar climatic conditions (the coastal zone along the Baltic sea) where the utilization ofwind energy could be to better advantage than in the rest of the territory. The heat from solar radiation (solar heat collectors for water heating) is reduced to such an extent in winter that it is practically unfit for use (Fig.4) Redundancy can easily be achieved by means of the electrical installation. In order to utilize night-time electricity heat accumulators must necessarily be installed which simultaneuosly can be applied to the solar heat collector or wind-turbine generator used for hot water production. The choice of capacity for the solar heat collector or wind-turbine generator could be made on the basis of the formula 6 provided it does not exceed the capacity of the electroheat installation forming a component part of CHSS for which they are kept redundant. The heat requirements of CHSS are covered by the heat installation utilizing solid or liquid fuels during periods of cold in winter. The Physico-Energetical institute of the Latvian Academy of Science is conducting experiments on the usage of solar heat collectors and wind-turbine generators to supply heat for a standardized prefabricated rural house by means of CHSS. One of the basic tasks of the experiment is to develop the program for CHSS management in order to ensure more efficient use of AES i.e. solar heat and wind energy. In practice one of these unconventional sources will be available in CHSS.(in this case CHSS will comprise the following three installations - electro heat installation, fuel heat installation and the installation utilizing AES) depending on the territorial location of the house. In the Baltic it is extremely convenient to utilize wind-turbine generators in close vicinity to the Baltic sea coast. However, either of the installations - solar heat collector and wind-turbine generator are simultaneously subject to testing at the experimental house and the CHS system comprises four heat installations,which permits to speed up the experiment. In the climatic conditions typical for the Baltic solar heat can cover about 10%, whereas wind energy can account for 20 - 25% from the total heat consumption of the house. The experiments on supplying heat for rural houses in Latvia,collecting heat from the upper layer of the soil with the help of heat pumps, demonstrated the advantage of the CHS system in which the base-load part of heat consumption chart for CHSS was covered by the heat pump installation utilizing night-time electricity, whereas the peak-load part was covered by the heat installation working on solid or liquid fuels in the periods of cold in winter. Heat pump installations are subject to the same regulations as regards their capacity restrictions (0.4 - 0.6 of the total CHSS capacity) in accordance with the network loading.

Biogas installations can fully meet the heat requirements of a rural house or some other building and the application of CHSS can well be justified providing the biogas installation has a low output and in the periods of cold it is necessary to resort to an extra heat source based on the conventional energy. Typical for the Baltic as well as for the rest of Northern Europe is the fairly high biogas consumption rate in winters for maintaining of the processes in biogas installation reactors.

Conclusions

In the climatic conditions of the Baltic a majority of alternative energy forms can successfully be utilized only in the combined heat supply systems. The meteorological data study is an issue of substantial significance for the right choice of the capacity of electrical installations involved in CHSS. Moreover, meteorological data processing for each form of alternative energy sources had certain peculiarities of its own. The automatic control of CHSS under these circumstances must ensure the maximum efficiency of AES utilization in heat installations with limited capacity.

REFERENCES

1. V.Zeberg, P.Shipkovs. The influence of climatic conditions on the selection of combined heat supply systems for animal farms In;Proceedings of the 4th International Conference"Physical Properties of Agricultural Materials, Volume 2 (1989) pp.907-911.
2. V.Zeberg, P.Shipkovs. Ermittlung der Parameter kombinierter Wärmeversorgung landwirtschaftlicher Betriebe in Abhängigkeit von meteorologischen Daten // 10.Internationaler Kongress für Landwirtschaftstechnik (CIGR): Zusatzband.-Budapest,1984.-S.232-235.

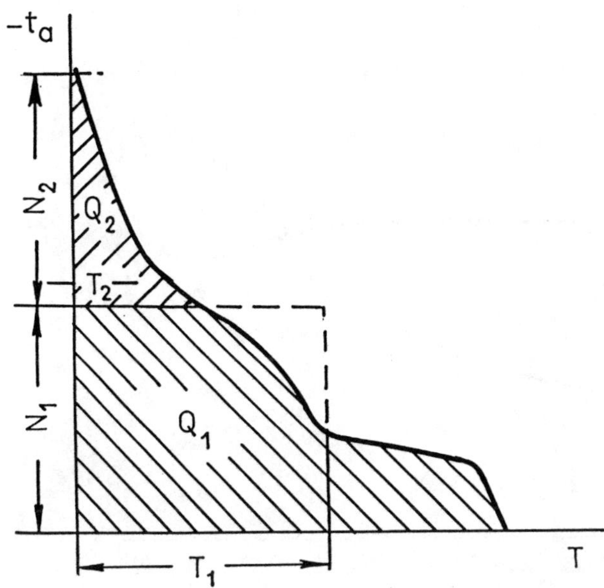

Fig. 1. - Outside air temperature duration curve (t_a) for the Baltic region in combination with characteristics of CHSS consisting of two installations

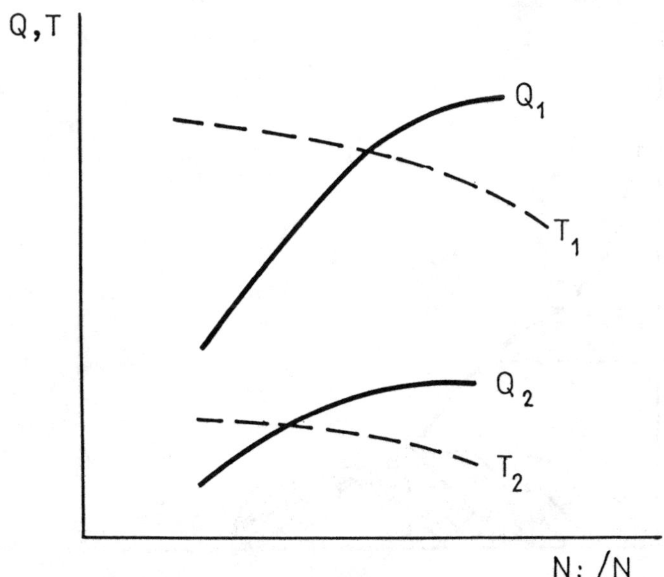

Fig. 2. - Determination of energy output Q and the number of yearly utilization hours T for combined heat supply systems

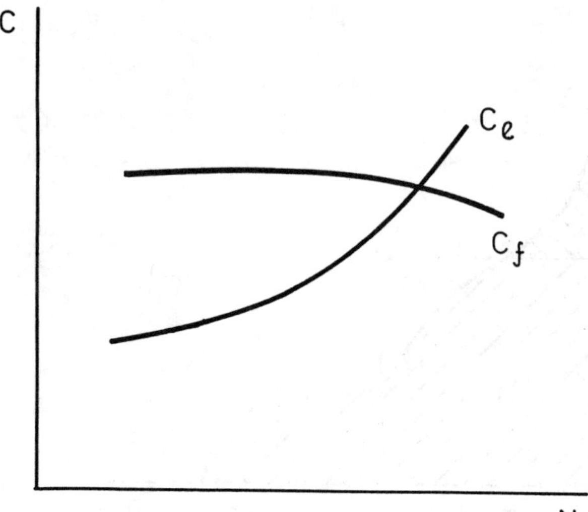

Fig. 3. – Economical characteristics for night-time electricity supply (C_e) and fuel supply (C_f) plotted vs CHSS capacity

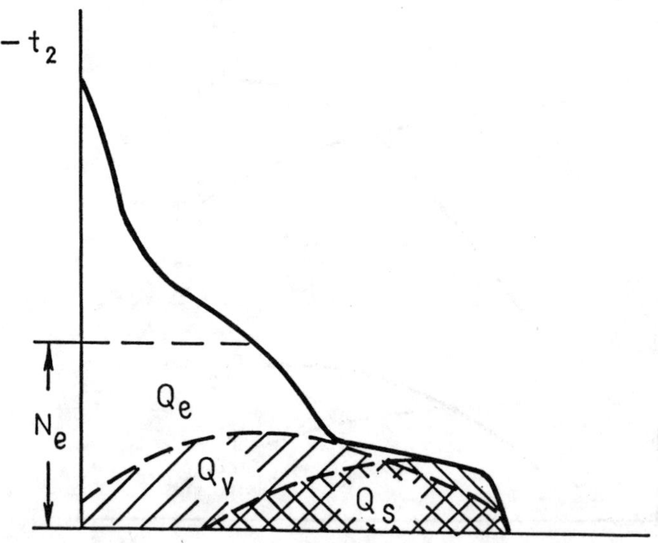

Fig.4. – Outside air temperature duration curve in combination with heat energy output charts by electroheat installation Q_e , wind-turbine generator Q_v and solar heat collector Q_s in CHSS

DESIGN AND OPERATION OF A SOLAR ENERGY ASPHALT HEATING SYSTEM

Chang Song-ying and Ma Guang-yuan
Solar Energy Research Lab., Zhengzhou Institute of Tech.
Zhengzhou, Henan 450002, People's Republic of China

Qiao Ping
Meng Country Road Admimistration Authorities
Meng County, Henan, People's Republic of China

Robert M Wirtshafter
Center for Energy and the Environment
University of Pennsylvania, Philadelphia, PA 19104, U.S.A.

Zhou Guan Lu
Jiaozuo City Transportation Bureau
Jiaozuo, Henan, People's Republic of China

ABSTRACT

This paper introduces a solar and electrical heating system which is used to heat asphalt for road construction. The system includes the use of a solar greenhouse, concentrating parabolic collectors, and electric heating. The basic design, the description of each component, operating statistics, and an economic evaluation are provided.

1. INTRODUCTION

A key element of Chinese development plans requires the vast improvement in the Chinese network of highways. China currently has less than 200,000 km of paved highway.[1] By comparison, the United States in 1984 had 5,200,000 km of non-urban paved highways.[2] The Chinese road network is inadequate to move the existing amounts of goods and services across China. The transportation bottleneck is further exacerbated by the limited amount of rail capacity, a dominant portion of which is currently used to bring energy resources from the northern and western parts of the country to the population centers along the south-eastern coast.

Most of the responsibility for maintaining and improving highways falls to the county officials. Each of China's 2000 counties has on average only 100 km of paved highway. Given this modest figure, each county cannot and does not possess modern road paving and maintenance equipment. In general, self-contained asphalt heating trucks are not available in most areas of China.

Meng County, located in Henan Province, is typical of many of the semi-rural counties found all over China. Meng County has only 110 km of paved highway.

Prior to the development of the solar heating system described herein, the preparation of asphalt for road resurfacing utilized traditional methods.

<u>The Traditional Preparation of Asphalt in Rural China</u>

Each county in China is given an annual allotment of asphalt, which is stored until needed. Asphalt is delivered from primary manufacturing and distribution areas to each county in large insulated tank trucks, typically holding 3 tons per truck. The asphalt is dumped into an open pond where it is stored until needed. The asphalt normally solidifies in the pond turning into a hard and sticky solid. When asphalt is needed, chunks of it are manually transferred from the holding pit to the heating tank or open cauldron. It is often necessary to heat the shovels used to dig out the asphalt in order to be able to dig it out.

Heating the asphalt to the required temperature of 160°C presents several problems.

1) The heating of small amounts of asphalt is inefficient, especially using solid fuels. Typically an efficiency of 3 to 5 percent is achieved, so that 400 to 700 kg of coal is required to heat one ton of asphalt. (The Chinese standard of 213 kg/ton of asphalt which was established by the Ministry of Transportation is seldom achieved). Liquid fuels are in short supply and therefore used exclusively for transportation. Natural gas is almost non-existent in rural China. Fuelwood, traditionally used in some areas, is too scarce in most of China. For many areas, coal is either not available, difficult to obtain, or too expensive.

2) Heating asphalt by direct burning of solid fuels is also difficult to control. Cold asphalt is a poor conductor of heat. Asphalt must be heated slowly. Too hot of fire will result in the coking of the asphalt nearest the heat source. This layer further reduces the heating efficiency.

3) Heating of asphalt can be dangerous. Open burning of asphalt creates air pollution from the fuel and asphalt burning. The water contained in the asphalt can also create a dangerous situation when water, which is heavier than asphalt, sinks to the bottom of the heating vessel and is heated. As this water boils to the surface it can violently spew hot steam and asphalt on nearby workers.

The difficulties inherent in the traditional means of preparing asphalt are a major obstacle to improving road conditions in Meng County. Unfortunately, the modern approaches to heating asphalt are unavailable to Meng County. No liquid fuels are available for direct-firing of asphalt at the construction site. The scale of county operations does not permit them to construct a large solid-fuel boiler to indirectly heat the asphalt. The reliance on coal to heat the asphalt from solid stage to 160°C means questionable fuel supplies, lower productivity, and dangerous working conditions. Under these circumstances, a solar greenhouse, concentrating collector, and electric heating system was designed and constructed in 1987.

2. THE DESIGN OF THE SOLAR ENERGY ASPHALT HEATING SYSTEM

The Solar Greenhouse Asphalt Pond

The pond shown in Figure 1 is designed to hold the asphalt in liquid form without requiring supplemental heating. It utilizes a simple insulated masonry building (80 cm thick with 10 cm of insulation) covered by a south facing glass roof. The design provides a reasonable tradeoff between efficiency and cost.

The solar collector is about 160 m^2, sloped so that heat gain is maximized during the May to October road construction period. A 2 cm thick insulation curtain is used to reduce heat losses at night. The pond, sized to hold the 400 tons of asphalt used annually by Meng County, can hold 2.5 meters of asphalt. The air temperature within the structure is normally 40 °C above the outside ambient temperature. This heat differential is adequate to melt the top 10 cm of asphalt so that it can flow to the primary heating devices.

There are large diameter pipes at the base of the greenhouse wall to draw the melted asphalt out. The greenhouse is built at a higher elevation than the rest of the system to allow gravity to assist in maintaining the flow. Electricity is used initially to free the flow through the pipes. As soon as the hot asphalt can freely flow through the entire pipe and maintain a certain flowrate, then the electricity can be turned off. On sunny days in the summer about 20 tons of hot asphalt at a temperature of 60-70 °C can be drawn out of the pond daily.

The Cylindrical Parabolic Concentrating Collectors

The greenhouse can only raise the asphalt to around 65 °C. In order to raise the temperature using solar energy, it is necessary to utilize a concentrating collector. For this application, cylindrical parabolic concentrating collectors shown in Figure 2 were selected. This technology was selected because this type of system can be cheaply built from equipment available in China.

The collectors were manually constructed in small sections 4.5 meters long and 3.6 meters wide. The collecting angle is about 193 degrees. The collecting area of each section is about 16 M. The focal length of each collector is 800 mm. The concentrating ratio is about 14.2:1. The arc length of the cylinder is about 4.22 m. The base material of the collectors is made of galvanized iron sheet reinforced by a rolled-angle steel frame. Vacuum aluminium plating film was glued carefully on the surface as the reflecting material. The solar energy is focused on a 80 mm copper pipe that runs the length of the collector. The asphalt flows through the pipe and raises the temperature of the collector by approximately 20 °C.

In order for the collector to be effective, the entire collector must rotate in order to track the sun's position. A motor and governor system is used to drive the individual collectors to rotate around the accepting pipe, which is positioned at the focal line. A hydraulic transmission system is used to control the collector inclination. A multi-sensor sun position detector provides sun position information. The system tracks the sun on a continuous basis, so that the collectors reposition themselves every few minutes.

There are a number of issues that have yet to be established regarding the use of parabolic cylindrical collectors. The principal issue relates to the question of cost-effectiveness which will be addressed in more detail in a later section of this paper. There are also a number of technical issues, among the most important of these are the following:

1) Precise construction of the collectors is required in order to achieve the maximum heat gain. Can these low-cost, hand-made collectors operate effectively? What is the tradeoff between precision and cost-effectiveness?

2) The diameter of the acceptor pipe influences the temperature of the asphalt. A larger diameter pipe would not raise temperatures as high as a smaller one. However, a larger pipe would reduce the need for precision in focusing the collector and make asphalt flow more easily.

3) What level of control is justified for tracking of the collectors? Should a manual system of adjustment be used or is an automatic control strategy which more precisely adjusts the collectors cost-effective?

Electrical Heating Asphalt Tank And Other Electrical Heating Devices

Asphalt leaving the parabolic collectors has reached a temperature of around 85 °C. In order to reach the desired temperature of 160 °C, the asphalt is transferred to holding tanks with electric heating coils, see Figure 3. The four tanks, each measuring 2 meters deep by 1.5 meters wide and 1.6 meters long, are made of 4 mm steel, surrounded by 60 mm of perlite insulation. The tanks are design to hold the equivalent of two days of asphalt. The heating capacity is 40 kW per tank which is enough to heat 10 tons of asphalt per day should there be no solar contribution.

The holding tanks are designed to gradually raise the temperature of the asphalt to 160 °C. At 90 °C, water which is heavier than the asphalt falls to the bottom of the tank, where it is drained from the system. The elimination of water reduces the heat require for water vaporization and eliminates the threat of asphalt eruptions.

In order to ensure that solidified asphalt can flow through the entire system, electric wires have been wrapped around all of the pipes. Approximately 140 kW capacity of wire are utilized. In an emergency, this capacity could guarantee flow within thirty minutes.

Ideally the tanks hold the asphalt at around 90 °C and then bring to 160 °C only the amount required by the work crews at that time. The poor quality of the Chinese communication system does not yet make it practical to establish an exact schedule for when and where asphalt will be needed. This forces the plant to keep a larger amount of asphalt at the higher temperatures.

3. ECONOMIC ANALYSIS

The solar asphalt system operating in Meng County clearly has advantages over the traditional approach, particularly with respect to improvements in working conditions. There remains the fundamental question as to economic viability of the solar system as a whole and to appropriateness and cost-effectiveness of the individual components. Given the scarcity of electric power in China, the choice of electric supplemental energy is an interesting one.[3]

Cost Estimates

Cost estimates for the different components were obtained based on actual costs. Costs for products in China vary from location to location and from one instance to the next, and are not necessarily a good measure of future costs. In addition, the prices for coal, electricity, and labor are government controlled and do not reflect the true social costs. For this reason, we have tried to present two assessments, one analyzing the finances as seen by the Meng County officials constructing the facility, and the other reflecting the societal opportunity costs.

The construction costs are presented in Table 1. It is difficult to assign almost fifty percent of the expenses to one of the three individual components. We have allocated the costs based on the best estimate of individual component responsibility. The largest of these common expenses is the cost of a computer data acquisition system (DAS). While the system does help control the system, it is also included for research purposes. A simpler control system without DAS capability might lower the overall costs by ten percent.

Performance Data

The data acquisition system has been installed to collect the performance values. As of now, however, only generalized information on performance is available. The results indicate that the greenhouse raises temperatures from 15 °C to 65 °C, a 50 °C increase. The parabolic collector raises the temperature from 65 °C to 87 °C, a 22 °C increase. Electric heating is needed to increase the temperature from 87 °C to 160 °C, a 73 °C increase.

There exists only scant information on the energy consumption of the new plant. The consumption of the electricity used for heating the holding tanks was provided for a typical day. This figure does not include electricity used to position parabolic collectors, run the data acquisition/controller, and heat piping clogged with solidified asphalt. While the latter usage has a high electric consuming potential of 140 kW, the actual use of the entire capacity is limited to a couple of times a year for less than one half hour at a time. There were more frequent, yet unrecorded requirements for heating of an isolated clogged pipe.

The economics of the entire system are shown in Table 2. From the analysis, it appears as though the system is cost effective using the Chinese accounting approach. This approach uses a low cost recovery factor of 0.0035 per month or 0.042 per year. Under these circumstances, the entire system saves 9.8 Yuan per ton of asphalt. Using Western accounting procedures which include the interest charges on the capital investment, indicates that the plant is not cost-effective. The payback assuming 400 tons/year is 16 years. The low output to capacity ratio (400 ton/yr:10 tons/day) affects the economics. If the plant were to produce 800 tons/year. The payback period would be nearly halved.

The economics of the individual components, the solar greenhouse, the parabolic collectors, and the electric heating tanks, have been evaluated in Tables 3, 4, and 5. Data were only available on the overall efficiency of using coal for heating asphalt, therefore we assumed that coal savings was uniform across the temperature range of 15 °C to 160 °C. The results of this approximation are that the greenhouse is cost-effective using both Chinese and Western accounting. This is particularly true if a simplified system without the expensive controller were installed. Even at the low capacity of 400 tons/year, this greenhouse would have a payback of under 5 years.

The economics of the parabolic collectors and the electric heating tanks are not as favorable. These components still merit consideration largely because the official prices of coal and electricity used in the analysis are not representative of the true costs. The government supplies a limited amount of coal to industries at the official price. If a firm requires additional supply, they must secure it at the market price. In addition, they must contract to have the coal transported from the mine to the site. Given these conditions, the supply of coal to small, rural enterprises is neither guaranteed nor inexpensive. For this reason the substitution of coal with other sources, particularly solar may be warranted.

The substitution of electricity for coal is more complex. Surprisingly, using electricity for heating the asphalt is more efficient than directly burning coal. It takes 90 kwh to heat a ton of asphalt from 87 °C to 160 °C. Assuming that the least efficient electric plant consumes around 660 g/kwh, and that losses in electric transmission and in conversion to heat in the tanks are 25 percent and 20 percent, this converts to approximately 200 kilograms of coal per ton of asphalt (15 °C to 160 °C). It takes 400 kg of coal if the coal is used directly. The use of electricity makes sense because of the low efficiency of directly firing coal. It may be possible to develop a coal boiler system that produces steam to heat the asphalt. However, the heating requirements are so small, that it would be difficult to develop an efficient boiler/steam producing system.

Electricity supply in rural China is often interrupted due to limited generation and transmission capacity. It would seem that the asphalt plant is substituting one unreliable source for another. Considering that coal can at least be stockpiled, electricity, at face value, seems to be the wrong choice. Again, the pricing policies for energy supply in China, influence the decision process of the asphalt plant developers. Electricity prices for industry have always been subsidized heavily in China. This plant pays approximately 0.11 Yuan/kWh. The real cost of production is probably closer to the 0.3 Yuan/kwh normally charged residential customers. In those areas where local electric producers can charge market rates the price may be bid above .8 Yuan/kWh.[3] If electricity shortages become more acute, then an alternative to electricity will need to be found.

4. DISCUSSION

This prototype solar heating system has demonstrated an alternative to the traditional means of heating asphalt. The system has proven to be technically and economically superior to the old approach. The development of the system has uncovered the following potential improvements to the system:

1) The external placement of electric heating coils around the piping should be replaced with internal heating coils.

2) Covers should be built on the large heating tanks to reduce the heat loss.

3) The output from the system should be increased. This may require the combining of facilities from adjoining counties. It appears that the economics of the system will be greatly enhanced if output were doubled.

While closer monitoring is needed, the authors suggest that this type of solar heating system, particularly the solar greenhouse, might be utilized by similar asphalt preparation facilities worldwide.

REFERENCES

[1] The Chinese State Statistics Bureau reports that China had 962,800 km of highways, (this figure excludes city streets), A Statistical Survey in 1987, State Statistical Bureau, Beijing, PRC, p. 53. However, a World Bank report notes that only one sixth of the total road network is in fact paved, see World Bank, China The Transport Sector: Annex 6 to China Long-Term Development Issues and Options, World Bank, Washington D.C., 1985, p. 112.

[2] Table 1012, Statistical Abstract of the United States, 1987, from US Federal Highway Administration, Highway Statistics, Annual.

[3] For a description of power shortages in rural China see, Wirtshafter, "Decentralization of China's Electricity Sector: Is Small Beautiful?". World Development, forthcoming Vol. 18, No. 4, April 1990.

ACKNOWLEDGEMENTS

The authors wish to thank the Meng County construction and operating team, including Mr. Xu Jian-chao, Mr. Ren Shi-xin, Mr. He Cheng-zhu, Mr. Cao Yu-zhong and Mr. Yan Ji-yi, who provide background materials for this paper. We also give our thanks to Mr. Wang Le-xuan, Ms. Yang Man-xiang and Mr. Liu En-chen at Solar Energy Research Lab. of Zhengzhou Institute of Technology, who assisted us in the preparation of this paper.

TABLES

TABLE 1: COMPONENT COST OF THE SOLAR ASPHALT HEATING SYSTEM

Item	Unit	Amount	Unit Cost(yuan)	Total Cost(yuan)	
Common Costs					
Control Structure	Sq. meter	50	250	12,500	
Power Supply System			20,000	20,000	
Instruments, Control Device and DAS			50,000	50,000	
Relays		70	50	3,500	
Plumbing and Insulation	meter	300	40	12,000	
Subtotal Common Costs				98,000	
Greenhouse					
Structure	Sq. meter	160	300	48,000	
Electric heating	KW	40	50	2,000	
Subtotal Greenhouse with 30 % allocated common					79,400
Concentrating Collectors and Tracking System	Sq. meter	48	500	24,000	
Electric heating	KW	140	50	7,000	
Subtotal Parabolic Collectors with 20 % allocated common					50,600
Electric Heating System					
Structure	Sq. meter	50	250	12,500	
Electric heating	KW	160	50	8,000	
Motor and pump		4	350	1,400	
Subtotal Electric Heating with 50 % allocated common					70,900
Total				200,900	

TABLE 2: COST ANALYSIS FOR HEATING TON OF ASPHALT

Expenses	Cost per unit (Yuan)	units	Conventional System		Solar System		
			quantity	cost	quantity	cost	
Coal	0.045	kg					
low efficiency			400	18			
high efficiency			213	9.585			
Electricity	0.11	kWh	20	2.2	90	9.9	
Shovels and Sticks				1.4			
Gasoline	1	kg	0.5	0.5	0.5	0.5	
Labor	5	person/dy	3.5	17.5	0.8	4	
Lost Asphalt				4			
Maintenance				0.9		1.34	
Administrative				3.3		1.2	
TOTAL VARIABLE COSTS			39.385 to 47.8			16.94	
Capital Cost							
Investment				200900	Yuan		
Interest Rate				11.00%			
Lifetime				20	years		
Capital Recovery Factor (Western Analysis)				0.13			
Annual Payment				25228.15			
Output in tons per year				400.00			
CAPITAL COST PER TON (Western Analysis)				63.07		TOTAL COST	80.01 Western
Capital Recovery Cost (Chinese Analysis)				0.04			
Annual Payment				8437.80			
Output in tons per year				400.00			
CAPITAL COST PER TON (Chinese Analysis)				21.09		TOTAL COST	38.03 Chinese

TABLE 3: COST ANALYSIS FOR SOLAR GREENHOUSE ONLY

Expenses	Cost per unit (Yuan)	units	Cost Allocation Factor	Conventional System quantity	Conventional System cost	Solar System quantity	Solar System cost
Coal	0.045	kg					
low efficiency			0.37	400	6.67		
Electricity	0.11	kWh	0.1	20	0.22	90	0.99
Shovels and Sticks					1.4		
Gasoline	1	kg		0.5	0.5	0.5	0.5
Labor	5	person/dy	0.8	3.5	14	0.8	3.2
Lost Asphalt					4		
Maintenance					0.7		0.6
Administrative					same		same
TOTAL VARIABLE COSTS					27.49		5.29
Capital Cost							
Investment					79400	Yuan	
Interest Rate					11.00%		
Lifetime					20	years	
Capital Recovery Factor (Western Analysis)					0.13		
Annual Payment					9970.71		
Output in tons per year					400.00		
CAPITAL COST PER TON (Western Analysis)					24.93	TOTAL COST	
30.22							Western
Capital Recovery Cost (Chinese Analysis)					0.04		
Annual Payment					3334.80		
Output in tons per year					400.00		
CAPITAL COST PER TON (Chinese Analysis)					8.34	TOTAL COST	
13.63							Chinese

TABLE 4: COST ANALYSIS FOR PARABOLIC COLLECTORS ONLY

Expenses	Cost per unit (Yuan)	units	Cost Allocation Factor	Conventional System quantity	Conventional System cost	Solar System quantity	Solar System cost
Coal	0.045	kg					
low efficiency			0.163	400	2.93		
Electricity	0.11	kWh	0.1	20	0.22	90	0.99
Labor	5	person/dy	0.1	3.5	1.75	0.8	0.4
Maintenance					0.1		0.5
Administrative					same		same
TOTAL VARIABLE COSTS					5.00		1.89
Capital Cost							
Investment					50600	Yuan	
Interest Rate					11.00%		
Lifetime					20	years	
Capital Recovery Factor (Western Analysis)					0.13		
Annual Payment					6354.13		
Output in tons per year					400.00		
CAPITAL COST PER TON (Western Analysis)					15.89		TOTAL COST
17.78							
							Western
Capital Recovery Cost (Chinese Analysis)					0.04		
Annual Payment					2125.20		
Output in tons per year					400.00		
CAPITAL COST PER TON (Chinese Analysis)					5.31		TOTAL COST
7.20							
							Chinese

TABLE 5: COST ANALYSIS FOR ELECTRIC HEATERS ONLY

Expenses	Cost per unit (Yuan)	units	Cost Allocation Factor	Conventional System quantity	Conventional System cost	Solar System quantity	Solar System cost
Coal	0.045	kg					
low efficiency			0.467	400	8.4		
Electricity	0.11	kWh	0.8	20	1.76	90	7.92
Labor	5	person/dy	0.1	3.5	1.75	0.8	0.4
Maintenance					0.1		0.24
Administrative					same		same
TOTAL VARIABLE COSTS					12.01		8.56
Capital Cost							
Investment					70900	Yuan	
Interest Rate					11.00%		
Lifetime					20	years	
Capital Recovery Factor (Western Analysis)					0.13		
Annual Payment					8903.31		
Output in tons per year					400.00		
CAPITAL COST PER TON (Western Analysis)					22.26		TOTAL COST
30.82							Western
Capital Recovery Cost (Chinese Analysis)					0.04		
Annual Payment					2977.80		
Output in tons per year					400.00		
CAPITAL COST PER TON (Chinese Analysis)					7.44		TOTAL COST
16.00							Chinese

FIGURES

Fig. 1 Sketch of Asphalt Heating Solar Greenhouse

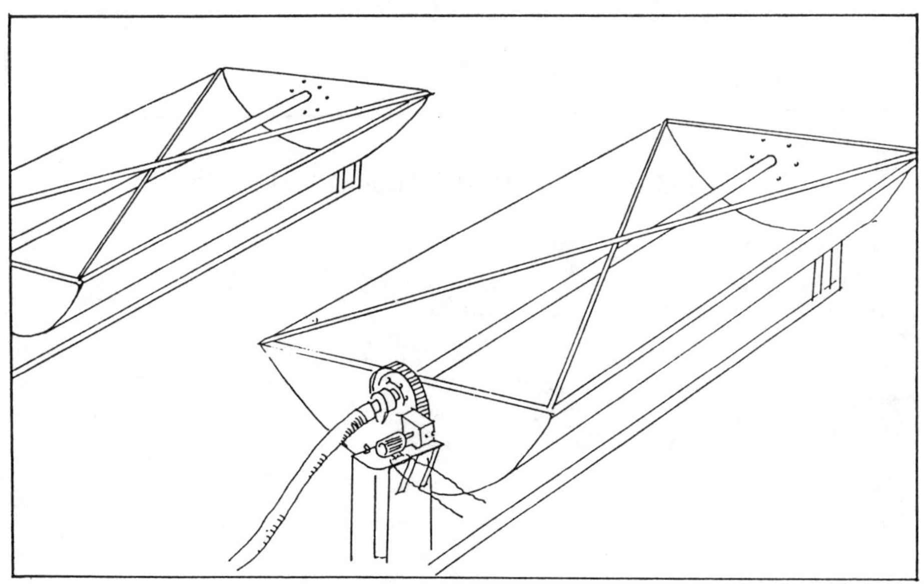

Fig. 2 Sketch of Parabolic Concentrating Collectors

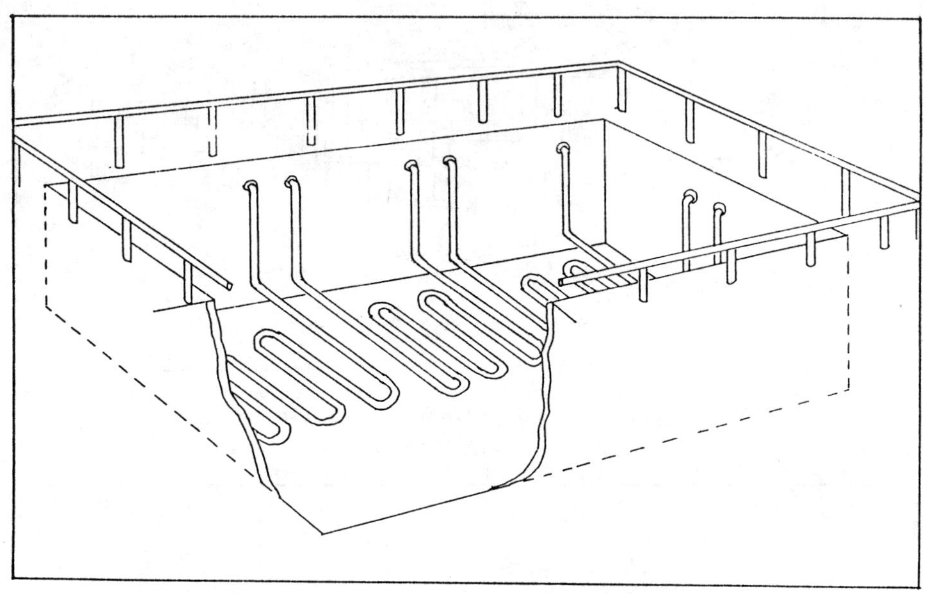

Fig. 3 Sketch of Electric Heating Tanks

ENERGY TRANSMISSION

THE INCLINATION EFFECT ON THE PERFORMANCE OF WATER FILLED HEAT PIPES

Ali Cetin Gurses & Levent Tezcan
Dokuz Eylul University
Eng. & Achitecture Faculty, Bornova-Izmir/Turkey

1. INTRODUCTION

The heat pipe idea has been proposed first in 1942 by Gaugler (1).Grover has improved the idea and introduced the well known typical heat pipe geometry in 1966 (2).In the heat transfer applications, heat pipe has some advantages as construction simplicity, perfect adjustibility, easier control and high heat transfer ability in low temperature differences .In addition to this features, heat pipes do not need an additional energy source as well. In it's convertional form (see Figure 1), heat pipe is a closed tube or chamber of different shapes whose inner surfaces are lined with a porous capillary wick. The wick is saturated with the liquid phase of a working fluid and the remainig volume of the tube contains the vapour phase.Heat applied at the evaporator by an external source vaporizes the working fluid in that section.The resulting difference in pressure drives vapour from the evaporator to the condencer where it condenses releasing the latent heat of vaporization to a heat sink in that section of heat pipe. Depletion of liquid by evaporation causes the liqid- vapour interface in the evaporator to enter into the wick surface and a capillary pressure is developed there.This capillary pressure pumps the condenced liquid back to the evaporator for re-evaporation.

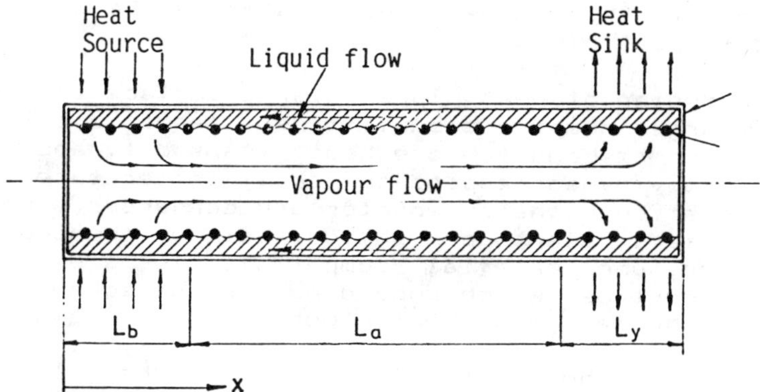

Figure 1. Standart heat pipe and the operation principle.

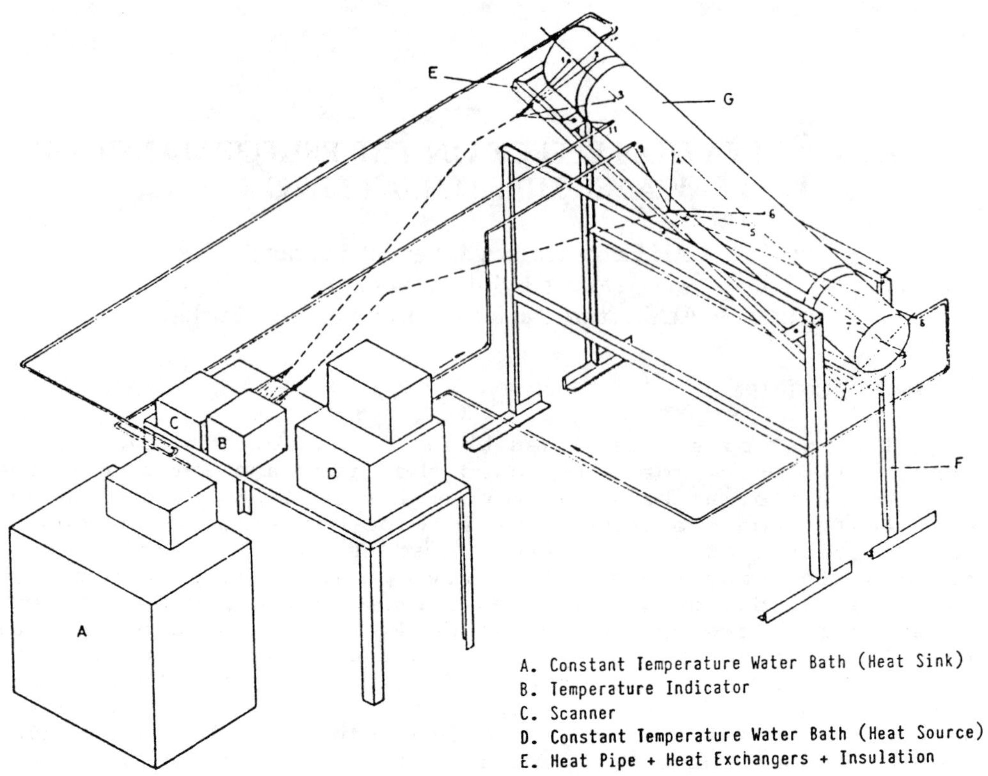

Figure 2. Perspective view of the Experimental set-up

Heat Pipes have been used in wide ranges of applications today. Heat recovery from exhaust gases and HVAC systems, Solar energy, geothermal energy, nuclear and isotope reactors are some examples of these typical fields. Cryogenic liquids, liquid metals, various gases or pure water can be used as working fluid in heat pipes.

In the design of heat pipes , large surface tension correspond to larger pumping capabilities, large latent heat of vaporazition correspond to more heat transport, and large thermal conductivity correspond to small temperature drop across to wick at both the evaporator and condenser side. Meanwhile entrainment, start up dynamics, boiling phenomena, working fluid and tube material compatibilities are the other important properties which should be considered and investigated during the design of heat pipes.

According to the type of application, the following parameters should be looked for suitable working fluid selection.

a- Compability between wick and tube material.
b- Good thermal stability
c- Wettability wick and tube materials.
d- Vapour pressure should not be high in operating temperature range.
e- High latent heat.
f- High thermal conductivity.
g- Low liquid and vapour viscosities.
h- High surface tension.
i- Acceptable freezing or pour point.

The selection of working fluid should also be based on some thermodynamic consideration such viscous, sonic, capillary, entrainment and nucleate boiling limitations which are concern with the heat flow occuring with in the heat pipe.Meantime , it is necessary for the working fluid to wet the con tainer and wick material. Another important part of heat pipe is wick side and many selection factors of wick materials are closely linked to the properties of working fluid. Today homogenous wicks takes many forms; meshes, foams, felts, fibres , sinters, grooves and arterial wicks. Some basic expected properties from a suitable wick material can be summarized as follows:

a- The necessary flow passages for the return of condensed liquid.
b- Surface pores at the liquid-vapour interface for the development of the required capillary pumping pressure.
c- A heat flow path between the inner wall of the container and the liquid-vapour interface.

The function of the container is to isolate the working fluid from the outside environment. It has to be leak proof, maintain the preesure differential across its walls,and enable to transfer of heat to take place into and from the working fluid. Selection of container material depends on several factor. These are as follows:

a- Compability (Both with the working fluid and the environment)
b- Strength to weight ratio
c- Thermal conductivity
d- Ease of fabrication, including weldability, machineability, ductility
e- Porosity
f- Wettability

2. PURPOSE OF THE RESEARCH

Flat plate solar collectors are one of the main devices for effective solar energy collection and there is an unnegligiable manufecturing potential in many parts of the world. In this research, it was investigated the suitibility and applicability of heat pipes in flat plate solar collector design, if it is achieved to good performance and manufecturing feasiblity,the heat pipe assisted solar flat plate collectors would have the foolowing additional advantages:

a- Freezing problem would be eliminated.
b- Under the consideration of one way working in heat pipe, there would be no night time heat losses from the collector.

With regard to the mentioned reasons, the following basic subjects was investigated during this research.

1- A theoretical design method was developed for heat pipes which use water as working fluid.

2- The thermal performance of one layer brass wick installed and water filled heat pipe was experiminially investigated due to:

a- Heat sink tempratures
b- Tilt angle of heat pipe

3- Experimental and theoretical results was compared and discussed

4- The feasibility of heat pipe assisted flat plate solar collectors was investigated.

3. THEORITICAL ANALYSIS

In order for the heat pipe to oparete the maximum capillary pumping head $(\Delta P_c)_{max}$ must be greater than the total pressure drop in the pipe.

$$(\Delta P_c)_{max} = \Delta P_l + \Delta P_v + \Delta P_g \tag{1}$$

If the condition defined in equation (1) is not met, the wick will dry out in the evaporator region and the pipe will not operate.

During the start up and with certain high temperature heat pipes the vapour velocity may reach sonic values. In such cases compressibility effects must be taken into account. So, sonic conditions set one limit to the maximum possible heat transport capability of a heat pipe. Other limits, are set, at low temperatures by viscous forces, and at increasing temperatures limits arise due to entrainment of working fluid in the wick by the vapour stream, by insufficient cappilary head and by evaporator burn out. During the theoritical analysis of heat pipe, these limitations was caltulated with the following equations.

Dry-out limit (3),

$$(Q\ L)_{k,max} = \int_{0}^{L_t} Q.dx \tag{2}$$

$$Q_{k,max} = \frac{(Q\ L)_{k,max}}{0.5\ L_y + L_a + 0.5\ L_b} \tag{3}$$

Sonic velocity limit (4) ,

$$Q_{s,max} = A_b\ \rho_o\ \lambda \left[\frac{\gamma_b\ R_b\ T_o}{2(\gamma_b + 1)}\right]^{1/2} \tag{4}$$

Entrainment limit (5),

$$Q_{e,max} = A_b \cdot \lambda \left[\frac{\sigma \cdot \rho_b}{2 \cdot r_{h,s}}\right]^{1/2} \tag{5}$$

Boiling limit (6),

$$Q_{b,max} = \frac{2\pi\ L_e\ k_e\ T_b}{\lambda \cdot \rho_b\ Ln(r_i/r_b)} \left(\frac{2\sigma}{r_n} - p_k\right) \tag{6}$$

The heat transfer capacity of a heat pipe is the function of three thermal resistances,

R_1= Resistance between the heat source bulk temperature T_{so} and the otside heat pipe wall temperature T_{se} at the evaporator section.

R_2= Resistance between the outside heat wall temperature T_{se} at the evaporator section and the outside heat pipe wall temperature T_{sc} at the condenser section

R_3= Resistance between the outside heat pipe wall temperature T_{sc} at the condenser section and the heat sink temperature T_{si}.

If the heat pipe must transfer an amount of heat Q between two temperatures T_{so} & T_{si} then the following must hold.

$$\Sigma R < \frac{T_{so} - T_{si}}{Q} \tag{7}$$

$$R_1 + R_2 + R_3 \leq R$$

Resistances R_1 and R_3 can be easily determined from well known heat transfer design formulae. The maximum acceptable thermal resistance of heat pipe (R_2) can be determined as follows,

$$R_2 = \frac{T_{so} - T_{si}}{Q} - (R_1 + R_3) \tag{8}$$

The correlations between heat sources, heat sink and sur face temperatures in equation (8) would be as follows,

$$T_{se} = T_{so} - (Q\ R_1) \tag{9}$$

$$T_{sc} = T_{si} - (Q\ R_3) \tag{10}$$

$$T_b = \frac{T_{se} + T_{sc}}{2} \tag{11}$$

Under the assumption of,the wick material would be saturated by working fluid,the effective wick thermal conductivity was calculated with the following equation:

$$k_e = k_s \frac{(k_s - k_f) - (1-\varepsilon)(k_s - k_f)}{(k_s - k_f) + (1-\varepsilon)(k_s - k_f)} \tag{12}$$

The purity of working fluid is another important factor in heat pipe design. The amount of working fluid, which would befilled into the pipe can be determined as foolows,

$$m = A_b\ L_t\ \ {}_b + A_w\ L_t\ \rho_s \tag{13}$$

The heat pipe must not be filled with the working fluid, unless the container fully evacuated from the uncondensed gases and cleaned from all harmfull residues. This process can be done with an powerfull and sensitive vacuum pump.

4. EXPERIMENTAL STUDIES

After the manifacturing of the prototype heat pipe, the rest apparatus shown in Figure 2 was installed in the "Heat Transfer Laboratory of D.E. University". After the cleaning and evacuation; the heat pipe filled with 28 cc distilled water. The condenser side equipped with extruded aluminyum fins Brass made, 18 mesh/cm wick was fitted into the heat pipe.The dimensions of the heat pipe which was used during the experiments was as follows:

Inside /Outside diameter of tube : 15/18 mm
Total length of the heat pipe : 1930 mm
Condenser length (Finned side length) : 420 mm
Outside fin diameter of the condenser: 400 mm
Root diameter of the fin : 200 mm

During the experiments, temperature measurements continuoisly recorded in 11 different locations. The mass flow rates of water, which was circulated between the heat exchangers (see Fig. 2) and the water baths was controlled with measuring of water weight.

During the experiments the inclination effect on water flow rate was neglected.(This was around of ± 2%). Before the experiments thermocouples, scanners recorders was calibrated.

The experiments was repeated for 15°, 30°, 45°,60°,75°, 90° inclination angle of the heat pipe. Test results was recorded after the observation of stable conditions in each experiment.

5. CONCLUSION

After a several sets of experiments, it was observed the effective heat transfer rate of water filled heat pipe depends on the inclination angle. The effective heat transfer rate increases with the angle, and reaches a maximum value in the vertical position. The minimum heat transfer rate was always obtained in the horizontal position. The instantaneous temperature and inclination angle variations did not create any problem on the response of heat pipes to new working conditions. In Fig.3 the experimental effective heat transfer rates was shown due to constant condenser but various evaporator side temperatures. In Fig. 4, the experimental heat transfer rates of the heat pipe were shown with a constant condenser-evaporator temperature difference for various heat sink temperatures. In Fig. 5, the most important factor effecting heat transfer capacity is capillary pumping limit up to 45° tilt , but for higher inclinations the capillary and entrainment becomes together effective on heat transfer rate.

As a conclusion it can be said that, inclinations of between 45° to 90° tilts are the most effective range, for water filled gravity assisted heat pipes.

NOMENCLATURE

A_b : Vapour cross sectional area ,m^2
A_w : Wick cross sectional area, m^2
d_b : Vapour volume radius, m
F : Frictional coeficient, $(N/m^2)Wm$
g : Gravitational constant, m/s^2
K : Wick permability, m^2
k : Thermal conductivity, W/m°C
L : length, m
P : Pressure drop,
P : Capillary preesure,
Q : Heat transfer rate, W
QL : Capillary heat transfer factor, Wm
R_b : Universal gas constant, J/kg°C
r : Radius, m

$r_{h,s}$: Hydraulic radius for liqid flow, m
R : Thermal resistance, °Cm²/W
T : Temprature, °C
T_o : Vapour temperature,°C
T_b : Internal operation temperature of heat pipe, °C
γ : Specific heat ratio, -
ϵ : Wick porosity, -
λ : Latent heat of vaporization, kJ/kg
μ : Dynamic viscosity, Ns/m²
ρ : Density, kg/m³
σ : Surface tension, N/m
ϕ : Inclination angle, ° (angle between the horizontal surface and the pipe axis)

Subscripts

b : Vapour
w : wick
s : Liquid
e : Wick material
f : Solid phase
t : Total
cap,max : Maximum capilarity
g : Gravity
k : Capillary
e,max : Entrainment limit
k,max : Capillary limit
s,max : Sonic limit
i : Inside
se : Evap. side
sc : Cond. side
1 : Evap. section
2 : pipe
3 : Cond. section
si : sink
so : source

REFERENCES

1. Gaukger, R.S., "Heat Transfer Device", U.S Patent 2350348, Appl.21, Dec.,1942.
2. Grover, G.M., "Evaporation-Condensation Heat Transfer device", U.S. patent 3229759, Appl.2 Dec.,1963.
3. Chi,S.W., "The Heat Pipe Theory & Practice", Hemisphere Publishing Coopperation,1976.
4. Levy,E.K., "Theoretical Investigations of heat pipes operating at Low Vapour Temp.",J.Eng.,Vol 90,P 547-552 1968.
5. Kemme,J.E.,"High Performance Heat Pipes",Proc 1967 , Thermoionic Conversion Specialist Conference, Palo Alto California,1967.
6. Marcus, B.D.,"Theory and Design of Variable Conductance Heat Pipes", Nasa CR-2018, 1972
7. Dunn,P.,Reay,D.A.,"Heat Pipes",Pergamom Press,1976

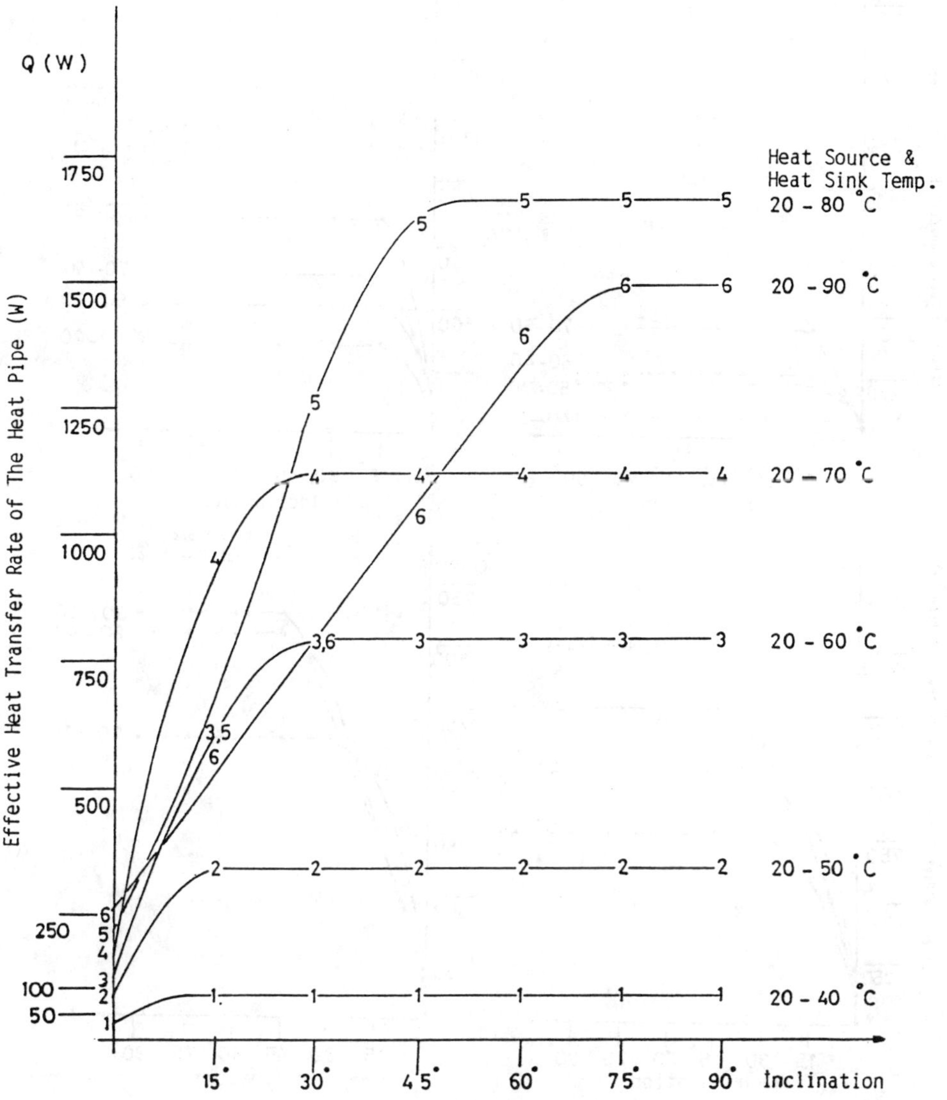

Figure 3. The experimental effective heat transfer rates of the heat pipe due to constant condencer but variable evaporator side temperatures.

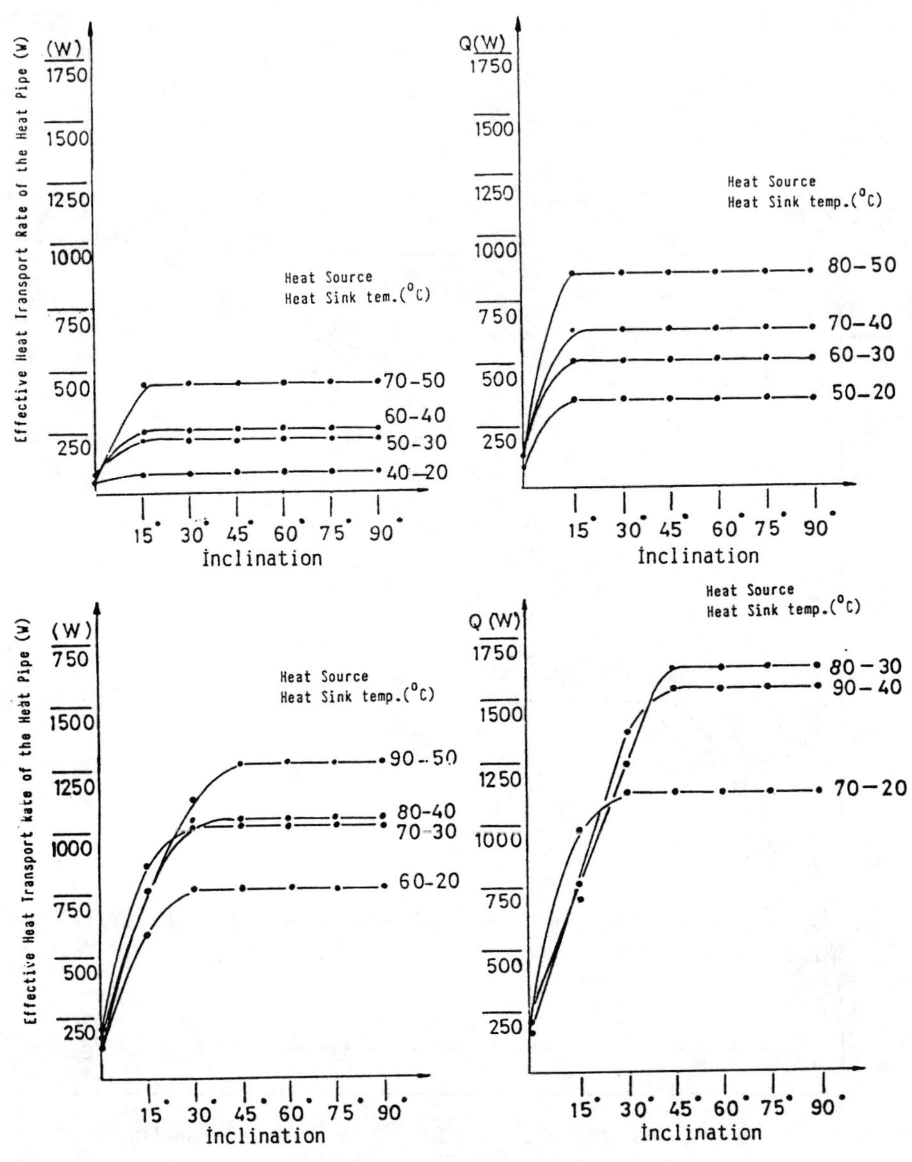

Figure 4. The effective heat transfer rates of the heat pipe due to variable evaporator (heat source), and Condenser (heat sink) temperatures.

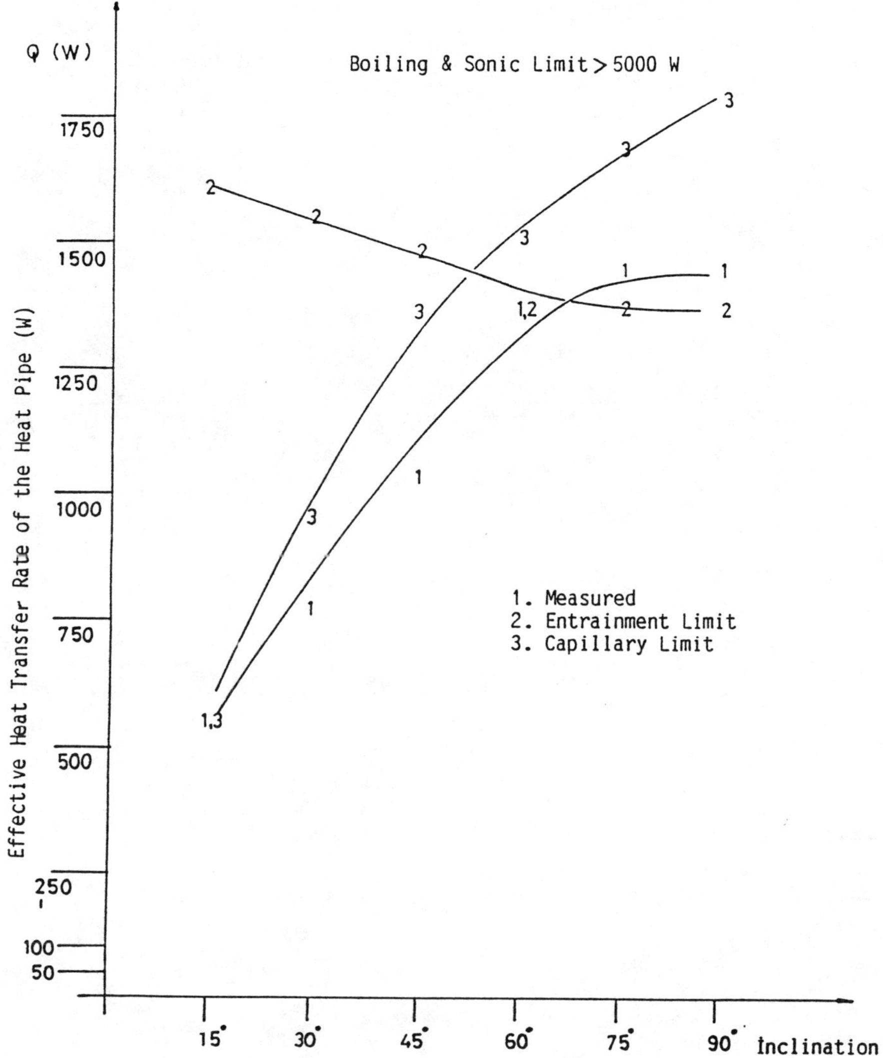

Figure 5. The comparison of the measured and calculated heat transfer capacity of the prototype heat pipe for 90-20 °C heat source-sink temperatures.

SONIC IMPULSES, THE AVAILABLE CLEAN ENERGY

Casin Popescu
Neo Sonic Power Company, 7800 Freiburg 73, Sautierstr., West Germany

SUMARY

It is known that the theory of sonic engines, i. e. when the transmission of energy is effected by means of a liquid medium through pipelines in which flow both the energy as sonic waves (or sonic output) and the resp. fluid, as well as the reckoning of sonic generators, haven't yet been formulated. With the aim to realize an immediate practical application of a new and very interesting theory in this field, we intended to briefly present in this exposé the theory of sonic (oscillating) engines, used as sonic energy consumers in a particular case: the hydrosonic attenuators, giving at the same time the calculation formulae for a new generation of sonic generators wich don't deliver energy (Qo = 0) and which thus provide only sonic energy, the so-called neosonic generators. From the exposed theory it results that the hydrosonic engines give a maximum output when they work under resonance conditions, thus a method was developed for setting up a diagram especially imagined to grasp at first sight, synoptically, their working conditions during resonance. This theory may be generalized and applied to all types of such engines, comprising also those with which the energy transmission is effected through solid mediums.

1. Introduction

Sonicity represents a new science, first created, developed and applied by the British scientist of Romanian origin George Constantinescu (1). Its aim is to transmit power by means of periodic forces and movements, transmitted as elastic waves, across liquids, solids and gases. Sonicity, other than the other sciences, is based on the phenomenon of compressibility of liquids -which actually was neglected in all hydraulic calculations- and on the time component, which plays a very important part.

In the field of sonicity are known: the transmissions, the generators and the sonic engines, the sonic pumps, the sonic hammers and perforating devices, the sonic injectors, the sonic sirens and the radar, the "Torque Convertor", the sonic boring and particularly the ultrasonic waves and their applications.

2. The hydrosonicity

By **hydrosonicity** one understands the utilization technique of the undulatory and oscillating movement of liquids for the transport of energy, its converting and the operating of mechanical devices taking into account the compressibility of liquids.

Two fundamental notions are attributes of the hydrosonicity:

the sonic output **q**
the sonic pressure **p**

By definition the sonic output **q** is the oscillatory part of the output, equal with the difference between the instantaneous value **q'** and the average value **Qo** of the output

$$q = q' - Qo \quad (1)$$

Also, the sonic pressure **p** is the oscillatory part of the pressure, equal with the difference between the instantaneous value **p'** and the average value **Po** of the pressure

$$p = p' - Po \quad (2)$$

We can demonstrate that the sonic pressure **p** is expressed by

$$p = \rho c u \quad (3)$$

ρ is the density of the liquid
c is the celerity of the waves
u is the oscillatory speed of the liquid

$$u = U \sin \omega (t - \frac{x}{c}) \quad (4)$$

Thus, il results that the sonic pressure generated by a sonic wave of the formula (4) is proportional to the density of the liquid ρ, to the celerity of the waves c and to the speed u. Consequently, it is also a priodic function of the time t.

It is noticed that the sonic pressure is alternatively positive and negative and that its average value in a point at random is zero.

Thus, the pulsations are the manifestation of the sonic waves and we shall use further sonic dimenions for designating and discussing these phenomena.

3. The common generators of sonic waves

All piston pumps are generators of sonic waves. Actually the output of these pumps varies at every moment, getting periodically the same value. If **q'** is the instantaneous output, a periodic time function, **Qo** the average constant output, the relation (1) is written

$$q = q' - Qo \quad (5)$$

thus, the piston pump delivers a sonic output **q** (Fig. 1), expressed by this relation (5).

Also, **p'** being the instantaneous pressure, periodic time function, and **Po** the constant average pressure, the relation (2) is written

$$p = p' - Po \quad (6)$$

p being the sonic pressure developed by the pump.

If we want to find the average value of the sonic output **q** during a period **T** we get

$$\frac{1}{T}\int_o^T q dt = \frac{1}{T}\int_o^T q' dt - Q_o = 0$$

Thus, the average value of the sonic output is zero, what proves that it isn't properly a so-called output, but actually an oscillatory shifting.

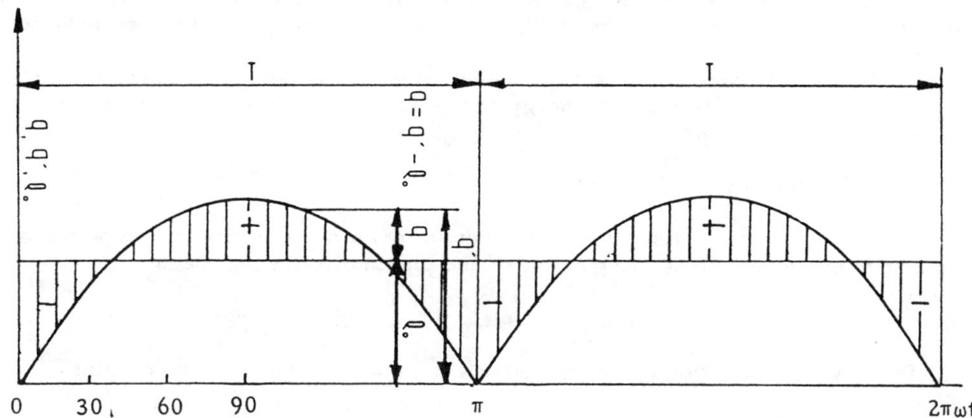

Fig. 1 **Monocylindric pump with double effect. q'** = instantaneous output, **q** - sonic output (hachured), **Qo** = the average constant output, **T** = the period, ω = the pulsation

Also, for the average value of the sonic pressure we have

$$\frac{1}{T}\int_o^T p dt = \frac{1}{T}\int_o^T p' dt - P_o = 0$$

Thus, the average value of the sonic pressure is also zero.
Let us write the relations (5) and (6) as follows

$$q' = q + Qo$$
$$p' = p + Po$$

and multiplicate them. One obtains

$$p'q' = pq + P_oQ_o + pQ_o + qP_o \qquad (7)$$

The result **p'q'** represents the instantaneous power. Let us look for the average value of the power during a period **T**

$$\frac{1}{T}\int_o^T p'q' dt = \frac{1}{T}\int_o^T pq dt + P_oQ_o \qquad (8)$$

The relation (8) demonstrates that the power delivered by the pump consists of two terms. The first is the sonic power and the second one the necessary hydraulic power. Thus, any piston pump delivers a necessary power, which corresponds to the average output **Qo** under the pressure **Po** and sonic extraneous power.

It is this sonic power which generates the incriminated mechanical effects, as shocks, vibrations, shudderings etc. Thus, the aim any attenuator should be to dissipate this kind of energy.

One may notice that a sonic generator, of the kind which was built by the inventor, doesn't deliver energy (**Qo** = 0), as it was expected, but delivers only sonic energy.

On the other hand, a centrifugal pump which gives an instantaneous constant output **q'**, doesn't deliver sonic energy and is thus free of the particular disadvantages of the piston pumps.

3. 1. The monocylindric pump

A monocylindric piston pump, with double effect, stemming in a pipeline a liquid stream, has at the starting-point of the pipeline following speed

$$u = |U \sin \omega t| \qquad (9)$$

This periodic function can be developed in Fourier series and one obtains

$$u = \frac{2U}{\pi} - \frac{4U}{\pi} \sum_{1}^{\infty} \frac{1}{(2k-1)(2k+1)} \cos 2k\omega t \qquad (10)$$

k being a natural number.

The instantaneous output **q'** will be, **S** designating the secion of the pipeline

$$q' = uS$$

and the sonic output

$$q = -\frac{4US}{\pi} \sum_{1}^{\infty} \frac{1}{(2k-1)(2k+1)} \cos 2k\omega t \qquad (11)$$

Thus, this sonic output results from the superposition of an unlimited number of harmonious waves.

The sonic pressure will be

$$p = -\frac{4\rho c U}{\pi} \sum_{1}^{\infty} \frac{1}{(2k-1)\,2(k+1)} \cos 2k\omega t \qquad (12)$$

It es easy to find out that only the fundamental wave delivers 95% of the sonic energy turned out by the pump. Thus, one may keep in mind only the fundamental wave and obtain

$$q_1 = -\frac{4US}{\pi} \frac{1}{1.3} \cos 2\omega t \qquad (13) \qquad\qquad p_1 = -\frac{4\rho c U}{\pi} \frac{1}{1.3} \cos 2\omega t \qquad (14)$$

3. 2. The duplex pump

A duplex pump, with double effect, comprises two cylinders set in a parallel direction, the handles being offset at 90°. For the fundamental wave one gets

$$q_1 = - \frac{8US}{\pi} \frac{1}{3.5} \cos 4\omega t \qquad (15) \qquad\qquad p_1 = - \frac{8\rho cU}{\pi} \frac{1}{3.5} \cos 4\omega t \qquad (16)$$

3. 3. **The triplex pump**

A triplex pump, with double effect, comprises three cylinders set together, the handles being offset at 120°. For the fundamental wave one gets

$$q_1 = - \frac{12US}{\pi} \frac{1}{5.7} \cos 6\omega t \qquad (17) \qquad\qquad p_1 = - \frac{12\rho cU}{\pi} \frac{1}{5.7} \cos 6\omega t \qquad (18)$$

4. **The oscillator**

The oscillator is a mechanical device, mainly consisting of a mass **m** joined to a spring. If a periodic force **F** cos ωt is applied, directed along the symmetrical axis of the system, the mass **m** starts moving, performing periodic oscillations. The equation of the movement is the following

$$m\ddot{x} + r\dot{x} + kx = F \cos \omega t$$

All terms represent forces. The first **mx** represents the force of inertia, the second **rx** the friction and the third **kx** the elastic force. **F** is the amplitude of the excitation. **x** represents the elongation, namely the distance at which one finds at any moment the mass **m** in relation to its position at rest.

The mechanical impedance of the system Z_M is

$$\sqrt{r^2 + (m\omega - \frac{k}{\omega})^2} \qquad (19)$$

The amplitude of the speed is the following

$$V = \frac{F}{\sqrt{r^2 + (m\omega - \frac{k}{\omega})^2}} \qquad (20)$$

Subsequently the amplitude of the oscillations is

$$X = \frac{V}{\omega} \qquad (21)$$

and the course

$$2X$$

The elastic force which attracts the spring is equal to

$$F_e = kX = k\frac{V}{\omega} \qquad (22)$$

Without stressing the importance of this well-known device, we only want to point out the fact that the energy supplied for the oscillator is solely used by friction, while the kinetic and the potential energies are always recovered, as a periodical exchange between mass and spring takes place.

5. The hydrosonic attenuator

Any piston pump has to be provided with a pulsation attenuator. With this device the pump gets noiseless and innocuous.

The author constructed a hydrosonic attenuator for the Canadian "Duplex" pumps used in the oil industry, the schema of which -imposed by the necessity to keep free the delivering pipeline for the passage of the spokeshave- is represented below (Fig. 2).

Close to a piston pump, which delivers a liquid in a pipeline with a diametre D, was connected a by-pass having the same diametre D, a derivation which may be isolated by means of two taps.

Inside the derived pipeline there is a hollow piston moving freely, backed-up by a spring. As a matter of fact, the course is limited by two block stops. The piston, having the shape of a winding off bobbin, is cut by an axial convergent-divergent channel, which allows the draining of the liquid flowing in the derivation. The device functions as follows:

After running the pump, the lower tap on the derivation must be completely turned on, and then, gradually, also the upper tap. Thus, the liquid stream is divided, one part flowing through the derivation. Under the action of the sonic waves which join the stream, the hollow piston starts oscillating. One gets an optimum output, owing to the fact that the pulsations generated by the piston pump are absolutely absorbed and the rhythmic noise of the strokes of the piston pump propagated along the pipeline is suppressed.

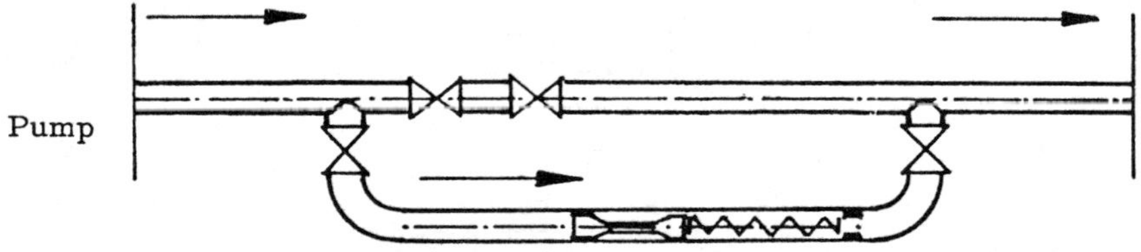

Fig. 2. Schema of the hydrosonic attenuator

5. 1. The theory of the hydrosonic attenuator

The described device is an oscillator running by means of the sonic energy developed by the pump. Thus, it can be assimilated to a sonic engine, having in view that it consumes a sonic energy which it converts into friction action, to overpower the resistance to the movement opposed by the liquid medium in which it is immersed.

Thus, a sonic power transmission takes place, comprising:

a) A sonic generator, namely the piston pump

b) A gearing line, namely the delivering pipeline

c) A sonic engine, the oscillator, connected to the gearing line, close to the pump.

We shall transform these data in sonic terms.

The generator (the piston pump) develops a sonic output $\overline{\mathbf{Q}}$ and sonic pressure $\overline{\mathbf{P}}$. We designate in this way the resp. complex dimensions, while **Q** and **P** will represent the resp. amplitudes.

The gearing line is characterized by a characteristic impedance, which is a real quantity

$$z = \frac{\rho c}{S} \qquad (23)$$

Besides, one has

$$\overline{P} = z\overline{Q}$$

The sonic engine (the oscillator) is characterized by a sonic impedance Z

$$\overline{Z} = R + j(L\omega - \frac{1}{C\omega}) = R + jK \qquad (24)$$

The value of the sonic impedance being

$$Z = \sqrt{R^2 + K^2} \qquad (25)$$

The equivalent schema of the sonic transmission is the following (Fig. 3):

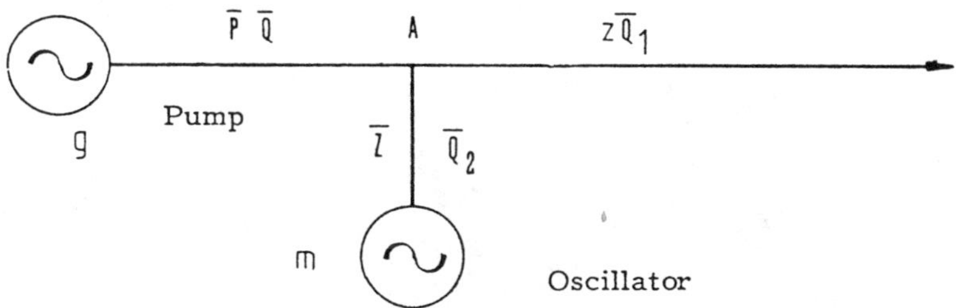

Fig. 3. Equivalent schema of the sonic transmission

At the point A the sonic output of the pump is split up and thus we have

$$\overline{Q} = \overline{Q}_1 + \overline{Q}_2 \qquad (26)$$

designating with $\overline{\mathbf{Q}}_1$ the sonic output circulating in the direct branch and $\overline{\mathbf{Q}}_2$ the sonic output absorbed by the engine (the oscillator).

At the junction point **A** the sonic pressure $\bar{P}$ gets the value

$$\bar{P} = z\bar{Q}_1$$

in relation to the gearing line. $\bar{P}$ and $\bar{Q}_1$ are thus in phase, as **z** is a real quantity.

In relation to the sonic engine, this sonic pressure gets the value

$$\bar{P} = Z\bar{Q}_2$$

and having in view that just at point **A** the sonic pressure can have only one and the same value, one gets

$$\bar{P} = z\bar{Q}_1 = Z\bar{Q}_2 \quad (27)$$

It is the second relation, necessary for the determination of the value of each sonic output of both Q_1 and Q_2 and consequently the value of **P.**

Thus, the sonic pressure is the same along the whole pipeline, what proves that the effect of the attenuator is felt right from the pump itself.

5. 2. **Functioning diagram**

In order to study the way the hydrosonic attenuator functinos, we set up the diagram represented below (Fig. 5)

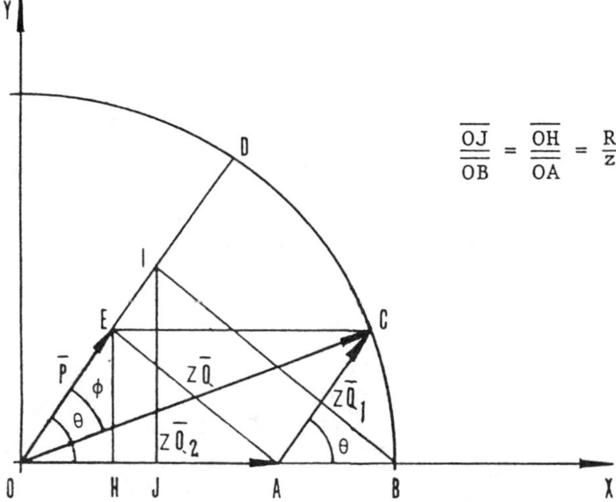

Fig. 5. Functioning diagram of the attenuator

Knowing the basic elements of the oscillator, one may set up the triangle **OJI** which represents the sonic impedance of the oscillator (Fig. 4). The $\overline{OJ}$ side corresponds to the resistance **R** and the $\overline{JI}$ side to the reactance **K.** The triangle **OHE** correspond subsequently to the sonic pressure $\bar{P}$, the amplitude **P** of which is equal to $\overline{OE}$. The **OH** side corresponds to RQ_2 and the $\overline{HE}$ side to KQ_2.

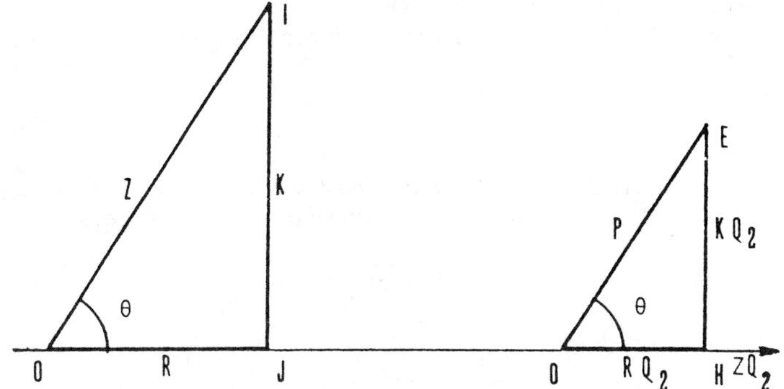

Fig. 4. Sonic impedance of the oscillator

Thus one gets

$$P = Q_2\sqrt{R^2 + K^2} = ZQ_2 \qquad (28)$$

With this diagram one can represent different dimensions, namely of the sonic pressures and outputs. In order to apply only one scale, that of pressures, we shall multiply all sonic outputs with the sonic constant and real impedance (23)

$$z = \frac{\rho c}{S}$$

which is actually the characteristic impedance of the gearing line, thus we shall have to represent at the chosen scale the dimensions $\mathbf{z\bar{Q}}$, $\mathbf{z\bar{Q}_1}$, $\mathbf{z\bar{Q}_2}$, expressed in pressure units.

In order to represent the complex sonic dimensions, we shall apply a system of rectangular axes **OX** and **OY**. **OX** will mark the origin of the phases and the direction of the real quantities.

The sonic output of the pump is constant, being invariably connected to the average output **Q** which is constant and consequently the radius **OC** being equal to **zQ** one draws the circke **BCD** having its centre at **O** and the radius equal to **OC.**

The sonic pressure delivered by the pump is equal to

$$P = zQ$$

z being the characteristic impedance of the pipeline (23).

It results that the amplitude of the sonic pressure **P** is represented at the scale chosen by **OB, OC** or **OD.**

Let's draw the triangle **OHE** which fixes the direction **OD** of the complex sonic pressure $\mathbf{\bar{P}}$.

The parallelogram **OACE** noticed on the figure, gives:

OE which represents the absorbed sonic presure $\overline{\mathbf{P}}$
OA which represents the sonic output $\overline{Q}_2$
AC which represents the sonic output $\overline{Q}_1$
OC which represents the sonic output $\overline{Q}$

Thus, the triangle **OAC** gives the following relation, which is the first condition to be observed (26)

$$\overline{Q} = \overline{Q}_1 + \overline{Q}_2$$

The **AC** side being parallel to **OE**, it means that the sonic output $\overline{Q}_1$ is in phase with the sonic pressure $\overline{P}$ and these two sides being equal, one gets

$$\overline{P} = z\overline{Q}_1$$

On the other hand one has (28)

$$\overline{P} = \overline{ZQ_2} = z\overline{Q}_1$$

Thus, one gets the second necessary condition which together with the first one (26) defines the values of $\overline{Q}_1$, $\overline{Q}_2$ and $\overline{P}$.

Thus, it is evident that the sonic pressure which initially had the dimension P = $\overline{OD}$ is reduced to P = $\overline{OE}$ following the effect of the attenuator.

Projecting the triangle **OAC** on the direction **OD** one obtains

$$\frac{PQ}{2}\cos\phi = \frac{PQ_2}{2}\cos\Theta + \frac{PQ_1}{2} \qquad (29)$$

what means that the sonic power delivered by the pump is partially consumed by the oscillator and a certain residue is conveyed on the line.

The diagram gives the opportunity to check the behaviour of the attenuator under various conditions.

Let's suppose, for instance, that the sonic output $\overline{Q}_2$ diminishes; the sonic output $\overline{Q}_1$ as well as the sonic pressure $\overline{P}$ increase. When the sonic output $\overline{Q}_2$ is cancelled, the sonic output $\overline{Q}_1$ becomes equal to $\overline{Q}$ and the sonic pressure attains a maximum. This state corresponds to the short-circuited attenuator.

On the other hand, let's suppose that the sonic output $\overline{Q}_2$ increases and attains the maximum $\overline{Q}$ which corresponds to $\overline{Q}_1 = 0$, namely to the situation in which the output of the pump passes through the derivation. In this case the sonic pressure $\overline{P}$ increases and becomes **OI.**

It is thus evident that the minimum sonic pressure is obtained for a certain relation between the sonic output $\overline{Q}_2$, absorbed by the oscillator and the total sonic output $\overline{Q}$.

Let's now examine the influence of the dephasing angle θ on the behaviour of the attenuator.

The angle θ is defined by the relation

$$\operatorname{tg}\Theta = \frac{K}{R} \qquad (30)$$

If we put in equation the elements of the figure, we get

Equation of the circle $x^2 + y^2 = r^2$
Equation of the right **JI** $x = a$
Equation of the right **OD** $y = mx$

r designating the radius of the circle, **a** the lenght $\overline{OJ}$ and **m** the angular coefficinet of the right **OD**, thus $m = tg\theta$.

In this way, for the lenght $\overline{OE}$, representing the amplitude of the absorbed sonic pressure **P** one gets

$$P = ra\sqrt{\frac{1 + m^2}{m^2a^2 + (a+r)^2}} \qquad (31)$$

In order to find the minimum of **P**, considered as function of **m**, the equation

$$\frac{dP}{dm} = 0$$

gives us the condition $m = 0$ and we find

$$P = \frac{ra}{a+r} \qquad (32)$$

If we look for the geometric site of point **E**, the extremity of the vector $\overline{P}$, we find the equation of an ellipse

$$\frac{x^2}{r^2b^2} + \frac{y^2}{r^2} = 1 \qquad (33)$$

r being the radius of the circle and

$$b = \frac{a}{r+a}$$

For $x = 0$ one finds $y = r$, the maximum value of the sonic pressure $P = r$. For $y = 0$ one gets

$$x = rb = \frac{ra}{a+r}$$

and we find anew the minimum value of the sonic pressure, which corresponds to the case where the reactance K is zero, namely to the resonance of the oscillator.

And for $m = 0$ one gets

$$Q_2 = \frac{r}{a+r}Q \quad \text{and} \quad Q_1 = \frac{a}{a+r}Q \qquad (34)$$

relations determining the repartition of the sonic outputs, for which the sonic pressure is minimal.

5. 3. Efficiency of the attenuator

The efficiency of the attenuator is expressed by the relation

$$\eta = \frac{N-N_r}{N} \qquad (35)$$

where

N is the developed sonic power

$\mathbf{N_r}$ is the sonic residual power.

But

$$N = \frac{PQ}{2} \qquad N_r = \frac{P_a Q_1}{2}$$

and one gets

$$\eta = 1 - \frac{P_a Q_1}{PQ}$$

On figure 5 one notices that

$$\frac{Q_1}{Q} = \frac{P_a}{P}$$

and it results

$$\eta = 1 - (\frac{P_a}{P})^2 \qquad (36)$$

It is the expression which allows us to calculate the efficiency of the attenuator, measuring the realtion P_a/P between the amplitude of the absorbed sonic pressure and the amplitude of the initial sonic pressure.

Referring to the expression (31) of the absorbed sonic pressure, one gets

$$\eta = \frac{r^2 + 2ar}{m^2 a^2 + (a+r)^2} \qquad (37)$$

relation which expresses the efficiency of the attenuator, reported to **m,** namely in function of the reactance of the oscillator.

It is easily seen, that the maximum efficiency takes place in case of **m** = **0,** namely when the oscillator works as resonator.

This proves the necessity to make the device work under resonance conditions.

6. Study of tje hydrosonic attenuator submitted to resonance

For the resonance we can't apply the functional diagram of the hydrosonic attenuator, set up for hydrosonic study, because the drawn geometric figures disappear when the angles ϕ and θ are zero.

Thus, we have to set up a new diagram, which should represent the characteristic hydrosonic elements of the oscillator under resonance conditions.

In that case the sonic impedance (24) becomes

$$Z = R \qquad (38)$$

namely a real quantity, equal to the sonic resistance of the oscillator.

Also, between the sonic outputs and the sonic pressure there are fundamental relations, namely

$$Q = Q_1 + Q_2 \quad (39)$$

$$P = zQ_1 = ZQ_2 = RQ_2 \quad (40)$$

All these values are scalar.

In the previous study we found in this case that the minimum sonic pressure is

$$P = \frac{ra}{a+r} \quad (41)$$

expression in which **r** represents the initial sonic pressure

$$r = zQ \quad (42)$$

z being the characteristic impedance of the pipeline and **a** a parameter defined by the relation

$$a = RQ = \frac{2N}{Q} \quad (43)$$

R being the sonic resistance of the oscillator.

We also found the expression of the sonic outputs Q_2 and Q_1

$$Q_2 = \frac{r}{a+r} Q \quad (44)$$

$$Q_1 = \frac{a}{a+r} Q \quad (45)$$

Starting from these relations we set up the diagram (Fig. 6) drawn below, which represents the sonic pressure, the sonic outputs and the sonic power delivered by the pump **N** for the oscillator **N'** and consumed by the oscillator **N"**.

We marked by **x** the variable relation

$$x = \frac{Q_2}{Q} \quad (46)$$

between the sonic output got by the oscillator and the total, constant sonic output **Q**. Thus, the relationship **x** varies between zero and one.

The ordinates of the diagram are:

-the sonic outputs **Q**, Q_1 and Q_2

-the sonic pressure **P**

-the sonic power **N, N'** and **N"**

-the velocity **V** and other parameters

The resp. equations are the followings:

The total sonic output Q. Which being constant, is graphically represented by a right **AB** parallel to **Ox.**

The sonic output Q_2

$$Q_2 = xQ$$

Is represented by the diagonal of the square which passes across the origin, namely OB

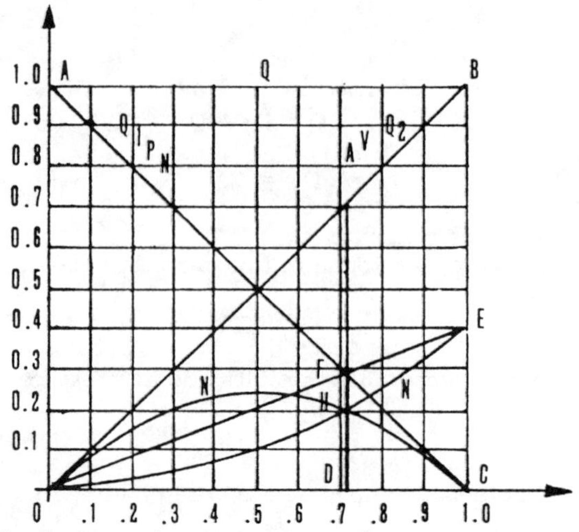

Fig. 6. Diagram of the hydrosonic attenuator under resonance conditions

x represents the relation Q_2/Q
AB represents the constant sonic output **Q**
AC represents the sonic output Q_1
OB represents the sonic output Q_2
AF represents the sonic pressure $P = zQ_1$
OE represents the sonic pressure $P = RQ_2$
AFE represents the sonic power delivered by **N**
DF represents the minimal sonic pressure
OHC represents the parabola of the sonic power delivered by **N'**
OHE represents the parabola of the consumed sonic power **N"**
OD represents the abscissa of the functioning point
HF represents the residual sonic power

The sonic output Q_1

$$Q_1 = (1 - x)Q$$

is represented by the diagonal **AC**

The sonic presssure P

According to the relation (40)

$$P = zQ_1$$

z being a constant, the right **AC** represents also, at another scale, the sonic pressure **P** up to the point **F.** The rising branch which corresponds to the relation

$$P = RQ_2 = xRQ$$

will be represented by a right **OE** the angular coefficient of which is lower than that of the right **OB,** because **R** must be smaller than **z.**

For $x = 1$ one gets $P = RQ = \overline{CE}$ and consequently

$$\frac{\overline{CE}}{\overline{CB}} = \frac{R}{z} \qquad (47)$$

Thus, the broken line **AFE** represents the variation of the sonic pressure **P**, function of **x** with the minimum $\overline{DF}$. The abscissa of the point **D** is determined by the equation

$$ZQ_1 = RQ_2$$
$$z(1-x)Q = xRQ$$

the solution of which is

$$x = \frac{z}{R+z} \qquad (48)$$

Delivered sonic power N

The sonic power delivered by the piston pump is

$$N = \frac{PQ}{2} = \frac{zQ_1Q}{2} = \frac{1}{2} zQ^2(1-x) \qquad (49)$$

It is represented at the resp. scale by the right **AC** up to the point **F**. From this point the relation becomes

$$N = \frac{PQ}{2} = \frac{RQ_2Q}{2} = \frac{1}{2} RQ^2x \qquad (50)$$

and the power delivered by the pump is represented at the scale of the powers by the right **FE**. The sonic power delivered, passes through a minimum represented by the segment $\overline{DF}$.

Sonic power delivered to the oscillator N'

The sonic power delivered to the oscillator is, of course,

$$N' = \frac{PQ_2}{2} = \frac{zQ_1Q_2}{2} = \frac{1}{2} zQ^2x(1-x) \qquad (51)$$

This parabola passes across the points **O** for $x = 0$ and **C** for $x = 1$ and it is tangential to the diagonals **OB** and **CA**. The peak is determined by the equation

$$x = 1-x \qquad x = \frac{1}{2}$$

and it is easy to notice that

$$N'_{max} = \frac{N_{max}}{4} \qquad (52)$$

Thus, the oscillator can get at the utmost a quarter of the maximum sonic power delivered by the pump.

Consumed sonic power by the oscillator N"

The sonic power comsumed by the oscillator is

$$N'' = \frac{1}{2} rV^2 \qquad (53)$$

r being the mechanical resistance opposed to the movement of the oscillator immersed in the liquid medium and **V** the amplitude of the oscillator's velocity.

Between the mechanical resistance **r** and the sonic resistance **R** one finds the following relation

$$r = R\,(S - s)^2 \qquad (54)$$

(S-s) being the plain section of the piston.

The velocity is connected to the sonic output Q_2 by the relation

$$V = \frac{Q_2}{S-s} = x\,\frac{Q}{S-s} \qquad (55)$$

and it results

$$N'' = \frac{1}{2} rV^2 = \frac{1}{2} r\,\frac{Q_2^2}{(S-s)^2} = \frac{1}{2} RQ_2^2 = \frac{1}{2} RQ^2x^2 \qquad (56)$$

It is the equation of a parabola having its peak in **O** and the axis **OA.** For **x**=**1** one states that this curve corsses the point **E,** the extremity of the right **OE.**

Determination of the functioning point

Evidently, the functioning point of the attenuator is attained when the power delivered to the oscillator is equal to the power consumed by the latter

$$\frac{1}{2} RQ^2x^2 = \frac{1}{2} zQ^2(1-x)x \qquad (57)$$

equation which gives the seeked solution

$$x = \frac{z}{R+z} \qquad (58)$$

Referring to the relation (48) one notices that it is similar to the value found for the abscissa of the point **F.** The intersection points **H** and **F** have obviously the same abscissa.

It should be noted that in these two points (**H** or **F**) the fundamental relations (39) and (40) are fulfilled.

Velocity of the oscillator

The amplitude of the velocity **V** is given by the relation (55)

$$V = \frac{Q_2}{S-s} = x\,\frac{Q}{S-s} \qquad (55)$$

and it is represented at the scale of velocities by the right **OB.**

Thus, the maximum amplitude for **x** = **1** is

$$V_{max} = \frac{Q}{S-s} \qquad (59)$$

Acceleration of the oscillator

The amplitude of the acceleration is the following

$$A = V\omega = x \frac{Q}{S-s} \omega \qquad (60)$$

and is represented at the scale of the accelerations by the right **OB.**

Amplitude of the oscillation

The amplitude of theoscillationis given by the relation

$$X = \frac{V}{\omega} \qquad (61)$$

and is also represented by the right **OB,** because **ω** is a constant.

The course of the oscillator

The course of the oscillator is **2X** and is represented also by the right **OB.**

The friction force

The oscillator in movement opposes a friction force expressed by the relation

$$F_f = rV = R(S-s)^2 \frac{Q_2}{S-s} = RQ_2(S-s) \qquad (62)$$

which is thus represented, at the scale of forces, by the right **OE.** During resonance, this force is directly opposed ti the active force which is represented at the same scale by the right **AC** with the following equation

$$F_a = \Gamma(S-s) = zQ_1(S-s) = (1-x)zQ(S-s) \qquad (63)$$

When the equilibrium is reached, which corresponds to the functioning point, one gets

$$F_f = F_a = \overline{DF} = \frac{R}{R+z} zQ(S-s) \qquad (64)$$

Efficiency of the attenuator

The efficiency of the attenuator is the following

$$\eta = \frac{\overline{OA} - \overline{HF}}{\overline{OA}} = 2x - x^2 \qquad (65)$$

This efficiency expresses the rate at which the attenuator reduces the destructive sonic energy propagated along the pipeline.

Residual sonic force

The residual sonic force $\overline{HF}$ is expressed by

$$\overline{HF} = \frac{1}{2} zQ^2(x^2-2x+1) \qquad (66)$$

which is cancelled only for x = 1, a condition which cannot be achieved.

One notices that the performance expressed by the relation (65) increases rapidly and attains the value = 0,96 for x = 0,8. Thus, if the oscillator is able to absorb 80% of the sonic output, the performance of the attenuator reaches 96% and the residual sonic force which can be recovered, is reduced to 4%.

7. Experiences

The test effected by the Oil Ministry of Romania and conducted by the author for the study of the functioning of the hydrosonic attenuator and its behaviour in the pumping installations, as well as the tests on the assembly-bench recently effected in the Federal Republic of Germany (1985), have absolutely confirmed the above presented theory. In the practical field -on condition that the oscillator should be well projected- the results of these experiences led to four essential conclusions:

1. The optimum performance of the hydrosonic attenuators mounted in the aspiration pipeline is situated close to the unit; namely, out of 17 different cases (trilex pump, compressing pressure 100; 250; 300; 305 bar, aspiration pressure -o,1; 1,0; 3,6 bar number of turns/minute 199; 201; 400; 500, fluid, salted water), for 11 cases the oscillograms gave performances comprised between 0,992 and 0,9989.
2. The optimum performance of the hydrosonic attenuators functioning behind the pump can reach:
   a) in the circuit pumping installations, values beyond 0,95
   b) in the on the line pumping installations, values beyond 0,97
3. The hydraulic ram shocks resulting from the sudden stopping of the pump or of another leading device of the flowing are also attenuated.
4. If the hydrosonic attenuator functions in the aspiration pipeline, the cavitation disappears, because the required net positive suction head goes on being inferior to the available net positive suction head, owing to the action of the oscillator.

The conclusions 3 und 4 are logic consequences of the relations (27) expressed above.

8. An ignored source of sonic energy

Instead of the pump, one can imagine a sonic generator of the type of those delivering only sonic energy and which, by a suitable transmission, are able to transmit to a free suspended oscillator on a gliding line periodic impulses, having a constant dimension and frequency. Let's suppose, for instance, that the oscillator consists of a hollow cylinder on which is winded an isolated metallic wire performing regular, alternative two-way movements, so as to pass across the same positions with the same velocities as in front of a magnet. Thus, one can imagine a parametric oscillometre, namely an electric engine with alternating movements, functioning according to the principle of the electromechanic resonance

Here we have a new source of clean (natural) energy, which may be used on industrial scale by means of the sonic generators.

8. 1. The neosonic generator

A liquid submitted to compression accumulates during this time a certain quantity of potential elastic energy, which it delivers solely as cinetic energy, at the moment of the total decompression. I call **neosonic generator** any apparatus able to transmit almost entirely to the exterior this kind of decompression energy, just at the moment when the volume of decompressed liquid regains its initial volume (prior to the compression), corresponding actually to the moment when the energy delivered by the liquid attains its peak. Thus, in principle, the most unsophisticated neosonic generator consists of a limited volume of liquid which is alternatingly compressed and decompressed in a tank, one wall of which being a membrane outfitted with a plug controlling the functioning of the membrane. This plug provides at the same time the necessary resistance in order to keep the liquid compressed under the required pressure, up to the moment when one wishes to decompress it. Beside this function of resistance to keep the liquid in compressed state, the plug has furthe two functions:

1. it starts by instantaneous release the decompression of the liquid at the due moment;

2. it restores the membrane to its initial position, just when the liquid regains its initial volume. This condition is absolutely necessary, because after this moment the decompression energy of the liquid is dissipated in the intermolecular space and ceases to act on the membrane with its maximal intensity.

If the pressure necessary to compress the liquid in the tank is given by a column of liquid (the same as that in the tank), the tank should be equipped with an absorption (or evacuation) system of the excess of liquid introduced in the tank in order to achieve the required compression. In such a case, the functioning of the absorption (or evacuation) system should be also controlled by the plug, because the elimination of the liquid in excess must take place when the membrane recovers its initial position from its position of maximum admitted deformation.

In the center of the membrane is fixed in a perpendicular position to it a steel ram-cylinder. The latter informs the exterior of the energy assumed by the membrane, but also of the centering of the plug.

The alternating conpressing and decompressing liquid, like a liquid ressort -which evidently doesn't observe the law of Hooke, which moreover was invalidated since long by the experiences of Bach & Schule- must behave as a sonic capacity, in order to give the above described generator the possibility to deliver a sonic energy of required value. This implies that the interior lenght of the tank in the direction in which the decompression force (the impulses) acts on the membrane should be short if compared with the lenght of the sonic wave. Under the mentioned conditions, the membrane acts as sonic condensator.

From the theoretic point of view, the functioning of the so-called neosonic generator is similar to that of the already known sonic generators with membrane. Yet, the main difference between these generators and all their predecessors is constituted by the functioning principle of the above defined membrane, owing to which a natural phenomenon particular to liquids is put forward, namely: the sonic elastic reaction of an impulse liquid is more intense than the force (pressure) generating it.

Of course, in order to be able to apply this principle, the deformation of the membrane must be kept in the elastic field of very small deformations -at a given frequency- and the transmission to the exterior of the sonic power should be done in this field.

If R_{liq} is the sonic reaction of the liquid, and F_i(=F/t") the impulsion force applied on the liquid, ψ a coefficient without dimension which defines the size of the sonic reaction in relation to F_i, one demonstrates that

$$R_{liq} = \psi F_i \qquad (67a)$$

where

$$R_{liq} = \psi P \qquad (67b)$$

having in view that any force acts on a liquid like a pressure P (dP=dF/dA, where A is the surface of the liquid on which acts the force F).

The value of the coefficinet ψ is given by

$$\psi = 3\rho c^2 b \qquad (68a)$$

where

ρ is the density of the liquid (Kg sec²/m⁴),

c the celerity of the sound (m/sec)

b the compressibility coefficient (m²/kg).

Under a more general form, the relation (68a) is written, adopting for c the formula suggested by the American scientists Hueter and Bolt (2)

$$\psi \simeq 3\rho(4,3 \frac{c_{gas}}{1-s})^2 b \qquad (68b)$$

Taking in consideration the relations (68a), (68b), the relations (67a) and (67b) are written

$$R_{liq} = 3\rho c^2 b F_i \qquad (69a)$$

and

$$R_{liq} \simeq 3\rho(4,3 \frac{c_{gas}}{1-s})^2 bP \qquad (69b)$$

The demonstration of the relations (67a), (67b) and (68a), as well as the physical interpretation of the coefficient go beyond the limits of this exposé and the reader eventually interested in these problems can find the theoretical details in a supplement on this matter, which will be separately published.

The builder of the first hydrosonic monophasic generators, Geroge Constantinescu, states -quite surprised- that he succeeded in transmitting, by means of water, to the atmospheric pressure a sonic power of about 3,5 kg/cm² (3).

As a matter of fact, with

ρ = 101,79 kg sec²/m (the density of water at 20°C),

c = 1481 m/sec (the sound celerity in fresh water at 20°C),

b = 0,000000005 m²/kg,

and

P = 1,0336 kg/cm²,

the relation (67b) gives:

$$R_{liq} = 3,46 \simeq 3,5 \text{ Kg/cm}^2$$

That proves that the neohydrosonic generators are able to catch and to transmit almost integrally especially the sonic (specific) reaction of water and of liquids in general.

In order to project a neosonic generator, one must know at first the exact value of the compressibility coefficient of water for any worging pressure and temperature, in order to establish the volume of the compressed water, Δv, which can be determined only by experiences.

To clear up this difficulty, a general formula was elaborated for the reckoning of the compressibility $b_{p,t}$ of water:

$$b_{p,t} = \frac{b}{\sqrt{1+\alpha t}} e^{b(p+1)/\sqrt{1+\alpha t}} \frac{\rho_t^2}{(\chi\rho_p)^2} \qquad (70)$$

where

b is the compressibility coefficient of fresh water (under isothermic conditions), taken equal to -0,00005 cm^2/kg,

α is a constant equal to 1/273,

p, t, ρ_t, ρ_p represent the pressure, the temperature and the resp. densities of water, for which one reckons the corresponding values of $b_{p,t}$ and

χ the ration of the specific heats. By means of the known formula

$$\frac{\rho_p}{\rho_o} = 1+0,00005(p-1)$$

one can calculate ρ_p where ρ_o is the density of water under the pressure of 1 atm.

The values obtained with the formula (70) correspond to those given by the International Critical Tables.

When the value of $b_{p,t}$ is known, by means of the known formula

$$b = \frac{1}{v_1} \cdot \frac{v_1-v_2}{p_2-p_1}$$

one gets the seeked volume

$$\Delta v = v_1 - v_2$$

for which the deformation of the membrane and the functioning of the neohydrosonic generators should be calculated.

8. 2. Remak 1. The sonic reaction of the sea-water is the basis of the formation of waves under the impuse of Aeolian forces.

8. 3. Remak 2. The paradox of Bergeron is better explained by the actual sonic reaction force of water, than by the principle of equality between action and reaction, the rightness of which is moreover doubted since a long time.

8. 4. Remark 3. The characteristic sonic reaction of liquids is at the origin of several natural phenomena which are still unsatisfactorily explained, as for instance the ejaculation of certain secretions and particularly of sperm.

8. 5. Remark 4. Among these phenomena which are not well defined, we must mention also the hydraulic ram. For all these natural cases, which by analogy can be considered sonic generators, the membranes are automatically formed and function like ideal membranes, perfectly balanced on the critical positions, which transmit to the exterior the sonic energy, precisely at the moment when the volume of the compressed liquid attains in its expansion on the direction of the exterior force, the initial volume.

9. Conclusions

The conclusions of this exposé are the followings:

1. The hydrosonic engines are the best suitable to be used as attenuators of pulsations and noises in the pumping installations or similar ones. They are also recommended as energy supplying devices for various practical aims.

2. The experiences very recently realized by the author in the Federal Republic of Germany, with two hydroneosonic genreators -investigated models- symetrically set up in relation to a bobbin placed in a magnetic field, in such way as to be able to perform rhythmic movements (forward and backward) under parametric resonance conditions resulting from the impulses of the generators, showed that:

a) the sonic reaction of the water to impuses is more intesive than the reaction resulting from the principles of the mechanics of Newton.

b) it is possible to build hydroneosonic generators at industrial scale, for obtaining different forms of energy: electric, mechanical, calorific, associating them to transmissions or to adequate mechanical systems.

Appreciations

The research works presented in these studies (4) have been realized alone by the endeavours of the author. In order to be able to resume his research works and to go beyond the theoretical stage, the author would absolutely need to cooperate with an interested research or industrial institute.

References

1 George Constantinescu, **Theory of Sonics. A treatise on Transmission of Power by Vibrations,** Vol. I, London 1918.

2 Theodor F. Hueter, Richard H. Bolt, **Sonics,** New York/London 1955, p. 429.

3 **Sonics.** "The Transaction of the Society of Engineers" (Inc.) Abbey House, Westminster S.W.1, pp. 69-102, June 1959. Reprinted by the Lewes Press Wightman & Co. Ltd. Friars Walk, Lewes, Sussex 1959.

4 Having in view that the neosonic generator and the oscillomotor are under way to be patented, owing to certain formalities imposed by the rules, it wasn't possible to present more practical and theoretical details. In what concerns the hydrosonic attenuator, it was considered useless to reproduce the oscillograms or the practical characteristic found during the tests, this device being known and applied from the practical point of view. The readers interested in these problems may contact the author to get more details.

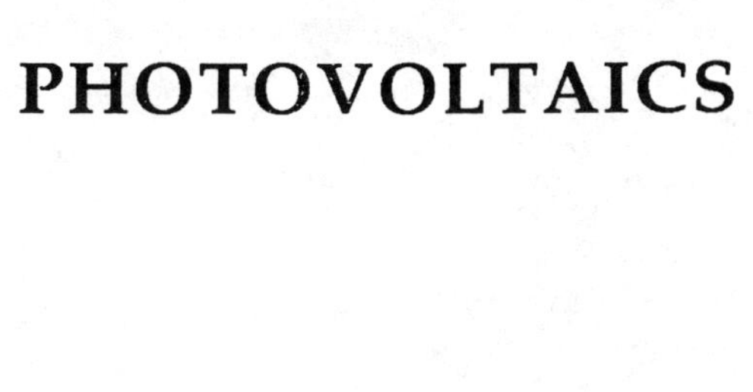

PHOTOVOLTAICS

PROGRESS IN THIN FILM SOLAR PHOTOVOLTAIC TECHNOLOGIES

Harin S. Ullal and Kenneth Zweibel
Solar Energy Research Institute
1617 Cole Boulevard, Golden, CO 80401, U.S.A.

Abstract

This paper focuses on the rapid recent advances made by thin film solar cell technologies, namely, amorphous silicon, copper indium diselenide, and cadmium telluride. It also indicates the several advantages of thin films. Various consumer products and power applications using thin film solar cells are also discussed. The increasing interest among the utilities for PV system applications is also elucidated.

1. INTRODUCTION

The widespread utilization of photovoltaics (PV) for large-scale applications requires higher efficiency, lower-cost, and increased reliability of the devices. Accordingly, much of the research and development in the past decade has focused on high-efficiency, low-cost thin film solar cells. Other PV technologies, such as crystalline silicon, polycrystalline silicon, sheet and ribbon silicon, gallium arsenide, and concentrator solar cells are also being investigated as PV options by several research organizations [1].

This paper focuses on the rapid advances made by thin film solar cells, such as amorphous silicon (a-Si), copper indium diselenide ($CuInSe_2$, CIS), and cadmium telluride (CdTe) in the past few years. Figure 1 shows the progress of these thin film solar cell technologies.

Advantages of Thin Films

The main advantage of thin films is the minimum amount of material requirements. For example, in the crystalline silicon technology the film thickness of the cell is roughly 300 microns, whereas in the case of the thin film solar cells the film thickness is 1-3 microns. This is primarily due to the high optical absorption of thin film materials. This reduced material requirement results in considerable cost savings. In addition, several low-cost, high throughput and scalable methods such as plasma enhanced chemical vapor deposition, electrodeposition, spraying, sputtering and selenization are available for fabricating the thin film solar cells. Also, low-cost substrates such as soda lime glass and plastic are used in large-area modules. Module fabrication is further simplified by monolithic interconnection of cells during the actual fabrication process. Thin film solar cells can also be used in both single and multijunction configurations with 15% – 20% efficiency ranges expected for optimal cell designs. Finally, the use of glass as the top encapsulant in a superstrate structure eliminates the problem of degradation of polymers such as EVA and PVB since they are not exposed to sunlight.

2. AMORPHOUS SILICON

For the past several years, since the first a-Si solar cell was reported by Carlson and Wronski in 1976 [2], there has been a great deal of progress in improving the efficiency of a-Si devices, fabricating large-area modules, and increasing the infrastructure for research and development. To-date, 26 groups around the world have reported efficiency in excess of 10% for small-area devices. The highest efficiency a-Si solar cell is a multijunction device with a conversion efficiency of 13.3% (active area) that has been fabricated by Energy Conversion Devices (ECD). The bandgaps of the three light absorbing layers of this device are 1.7 eV, 1.7 eV and 1.45 eV, respectively. The top two cells are silicon based, while the bottom cell is a silicon-germanium alloy. ECD has also fabricated a 8.4% efficient square foot multijunction module using a silicon-germanium alloy. In the U.S. alone there are at least seven (7) companies that are actively involved in taking the a-Si technology from the lab to the market place. Table I summarizes the major players in this area.

In terms of technical performance, ARCO Solar has fabricated a 9.4% efficient semitransparent single junction a-Si square-foot module with a white back reflector. Most recently, Chronar has fabricated the world's largest 2.6 ft $\times$ 5 ft single junction monolithic module (Fig. 2) with a power output of 61.76 W tested outdoors at an insolation of 1043.93 W/m^2, which has been verified by SERI. The module parameters are I_{sc} = 1.7 amperes, V_{oc} = 57.9 volts, FF = 0.628, and Eff = 5.22%. There are 77 cells connected in series in this module, which is part of a 10 MW – Eureka project [3].

The most critical issue facing the a-Si technology is stability. Ever since the first report of the so-called "Staebler-Wronski" effect was published in 1977 [4], researchers have aggressively addressed this problem. Although progress has been made over the years [5], the average degradation observed in actual field testing is approximately 20% [6,7], whereas laboratory devices exhibit a 10%-15% degradation using multijunction structures. The use of multijunction device structure represents an acceptable engineering solution [8]. Most manufacturers of a-Si are engaged in research on multijunction modules which may result in cost effective thin film a-Si modules.

3. COPPER INDIUM DISELENIDE

Copper indium diselenide ($CuInSe_2$, CIS) is presently the leading thin film PV material in terms of efficiency and reliability for terrestrial PV applications [9-12].

The first 10% efficient thin film $CuInSe_2$ solar cell was reported by R. A. Mickelsen and W. S. Chen of Boeing in 1982 [13]. Since then ARCO Solar, the Institute of Energy Conversion at the University of Delaware, International Solar Electric Technology, the Solar Energy Research Institute, and the University of Stuttgart have all reported efficiency over 10%. In the past few years, the rate of progress has improved dramatically at both the cell and module level. Table II lists the various deposition methods for the growth of $CuInSe_2$ films. Innovative cell design, directed at improving $CuInSe_2$ cell blue response, was first proposed by Choudary et al. [14] of ARCO Solar. It was reduced to practice by Potter et al. [15], who increased the cell efficiency to 12.5%. Further addition of Ga (< 10%) has enhanced the cell efficiency to 14.1% (active area) for a 3.5 cm^2 device also made by ARCO Solar [11]. Boeing has also improved their cell efficiency to 12.9% (active area) by the addition of Ga [16]. The Ga content in their device is about 27%. The most significant improvement was in the V_{oc} of 550 mV, which is a major improvement for this technology. Projected efficiencies for $CuInSe_2$ solar cells are in the range of 15%-20% [17,18].

Substantial technical progress has been made by ARCO Solar in the area of $CuInSe_2$ module fabrication. Using the sputtering/selenization method [19,20], efficiency of 11.1% for a square-foot module has been verified at SERI, and 9.1% for a four-square feet module shown in Fig. 3 with a power output of 35.8 W has been reported so far [21]. Tested outdoor for 240 days at SERI under natural sunlight under both load and open-circuit conditions, ARCO Solar's $CuInSe_2$ modules have demonstrated excellent reliability [22], as is shown in Fig. 4. This is a major accomplishment for the thin film $CuInSe_2$ technology.

4. CADMIUM TELLURIDE

Cadmium telluride normally referred to as the "darkhorse" of thin film solar cells, has shown considerable improvement in performance in the past few years. In fact, it was under development before a-Si and $CuInSe_2$ and made some advances in the 1970s and early 1980s.

Y.S. Tyan and E.A. Perez–Albuerne of Kodak were the first to report 10% thin film CdTe solar cells in 1982 [23]. Since then several groups such as Ametek, ARCO Solar, BP Solar, Georgia Institute of Technology, Jet Propulsion Laboratory, Monosolar, International Solar Electric Technology, Photon Energy, SOHIO, Southern Methodist University, and Matsushita Battery have reported efficiencies of over 10%. A number of methods are used for depositing the CdTe thin-films, and are listed in Table III. The most promising low-cost approaches are electrodeposition and spraying. Screening printing has had limited success so far, due to limitations in module processing [24].

One of the key technology issues for CdTe devices is the contact stability. Ametek has circumvented this problem with a novel cell design [25,26]. In this device structure, the undoped CdTe film is sandwiched between n-CdS and p-ZnTe. There is no direct metal contact to the high resistivity CdTe films. The device structure and band diagram is shown in Fig. 5 and 6, respectively. The thin film CdTe in this case is deposited by electrodeposition, a potentially low-cost technique. Using this method, Ametek has been successful in fabricating 11.2% efficient devices. Ametek has also tested their cells and modules under illumination and load, and they have reported no change in performance after 3000 hours of exposure under controlled indoor testing conditions.

Another potentially low-cost approach for fabricating large-area CdTe thin film modules is spraying. Since 1984, Photon Energy has been actively pursuing this method to manufacture product-sized photovoltaic modules [27]. In 1988, with DOE and SERI support, they made a 7.3% square-foot thin film CdTe module [Fig. 7]. They have also fabricated four-square-foot prototype CdTe modules. Photon Energy has very aggressive near-term goals to enter the PV power production markets. Their cost estimates suggest that they may potentially be one of the lowest cost module manufacturers. They do not appear to require the same levels of manufacturing scale-up (i.e., 10 MW) to achieve economies-of-scale. In fact, they claim that they will achieve low cost ($1-$2/W modules) at a production level of about 3 MW/year.

One of the critical issues hindering the CdTe technology is the prescence of cadmium. However, environmental issues are closely associated with "elemental Cd," but most of the processing of CdTe technology does not require elemental Cd. All processing steps can use Cd-based compounds, which are at least a 100 times safer than elemental Cd.

Manufacturers believe that Cd can be appropriately handled in the workplace. Another issue is the product disposal at the end of 30 years lifetime. A possible solution is to return the modules to smelters, who could separate out the various chemical compounds and recycle the Cd, Te, and/or glass. Several other options are also possible. It is important to realize that the amount of Cd, in these modules is small. For instance, on the basis of producing energy, a CdTe module will produce 1 MWh for each gram of Cd, while a coal plant inadvertent releases the same amount of Cd, i.e., 1 gram/MWh (Cd present in coal). A coal plant also produces 120 gram of arsenic per MWh, as well as numerous other pollutants which are not present in PV modules. As with any strategies, one must weigh the risks and the rewards of new options against the ones of the existing technologies they are aimed at displacing.

5. THIN FILM SOLAR CELL APPLICATIONS

Consumer Products

In the early 1980s, a number of consumer products based primarily on a-Si thin film solar cells were introduced to the market by Japanese companies. Among them, calculators, clocks, watches, etc., were the most popular products. Since then other interesting applications like battery chargers, semi-transparent car sunroofs, garden or patio lights, battery chargers for RV vehicles, and billboard lights have found increasing acceptance by consumers of the thin film solar cell based products. In fact, in 1988 a-Si products accounted for 40% of the world market share in photovoltaic shipment, primarily for consumer applications.

PV Power Applications

In the past few years, thin film solar modules have also been used for a number of solar power applications. Among them are water pumping, irrigation, street lighting, railroad signals, and small demonstration systems by utilities (4 KW - 55 KW). The most recent one is the Photovoltaics for Utility Scale Applications (PVUSA, Fig.8) projects. System sizes varying from 20 KW to 400 KW have been installed or will be installed next year at Davis, CA using a-Si, CdTe, and $CuInSe_2$ for evaluation and field testing [28,29]. Thin film solar cells are making rapid progress towards helping us solve energy and environmental problems as we enter the 1990s.

6. NOMENCLATURE

A	ampere(s)
ASI	ARCO Solar, Inc.
a-Si	amorphous silicon
a-Si:Ge	amorphous silicon germanium
a-Si:H	hydrogenated amorphous silicon
BP	British Petroleum
Cd	cadmium
CdS	cadmium sulphide
CdTe	cadmium telluride
CIS	copper indium diselenide
$CuInSe_2$	copper indium diselenide
DOE	U.S. Department of Energy
ECD	Energy Conversion Devices

Eff	Efficiency
EMT-1	Emerging Materials Technology - 1
EMT-2	Emerging Materials Technology - 2
EVA	Ethylene Vinyl Acetate
FF	Fill Factor
Ga	gallium
GSI	Glasstech Solar, Inc.
IEC	Institute of Energy Conversion
I_{sc}	short circuit current
ISET	International Solar Electric Technology
JPL	Jet Propulsion Laboratory
J_{sc}	short circuit current density
KW	kilowatt(s)
MW	megawatt(s)
MWh	megawatt-hour
PEI	Photon Energy, Inc.
PV	Photovoltaic(s)
PVUSA	Photovoltaics for Utility Scale Applications
Si	silicon
SMU	Southern Methodist University
Te	tellurium
UPG	Utility Power Group
USF	University of South Florida
US-1	Utility Scale - 1
V_{oc}	open circuit voltage
W	watt(s)

Acknowledgments

This work was supported by the U.S. Department of Energy under contract # DE-AC02-83CH10093.

7. REFERENCES

1. Five Year Research Plan 1987-1991, National Photovoltaics Program, U.S. Department of Energy, May 1987, DOE/CH10093-7

2. D. E. Carlson and C. R. Wronski, Appl. Phys. Lett., 28, 671 (1976)

3. E. S. Sabisky, Z. Kiss, F. Ellis, E. Eser, S. Gau, F. Kampas, J. van Dine, and H. Weakliem, Proceedings of the 9th E. C. Photovoltaic Solar Energy Conference, Freiburg, F.R. Germany, September 25-29, 1989

4. D. L. Staebler and C. R. Wronski, Appl. Phys. Lett., 31, 292 (1977)

5. H. S. Ullal, D. L. Morel, D. R. Willett, D. Kanani, P. C. Tyalor and C. Lee, Proceedings of the 17th IEEE Photovoltaic Specialists Conference, Kissimmee, FL, May 1-4, 1984

6. L. Mrig, Proceedings of the 24th Intersociety Energy Conversion Engineering Conference, August 6-11, 1989, Washington, D.C.

7. G. H. Atmaram and C. Herig, Proceedings of the 1989 American Solar Energy Annual Conference, Denver, CO, June 19-22, 1989

8. W. Wallace, J. Ohi, W. Luft, B. Stafford, and E. Sabisky, Proceedings of the 20th IEEE Photovoltaic Specialists Conference, Las Vegas, NV, September 26-30, 1988

9. H. S. Ullal, K. Zweibel, and R. L. Mitchell, MRS 1989 Fall Meeting, Boston, MA, November 27 - December 2, 1989; SERI/TP-211-3595

10. K. Zweibel, H. S. Ullal, and R. L. Mitchell, Proceedings of the 20th IEEE Photovoltaic Specialists Conference, Las Vegas, NV, September 25-29, 19889

11. K. Zweibel and H. S. Ullal, Proceedings of the 24th Intersociety Energy Conversion Engineering Conference, Washington, D.C., August 6-11, 1989

12. H. S. Ullal (Ed.), Proceedings of the Polycrystalline Thin Film Program Meeting, Lakewood, CO, August 16-17; SERI/CP-211-3550

13. R. A. Mickelsen and W. S. Chen, Proceedings of the 16th IEEE Photovoltaic Specialists Conference, San Diego, CA, September 27-30, 1982

14. U. V. Choudary, Y. H. Shing, R. R. Potter, J. R. Ermer, and V. K. Kapur, U.S. Patent No. 4,611,091, September 9, 1986

15. R. R. Potter, C. Eberspacher, L. B. Fabick, Proceedings of the 18th IEEE Photovoltaic Specialists Conference, Las Vegas, NV, October 21-25, 1985

16. W. E. Devaney, W. S. Chen, J. M. Stewart, and R. A. Mickelsen, IEEE Trans. on Electron Devices, February 1990, to be published

17. J. R. Sites, Proceedings of the 20th IEEE Photovoltaic Specialists Conference, Las Vegas, NV, September 26-30, 1988

18. K. W. Mitchell, Proceedings of the 9th E. C. Photovoltaic Solar Energy Conference, Freiburg, F.R. Germany, September 25-29, 1989

19. R. B. Love and U. V. Choudary, U.S. Patent No. 4,465,575, August 14, 1984

20. J. R. Ermer and R. B. Love, U.S. Patent No. 4,798,660, January 17, 1989

21. J. Ermer, C. Fredric, K. Pauls, D. Pier, K. Mitchell and C. Eberspacher, Proceedings of the 4th International Photovoltaic Science and Engineering Conference, Sydney, Australia, February 14-17, 1989

22. L. Mrig and S. Rummel, in H. S. Ullal (Ed.), Proceedings of the Polycrystalline Thin Film Program Meeting, Lakewood, CO, August 16-17, 1989; SERI/CP-211-3550

23. Y. S. Tyan and E. A. Perez-Albuerne, Proceedings of the 16th IEEE Photovoltaic Specialists Conference, San Diego, CA, September 27-30, 1982

24. K. Zweibel and R. Mitchell, "$CuInSe_2$ and CdTe: Scale-Up for Manufacturing," October 1989; SERI/TR-211-3571

25. P. V. Meyers and C. H. Liu, in Proceedings of the 8th E. C. Photovoltaic Solar Energy Conference, Florence, Italy, May 9-13, 1988

26. H. S. Ullal, Electronic Structure of Electrodeposited Thin Film CdTe Solar Cells, May 1988; SERI/TP-211-3361

27. S. P. Albright, B. Ackerman, and J. F. Jordan, in IEEE Trans. on Electron Devices, February 1990, to be published

28. S. Hester, Proceedings of the 20th IEEE Photovoltaic Specialists Conference, Las Vegas, NV, September 26-30, 1988

29. S. Hester, 1989 DOE/Sandia Crystalline Photovoltaic Technology Project Review Meeting, Albuquerque, NM, July 11-13, 1989, SAND89-1543

TABLE I

Amorphous Silicon

- ARCO Solar
- Chronar
- Energy Conversion Devices
- Glasstech Solar
- Solarex
- Utility Power Group
- 3M/Iowa Thin Films

TABLE II
$CuInSe_2$ Processing

- Coevaporation
- Electrodeposition/selenization
- Electron beam/selenization
- Hybrid evaporation/sputtering
- Sputtering/selenization
- Reactive sputtering
- Close space vapor transport
- Metal organic chemical vapor deposition
- Sputtering/laser assisted annealing
- Sputtering/rapid isothermal processing
- Spraying

TABLE III
CdTe Processing

- Electrodeposition
- Spraying
- Close space vapor transport
- Screen printing
- Chemical vapor deposition
- Hot wall evaporation
- Ion assisted evaporation
- Laser assisted evaporation
- Thermal evaporation
- Sputtering
- Sputtering/laser assisted annealing
- Molecular beam epitaxy
- Metal organic chemical vapor deposition

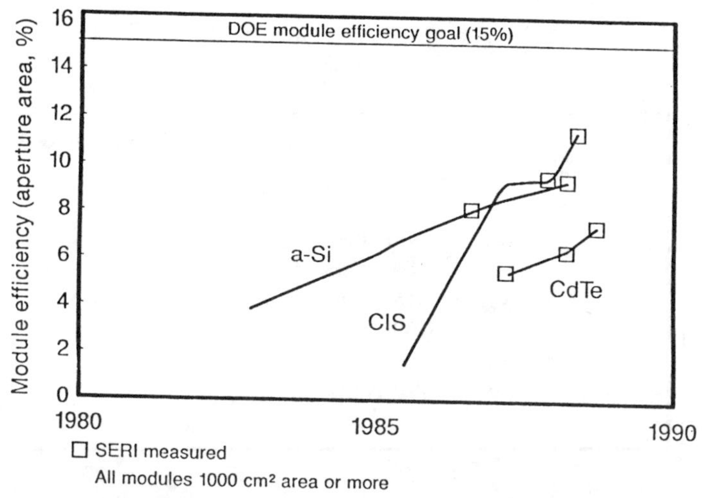

Figure 1 Progress in thin film solar photovoltaic module technologies

Figure 2 A 2.6 ft x 5 ft very large a-Si module fabricated by Chronar with a power output of 61.67 W

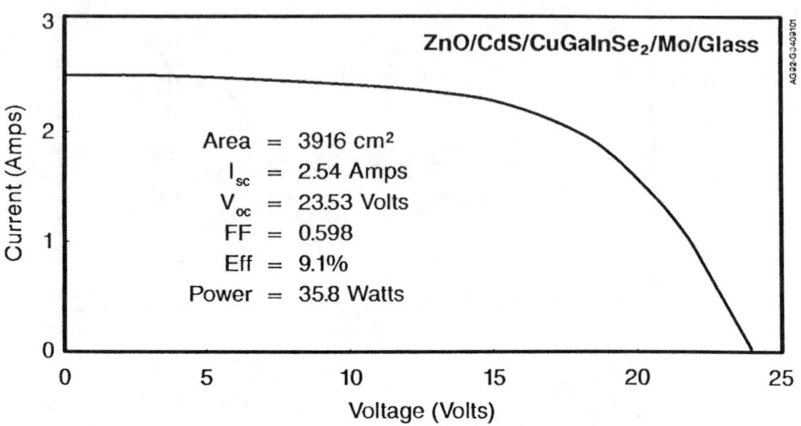

Figure 3 Light I - V characteristics of a 3916 cm^2 $CuInSe_2$ ARCO Solar power module with an output of 35.8 W

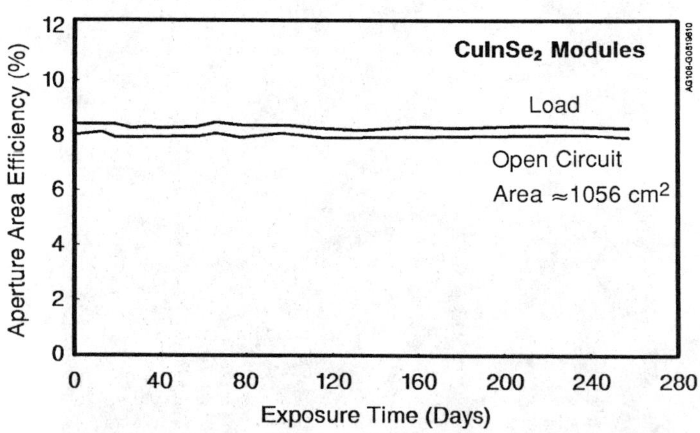

Figure 4 Stability performance of $CuInSe_2$ modules tested outdoors at SERI under load and open circuit conditions

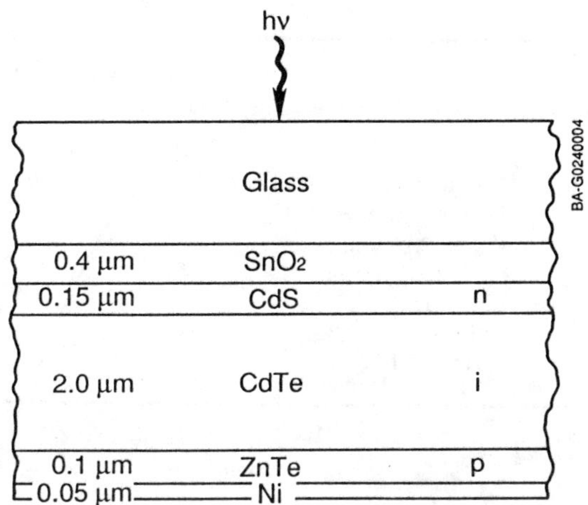

Figure 5 Solar cell structure of a n-i-p thin film CdTe solar cell with a device configuration of Glass/SnO_2/CdS/CdTe/ZnTe/Ni

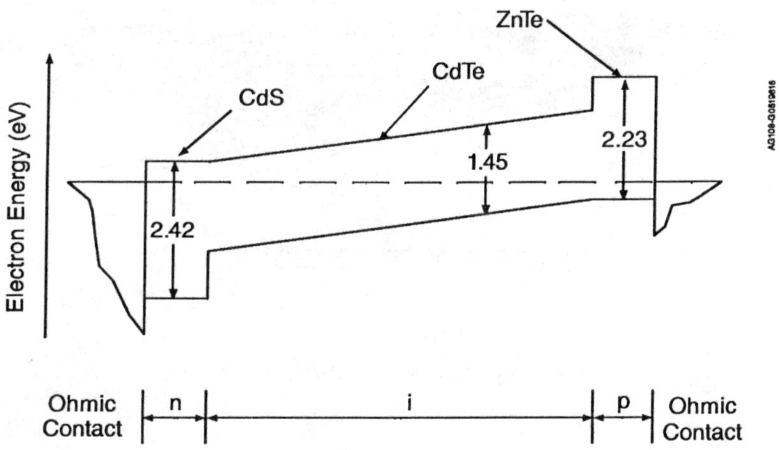

Figure 6 Simplified band diagram of a n-i-p thin film CdTe solar cells

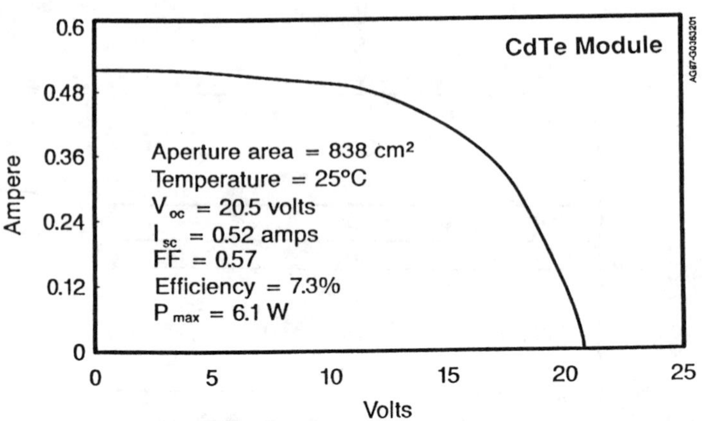

Figure 7 Light I - V characteristics of a 838 cm^2 CdTe Photon Energy power module

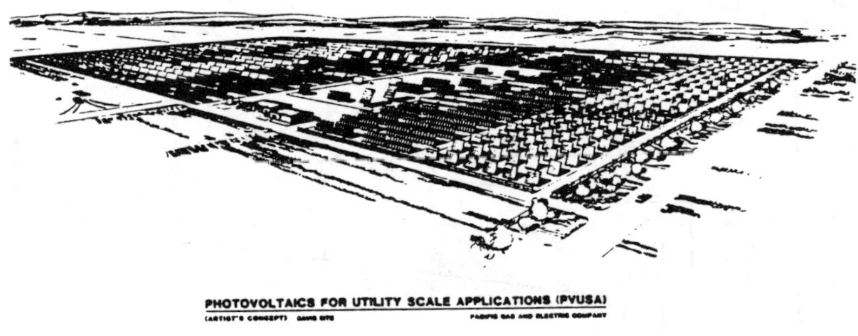

Figure 8 Artist rendition of the PVUSA project at Davis, CA

GALLIUM ARSENIDE THERMOPHOTOVOLTAIC SYSTEMS AND CONVERTER TEMPERATURE

Dr. R.C. Neville
College of Engineering and Technology, Box 15600,
Northern Arizona University, Flagstaff, Arizona 86011, U.S.A.

Abstract

The performance of thermophotovoltaic systems utilizing gallium arsenide photovoltaic cells to produce electrical energy and a Carnot cycle to produce thermal energy is theoretically examined. The principle variable is the temperature of the central converter, which is varied from 1000 to 4000 degrees Kelvin. Several assumptions are made concerning the construction of the thermophotovoltaic converter and the photovoltaic cell as well as the junction temperature of the photovoltaic cells. Overall system efficiencies between 30% and 40% are obtained and it is determined that maximum overall system efficiencies are obtained for converter temperatures between 2500 and 3500 °K depending on system design and photovoltaic cell junction temperature.

1. INTRODUCTION

A thermophotovoltaic system produces both electrical and thermal energy. A typical gallium arsenide thermophotovoltaic system is constructed of a number of parts (Figure 1). The entire operating system is enclosed in a vacuum bottle. This is to reduce the conduction and convection of thermal energies from one portion of the system to another. The vacuum bottle has an optical input port set into the upper (illuminated) side. This port has a lens which focuses solar insolation upon the converter. The converter is a spherical object placed at the center of the thermophotovoltaic system, constructed of a high temperature material capable of acting as a grey or black-body radiator and is supported on a thin, rigid stalk (not shown in Figure 1) which is fabricated of a low thermal conductivity material capable of high temperature operation. The converter radiates secondary photons towards the inner walls of the thermophotovoltaic system.

The inside surface of the vacuum chamber is coated with a layer of photovoltaic cells backed by a heat exchanger. Between the photovoltaic cells and the vacuum is an optical filter which serves to "condition" that photon spectrum which reaches the photovoltaic cells. Behind the photovoltaic cells is a heat exchanger. External to the thermophotovoltaic system shown in Figure 1 there is an optical concentrator which serves to provide a sufficiently high density solar insolation.

The thermophotovoltaic system operates as follows. Solar photons are concentrated by the external concentrator and focused on the converter by the optical input port, thus heating the converter. In turn, the converter reradiates secondary photons. The spectral distribution of the secondary photons depends on the temperature and nature of the converter. This spectral distribution can be modified, by adjusting the temperature of the converter, in order to provide a better match of the secondary photon distribution to the energy gap of the semiconductor--better than the spectral distribution in ordinary sunlight. These secondary photons leave the converter and impinge on the walls of the vacuum container. These walls are lined with photovoltaic cells which then convert the energy in the secondary photons to electrical energy. The conversion efficiency of the photovoltaic cells depends on the energy of the photons and its relationship to the energy gap of the semiconductor.

The best energy conversion efficiencies are obtained for photon energies slightly in excess of the energy gap of the semiconductor. To improve overall efficiency the photovoltaic cells are "fronted" by an optical, band-pass filter. This filter passes, from the converter to the photovoltaic cells, a limited energy range of photons. photons with energies from that of the semiconductor energy gap to some 10% higher. All other photons are reflected back to the converter. This assists in keeping the converter at its desired temperature.

As a further optical improvement, the optical input port has a optical filter, affixed to the vacuum side. This is an infrared reflecting filter and its purpose is to reflect secondary infrared photons back to the converter. This filter introduces a slight loss in the solar input, but the loss in solar infrared photons is more than made up for by the retention of infrared secondary photons which would otherwise be lost out of the input port.

Finally, backing the photovoltaic cells is a heat exchanger. This is needed to keep the photovoltaic cell junction temperatures low enough to insure their efficient operation. The heat so removed is made available, along with the electrical energy output from the photovoltaic cells, to energy consumers--hence, the name thermophotovoltaic.

2. INVESTIGATION

Goal of Study

The overall system energy output of a thermophotovoltaic system depends on the energy conversion efficiency of the photovoltaic cells. The electrical energy output is directly dependent on this efficiency. The thermal energy available for potential output depends on the electrical efficiency according to:

$$E_{TA} = E_A \ (1 - n_E/100) \qquad (1)$$

where E_{TA} is the available thermal energy, E_A is the total energy available (in photons) to the photovoltaic cells and n_E is the efficiency of the photovoltaic cells. There is a small contribution of thermal energy due to conduction via the stalk. However, due to the construction of the stalk this may be ignored and because of the vacuum there is no other conduction/convection transfer of thermal energy to the photovoltaic cells from the converter.

The energy conversion efficiency of the photovoltaic cells depends on the nature of the semiconductor employed in the photovoltaic cells, upon the spectral distribution of the secondary photons reaching the photovoltaic cells (controlled by the optical band-pass filter) and on the concentration of the photons. Both the spectral distribution and concentration of secondary photons depend on the temperature and nature (black-body, grey-body, etc.) of the converter surface.

This paper is a theoretical study of the performance of a typical thermophotovoltaic system employing gallium arsenide as the semiconductor for the photovoltaic cells. The converter is assumed to be a black-body. The variables to be studied are the temperature of the converter, the spacing of the solar cells from the converter surface, the impurity concentration of the photovoltaic cell substrates and the junction temperature of the photovoltaic cells.

Theory and Assumptions

This section treats the general background for the study. It starts with the initial energy input--concentrated solar insolation. The optical input port is assumed to be ideal with the exception of the infrared reflecting coating. This coating exists to insure that secondary photons are retained within the vacuum cavity. Study of Figure 2 [1] indicates that, if the filter reflects all photons with a wavelength in excess of two microns and passes all other photons, some eight percent of the incoming solar insolation is lost.

These input photons are focused on the converter and act to heat it to its designed operating temperature. The converter, assumed to be a black-body radiator, emits secondary photons towards the photovoltaic cells which line the inner walls of the vacuum chamber. A fraction; dependent on the temperature of the converter and the ratio of the radius of the optical input port/photovoltaic cells, rs, to the radius of the converter, rb; are lost through the optical input port (the radii are taken to start from the center of the converter).

The operating converter temperatures to be examined in this study are given in Table I and range from 1000 to 4000°K. Given this temperature and using the Stefan-Boltzmann law for the power radiated from a black-body:

$$P_{rb} = 5.64 \times 10^{-12} T^{4} \text{ watts/cm}^{2} , \qquad (2)$$

where T is the absolute temperature in degrees Kelvin; the necessary degree of concentration, Cm, of the solar input to balance secondary photon losses from the converter through the optical input port can be determined. Three ratios of the input port radius to the converter radius, rs/rb are assumed, two, three and four. This minimum concentration (note that under these conditions there is no power available to the photovoltaic cells) is also provided in Table I.

In any realistic system the vast majority of the secondary photons fall upon the optical band-pass filters fronting the photovoltaic cells. These filters are assumed to be perfectly reflecting for photons in the energy ranges:

$$E \leq Eg \quad \text{and} \quad E \geq 1.1\ Eg \qquad (3)$$

where E is the photon energy and Eg is the energy gap of the photovoltaic cells. Within the narrow range of energies left, the band-pass filters are considered to be 95% transmitting.

The energy of those secondary photons which reach the gallium arsenide photovoltaic cells is converted, with some efficiency, into electrical energy. In order to provide optimum performance, considering the heavily concentrated insolation to the photovoltaic cells, their construction is of an inverted type (Figure 3). The down (n-type) region is doped to ten to the twentieth per cubic centimeter. The substrate of these photovoltaic cells is variably doped at the junction--Na varies from ten to the fourteenth to ten to the sixteenth per cubic centimeter. The optical surface substrate concentration is taken to be ten to the nineteenth per cubic centimeter. This variable doping produces an electric field in the substrate which enhances the photocurrent as well as reducing the effective series resistance of the photovoltaic cells. The equivalent circuit for the photovoltaic cells and load is given in Figure 4.

If the load resistance of this photovoltaic cell circuit is optimized for maximum power delivery, then it can be shown that [2, 3] the maximum power delivered by the photovoltaic cells is given by:

$$\frac{Pmax}{A} = \frac{\{J_{ph} + J_s\}\{qV_D/kT_J\}\{V_D\}}{qV_D/kT_J + 1 + K} \frac{\{1 + K/2\}}{1 + K}, \quad (4)$$

where J_{ph} is the photocurrent density of the photovoltaic cells, J_s is the saturation current of the cell junction, q is the positronic charge, V_D is the photovoltaic cell junction voltage for maximum power transfer, k is Boltzmann's constant, T_J is the absolute temperature, K is the loss factor and A is the photovoltaic cell area.

The junction voltage for maximum power transfer and the loss factor are related by the following equations:

$$\{qV_D/kT_J + 1 + K\} e^{qV_D/kT_J} = \frac{J_{ph} + J_s}{J_s} \{1 + K\}, \quad (5)$$

and

$$K = 2r\{J_s A\}\{q/kT_J\} e^{qV_D/kT_J}, \quad (6)$$

where r is the photovoltaic cell internal resistance.

The photocurrent density of the photovoltaic cells is computed using the effective secondary insolation at the photovoltaic cell radius, rs, deducting five percent for transmission losses through the band-pass filters, determining the photon density using photon energy equals Planck's constant times frequency and assuming one electron-hole pair generated per photon. This value is then reduced by a percentage based on the lifetime of gallium arsenide (see Table II).

The saturation current density of the photovoltaic cells is taken to be that for minority carrier injection and is given by:

$$J_s = qn_i^2 \left\{ \frac{1}{Na}\sqrt{\frac{D_n}{\tau_n}} + \frac{1}{Nd}\sqrt{\frac{D_p}{\tau_p}} \right\}, \quad (7)$$

where n_i is the intrinsic carrier concentration of gallium arsenide, D_n and D_p [5] are the minority carrier diffusion constants on the n and p sides of the photovoltaic cell junction, Na is the impurity carrier concentration on the p-side of the junction, Nd is the impurity concentration on the n-side of the junction and τ_n and τ_p are the minority carrier lifetimes on the p and n sides of the junction.

The series resistance, r, of the photovoltaic cells is determined using:

$$rA = \frac{xn}{q\mu_n Nd} + \frac{1}{q}\int_o^{xp} \frac{dxp}{\mu_p P} \text{ ohm-cm}^2 , \tag{8}$$

where xn is the thickness of the n-region (100μm) and xp is the thickness of the p-region (20μm) of the photovoltaic cells.

The efficiency of solar-to-electrical energy conversion by the solar cells is given by:

$$n_E = \frac{Pout}{Pin} \times 100\% , \tag{9}$$

where Pout is obtained from Equation 4 and Pin is the effective insolation at the solar cells.

The power available to the heat extractors, P_{TA}, which back the photovoltaic cells is:

$$P_{TA} = Pin - Pout + .05\ Pin , \tag{10}$$

where the final term in Equation 10 is due to the energy absorbed by the band-pass filter. Assuming the heat extractors are delivering their thermal energy to a load using a Carnot cycle, the actual thermal power delivered by the system is:

$$P_T = n_{Th}\{P_{TA}\}/100 , \tag{11}$$

where:

$$n_{Th} = (1 - T_c/T_h)100 . \tag{12}$$

Here, T_c is the cold energy reservoir temperature and is assumed to be 300°K. The high temperature reservoir temperature is taken to be that of the photovoltaic cell junction. This study assumes three photovoltaic cell junction temperatures, T_j, (see Table II).

Finally, the overall efficiency of the thermophotovoltaic system is obtained by assuming an input power of 100 watts at the entrance of the optical input port, deducting the losses at the input port, the amount of energy lost via secondary photons exiting the input port and the losses in the band-pass filter covering the photovoltaic cells. The output power of the photovoltaic cells is now computed. The input power to the thermal energy extractors is given by Equation 10. The thermal energy output is now computed. We end with three possible efficiencies: the overall efficiency computed by summing the thermal output power and the electrical output power, the electrical efficiency, and the thermal energy efficiency.

Results

From the data in Table II, the mobilities of gallium arsenide [5] and the dimensions and doping levels of the photovoltaic cells, we can compute the electrical efficiency of the photovoltaic cells. These data are presented in Figures 5-7 Here the photovoltaic cell efficiencies are given as a function of the converter temperature, Tb. The ratio of the photovoltaic cell radius, rs, (in the system) to the converter radius, rb, in these figures is taken to be two and the parameter in the figures is the junction temperature of the photovoltaic cells. Table III presents the ratio of the photovoltaic cell efficiencies for system configurations of rs/rb equal to three and four as a function of the photovoltaic cell junction temperature and the converter temperature.

Table IV presents the operating efficiency of the thermal heat extraction mechanism as a function of the photovoltaic cell junction temperature assuming a Carnot cycle.

Assume an input power level of 100 watts of concentrated solar energy to the optical input port of the thermophotovoltaic system. As stated earlier, some eight watts will be rejected by the infrared filter on the input port, leaving 92 watts as the input power to the converter. This power is reradiated towards the optical input port and the photovoltaic cells lining the inside wall of the system.

Now, make an assumption regarding the apportioning of this power radiated by the converter. If the ratio of the photovoltaic cell area to input port area is that given in Table V, Case A, half of the converter radiated power will be lost out of the input port and half will be absorbed by the photovoltaic cell/thermal energy extraction mechanism on the inner walls of the system. At the same time, the degree of initial optical concentration (outside of the thermophotovoltaic system proper) will be doubled from that listed in Table I. As we will see this does not yield a particularly high system efficiency. A considerable improvement in overall system efficiency is experienced if seven times as much power is absorbed by the photovoltaic cell/thermal energy extraction mechanism as is lost through the input port. In this case, Case B, the required initial concentration is eight times that indicated in Table I and the area ratios must be four times those given in Table V.

This study now proceeds by assuming the two cases considered above (see Table VI for details), computing the actual output power and determining the percentage of output power divided by input power--the system overall efficiency. These data are presented in Figures 8-10 where the overall system efficiency,

n_T, is plotted versus the converter temperature, Tb, with junction temperature, T_J, as a parameter. For these curves the rs to rb ratio is taken as two. Table VII presents the ratio of the efficiency for rs/rb equal to four to the efficiency for rs/rb equal to two.

Thus far, the theoretical performance of a thermophotovoltaic system has been obtained. In the next section we discuss the performance limitations and operating converter temperature limits which result from this performance.

3. DISCUSSION AND CONCLUSIONS

Consider the operating efficiency of the gallium arsenide photovoltaic cells. With the large band gap of gallium arsenide, this efficiency exceeds forty percent high sufficient photocurrent is available. The data in Figures 5-7 indicate that the efficiency is roughly constant once the converter temperature reaches 2000°K for junction temperatures of 300°K and 400°K. For junction temperatures of 500°K a converter temperature of at least 2500°K is required. This condition applies independently of the substrate impurity concentration. While the curves in Figures 5-7 are for rs/rb equal to two, Table III indicates that the converter temperature range for rs/rb equal to three and four is also restricted to the same temperatures.

The thermal output efficiency (Table IV) depends only upon the junction temperature of the photovoltaic cells, since the cold junction temperature has been taken to be 300°K. This thermal efficiency is sharply higher for a photovoltaic cell junction temperature of 500°K than for one of 300°K. Study of Figures 5-7 indicates that the photovoltaic cell's efficiency is higher for 300°K than for 400 or 500°K junction temperatures.

Overall system efficiencies are provided in Figures 8-10 as a function of the converter operating temperature, Tb with junction temperature, T^J, and Case (Table VI) as parameters. Note that in Case A the radiated secondary photon power lost through the optical input port is equal to the radiated secondary photon power which is absorbed in the band-pass filter, the photoelectric cells and the thermal energy extraction mechanism. In Case B the photoelectric cells, thermal energy extraction mechanism and band-pass filter absorb seven times the power which is lost through the optical input port. Note that overall efficiency is much higher for Case B than Case A and that the efficiencies are higher for photovoltaic cell junction temperatures of 300°K than for junction temperatures of 400°K or 500°K. The data of these figures indicate that increasing the converter temperature increases the overall operating efficiency, however, the bulk of the increase occurs for low converter temperatures. Thus, these data indicate that the optimum converter range of converter temperatures is above 2500°K.

At a 2500°K converter temperature the efficiency of a thermophotovoltaic system employing a junction temperature of 300°K: a radii ratio, rs/rb, of two; Case B power conditions and a substrate impurity concentration of ten to the fourteenth is approximately 39%. Increasing the substrate impurity concentration by an order of magnitude improves the efficiency to 40%. Comparing these efficiencies to a similar Case A situation with efficiencies of 22% and 24%, respectively indicates that Case B is a much better producer of energy. Further note that at 4000°K the respective efficiencies are 42% and 43% for Case B and 24% and 25% for Case A. Table VII provides the input necessary to consider efficiencies for rs/rb equal to four. Note that the difference is but a few percent.

There is a potential (major) problem. In Table I there is a listing of the minimum optical concentration required of the optical concentrator external to the thermophotovoltaic system. This value is large for higher values of the converter operating temperature. For Case A conditions, this concentration must be doubled, reaching values of 4,754 for Tb = 4000°K and rs/rb equal to two. For case B this concentration must be eight times the value in Table I, reaching values in excess of 19,000 for Tb = 4000°K and rs/rb equal to two. Now, the concentration system does not have to concentrate down to a point since we have a focusing lens in the optical input port and the converter is of finite dimensions. Never-the-less there are physical limits on the degree of concentration, partly imposed by the necessity of tracking the sun accurately and partly by the large degree of concentration. If we arbitrarily impose a maximum value of concentration of 3,000, this limits the converter operating temperature to a maximum of 2500°K for rs/rb equal to two, 3000°K for rs/rb equal to three and 3500°K for rs/rb equal to four for case B and for Case A to converter temperatures of 3500°K for rs/rb equal to two and to 4000°K for rs/rb equal to three and four.

Considering the improvement of thermophotovoltaic system efficiency with increasing converter temperature, we can consider the converter temperatures at the end of the preceding paragraph to be the ideal operating temperatures and derive Table VIII. In this table, the overall thermophotovoltaic efficiency is provided as a function of rs/rb with substrate doping, Na, and junction temperature, T_j, as parameters.

Study of Table VIII indicates that 40% overall system efficiencies are possible, a significant improvement over current commercial concentrator systems. However, we have another factor to consider. The system output is composed of both thermal and electrical energies. When these two power types are compared, we see that: 1) electrical energy is consistently more portable and 2) electrical energy is more efficiently converted to other forms of energy. These considerations clearly favor low photovoltaic cell junction temperature operation. On-the-other-hand, note that the ultimate energy consumer may have a preference as to the type of energy. In considering practical situations, we find that a residence in North America requires a power level of 5,000 watts--some 60-80% of which is for heating/cooling [6]--, a typical commercial laundry requires significant amounts of heat energy, while a bauxite-aluminum plant needs massive amounts of electrical energy. When we consider the data in Table VIII, the percentage of the output which is thermal, under the conditions detailed in Table VIII, is delineated in Table IX.

Exactly which system configuration (rs/rb, T_J, Na and Tb) is used, depends, then, on the mixture (thermal/electrical) of power desired by the consumer. Two other factors should be included in making the ultimate decision. One of these is the maintenance factor. Considering the large optical concentrations employed and the high operating temperatures the maintenance of a thermophotovoltaic system is going to be exacting. The other factor is internal. The temperature of the converter is high (2500 to 4000°K) and this will place a considerable stress on the input port infrared filter, the band-pass filter and the thermal energy extraction mechanism. This stress is reduced for larger values of rs/rb at some cost in operating efficiency.

Consumer requirements as to energy type, stresses on the optical components and maintenance considerations are important in a practical sense. Assuming that these conditions can be met, it appears that overall thermophotovoltaic system efficiencies in excess of 40% are theoretically obtainable. It would seem that the thermophotovoltaic system has a bright future.

REFERENCES

[1] R. C. Neville--"Solar Energy Collector Orientation and the Tracking Mode", Solar Energy 19 (1977).

[2] R. C. Neville--Solar Energy Conversion: The Solar Cell, Elsevier Scientific Publishing Co., Amsterdam, (1978) Chapter VI.

[3] R. C. Neville--"Photovoltaic Cells for Thermophotovoltaic Systems", 5th Miami International Conference on Alternative Energy Sources, Miami, Fl (December 1982).

[4] S. M. Sze--Physics of Semiconductor Devices, Wiley Inter-Science, N.Y., N.Y. (1981).

[5] H.C. Cases Jr. and M.B. Panish--Heterostructure Lasers, Academic Press, N.Y., (1978).

[6] R. C. Neville--"Solar Energy and the Residence-Some Systems Aspects", Solar Energy 19 (1977).

Table I: Converter operating temperatures, Tb, and the required degree of optical concentration to match secondary photon losses through the input optical port, Cm. Here, rs/rb is the ratio of the system photovoltaic cell radius to the converter radius.

rs/rb	Tb (°K) 1000	1500	2000	2500	3000	3500	4000
	Cm						
2	1.4	12.3	85.3	258	658	1289	2377
3	.6	5.5	37.9	114	293	573	1056
4	.03	3.1	21.3	64.4	164	322	594

Table II: Junction temperatures, T_j; carrier lifetimes, τ; substrate junction impurity concentrations, Na; saturation current densities, J_s and photocurrent densities, J_{ph}.

Na (cm^{-3})	T_j (°K) 300	400	500
	τ (μ-sec.)		
10^{14}	10^{-2}	$.56 \times 10^{-2}$	$.36 \times 10^{-2}$
10^{15}	10^{-3}	$.56 \times 10^{-3}$	$.36 \times 10^{-3}$
10^{16}	10^{-4}	$.56 \times 10^{-4}$	$.36 \times 10^{-4}$
	J_s (μA/cm²)		
10^{14}	7.38×10^{-10}	3.73×10^{-3}	1.2×10^{2}
10^{15}	2.30×10^{-10}	1.17×10^{-3}	3.75×10^{1}
10^{16}	6.50×10^{-11}	3.30×10^{-4}	1.06×10^{1}

J_{ph} (A/cm²)

Tb (°K) rs/rb	1000	1500	2000	2500	3000	3500	4000
2	5×10^{-5}	1.35×10^{-2}	.238	1.3	4	9	17
3	2.2×10^{-5}	6×10^{-2}	.106	.578	1.78	4	7.56
4	1.25×10^{-5}	3.38×10^{-3}	5.94×10^{-2}	.325	1	2.25	4.25

Table III: Ratio of operating efficiencies for rs/rb = 3 and 4 as a function of T_J and Tb.

	$n_E(x)/n_E(2)$					
rs/rb=x		3			4	
T_J (°K)	300	400	500	300	400	500
Tb (°K)						
1000	.95	.92	.55	.94	.84	.32
1500	.96	.94	.87	.94	.90	.79
2000	.97	.95	.90	.95	.91	.79
2500	.97	.95	.93	.95	.92	.82
3000	.97	.96	.93	.95	.92	.84
3500	.97	.96	.93	.95	.92	.86
4000	.98	.96	.92	.96	.93	.87

Table IV: Thermal output efficiency as a function of junction temperature.

T_J (°K)	300	400	500
n_T (%)	0.0	25.0	40.0

Table V: The ratio of the area of photovoltaic cells to optical input port area for equal amounts of power supplied to the photovoltaic cells and to the optical input port (ie., lost). Variables are the converter temperature, Tb, and rs/rb.

	Area Ratio						
Tb (°K)	1000	1500	2000	2500	3000	3500	4000
rs/rb							
2	1699	57	23	13	10	9	9
3	1100	55	23	13	10	9	9
4	285	55	23	13	10	9	9

Table VI: The power reaching the photovoltaic cells, Ps, the power lost out of the input port. Pio, and the thermal power absorbed by the band-pass filter, Pbf.

	Case A	Case B
Ps (watts)	43.7	76.475
Pio(watts)	46	11.5
Pbf(watts)	2.3	4.025

Table VII: The ratio of the overall thermophotovoltaic system efficiency for rs/rb = 4 to the efficiency for rs/rb = 2.

Substrate Doping, Na (cm^{-3})	10^{14}			10^{15}			10^{16}		
Junction Temperature, T_J (°K)	300	400	500	300	400	500	300	400	500
Ratio	.95	.95	.96	.95	.96	.97	.96	.97	.98

Table VIII: Best overall thermophotovoltaic system efficiency as a function of Case, rs/rb, Na, and T_J.

Case A

n_T

Na (cm^{-3})		10^{14}			10^{15}			10^{16}		
T_J (°K)		300	400	500	300	400	500	300	400	500
rs/rb	T_c (°K)									
2	3500	23	24	24	23	24	24	19	22	24
3&4	4000	23	24	24	24	24	24	20	23	24

Case B

n_T

Na (cm^{-3})		10^{14}			10^{15}			10^{16}		
T_J (°K)		300	400	500	300	400	500	300	400	500
rs/rb	T_c (°K)									
2	2500	39	40	40	39	41	41	33	38	40
3	3000	39	40	40	38	40	41	33	38	40
4	3500	39	40	41	38	41	41	33	38	41

Table IX: The percentage of system output which is thermal energy as a function of Case, rs/rb, Na and T_J.

Case A

n_T

Na (cm^{-3})		10^{14}			10^{15}			10^{16}		
T_J (°K)		300	400	500	300	400	500	300	400	500
rs/rb	T_c (°K)									
2	3500	0	30	60	0	30	58	0	34	61
3	4000	0	30	61	0	30	59	0	34	61
4	4000	0	31	63	0	31	60	0	35	62

Table IX (continued)

		Case B n_T								
Na (cm^{-3})		10^{14}			10^{15}			10^{16}		
T_J (°K)		300	400	500	300	400	500	300	400	500
rs/rb	T_c (°K)									
2	2500	0	35	70	0	34	66	0	38	68
3	3000	0	35	70	0	34	62	0	38	68
4	3500	0	35	70	0	33	63	0	36	64

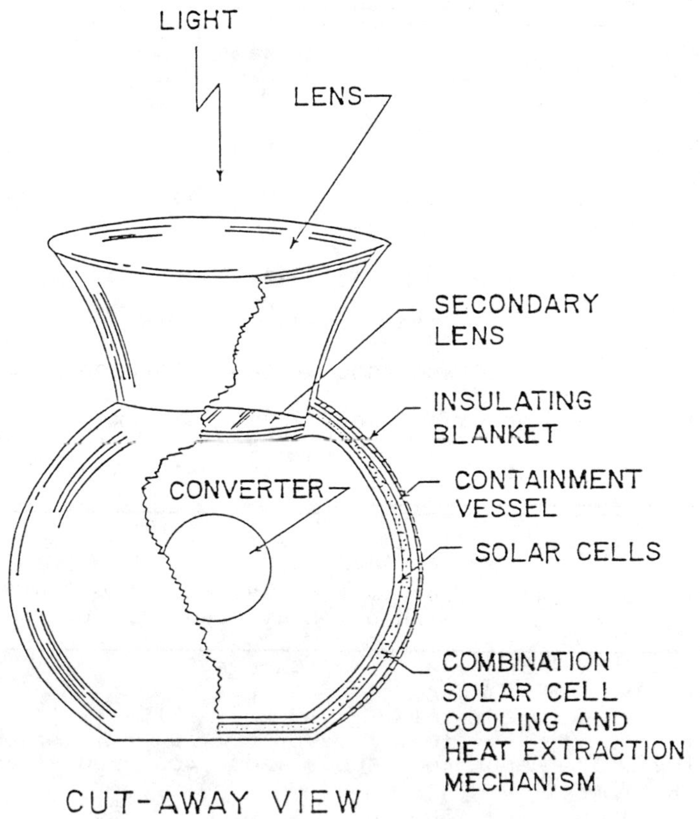

Figure 1: A typical thermophotovoltaic system. Note that the converter is supported by a thin, high thermal resistance, stalk (not shown here) and the internal volume is a vacuum.

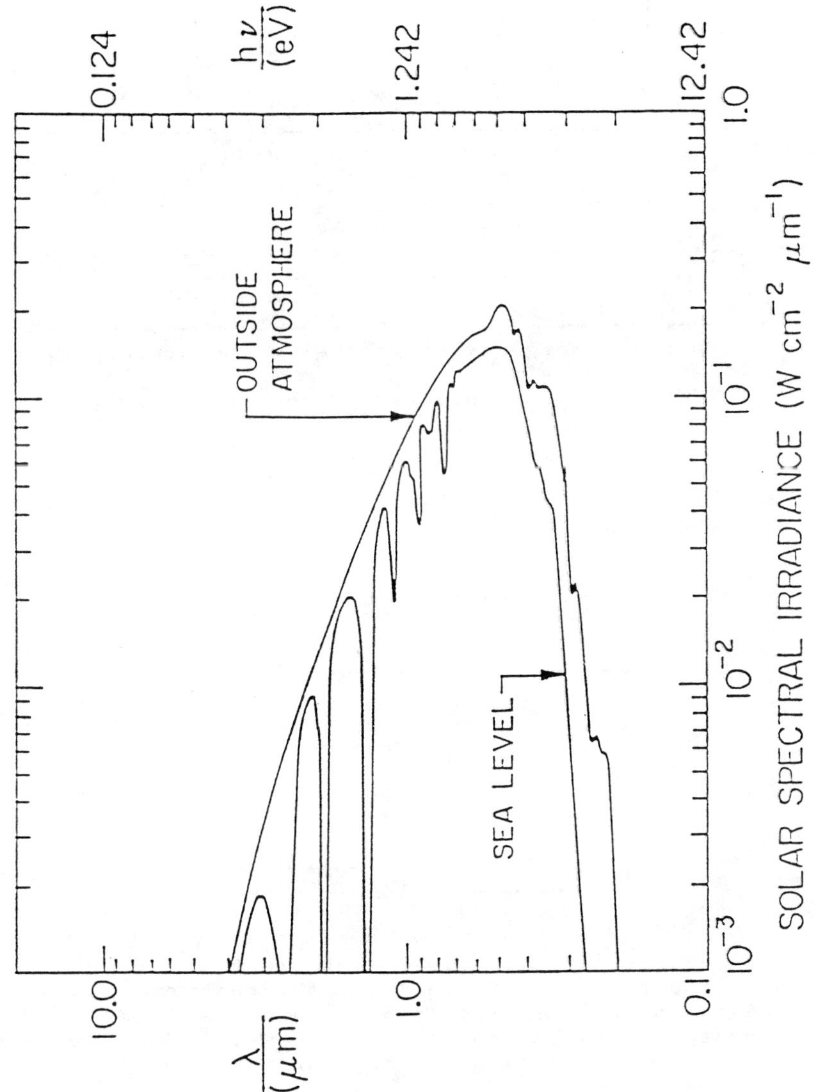

Figure 2: The solar spectral irradiance for AM0 and AM1 conditions [1].

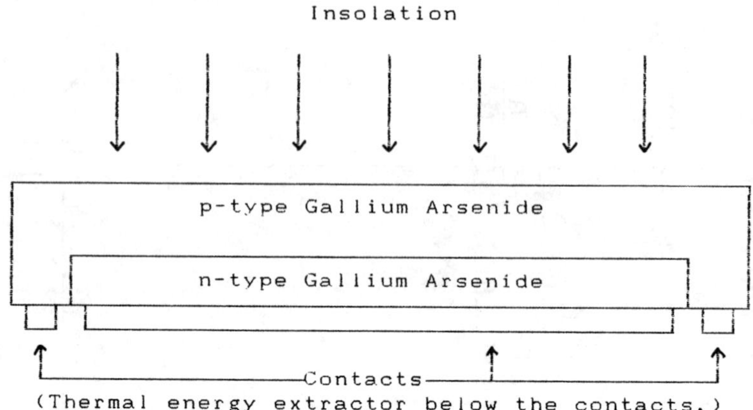

Figure 3: An inverted photovoltaic cell.

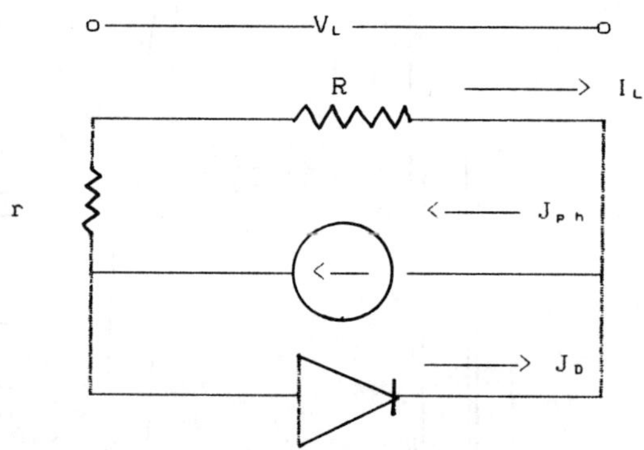

Figure 4: Electrical equivalent circuit.
R is the load resistance, r is the photovolotaic cell resistance, V_L is the load voltage, I_L is the load current, J_{ph} is the photocurrent density and J_D is the diode junction forward current density.

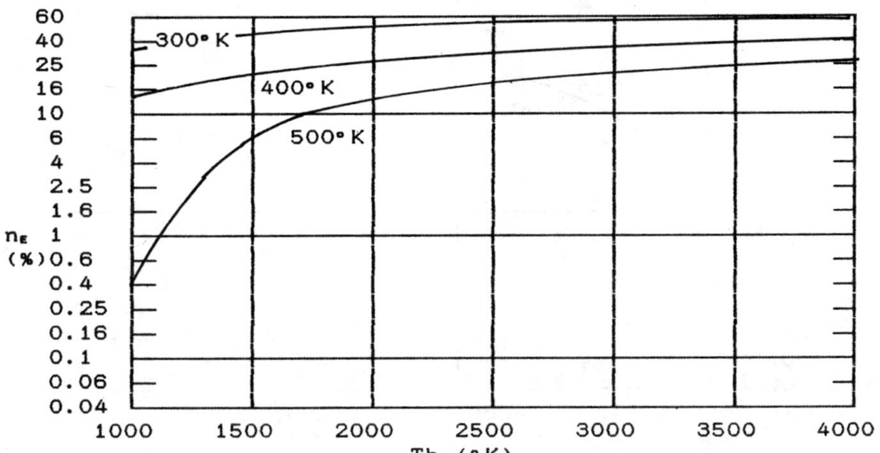

Figure 5: Photovoltaic cell efficiency as a function of the operating temperature with T_J as the parameter and for a substrate impurity concentration of ten to the fourteenth per cubic centimeter at the junction. rs/rb = 2.

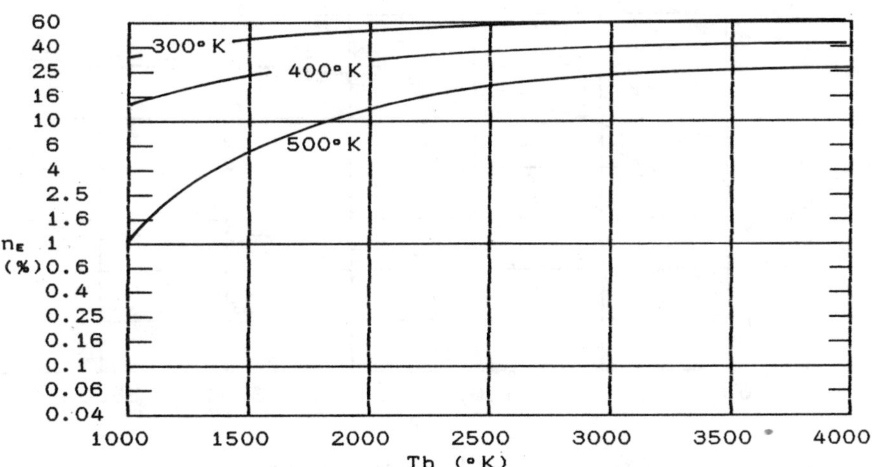

Figure 6: Photovoltaic cell efficiency as a function of the operating temperature with T_J as the parameter and for a substrate impurity concentration of ten to the fifteenth per cubic centimeter at the junction. rs/rb = 2.

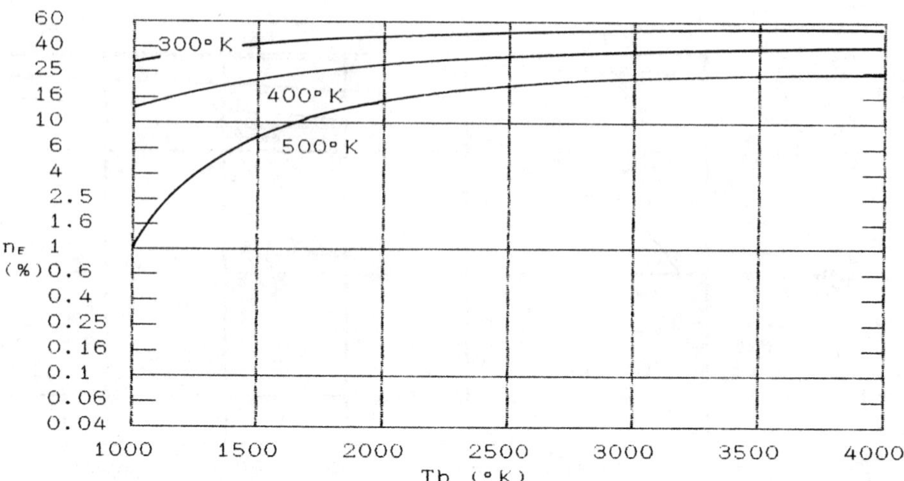

Figure 7: Photovoltaic cell efficiency as a function of the operating temperature with T_J as the parameter and for a substrate impurity concentration of ten to the sixteenth per cubic centimeter at the junction. rs/rb = 2.

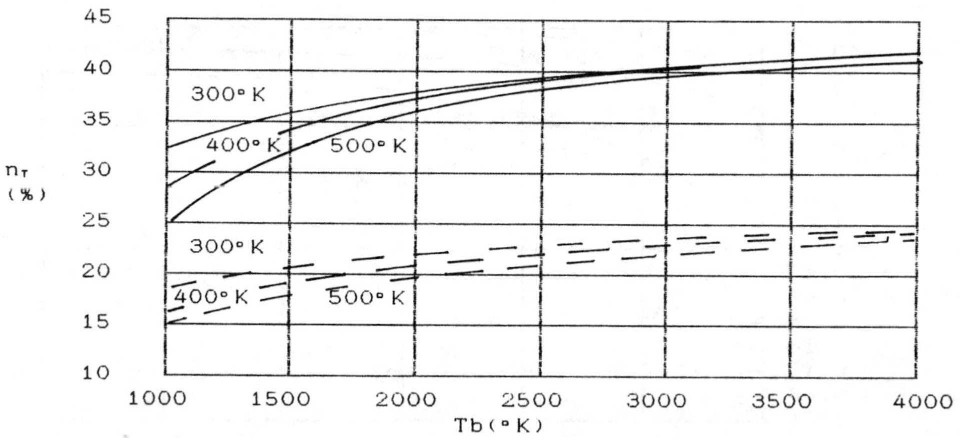

Figure 8: The overall thermophotovoltaic system efficiency, n_T, as a function of converter temperature, Tb, with the photovoltaic junction temperature as a parameter. Dashed lines are for Case A and solid lines are for Case B (see Table VI). rs/rb = 2 and the substrate impurity concentration is ten to the fourteenth per cubic centimeter.

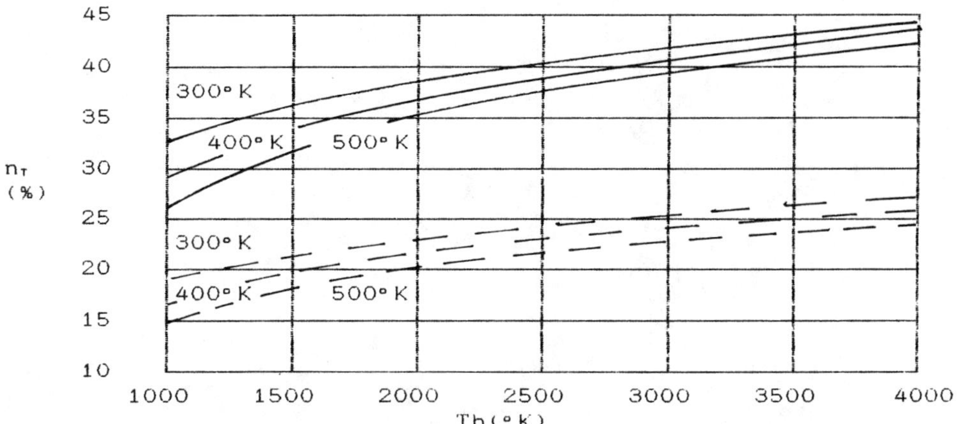

Figure 9: The overall thermophotovoltaic system efficiency, n_T, as a function of converter temperature, Tb, with the photovoltaic junction temperature as a parameter. Dashed lines are for Case A and solid lines are for Case B (see Table VI). rs/rb = 2 and the substrate impurity concentration is ten to the fifteenth per cubic centimeter.

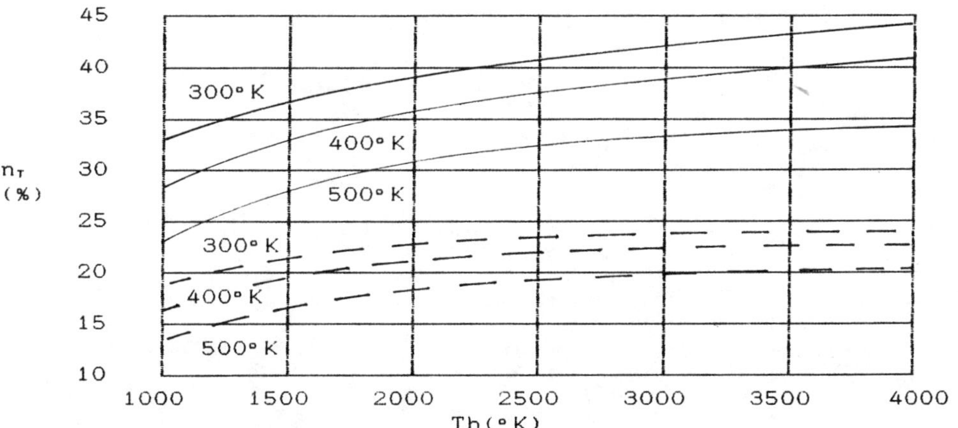

Figure 10: The overall thermophotovoltaic system efficiency, n_T, as a function of converter temperature, Tb, with the photovoltaic junction temperature as a parameter. Dashed lines are for Case A and solid lines are for Case B (see Table VI). rs/rb = 2 and the substrate impurity concentration is ten to the sixteenth per cubic centimeter.

PHOTOVOLTAIC CONVERSION AND STORAGE OF SOLAR ENERGY

Jose L Perez-Mira
University of Murcia, Chemical Engineering Dept .
c. Norte, No 5, Valle Escombreras, Cartagena, Murcia, Spain

Historical Note

In 1887, Heinrich Hertz, who is best known for his discovery of radio waves, observed that his receiving spark gap would fire more readily when the gap was illuminated with ultraviolet light. He did not understand why this should happen. In 1888, however, William Hallwachs explained the phenomenon, announcing that the ultraviolet light caused negative electricity to be emitted from the metal balls of the gap and that this electricity gacilitated the sparking. He proved his point by connecting a polished zinc plate to an electroscope, charging the combination, and exposing the surface of the plate to ultraviolet light. When the plate and electroscope were positively charged, the ultraviolet light had no effect; but when they were negatively charged, the ultraviolet light discharged the combination (thus, negative electricity passed from the plate into the surrounding air). This is the first recorded observation and explanation of the photoelectric effect, which is sometimes called the Hallwachs effect. In 1899 Lenard and Thomson identified Hallwachs negative electricity as a stream of electrons.

In 1873, the photoconductivity of selenium was discovered by Willoughby Smith and by May. The photovoltaic effect in selenium was separately discovered by Adams and Day in 1876 and by Fritts in 1884. Becquerel's later discovery (1899) concerned electrolytic devices. He observed that a voltage appears between two similar electrodes immersed in an electryte, such as lead nitrate, when one of the electrodes is exposed to light.

A copper-oxide photovoltaic cell, known by the trade name Photox cell and developed by Bruno Lange, was introduced during World War I and marketed by Westinghouse. Albert Einstein received the Nobel Prize in 1905 for his explanation of the photoelectric effect.

Future of Photoelectricity

The substantial progress already made in application of photocells and solar cells in both casual and sophisticated electronics points to expanding use of these devices in commercial, scientific, industrial, military, and household areas. Some of the fields in which it is easy to envision further of photoelectricity are automatic control; crime detection and prevention; identification, sorting, and grading; counting; communications; safety measures; highway traffic management; pollution measurement and control; medical technology and sports and amusements.

One of the most engrossing prospects is that of harnessing sunlight for the production of electrical energy. Experts are fairly confident that at least part of the electricity required in the future will be supplied by rooftop mounted photovoltaic panels.

It is estimated that the earth receives approximately 126 trillion horsepower (94 trillion kilowatts) from the sun each second; and if this enormous and unlimited energy could be more fully utilized, our dependence upon fossil fuels might be dramatically reduced. In this direction it shows an early experimental electric automobile which is sun powered. Here, the large panel seen atop the Baker car (1912) contains 10,640 silicon photovoltaic cells. Its output converts sunlight into enough electricity to keep charged the storage batteries that run the car.

Since then, this research and development have been stepped up considerably. This acceleration has been occasioned in great part by shortages of crude oil. Government and industry both have feasibility studies under way, and in 1979 President Carter called for 26 percent of US energy needs to be supplied by solar means by the year 2000. Several energy systems have been suggested to reduce or even to end our dependence on oil and other fossil fuels. Under serious consideration are nuclear solar-steam, solar-hydrogen, geothermal, hydroelectric, wind, and tidal systems. Of these, various solar systems are attractive, since the sun is a perpetual source of energy. And of the solar systems, the photovoltaic method seems most desirable if it can be made economically feasible, since it converts solar energy directly into electricity, is pollution free, and generates no harmful wastes.

Many practical problems remain to be solved, however, before photovoltaic systems can become cost competitive with coal-fueled, oil-fueled, and nuclear systems for large blocks of power. Authorities assert that solar-cell-produced energy costs ten times as much as that produced from coal. At present, photoelectric power also demands a high capital investment. In the direction of improvement of the silicon solar cell additives (such as chlorophyl) are being tested and promise to increase the efficiency of this device .

More efficient storage batteries will be needed, and some improvement must be planned for dc-to-ac converters for use in conjunction with solar cells. One expectation is that by 1992 improved photoelectric energy systems will supply electricity at the rate of 70 cents; per peak watt, in terms of 1980 dollars. It has been estimated that 1,000 square feet of solar cells will be required to power the average house. But, because direct solar power is unavailable at night and on cloudy days, it might require a conventional-system backup. There are two approaches to photovoltaic power: Central-plant and individual installations. It has even been suggested that during the day when occupants are away from home and do not need the service, a residential installation might supply power, for which the local utility would pay, into the grid; at night, the customer would pay for power from the local utility.

Photovoltaic Cells

Semi-conductors have the useful property of turning sunlight directly into electric current. It is a trick that is being put to commercial use in solar-powered calculators, refrigerators, and of course satellites.

Research and development have been stepped up considerably in University of Murcia (Spain), and photovoltaic cells are not expensive for large-scale power generation.

Solar Cell

Generally speaking, a solar cell is a heavy-duty photovoltaic cell; that is, any self-generating cell that can produce usefully high voltage and current when exposed to sunlight. A typical solar cell, however, is a silicon cell; this type of photovoltaic cell delivers the highest output for a given light intensity. A solar battery is a dc power source made up of several solar cells connected in series or parallel, or both, to deliver useful amounts of power when illuminated by sunlight. Such heavy-duty photoelectric batteries are used in space satellites, control devices, emergency telephone power supplies, portable radios, and other places

Range of Characteristics of Photovoltaic Cell

Output current. -Varies with active area, illumination, and load resistance. The typical 100-fc short-circuit value for selenium is 12 μA to 77ø μA; for silicon 5 mA to 36 mA; for silicon solar-power assemblies, 18 mA to 500 mA.

Output voltage. -Open-circuit. Directly proportional to active area and illumination. Typical 100-fc value for selenium is 0.2 V to 0.45 V; for silicon, 0.3 V to 1.5 V; for silicon solar-power assemblies, 0.4 to 12 V.

Power output. -Varies with active area, illumination, and load resistance. For a given illumination and active area, the power output is maximum when the load resistance equals the internal resistance of the cell. When the cell output is pure dc, the power (in watts) is the simple product E.I, where E is the output voltage (in volts) and I the output current (in amperes).

Internal resistance. -For a given illumination, the dc resistance

$$R = E / I$$

where

E is the output voltage (volts) developed across the cell
I is the output current (amperes)
R is in ohms.

The resistance is lower in a silicon cell than in an equivalent selenium cell.

Peak spectral response. -For selenium, 5500 angstroms. For silicon, 8000 angstroms.

Solar Cell Systems Economical Performance

Most of the PV advanced research is aimed at photoelectrochemical cum storage devices to supply electricity either directly or through an intermediate hydrogen storage followed by a fuel-cell conversion.

There are also novel ideas which might further advance basic knowledge of energy and matter. We would like to conclude this brief assessment of the PV issue by mentioning a PV advanced research program being performed by the Cartagena Polytechnic University in Spain; a PV cum storage device consisting of an alkaline metal sheet imbedded in a transparent plastic matrix. Efficiencies in the 30 to 50% range have already been achieved and research is progressing towards 60% efficiency goal with polifacial devices.

International Trends

International trends headed by the USA focuses the spatial technological development, recently called, "Stars War". This potential direction of international technology makes sure the necessity of resorting to the only source for the planets: sun energy.

And it is also clear that in these conditions only one of the ways of conversion of solar energy is attractive: electrical photovoltaic capable of providing energy to the spatial electronical circuits.

In these conditions, with a view to the year 2000 and 2020, what a better study of photovoltaic Sun energy can be done?

Nuclear energy, magnetohydrodynamic, hydrogen fusion, combustible batteries will not be able to provide the energy required in the space. It is intended to get a photovoltaic cell which reaches an efficiency of 50%, carrying a system of electrical accumulation in the same cell, and it is also intended to apply this cell to three well defined fields like this:

a). For applications in the agricultural and cattle raising sectors;

b). For industrial applications like electrical energy of sun origin;

c). For transport applications such as electrical vehicles, in their air version, regarding as impossible the improvement of the wheeled vehicles, due to the high levels of sophistication acquired by these ones; nevertheless the support of air electrical vehicles could allow all the improvement of the wheeled vehicles to turbo and combustion engines and if the energy of electrical vehicle is sun, it is not necessary to comment anything more.

HIGH EFFICIENCY SOLAR CELLS

I.M. Tsidulko
Physical-Technical Institute
Tadjik SSR Academy of Sciences, Aini Street 299/1, Dushanbe, 734063, USSR

Abstract

A construction of a new type solar cells (SC) supplied with an inner electronic-optical concentrator is proposed, together with a modification of such a SC with an external microlens. Physical principles of the SC operation as well as some aspects of the SC fabrication technology are considered. It is shown that the efficiency of the SC having an electronic-optical concentrator, when being utilized in systems supplied with external concentrators of radiation, can be 49% more than that of the SC without a concentrator.

1. INTRODUCTION

One of the ways to increase the efficiency of the semiconductor solar cells is to reform the concentrated radiation. When growing degree of the radiation concentration K_c, the efficiency of the SC increases according to the logarythmic law ans, under the conditions of AM I, can achieve 35% if K_c= 1000 [1]. The further growing of K_c can be achieved by means of complicating the external optical systems, that increases their cost significantly. Moreover, when absorption of very concentrated radiation, within the SC a great quantity of thermal energy is disharged, the latter raising the work temperature of the SC and creates significant temperature gradients within the cell (under the conditions of compulsory cooling) as well as inner thermal stresses. As a result, the efficiency of the SC is sharply decreases and does not achieve the theoretical value. The thermal overloading of the semiconductor crystal results in degradation processes' precipitation an failure of the SC. That is why in practice the external radiation concentration degree is limited to the range of K_c=150÷ 200.Naturally,that under such conditions the limit theoretical values of the SC efficiency cannot be realized.

2. SC MODE OF OPERATION

Nevertheless,there is a possibility of the further raising for the radiation concentration degree within the reforming area without any complicating of the external concentrating systems.This possibility can be realized within the SC having a built-in radiation concentrator,which acts as an inner electronic lens [2], [3].

It is known,that in the usual Sc with a planar p-n junction the idling voltage U is defined from the formula

$$U=\frac{kT}{q}\ ln\ \left(\frac{J_p}{J_0}+1\right)\ , \qquad (1)$$

where j_p - photoelectric current density at the p-n junction,J_0 -saturation current density,q - the electron charge, T - the absolute temperature.The radiation concentration degree being increased,the photoelectric current grows linearly (up to K_c= 2500,when its saturation takes place) and the voltage at the p-n junction also becomes greater.When a stable radiation flow on the SC surface takes place the further increase of the photoelectric current density at the p-n junction is possible,if the form of the saturation layer is optimized and represented as a spherical or cylinder sector.In Fig.1(a) one can see the given SC with its cross-section,perpendicular to the p-n junction plane and parallel to the symmetry axis.The absorption layer I is encompassed by the energy band-gap isolating the semiconductor 2.In the narrow part of the absorption layer there is the p-n junction 3,with adjacent electrical contacts 4.Within the absorption layer an inner electrical field is created,pulling relatively to non-main charge carriers (NCC).Such field can be obtained.by means of the alloying admixture concentration gradient or by means of variation in the chemical composition for the semiconductor material,moreover,the wide energy-gap part of the absorption layer should be directed to the side of the radiation, as shown in Fig.1(b). NCC developed by the light are being gathered by the inner gathering field into the narrow part of the absorption layer, to the p-n. That is why, to eliminate the recombination losses of NCC in the process of their motion within the layer, it is necessary, that the diffusion-drifting NCC length L exceeds the linear dimensions of the sector in the sector in the motion direction. It is known, that NCC drifting length in the quasi-electrical field fprmed by the width gradient of the energy gap ∇E_g can be described by the formula

$$L_{dr} = \frac{|\nabla E_g| \mu \tau}{q} , \qquad (2)$$

and the NCC diffusion length will be

$$L_{dif} = \sqrt{D\tau} = \frac{\mu kT}{q} \tau , \qquad (3)$$

where μ - mobility, τ - life time, D - NCC diffusion coefficient, q - electron charge. That is why, for NCC diffusioning in the drift direction, the value L will be [4] :

$$L = \sqrt{L_{dif}^2 + \left(\frac{L_{dr}}{2}\right)^2} + \frac{L_{dr}}{2} . \qquad (4)$$

Thus, the maximum dimension l of the absorption layer in the direction of NCC motion (see Fig.1) should satisfy the condition

$$l \ll L \qquad (5)$$

The curvature radius of the absorption layer surface depends on the semiconductor material refraction factor and focus distance to the p-n junction, so that the lense obtained should ensure the primary localization of the radiation within the absorption layer. The focus of such a lens does not obligatory coincide with the p-n junction, if the sector shaping line slope allows to gather the reflected rays with the p-n junction area. Wide band-gap semiconductor encompassing the absorption layer should have the lattice period close to that of the absorption layer. This requirement is necessary in order to eliminate the non-correspondence deffects, arising at the joint of these semiconductors. The accuracy of the lattice period coincidence should be not worse than 0.5%. In this case the dislocation density proves to be small enough and the recombination of NCC at the heterojoint could be neglected.

3. SC EFFICIENCY

When meeting the above mentioned requirements all the NCC within the absorption layer will achieve p-n junction and generate photoelectric current. If the area of the layer receiving surface is designated as S_b, and the p-n- junction surface - S_s, then their relation $\xi = S_b/S_s$ shows how many times the photoelectric current density has increased at the p-n junction. Then the advantage in idling voltage and SC efficiency could be defined according to the formula

$$\frac{U_2}{U_1} = \frac{\ln(\xi\eta + 1)}{\ln(\eta + 1)} \approx 1 + \frac{\ln \xi}{\ln \eta}, \ (\eta = j_p/j_o), \quad (6)$$

However there exists a maximum possible degree of concentration ξ_{max} dependent on purely physical reasons. As the NCC concentration in the p-n junction area becomes higher than that upon the surface, the diffusion current j_{dif} appears which is directed towards the drift current j_{dr} and the inner NCC concentration process will continue until these currents will be equal. Knowing this, one can estimate the value $_{max}$. Such estimation was carried out earlier, in [2], where the expression for ξ_{max} was obtained

$$\xi_{max} = \frac{(\mu_n \Delta E_g)^2}{2qD_n} \quad (7)$$

Here ΔE_g -is the width variation of the band gap within the absorption layer between the ray-receiving surface and p-n junction, μ_n, D_n - electron mobility and diffusion coefficient, in the p-type layer, respectively. Let's estimate the value ξ_{max} for SC with the absorption layer, made on the base of solid solution $Al_yGa_{1-y}As$ ($0 \leq y \leq 0.8$). Substituting typical values of the semiconductor parameters into (7): μ_n = 4000 $cm^2/(B\ c)$, D_n=25 cm^2/s, ΔE_g=0.6 eV, we obtain that ξ_{max} is about 2×10^3. Such a degree of NCC concentration can be achieved, having made a spherical sector with area radius relation S_b and S_s equal approximately 45. In this case the ultimate possible idling voltage U_2 increase in comparison with the usual SC U_1 the values of photoelectric current and saturation current being typical, under conditions AMI (j_p = 20 mA/cm^2, $j_c=10^{-13}$ A/cm^2) according to (6) gives:

$$\frac{U_2}{U_1} \approx 1 + \frac{\ln (2 \times 10^3)}{\ln (2 \times 10^{11})} = 1.29$$

or 29% more, than in case of the SC having no concentration. If such a SC is put into the concentrated radiation flow with K_c=150, the optimal rejime from the point of view of the thermal loads - one can expect the efficiency increasing by 49%.

4. THE MATERIALS AND TECHNOLOGY

The main part of the proposed SC is the absorption layer, to which special requirements are put forward: the presence of an inner pulling field, that is strong enough, as well as large diffusion-drift length of the NCC. For its fabrication 3- and 4-component solid solutions of semiconductor materials of the type AlGaAs-GaAs, GaPAs-GaAs, InGaPAs-InP as well as Si and its compounds and the similar can be used from all the known and technologically developed. Within the above mentioned and similar solid solutions perfect epitaxial layers of the variable band -gap semiconductor can be created by means of liquid or gas epitaxy. Wide band-gap complexes of these compounds have the lattice parameters similar to narrow band-gap compounds and that is why can be used as an isolating semiconductor encompassing the variable band-gap layer. The fact that the band-gap width changing range coincides with the main part of the solar radiation spectrum in visual and near IR ranges, is favourable to utilization of those semiconductor compounds. Si, being the most technologically developed material, can also be used for these purposes. Among its advantages the large diffusion length of NCC in pure crystals, the possibility of creating an inner pulling field due to the admixture gradient, as well as creating a heterobarier Si-SiC should be noted.

Let's consider, for example, the construction of the SC made on the base of the solid solution of AlGaAs. The variable band-gap layer has the thickness l=26 mkm and the band-gap width variation ΔE_g=0.6 eV, so the gradient value $|\Delta E_g|$ will be about 230 eV/cm. The variable band-gap layer is improved by means of adding Si up to the level $10^{15} \div 10^{16} cm^{-3}$ and the NCC life time within it is about 10^{-7} s, while the diffusion-drift length of the carriers, defined by the formulas (2)-(4) will be about 900. Thus the condition (5) in the given case is fulfilled, so the NCC losses while their drift will be minimum. The radius of the maximum SC frontal section was within the limits 30 μm and the radius of the P-n junction $\sim$0.75 μm, so the possible concentration degree will be about ξ =1600. The surface curvature radius of the sector is about 50 μm.

As the dimensions of the microscopic SC do not exceed 100 micrometers it seems reasonable to fabricate the matrices of such SC during one technological cycle, e.g. as it is shown in Fig.2. The peculiarity of such a matrix is non-full encompassing of the graded-band layer by the layer of the wide band-gap semiconductor. Such a construction of the graded-band gap semiconductor layer will fulfill the function of the NCC concentrator under an additional condition: if the isoenergetic layers of the variable band-gap sector will be a part of the pspherical surface. In this case the direction of the band gap width gradient (and, consequently, the direction of the quasi-electric field) will conform with the direction towards the p-n junction and all the NCC will move towards it.

The SC matrix fabrication technology includes the following processes: forming of a microlens within the substrate,variable band-gap layer growing,forming of a V-shaped (or \/-shaped) grooves by means of selective chemical etching,filling the grooves with wide-zone semiconductor,diffusioning for creating p-n junction,fabricating the contacts. The shortcome of such a technology is the large quantity of operations together with difficulties related to co-axial orientation of the p-n junction with the microlens axis.

5. MATRIX-TYPE SC WITH A GRADIENT LENS

To decrease the quantity of the technological problems when fabricating matrix-type SC is possible when excluding the operations connected with the semiconductor microlens forming.The latter can be substituted by an external microlens.For example,planar gradient microlenses are known,which represent flat areas with a gradient refraction factor [5]. A glass matrix made of such microlenses having been jointed with a semiconductor matrix of SC (see Fig.3) will also act as a protective coating of the SC preserving it from the atmospheric impacts.Such a microlens can be fabricated using the technology of deep admixture electromigration [5].The jist of the technology is as follows.Upon the flat glass plate a mask is shaped by means of photolitography,in which open windows of the proper dimensions are left ,the latter forming the microlens apertures.Then the plate is immersed into the melted electrolyte,containing the salts of Tallium. When the external electric field is put in,the tallium ions penetrate into the glass,changing its refraction factor. Using this process it is possible to obtain a microlens with the focus spot diameter equal to several tens of micrometers. The radiation concentration degree of such a microlens can be several tens times.

The microscopic SC in the given case can have a flat ray-perceiving surface and pyramidal form of the band-gap layer.The dimensions of the ray-perceiving area should coincide with the focus spot dimensions.

To focus the radiation to the matrix-type SC it is possible to use microlenses of other constructions.For example, matrix-type difraction microlenses used in the work [6] for coherent composition of the radiation from the laser diodes' matrix are quite appropriate.Such microlenses consisted of the computer generated Frenell lenses (zone plates) obtained by means of an integral optics method.The theoretical efficiency of such zone plates is close to 100%,that excludes the possibility of radiation losses within them and can insure high efficiency of the SC.

6. CONCLUSION

The considered new principles of fabricating SC ,acting under the conditions of the concentrated radiation,give the possibility to increase their efficiency significantly.At the same time ,the realization of these principles in practice requires more complicated technology,close to the fabrication technology of the matrix element of the integral microelectronics and optics.This process is regular when trying to improve to the maximum possible degree the characte - ristics of the semiconductor devices and the increasing of the cost is regular as well.But in the case under consideration of the technology of the matrix-type SC fabrication can be justified by their high efficiency and the progress of the semiconductor technology allows to hope that the cost of such devices will decrease significantly in mass production.

REFERENCES

1. H.J.Hovel - Hovel materials and devices for sunlight concentrating systems "IBM J.Res;Developm." 1978,22.p-112-121.
2. I.M.Tsidulko - Solnetchniy element so vstroennim kontse ntratorom izlutchenia."Heliotechnika" 1989,N°1,s.20-23.
3. I.M.Tsidulko - Semiconductor solar cells with variable energy gap absorption layer. "Manila International Symposium on the Development and Management of Energy Re - sources,Jan.26-28,1989" Abst.of Papers,Manila,Philippines

4. G.V.Tsarenkov - "Fizika i technika poluprovodnikov"1975 9,s.253
5. B.A.Krasjuk, G.I.Korneev - "Opticheskie sistemi svjazi i svetovodnie datchiki.Voprosi technologii".M.- "Radio i svjas",1985, s.59-60.
6. J.R.Leger, M.L.Scott, W.B.Veldcamp - Coherent addition of Albaks lazers using microlenses and difrective coupling."Appl.Phys.Lett." 1988, 52,N°21,pp 1771-1773.

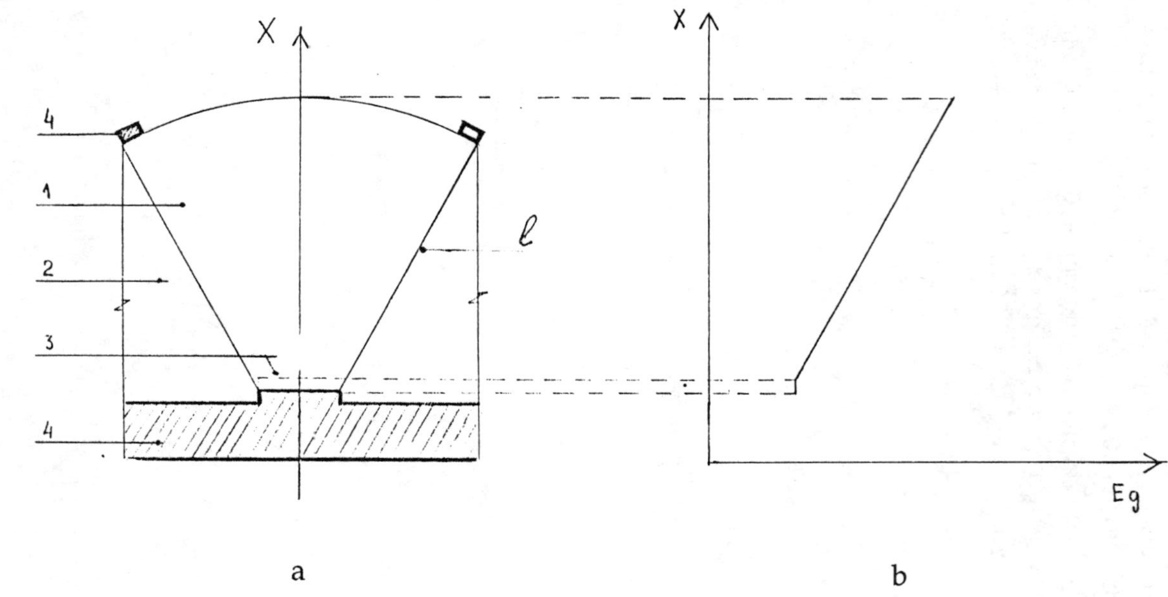

Fig.1(a). - Cross-section of the SC having an inner electronic-optical concentrator.

Fig.1(b). - Relationship of the band-gap width and the absorption layer thickness.

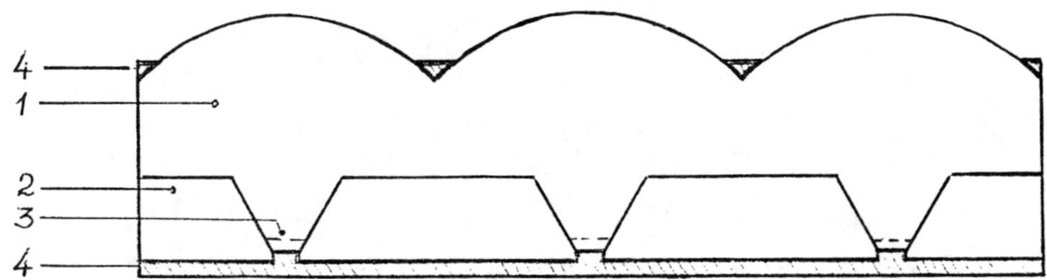

Fig.2. - Cross-section of the matrix for the SC having electronic-optical concentrators: (1) - variable band-gap absorbing layer;(2) - wide band gap semiconductor;(3) - p-n junction;(4) - electrical contacts

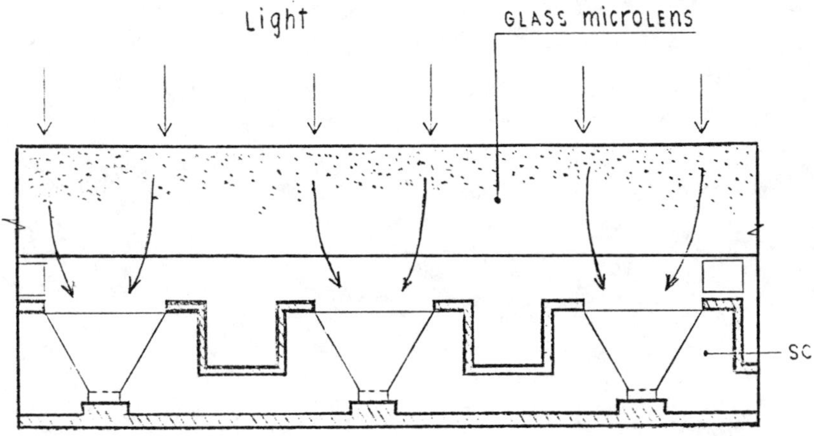

Fig.3. - Cross-section of the matrix for the SC having an external microlens. Hachures are used for the elcctrical contacts.

DEVELOPMENT OF PHOTOELECTRIC POWER STATIONS FOR THE PAMIRS REGION

I.M. Tsidulko
Physical-Technical Institute after S.U.Umarov,
Tadjik SSR Academy of Sciences, Aini Street 299/1, Dushanbe, 734063, USSR

1. INTRODUCTION

At present world attention is drawn to the problem of utilization of non-traditional energy sources (NES). It is excited, first of all, by the coming exhaustion of the organic fuel resources, limits of hydroresources and by the potential danger and high cost of atomic stations. An important factor orienting to NES utilization is the ecological one. It is of great significance for the regions which are not industrially developed but have a specific value as nature zones. That is why the energy supply strategy in such regions should base on maximum utilization of NES typical for these regions.

This paper considers some aspects within the problem of solar energy utilization in the Pamirs region,Tadjik SSR,by means of solar photoelectric and combined thermophotoelectric power stations.

2. POPULATION AND ENERGY

Tajikistan is a mountain republic where mountains from 1.5 up to 7.5 km high constitute 93% of the territory.Some regions, like Murghab,for example,are entirely situated highly in the mountains (more than 4000m).The population is concentrated mainly in the valley,but some 10% of it - which is hundred thousands of persons - reside in the high mountain regions.Its number quickly increases due to the greatest in the USSR natural accretion.This demographic factor raises economic problems related to population employment and food supply.That is why,an active peopling of the mountain and high mountain regions is preseen by the republic government for the nearest future as the first step of the mentioned problems solution.Consequently,in its turn,the problem of energy supply arises which cannot be solved by means of traditional energy sources utilization.First,the republic has not a sufficient reserve of an organic fuel,and its requirements are overcome at the expense of their importing from outside. Second,output of the local resources is disadvantageus due to almost inaccessibility of a great number of deposits.Besides , usage of an organic row material as a fuel can lead to ecological damage in this unique region and water balance change in the rivers of Central Asia (due to intensive melting of the Pamirs glaciers).Construction of big hydropower stations on the Pamirs rivers is problematic because of the difficulties in energy transmission under the mountain conditions, induced seismisity and so on. Small hydroelectric power stations are more perspective here but they are not able to meet all the energy needs for the debit of mountain rivers sharply drops in winter seasons.

At the same time, high mountain regions in Tadjikistan, particularly the Pamirs one, offer some unique potentialities to utilize solar energy. Highly in the mountains the density of solar radiation power axceeds 1100 W/m^2 and duration of sun radiance is more than 3000 hours per year. For example, there observed only 7 cloudy days a year in the zone of Kara-Kul lake, eastern Pamirs. That is why, mostly perspective to use in these regions are the installations transforming the solar radiation, such as photoelectric power stations (PEPS), combined P E P S (CPEPS), different air- and water-heating installations.

3. OBJECTS USING PEPS ENERGY

Analysis of the possibility to utilize solar electric installations in high mountain regions showed that the objects-users can be devided into several types according to their power consumption.

a). Low power installations (1-10 W) necessary as power sources for autonomic systems of the following types:
- TV broadcasting retransmitter with several kilometers radius of action, including reception and relaying of sputnik signals,
- a net of seismic and meteorologic sensors with telemetric transmission of information,
- road light-houses installed at the dangerous sections of mountain roads,
- power units for radio receivers.

b). Middle power installations (20-200 W) providing operation of domestic technics: TV-receiver, tape-recorder, refrigerator, room lighting and what conserns the country site, they can help a mechanization of some processes, such as shearing, milk separating e.t.c.).

c). More powerfull stations (300-1000 W) can be mostly usable ones. They can be used, in particular, at scools, medical aid stations, residental houses, cattle farms, hydrometeorological stations, geological parties and expeditions, tourist bases and campings, motels etc. It is reasonable, that such installations could produce hot water for economic purposes and would be supplied with a system following the sun. In this case the quantity of the produced power will be 30% higher than that of stationary systems of the solar stations.

According to preliminary estimations, at present the installations with the total power of about 10 MW are required to meet the power needs of the users. In spite of the fact, that the installations' cost is quite high (6-7 dollars per pcW), their utilization proves their value due to the high cost of the imported fuel and firing.

4. MODEL OF THE COMBINED PHOTOELECTRIC POWER STATION

The Tadjik Physical-Technical Institute has been working

for several recent years on the problem of creating such installations suitable for utilization in the Pamirs. For this purpose various assemblies and elements of the station have been tested under natural conditions. In particular, a model of the combined station has been developed, in which Si solar cells and radiation concentrators (Frenel lenses) are used, a general view of the station being represented in Fig.1. The model contains 16 sections of "solar element-concentrator", rigidly connected by means of a pyramidal aluminium frame (see Fig.2a). Such a construction has appropriate rigidity and small weight and allows to increase the number of the sections if necessary. The solar cells were installed in the carriers which had water cooling (Fig.2b). The efficiency of the solar element was rather low - under the conditions of AMO (the measurements were conducted using the solar radiation imitator). However, in the course of natural tests of the model the total efficiency of the solar elements went down to the level of 6% with respect to the power of radiation, incident to the installation. The radiation power losses are caused by a low efficiency of the Frenel lenses(0.6-0.7) as well as reflection losses within the glass plate,protecting the solar cells from the atmospheric effects.The thermal efficiency of the installation,computed on the base of cooling water temperature measurement data,was about 45%.Using such a model of the solar station,it is planned to carry out various tests for the solar cells and Frenel lenses to determine their degradation characteristics.At present the work towards developing a more powerful installation,containing 128 sections with the solar cells on the base of the silicon and gallium arsenide,have been begun at the Institute.

5. CONCLUSION

To supply energy for the human economic activity under the mountain and high mountain conditions,say,in the Pamirs,it is necessary to use the solar power.The utilization of the solar energy by means of photoelectric power stations proves its value even at present due to the high cost of the organic fuel and firing.Considering great distribution of the users it seems reasonable to produce installations having power up to 1 kW in order to supply the single users (with them).At present Tajikistan is a large potential market for saling such installations,the overall necessary power of which is 10 MW.

Fig.1. - General view of a combined photoelectric station model

a

b

Fig.2 (a) - Picture of PEPS model frame, at the top-Frenel's lenses, solar cells under them.

(b) - Picture of a solar cell under the Frenel's lens.

ON SOME CHARACTERISTICS OF THE $CdInSe_2/CdS$ - TYPE STRUCTURE

Sofia Melinte, Magda Gherghel, Gh. Calugaru, and D. Ursu
Politechnical Institute of Jassy, Physics Department,
Spl. Bahlui stg. 63, Jassy 6600, Romania

ABSTRACT

Photovoltaic solar cells based on thernary copper have a high efficiency and time stability, but the optoelectronic and structural properties of $CdInSe_2$ thin films are not completely known. We have studied this film by X-ray diffraction, correlating its structure with the temperature of the source used for film preparation. We have also determined the characteristic, parameters of the film and the I-V characteristics, whence we have deduced the values;

$$j_{sc} = 38mA/cm^2; V_{sc} = 0.45V; FF = 0.66; \eta = 11.2\%$$

1. INTRODUCTION

The $CuInSe_2/CdS$ - type solar cells have an excellent time stability and a conversion coefficient exceeding 10%/1.2. This type of cell has been intensely studied. The complete cell consists of one or two CuInSe2 films with different resistivity, on aluminium oxide substrate plated with a thin molibden film, then the CdS film/3/ aluminium grid. The cell repeatibility depends on a great number of parameters that interfere in the film preparation, and on the heterojunction thermal treatment. The authors have performed a study of some factors that influence the cell efficiency.

2. EXPERIMENTAL

A study of the vacuum evaporation have been performed, in order to choose the deposition method. The complex compounds dissociate in constituent elements and get evaporated at different rates during their evaporation. The constituent elements of $CuInSe_2$ are very different from the point of view of their evaporation parameters: evaporation temperature (corresponding to a vapour pressure of 10^{-2} torr) and evaporation rate calculated with the relation

$$N_e = 3.513 \times 10^{22} a_1 P_e / (M_e T_e)^{1/2} mol/cm^2 s \tag{1}$$

where N_e is the evaporating rate, M_e - the atomic weight, P_e - evaporation pressure at temperature T_e; a_1 - evaporation coefficient that is equal to 1 for a clean surface. In Tables I and II the deposition rates are given for each element, versus temperature and stoichiometrical structure of the $CuInSe_2$ thin film.

In Table I, one can see that Se has lowest evaporation temperature and the highest evaporation rate within the $CuInSe_2$ compound. Hence, it will be the first to evaporate, its evaporating ending when two elements begin to evaporate, in the flash method. One should expect that in the final CuInSe2 film, a selenium monolayer should be found on the substrate, but X-ray analyses showed absence of selenium. This fact is connected with Se reevaporation from the surface.

Still another difficulty appear selenium: its adherence on a clean substrate is null.

When analysing the parameters of $CuInSe_2$ constituents, one can see that, at the evaporation temperature corresponding to a vapour pressure of $p = 10^{-2}$ torr, the evaporation rates are close to each other, this fact determining us to obtain this film by the method of simultaneous evaporation from independent sources. Elements given in table II, taken in stoichiometric amounts, were vacuum evaporated from sources whose temperatures, measured with thermocouples, were kept constant at the corrresponding evaporation temperature. When producing the evaporators, their surface, temperatures and geometrical parameters were considered. The interdependence of these parameters is given by the relation (2)

$$\frac{T_1}{T_2} = \frac{d_1}{d_2}\left(\frac{L_2}{L_1}\right) \tag{2}$$

The cell was obtained by the deposition of a $CuInSe_2$ - type film, by using the multiply sources method by vacuum evaporation on an alumina substrate plated with molibden. The deposited film thickness ranged between (3.4 - 4) μm and it consisted of films of different resistivity.

The first film was of about 2.8μm and was deposited at a substrate temperature of 623°K and the second 0.9 μm thick film-for a substrate temperature of 723°K. Above this film, CdS was layed down at a substrate temperature of about 573°K and a substrate temperature of 1273°K /4/. The aluminium grid was formed by means of special masks. The photovoltaic parameters of the cell deposited as above were determined after a thermal treatment in air between (623 - 673)°K.

3. RESULTS AND DISCUSSION

3.1. X-Ray analysis

When the source temperature differs from the evaporation temperature of Cu, In, and Se, and the substrate temperature is kept constant and smaller than 623°K and 723°K respectively, in the obtained film of the selenium is noticed. this can be seen in Fig.1.

In Fig. 1. the coexistence of two phases is noticed, namely CuIn and $CuInSe_2$. Holding constant the temperatures and exceeding the evaporation and the substrate temperatures, i.e., 623°K and 723°K respectively, for those films of $CdInSe_2$, Fig. 3 shows that only one phase is obtained. The results are totalized in Table III.

Fig. 3 shows that the X-ray diffraction diagram contains an intense maximum at (112), also at present in Fig. 2 and other smaller maximes, which indicates that only one compound was obtained, $CuInSe_2$.

The X-ray analyses of the specimens showed that they contain the CuIn, $CuInSe_2$ phases; the content of $CuInSe_2$ increased when the sources temperature was increased (Fig. 3). The specimen (1) contained only CuIn, and the specimen (2) contained CuIn and $CuInSe_2$.

In summary, the conditions for sources temperature, substrate temperature, for multiply sources evaporation have been derived.

Analysis of our X-ray diffraction leads us to conclude that the dominant factor in multiply sources evaporation is the sources temperature and substrate temperature.

3.2 Electrical Parameters and Photovoltaic Effect

The film resisitivity was measured by measuring the resisitance R and film thickness $(\rho = R_{\blacksquare} \cdot d)$. The resistivity was $\rho = (0.5 - 0.8)\Omega cm$.

The charge carriers mobility in CdS was determined by Hall effect, a mean value μ = 3 $cm^2/V.s$ being found.

The dark current and the light current for a cell with A=(1.8 1.2) cm^2 is given in Fig. 4, whence we found that j_{sc} = 35 mA/cm^2, V_{sc} = 0.4 V; FF = 0.44 for an illumination of 100 $mW.cm^{-2}$, η=6.16%.

By reducing the cell area, one notices in the the I-V characteristics in Fig. 5 that, for A = (0.8 x 0.8) mm^2, one obtains Voc =0.45 V; FF = 0.66 for the same illumination, hence a conversion gain η= 11.2%.

4. CONCLUSION

In order to obtain highly efficient solar cells based on ternary copper, a permanent control is necessary of the deposition of the films forming the cell, for getting uniform films that should obey the stoichiometry of the final compound for both the p-type and n-type films.

The fact that high efficiency is obtained for small active areas shows the existence of imperfections, inhomogeneities in the film structure. Hence a proper thermal treatment is necessary.

REFERENCES

1. J.L. Scay and J.H. Wernich, in B.R. Pamplin (ed.), Thernary Chalcopyrite Semiconductors: Growth, Electronic Properties and Applications, Pergamon, Oxford, 1975.

2. R.A. Mickelson and W.S. Chen, Prec. 15th Photovoltaic Specialists Conf., Orlando Fla, May 12-15, IEEE, New York, 1981, p.800.

3. S. Melinte, A. Jeflea, M. Moise, "On Some Parameters Determining the Efficiency of Schottky Solar Cells Based CdS", Miami, Florida, U.S.A., Alternative Energy Sources VI, Vol. 2 (1985), p. 297.

4. S. Melinte, A. Jeflea, M. Moise and N. Mateescu, "The Diffusion Length of Minority Carriers in CdS Films used for Solar Cells" in J. of the Less-common Metals, 95 (1983) p.99.

Table 1

	Melting point T(°K)	Teat a 10^{-2} torr vapour pressure Te(°K)	Evaporation rate at $p=10^{-2}$ torr (mol/cm^2s)	Atomic weight (g)
Se	490	507	2.10^{17}	78.96
Cu	1356	1546	$1.4.10^{17}$	63.5
In	430	1225	1.10^{17}	100.0

Table 2: Stoichiometric Structure of $CuInSe_2$

Se	157.92g	47%
Cu	63.5g	18.9%
In	114.8g	34.1%
Mol	336.22g	100%

Table 3: X-Ray Analysis of $CuInSe_2$ Film

Crt. No.	Source temper. Se(°K)	Source temper. Cu(°K)	Source temper. In(°K)	Substrate temper. (°K)	Evap. rate mg/min	Deposited film phases
1.	503	1543	1173	593	20	CuIn
2.	508	1549	1223	623	20	CuIn, $CuInSe_2$
3.	573	1573	1233	(623-673)	20	$CuInSe_2$

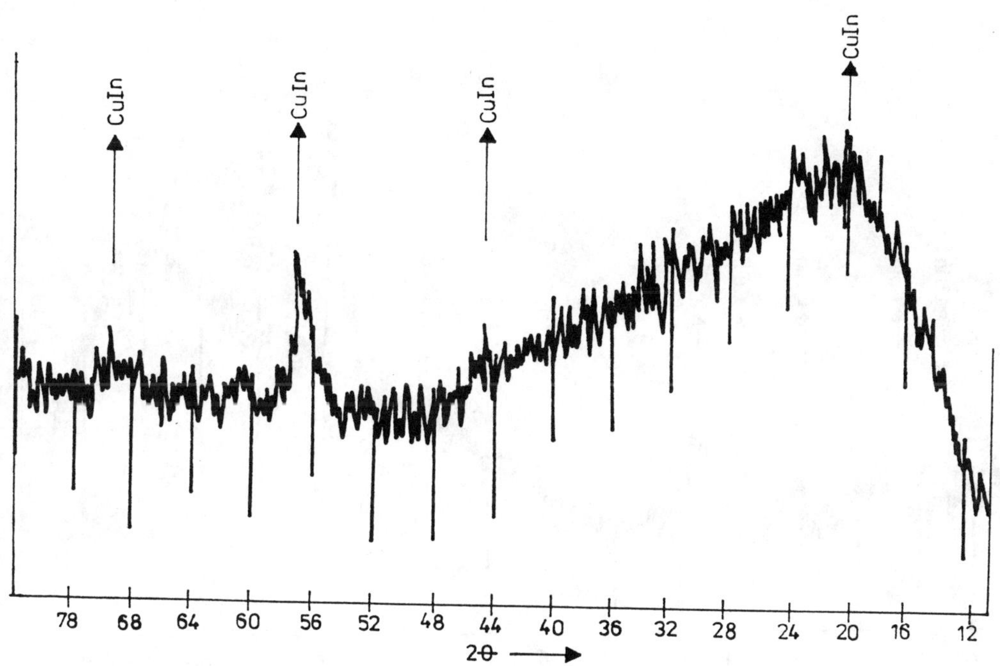

Fig. 1. X-ray Diffraction Pattern of $CuInSe_2$ Thin Film

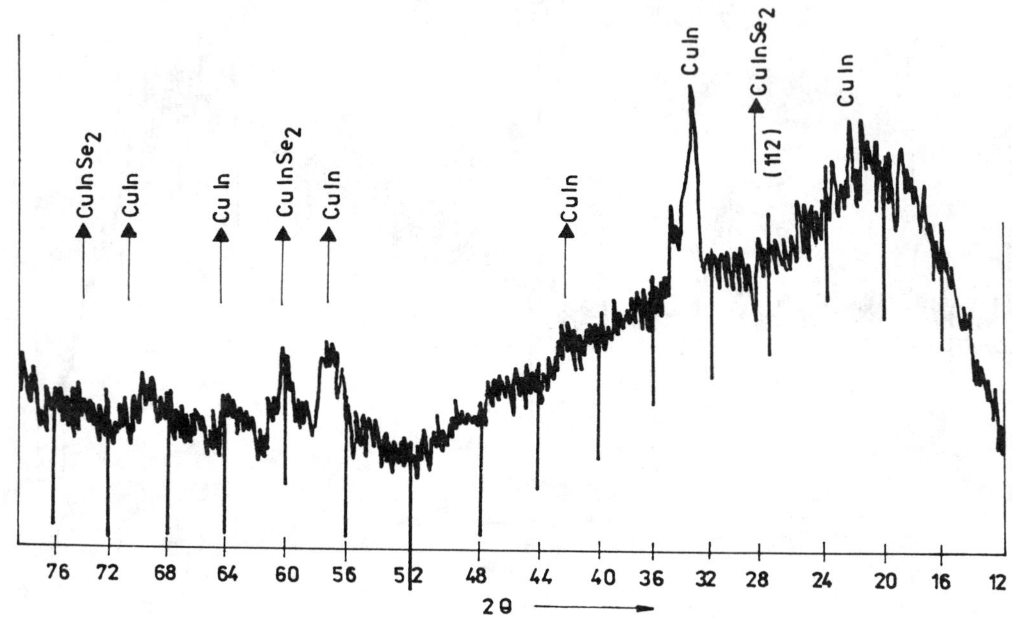

Fig. 2. X-ray Diffraction Pattern of $CuInSe_2$ Thin Film

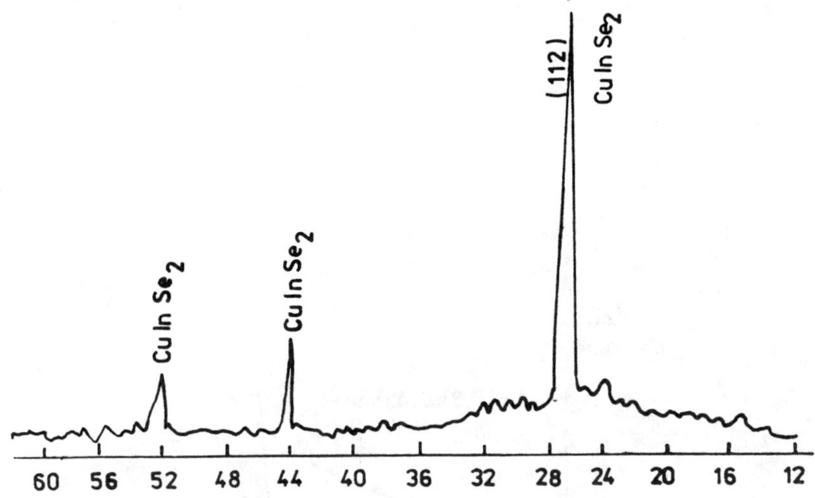

Fig. 3. X-ray Diffraction Pattern of $CuInSe_2$ Thin Film

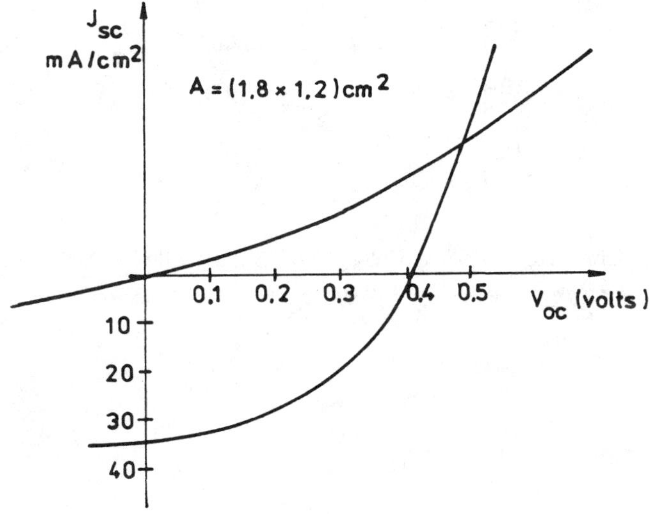

Fig. 4. J-V Characteristic of $CuInSe_2$/CdS Solar Cells in Dark and Under Simulated AM_1 Illumination

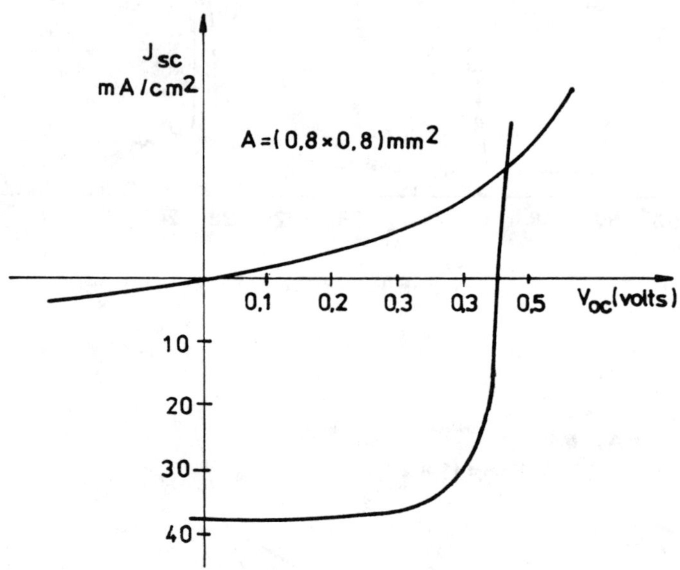

Fig. 5. J-V Characteristic of $CuInSe_2/CdS$ Solar Cells in Dark and Under Simulated AM_1 Illumination

ON THE PHOTOVOLTAIC PROPERTIES OF THE MIXED HETEROSTRUCTURES BASED ON CDS AND PHOTOSYNTHETIC PIGMENT

Madga Ghergel, Sofiaa Malinte, Gh. Calugaru, and D. Ursu
Polytechnical Institute of Jassy, Physics Department
63 Splai Bahlui Str., Jassy 6600, Romania

Abstract

The work presents the results of the experimental studies concerning the influence of both the impurities from the I_a and I_b groups [7] and the photosynthetic pigments [1-3,6] on the photoconducting properties of CdS- Cu_xS type photoconductive structures, obtained by liquid phase reaction method, with the aim of improving its photovoltaic properties.

Our experimental results led to the conclusion that both the photosynthetic pigments (chlorophyll) and the I_a and I_b group impurities (Na, Cu, Li, K) are able to improve the photoeffects at CdS- Cu_xS junction, by enhancing its photovoltages and photocurrents.

At the same time, by using the "progressive-analogical" modelling method, we have studied the influence of electric fields with different intensities and of the addition of the ascorbic acid on the photoconduction microcrystalline chlorophyll thin films included in a mixed heterostructure of the type SnO_x/CdS:Na- Cu_xS:Cu-Al- Chl/Cu.

1. INTRODUCTION

The CdS-Cu_xS type heterostructures obtained by the method of liquid phase reaction [4,5], present the advantages that their constituent photoconductive thin films can be obtained continuously and with a small energy consumption, in a relatively short time, having a good adherence to the substrate and a controllable stoichiometry [4, 5]. Their shortcoming consists in their photoconductive and photosensitive properties worse than that of the films produced by other methods (vacuum evaporation, sintering and dipping). Yet, their performances can be improved by controlling some physico-chemical parameters (temperature, concentration of the solution, impurities doping degree), that are directly influencing the chemical reaction kinetics and, hence, the film generation on the substrate.

With the purpose of improving the photovoltaic performances at the CdS-Cu_xS heterojunction, we directed our attention toward the use of some elements from the I_a and I_b groups (Na, Li, K) as dopants in CdS and Cu_xS films, respectively, taking into account that these impurities are cheaper and easier to find.

At the same time, on considering that the photosynthetic pigments (chlorophyll) show an intense absorption in the visible range (where the solar radiated energy is maximum), having a semiconduction activation energy of 1-2 eV, which corresponds to maximum efficiencies of the photovoltaic devices [4, 6], as well as the fact that the microcrystalline chlorophyll films show considerable semiconduction properties [1-3, 6], we proposed to study the influence of an Al-Chl-Cu type heterostructure, on the CdS-Cu_xS photojunction, with the aim of using the photosynthetic pigment thin films in producing mixed photovoltaic devices based on anorganic semiconductor materials. Simultaneously, we followed the influence of both electric fields of different intensities and the addition of the ascorbic acid ($C_6H_8O_6$) on the energy transfer and electron transport through the photosynthetic membrane of the chlorophyll molecules, with the object of improving their photoconduction performances [8].

2. EXPERIMENTAL. RESULTS AND DISCUSSION

The CdS-Cu_xS heterojunction was produced by the method of liquid phase reaction [4, 5] on glass substrates of (3x7) cm on which semitransparent SnO_x thin films were previously formed by the same method, these films representing the negative electrodes of the heterojunction [3, 5]. As the positive electrode, a copper grid covered later with a 0.2 μm thin Al film for avoiding the oxidation process was made by vacuum evaporation. Under the optimum conditions that we established in [5, 8], we obtained now the open circuit voltage V_{CD} = 190-250 mV and short-circuit current J_{SC}=8-8.5 mA/cm^2, for the heterojunction with CdS and Cu_xS films.

During our experiments, we found out that an improvement of the CdS-Cu_xS heterojunction photovoltaic parameters can be obtained by an optimum correlation between the sputtering temperature and the concentration of the solutions, these parameters being able to influence the equilibrium of the chemical reaction for the CdS and Cu_xS film formation [5].

For the purpose of improving the photoeffects of this type of photoelement, and increasing the heterojunction photovoltages and photocurrents, we sputtered above the CdS film, at a constant temperature of 120°C, a Na_2SO_4 solution with the concentration of 0.5-2%, thus obtaining a CdS:Na-Cu_xS structure. At the same time, we sputtered above the Cu_xS film obtained by the same method, at a constant temperature of 150°C, a $CuSO_4$ solution with the concentration of 0.5-2%, thus producing the Cu_xS:$CuSO_4$/Cu structure. A considerable improvement of the photovoltaic properties was seen at the CdS-Cu_xS heterojunction with the CdS and Cu_xS films doped at an optimum concentraion of Na_2SO_4 and $CuSO_4$ solutions, respectively, an amplification of both U_{CD} and J_{SC} being noticed: U_{CD} = 300 mV and J_{SC} = 8.7-9 mA/cm^2, respectively.

On considering that the photosynthetic pigments show an intense absorption in the visible range, where a maximum radiated solar energy exists, and that the chlorophyll microcrystalline thin films show remarkable photoelectric properties [6], an Al-Chl_{a+b+c}-Cu type heterostructure was made above the CdS:Na-Cu_xS:Cu photojunction, thus finally resulting a mixed SnO_x/CdS-CuS_x-Al-Chl_{a+b+c}/Cu type multijunction. Thin microcrystalline films were prepared by sputtering, electrodeposition and dipping methods, described in details in [6,8], having a thickness of 10-20 μm and a mean resistivity of 3.10^{11} Ωcm. For this mixed photoelement, we noticed an increase of both photocurrents and photovoltages, i.e. U_{CD} = 350-400 mV and J_{SC} = 10-10.5 mA/cm^2 (see Fig.1).

Taking into account that in the vegetal tissues there exist physiological gradients connected with the electrical gradients (hence stationary electric fields) and that the photoelectric voltages of these biological materials depend on the intensity of the metabolical processes, we decided to study the influence that an external electric field of different intensities (applied to biological materials from which the photosynthetic pigments ,were obtained) could have on the photoelectric behaviour of the Al-Chl-Cu heterostructure. Thus we applied to the vegetal samples that we used (from which we afterwards extracted the microcrystalline a and b chlorophylls), an artificial electrostatic field $\bar{E}$ directed downward, with intensities ranging between 75 and 60; 60-35; 35-7.5 V/cm. Aluminium plates of (35x65) cm and 10 μm thick multipolar platinum and wolfram electrodes were used.

During our experiments we noticed that thin chlorophyll films obtained from vegetal samples subjected to an average electric field $\bar{E}$–7.5V/cm, have higher photoeffects as compared to pigment films extracted from vegetal materials that were not subjected to electrical fields. At the same time, on considering that the chlorophyll molecules are complexed "in vivo" with proteins, enzymes and vitamins (mainly the ascorbic acid that is always present in green tissues), we supposed that a "progressive-analogical correlation" exists between the following parameters: electric field $\bar{E}$+ the concentration C of the ascorbic acid→proteic metabolism→photoconduction (volt-ampere characteristic).

With this aim in view, the vegetal samples used for chlorophyll extraction were subjected to an artifical electrostatic field directed downward, with intensities ranging between (75-60)V/cm; (60-35) V/cm; (35-7.5) V/cm. Aluminium plates of (25x35) cm and platinum and wolfram multipolar microelctrodes with a thickness of 10μm were used in the device. We established experimentally that thin chlorophyll films from the Al-Chl-Cu heterostructure obtained from vegetal samples subjected to an average electric field $\bar{E}$=7.5 V/cm showed higher photoeffects as compared with the films made of pigments extracted from vegetals not subjected to eletrostatic fields. We could also notice an increase of the photoeffects inthe microcrystalline chlorophyll films when adding in the pigment extract an ascorbic acid solution with the mean concentration of (0.5-2.5)%.

In Fig. 2 we give the volt-ampere characteristic J = f(V) for the Al-Chl-Cu heterostructure, where the microcrystalline chlorophyll films were obtained by treating the pigment in the electric field $\bar{E}$=7.5 V/cm with an addition of ascorbic acid in a concentraion of 1.5%. For these heterostructures, the mean values for U_{CD} and J_{SC} were U_{CD} =575 mV and J_{SC} = 25 nA/cm^2, respectively (Fig. 2).

For direct polarization (aluminium being the negative electrode) the dark current is higher than for the inverse polarization, the device showing rectifying properties. The photocurrent shows a contrary behavoiour, having small values at direct polarization and high values at inverse polarization (Fig. 2). Within the whole voltage range that we used, the direct dark current showed an exponential dependence on the voltage, and the inverse dark current depends on $V^{1/2}$, a fact that suggests the existence of a Schottky barrier at Al/Chl junction.

Taking into account that the photosynthetic pigments of the "chlorophyll reaction center" type are organized in two systems [6], P_{690} system (λ=690 nm) and P_{700} (λ=700 nm), we consider that these are permanently connected under the influence of photons, by an electron carrying chain. Thus, as the result of high hυ energy absorption, the P_{700} system emits electrons on an intermediate level n_3 of an unknown electronegative acceptor X (Fig. 3).

The P_{700} system regains its lost electron by means of an intermediate compound (suggested to be on the form: enzyme + vitamins C energetic level n'), the π electron excitation levels of the pigment molecules depending on the wavelength of the photon that enters the excitation resonance with the pigment molecule "reaction center", similarly with the optical pumping phenomenon produced in the LASER effect (Fig. 3).

At the same time, during the energy transfer from the chlorophyll molecule P_{690} to the molecule P_{700}, a photon excitation occurs of the elctrons from the magnesium atom (Mg^{++} included in pigment structure), this excitation determining two of the electrons of the atom to populate the higher energy levels, thus leaving the chlorophyll molecule, so that the free electrons with a high energy level can be transported (by means of enzymes and vitamins) through the photosynthetic membrane of the pigment molecules. On taking all this into account, we assign to chlorophyll molecules a p-type semiconduction behavior. We consider that, under the influence of the external electrostatic field $\bar{E}$, the positively charged states at the Al/Chl interface will attract electrons by reducing the space charge from the barrier layer and diminishing the barrier height (Schottky effect) (Fig. 4).

We also consider that both the electric field $\bar{E}$ and the ascorbic acid addition stimulate the protein and enzyme synthesis, resulting in the amplification of the electron transport through the photosynthetic membrane of the chlorophyll molecules, which renders evident a close correlation between the electric field, ascorbic acid, proteic metabolism and photoconduction mechanism in the photosynthetic pigment molecules.

Taking into account the influence of the ascorbic acid on the electron transport through the photosynthetic membrane, we suppose that the Q and X intermediate factors from the P_{690} and P_{700} systems have the form: enzyme + ascorbic acid (Fig. 3).

The experimental results obtained so far, made us come to the conclusion that both the impurities from the I_a and I_b groups (Na, Li, Cu, K), and the photosynthetic pigments can improve the photoeffects at the Cd-Cu_xS junction, by enhancing its photovoltages and photocurrents (U_{CD} = 350-400 mV; J_{SC} = 10-10.5 mA/cm^2).

We think that, under the influence of temperature (100-150°C), the Na_2SO_4 and Cu SO_4 thin films formed during the sputtering process give up the oxygen in the structural lattice of the heterojunction; oxygen diffuses and gets fixed at the CdS/Cu_xS interface, thus determining the generation of a rectifying junction. We also consider that during this process a special part is played by the two physico-chemical parameters, temperature and concetration of the doping solutions, the optimum values that we have found for concentrations being 0.75% for Na_2SO_4 and 1.2% for $CuSO_4$ solutions, respectively (Fig.1).

We consider that in the photovoltaic behaviour of this kind of mixed multijunction, the microcrystalline chlorophyll thin films bring a special contribution. This pigment traps photons and passes them over to different energy levels, by molecular induction or by inductive resonance [1,6], the excitation levels of π electrones of the chlorophyll molecules depending on the wavelength of the photon entered the excitation resonance with the pigment molecule.

We also remark, for this type of mixed multijunction, the SnO_x negative electrode being semitransparent, it is possible to illuminate the photoelement on its both faces (bilateral illumination), which increases the charge carriers collection gain in the Cu_xS film of the heterostructure.

In Fig. 5 we give the volt-amphere characteristic for such a mixed photoelement, where we see an improvement of the maximum values of the open circuit voltage U_{CD} and short circuit current J_{SC}.

We assigned the slight sudden increase of the absolute value of the photocurrent, on the volt-ampere characteristic, to the influence of the R' series resistance of the photoelement (R' being determined by both the semiconductor resistance and the contact resistance) that produces a sudden decrease of the current through the diode for small voltages: V<I R', according to the equation of the volt-ampere characteristic, where the series R' resistance of the photoelement is also considered:

$$I = I_s\left(e\frac{e(V - IR')}{\alpha kT} - 1\right) - I_L$$

where I_s is the saturation current, I_L is the photocurrent, $\propto$ is a constant of the semiconductor material, ranging between 1 and 2.

The experimental results showed that the photovoltaic properties of this kind of mixed photoelement are highly influenced by the duration of the alternative exposition to light and dark, the illumination period playing a special part in the processes occurring at the Al/Chl and Chl/Cu interfaces (Fig. 6).

3. FINAL CONCLUSIONS

Our experimental results made us come to the conclusion that both the photosynthetic pigments and the impurities of the I_a and I_b groups can lead to the improvement of the photoeffects at the CdS-Cu_xS interface, by increasing their photovoltages and photocurrents.

We think that the photovoltaic characteristics of this mixed multijunction are mainly determined by microcrystalline chlorophyll thin films. This pigment traps photons and passes them over to different energy levels, by molecular induction or by inductive resonance phenomena, the electrons excitation energy levels of the chlorophyll molecule depending on the wavelength of the atom entered the excitation resonance with the pigment molecule. By using the analogical modellation, we considered this mechanism as being similar to the optical pumping met at the LASER effect.

On taking into account the Schottky barrier at the metal/chlorophyll interface, we revealed the probability of electronic transfer between two neighbouring pigment molecules by quanto-chemical tunneling mechanism in the unielectronical approximation.

We also remarked that both the mean values of the electric field $\bar{E}$=7.5 V/cm and the mean concentration of the ascorbic acid of 1.5%, stimulate the protein synthesis within the proteic metabolism of the vegetal material, resuting in the intensification of the electronic transfer in the photosynthetic pigment molecule, thus rendering evident a close correlation between the electric field, the ascorbic acid concentration, the proteic metabolism and the photoconduction mechanism in the photosynthetic pigment molecule.

We consider that the explanation of the mechanisms of the energy transfer and electron transport through the photosynthetic membrane, as well as the study of the photoconductive properties of different forms of microcrystalline chlorophyll present a special fundamental and practical interest for the utilization of the photosynthetic pigment films in photovoltaic devices based on anorganic semiconductor materials.

REFERENCES

1. Tang, C.W., Albrecht, A.C., Mol. Liq. Cryst., 25,53 (1974).
2. Tang, C.W., Albrecht, A.C., J. Chem. Phys., 62, 2139 (1974).
3. Tang, C.W., Albrecht, A.C., J. Chem. Phys., 63, 953 (1975).

4. Besson, J., Plaques photovoltaiques au sulfure de Cadmium, L'onde electrique, 1. (1975).
5. Zet Gh., Gherghel Madga, Bul. I.P. Iasi, tom XXIX, fasc. 1-4, 87-91 (1983).
6. Zet Gh., Gherghel Madga, Bul. I.P. Iasi, tom XXX, fasc. 1-4, 101-106 (1984).
7. Bargale B.B., Pawar S.H., Indian J. Pure Appl. Phys., 18, 1, 15-19 (1980).
8. Gherghel Madga, Zet Gh., Ursu D., Vol. "Progrese in Fizica", 541, Institutul Central de Fizica, Bucuresti (1988).

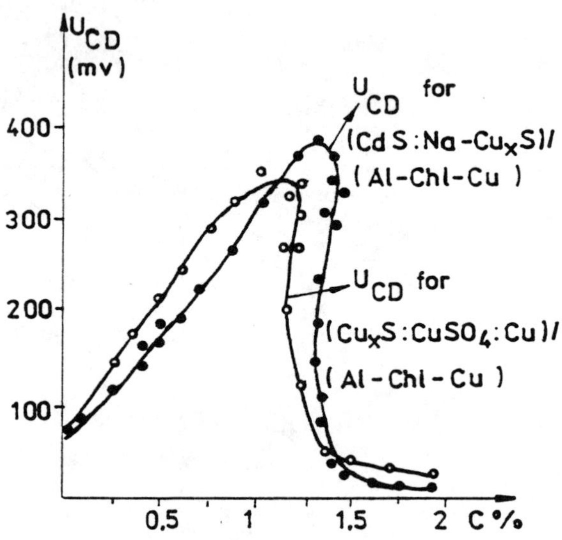

Fig. 1. Variation of photovoltage at room temperature for doped CdS and Cu_xS films

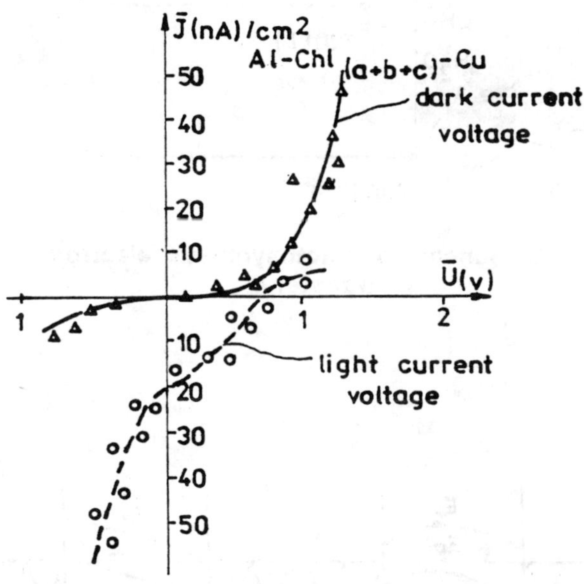

Fig. 2. The J-V characterisitic of Al-Chl$_{(a+b+c)}$-Cu

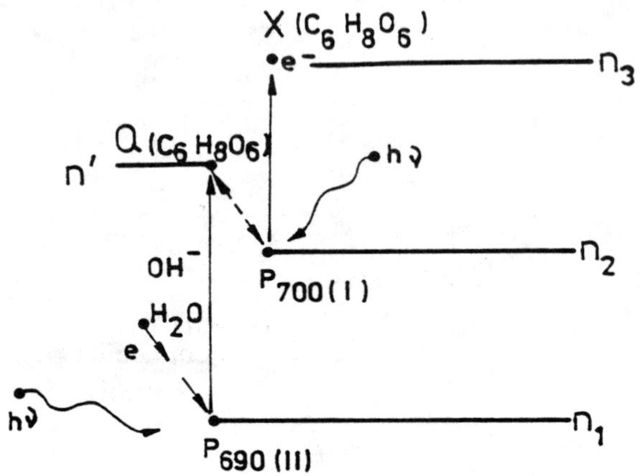

Fig. 3. Scheme of photosynthetic electron transfer

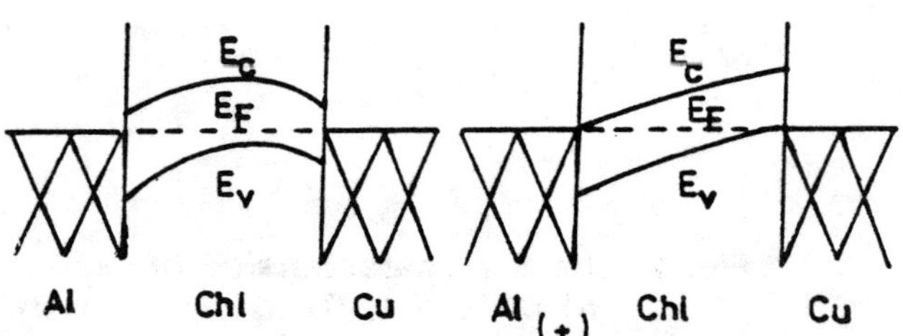

Fig. 4. Energy Scheme of Al-Chl-Cu (E_c - conduction band of Chl E_v - valence band of Chl and E_F - Fermi energy level)

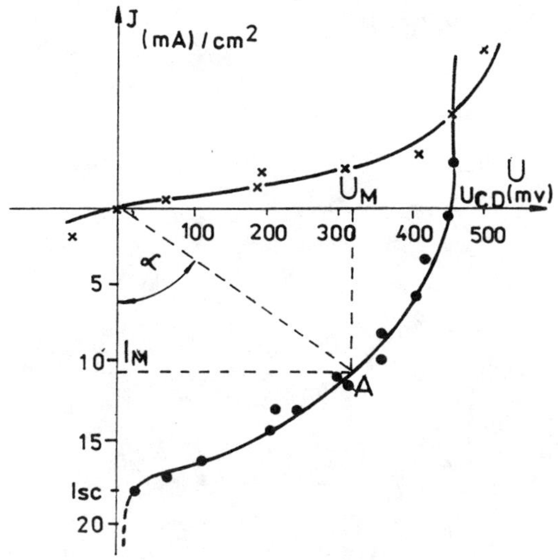

Fig. 5. The J-V characteristic for S_nO_x|Cds:Na-Cu_xS:Cu-Al-Chl|Cu cell

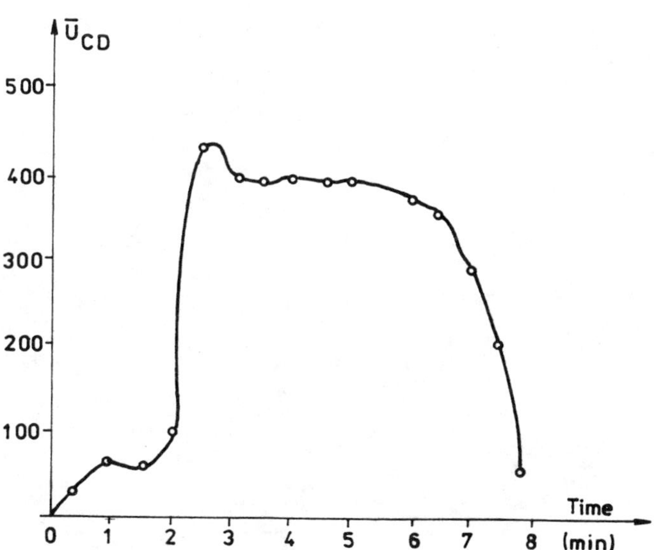

Fig. 6. Variation of photovoltage at room temperature, with illumination time

PHOTOVOLTAIC SOLAR WATER PUMPING IN BANGLADESH

M.A. Quaiyum
Bangladesh Atomic Energy Commission
P.O. Box 158, Dhaka-1000, Bangladesh

Abstract

In Bangladesh there is abundance of water during the rainy season while the country suffers from water shortages during the dry season. Hence, the rainfed agriculture requires to be supplemented by irrigation from surface and ground water. The traditional irrigation system is, however, inadequate. Four Photovoltaic pumping systems located at different latitudes of Bangladesh are described. Test results on the performance characteristics of three such systems are presented. The appropriateness and the social acceptability of solar pumping from users points of view are highlighted. It is concluded that further extensive testing is necessary before a decision is taken regarding adopting photovoltaic irrigation in Bangladesh, in a wider scale.

1. INTRODUCTION

Bangladesh forms the largest delta in the world and is situated approximately between 20.7°N and 26.8°N latitude, and 88.01°E and 92.75°E longitude. The great delta is flat throughout and stretches from near the foot-hills of the Himalays in the north to the Bay of Bengal in the south. Agriculture is the back-bone of Bangladesh economy contributing 47 per cent to the Gross National Product. Yet agriculture depends almost entirely on rain. In fact, during the rainy season there is abundance of water while the country suffers from shortage of water during the dry season when river flows are sharply reduced. The possibility of large scale storage of water is limited because of the flat terrain. Moreover, the rain fall is not uniformly distributed throughout the different regions of the country. Hence, rainfed agriculture is required to be supplemented by irrigation from surface and ground water. There are about 9 million hectares of arable land in the country. Only about 2.35 million hectares are irrigated with surface pumps, deep and shallow tube-wells and traditional means [1]. The rest of the agricultural land depends on seasonal rains only. Approximately 1,30,000 irrigation pumps have been sunk so far. Most of these pumps are diesel powered while about 11,000 are operated with electricity [2]. Diesel powered pumps currently in use are facing problems because of high cost, irregular supply and dearth of skilled manpower. It is in this context that Photovoltaic Pumping in Bangladesh will, from users' point of view, be discussed and test results on the performance characteristics of specific installations will be presented.

2. PRELIMINARY PERFORMANCE STUDY OF BUET

Hossain and others [3] reported preliminary performance study on a small solar photovoltaic pumping system consisting of an array of 280 circular silicon cells, a centrifugal pump and a D.C. motor. The test was done continuously on five days - one day in April, three days in May and one day in August, 1984 under the climatic conditions of Dhaka (Latitude = $23^{o}43'$N). The array was kept in a horizontal position on all days except on the day in April when it was tilted towards the south at an angle of 20° with the horizontal. Array and system overall efficiencies were based on the entire frontal area of the array.

On continuous testing of five days the maximum array power output has been indicated to be 144 W at a temperature of 45°C at a total array-plane solar irradiance of 1100 W/m^2. The rated nominal peak power was, however 250 Wp. Figure 1, 2, 3 and 4 show the performance characteristics of the system. The ranges of instantaneous array, sub-system and overall system efficiencies are 3-7%, 10-58% and 1-3% respectively for a total irradiance range of 250-1100 W/m^2 and total static head range of 2.28-4.11 m. It is concluded that the system would not be able to pump water through a total static head of more than 4.5m. The daily pumped volume of water through a static head of 3.43m is reported to be 34.3m^3 when the daily total insolation is 5.2 Kwh/m^2. The study recommends for extensive testing before considering solar photovoltaic irrigation in Bangladesh.

3. FIELD TEST AT MOUGACHI, RAJSHAHI

Wijlhuizen [4] reported performance study ona solar pumping system consisting an array of 1428 circular silicon cells, a submersible four stage centrifugal pump, an A. C. motor which operates at variable frequency and voltage, and an inverter. The test was done at a place named Mougachi (Latitude = $24^{o}30'$N) which is 19.3 Km north of Rajshahi town. The solar photovoltaic array consists of 6 in parallel and 7 in series connected modules making a total of 42. The total peak power (Wp) is 462 Watts. Gross array exposure area is 7.71 m^2 while cell area is 3.64m^2. Array inclination can be varied from 22° to 45° with the horizontal. The characteristics of discharge, static head and the efficiency of the pump motor combination are shown in Figure 5. The discharge is found to flactuate roughly between 0 to 2.5 l/sec at 5 m static head. The estimated maximum daily pumping rate on a monthly basis was found to be 67 m^3/day which occurred in the month of October. The average flow rate on monthly basis from January to December, 1985 was 47.79m^3/day.

4. FIELD TEST AT SAVAR

The solar module number is 35 (7 in series and 5 in parallel) consisting of 1225 circular silicon cells of diameter 10.16cm. The design maximum output of the panel is 1.4 Kw while actual measured output was 1.1 Kw. The panel was placed at an angle of 33° with the horizontal. The subsystem consists of a submersible 4 stage centrifugal pump powered with an A.C. motor and an inverter. The array is directly connected with the inverter through the main switch and the inverter, which in turn, is connected with submersible pump-motor set. The pump is made of high quality stainless steel. The motor powering the pump is high efficiency submersible motor, which, like the pump, is lubricated by the pump water. The inverter converts the D. C. power of the panel to 3-phase A.C. The efficiency of the inverter is 95%. The well is 45m deep and consists of riser, 15.24cm dia housing pipe, 5.08cm dia stainless steel strainer, 5.08cm blind pipe and a centraliser. The pump is placed at a depth of 15.24m and is connected with the inverter. There is a provision to change and fix the pump at desired levels at different depths.

Test boring was done upto a depth of 45.72m and an well was developed for installing the solar pump. In order to protect the solar array from hailstroms, movable wire mesh fixed on manually operated framed structure was provided. A water flow meter placed in the delivery line recorded the comulative values of the water volume lifted.

The measured solar irradiance at Dhaka by Bangladesh University of Engineering and Technology (BUET) was used as the value of solar insolation at Savar. This is reasonable, because Savar is situated only 40 Km north-west of Dhaka and the latitude of Dhaka ($23^{\circ}43'$N) does not vary significantly with that at Savar. Array power output was measured with the help of an ammeter and a voltmeter and array output was obtained at frequent intervals. No maximum power recording device was installed. Mercury-in-glass thermometers were used to record array front, array back and ambient temperatures. The total cost of the pump system including drilling & development of well and installation was US$ 29,520/00. The CIF cost of imported hardware was US$ 17,374/00. The installation cost including that of drilling and development of well was 41% of the total.

Continuous test run of the system was carried-out from January, 1985 to December, 1988 when the inverter want wrong. Typical performances of solar array, sub-system and the overall system are shown in Figures6 to 9.

The figures show the variations of instantaneous array power output, hydraulic power output, sub-system efficiency, array efficiency and the overall system efficiency against total solar irradiance in the plane of the array. The variation in water flow rate and solar irradiance with time on a bright sunny day is shown in Figure 8. A peak power of 1.1 Kw was recorded at array front temperature of 45°C and a total solar irradiance of 950 W/m^2 though the design peak power output is stated to be 1.4 Kw. Figure 9 shows the daily average array, sub-system and overall system efficiency. The nature of the curves in Figures 6 to 9 are very similar to that of Hossain et al and other works. It may be mentioned here that during the period of four years of operation as mentioned, earlier, except for some small conventional items for measurements, the water pumping system has been found to be maintenance free.

5. SOLAR PUMPING SYSTEM AT SYLHET:

A solar panel with a rated peak output of 2300 Watts to drive a centrifugal pump has been installed at a tea garden in Sylhet (Latitude = 25°N) for irrigation of its nursery. The system is being operated and the relevant data are being collected so as to establish the performance characteristics of the same.

6. CONCLUSION:

Of all the possible applications of solar energy its use for irrigation seems to be an ideal. The supply and demand for the energy are more or less closely matched. There is a natural relation between the availability of solar energy and the requirements of water for irrigation. The requirements grow during the periods of hot weather when the sun shines brightly and the solar array output is at its peak. A very small portion of arable land is irrigated in Bangladesh. The number of electric pumps is still less. The rural electrification programme is, in near future, not going to cover many isolated areas of the country. Hence, there is plenty of scope for experimentation in the use of alternative technologies such as photovoltaic solar water pumping. Systematic field test in this respect, has so far been very limited. Although the public may not immediately understand new technology and may be very apprehensive at the start, they are the best judges for the appropriateness and the social acceptability of any technology. They will in the long run accept a technology if it is reliable and has cost advantage, in terms of initial investment, operation and maintenance over the existing ones.

There is an ample scope for further electrification of irrigation pumps in Bangladesh. Photovoltaic pumping is an alternative to electrification. Hence, the work on photovoltaic pumping needs further field studies so as to generate sufficient data on the performance, cost economics and social acceptability of these pumps. These data will facilitate to draw a definite conclusion on the feasibility of applying solar pump for irrigation.

From the operating experience of BAEC it can be said that the initial investment and balance of the plant (BOP) components, like inverters and central control unit are, at present, the main problems in photovoltaic solar water pumping in Bangladesh. Solar panels have so far been found to be maintenance free. Hence, with the predicted reduced cost of @ $ 2/Wp by 1995 [5] and indeginous supply of BOP compenents such system may be viable in the future. It is, however, concluded that extensive field testing is necessary before a decision is taken regarding adopting photovoltaic irrigation in Bangladesh in a wider scale.

REFERENCES

1. Bangladesh Bureau of Statistics. Monthly Statistical Bulletin of Bangladesh. July, 1989.

2. Rural Electrification Programme, Bangladesh,1987-88. Rural Electrification Board Bangladesh, Dhaka, November, 1988.

3. Hossain, M.A., Helali, M., Mesbahuddin, A.K., Ahmed,S. "Preliminary performance study of a small solar photovoltaic pumping system." Paper presented at the International Conference on Solar and Wind Energy Applications. Beijing, China, 3 - 7 August, 1985.

4. Wijlhuizen, E. "Solar Pump; Mougachi." Christian Community Development (CCDB) Solar Energy Experimentation. Rajshahi Report, 1986.

5. Bentley, R.W. "Photovoltaic Solar Water Pumping as of 1988." Paper presented at Dhaka held in 31st October to 13th November, 1988.

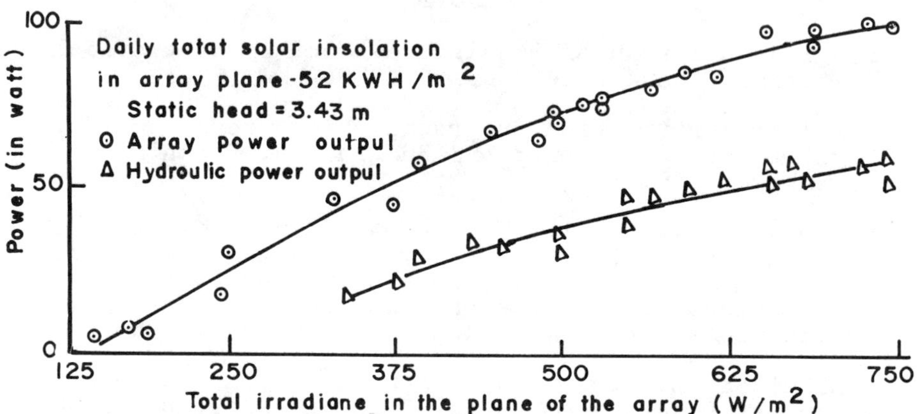

Fig. 1. Variation in array and hydraulic power outpul With total irradiance on 12 april 1984

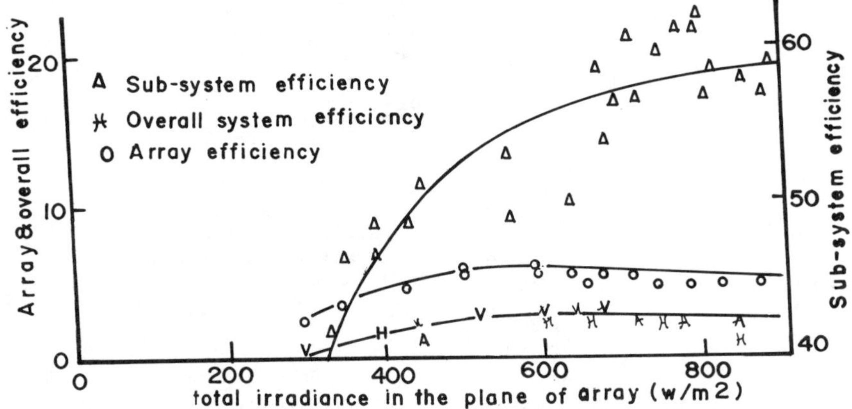

Fig.2. Variation in efficiencies with total solar irradiance on 12 april 1984

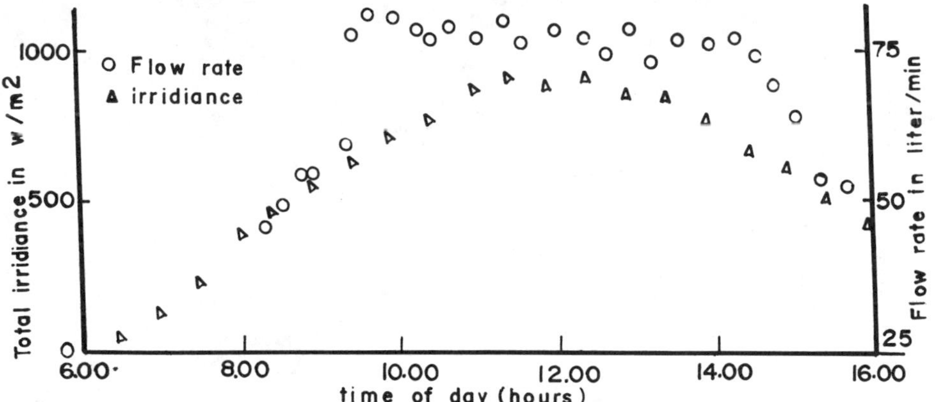

Fig. 3. Water flow rate and solar irradiance in the plane of the array on 12 april 1984

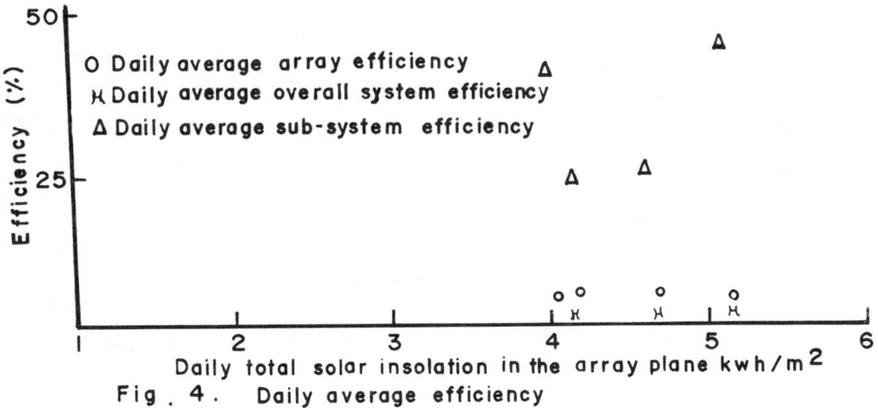

Fig. 4. Daily average efficiency

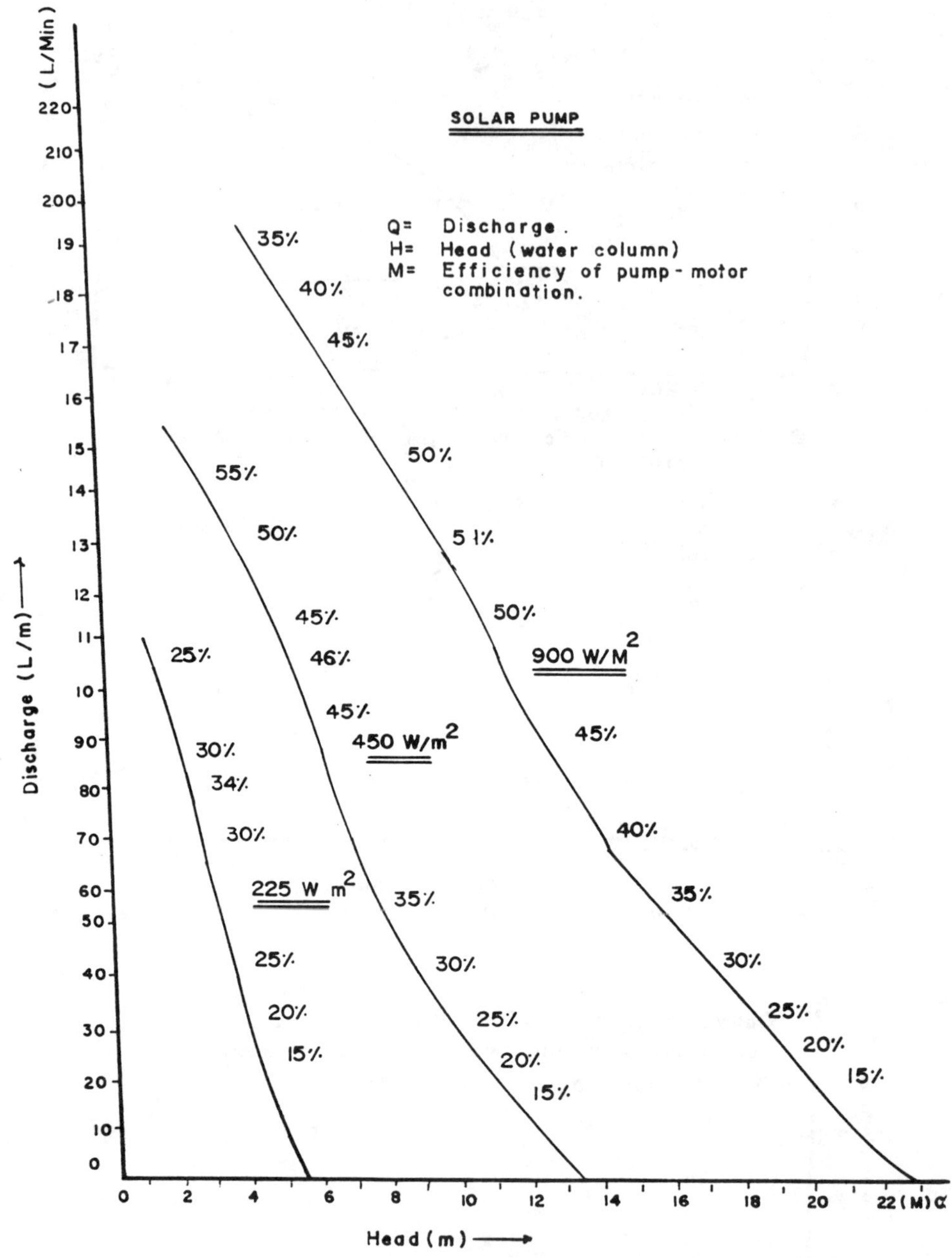

Fig. 5. H . M characteristic at dafferent radiation levels.

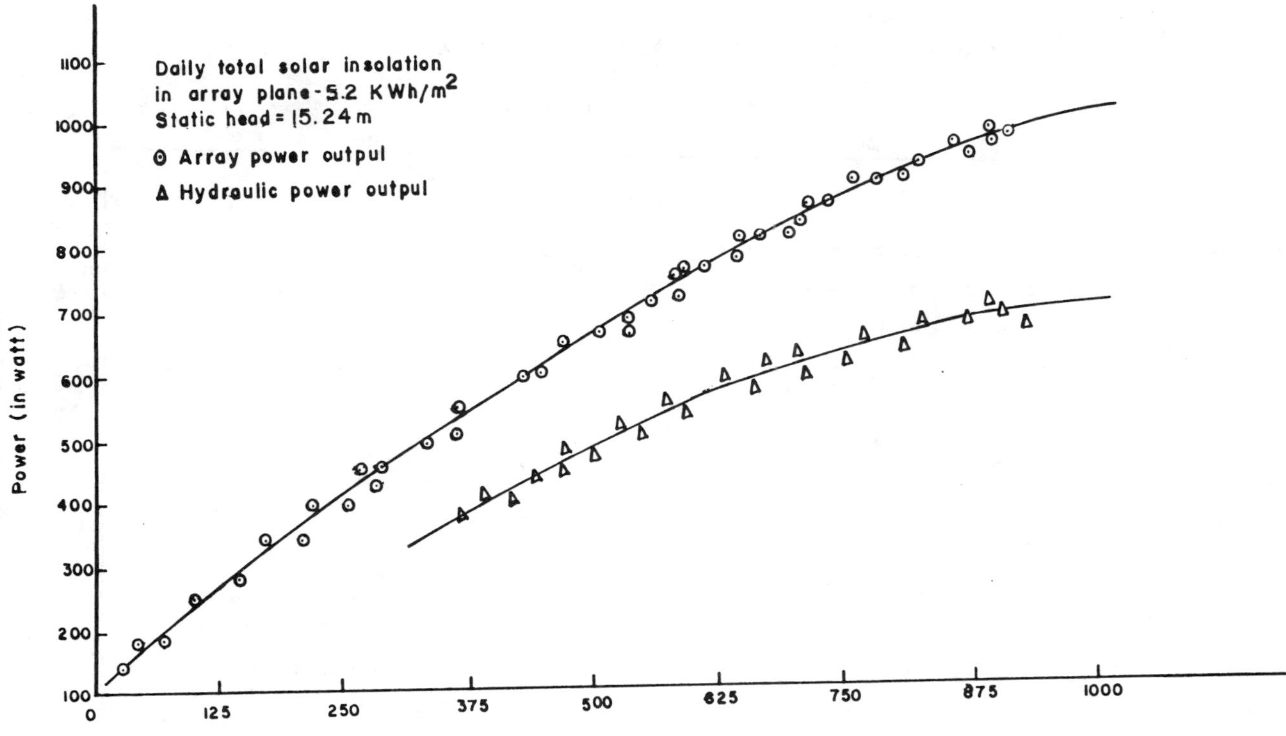

Total irradiance in the plane of the array (W/m^2)

Fig.6. Variation in array and hydraulic power output With total irradiance.

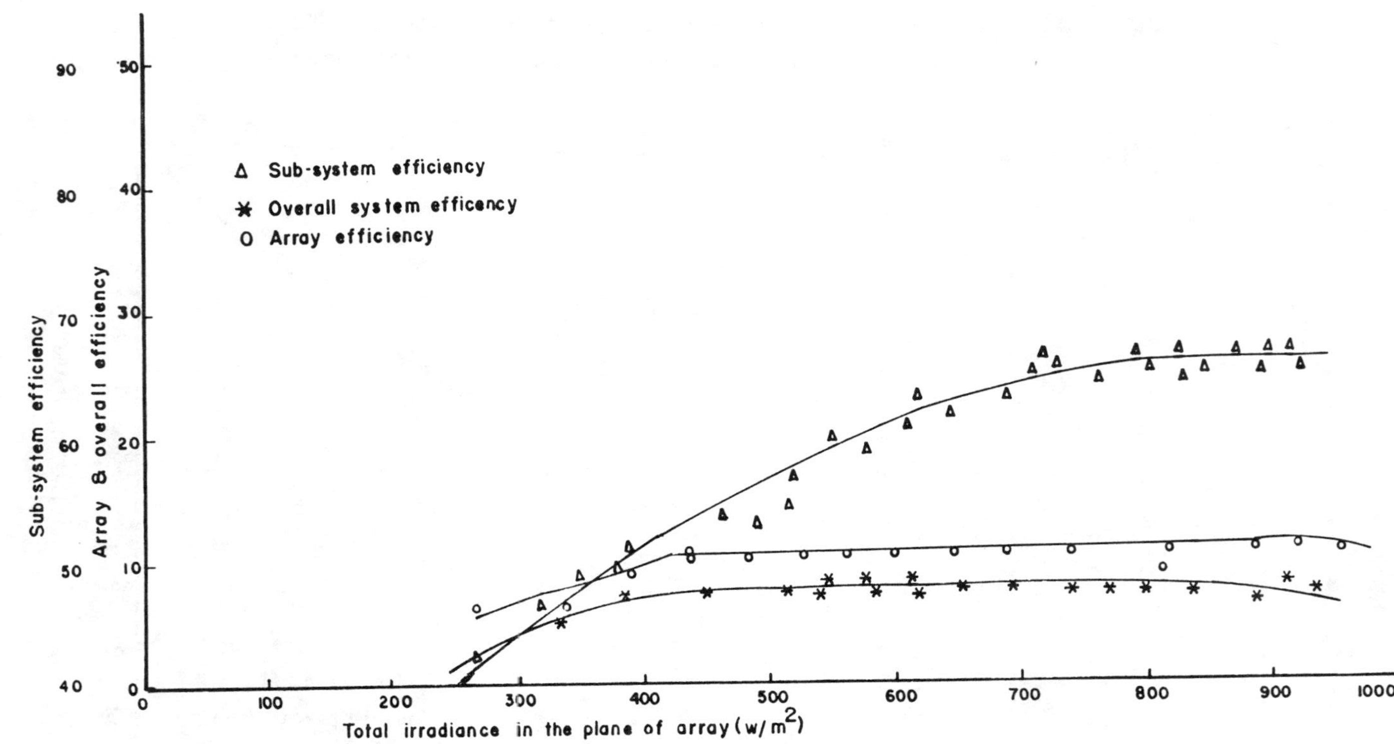

Fig. 7. Variation in efficiency with total solar irradiance.

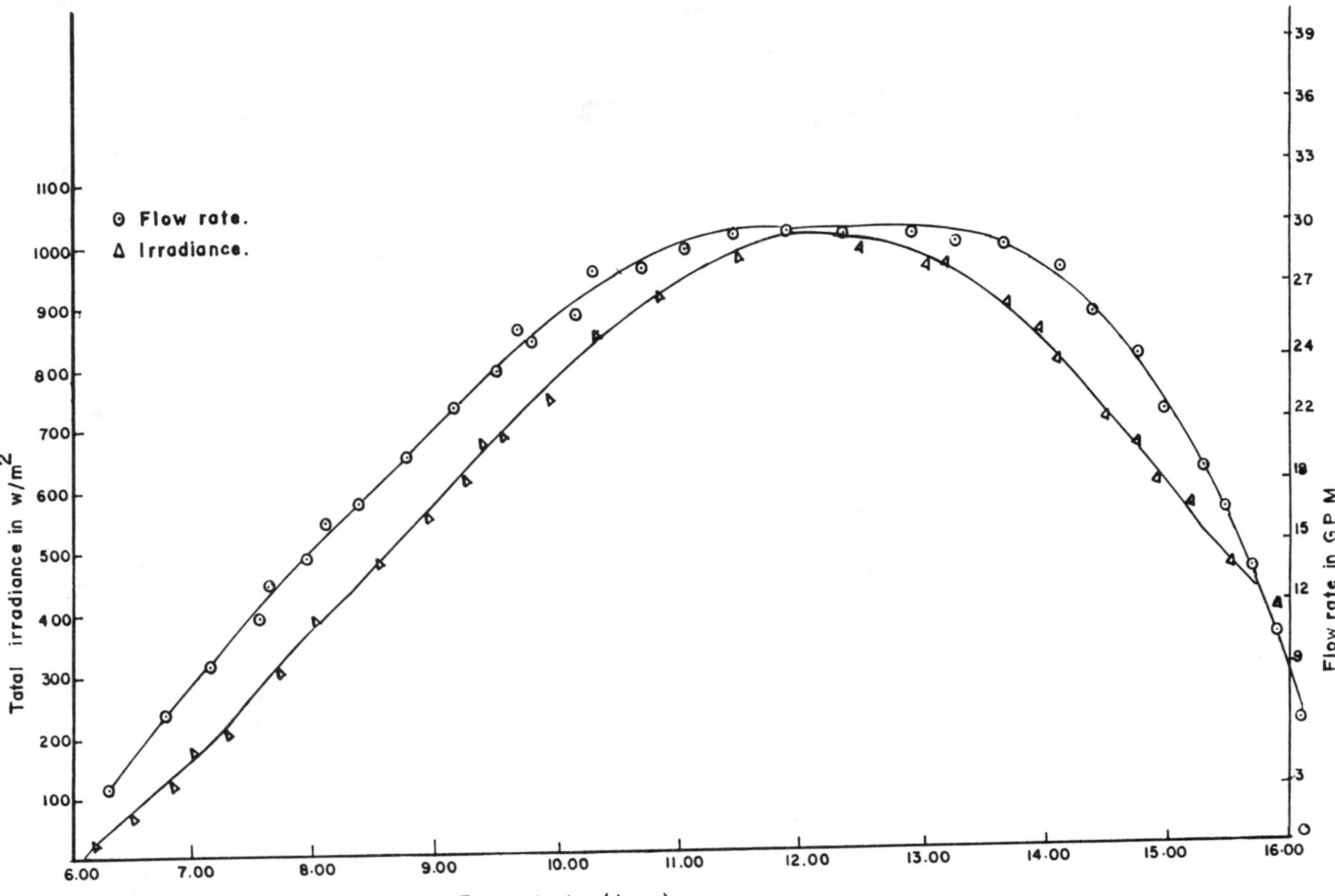

Fig : 8 . Water flow rate and solar irradiance on the plane of the array.

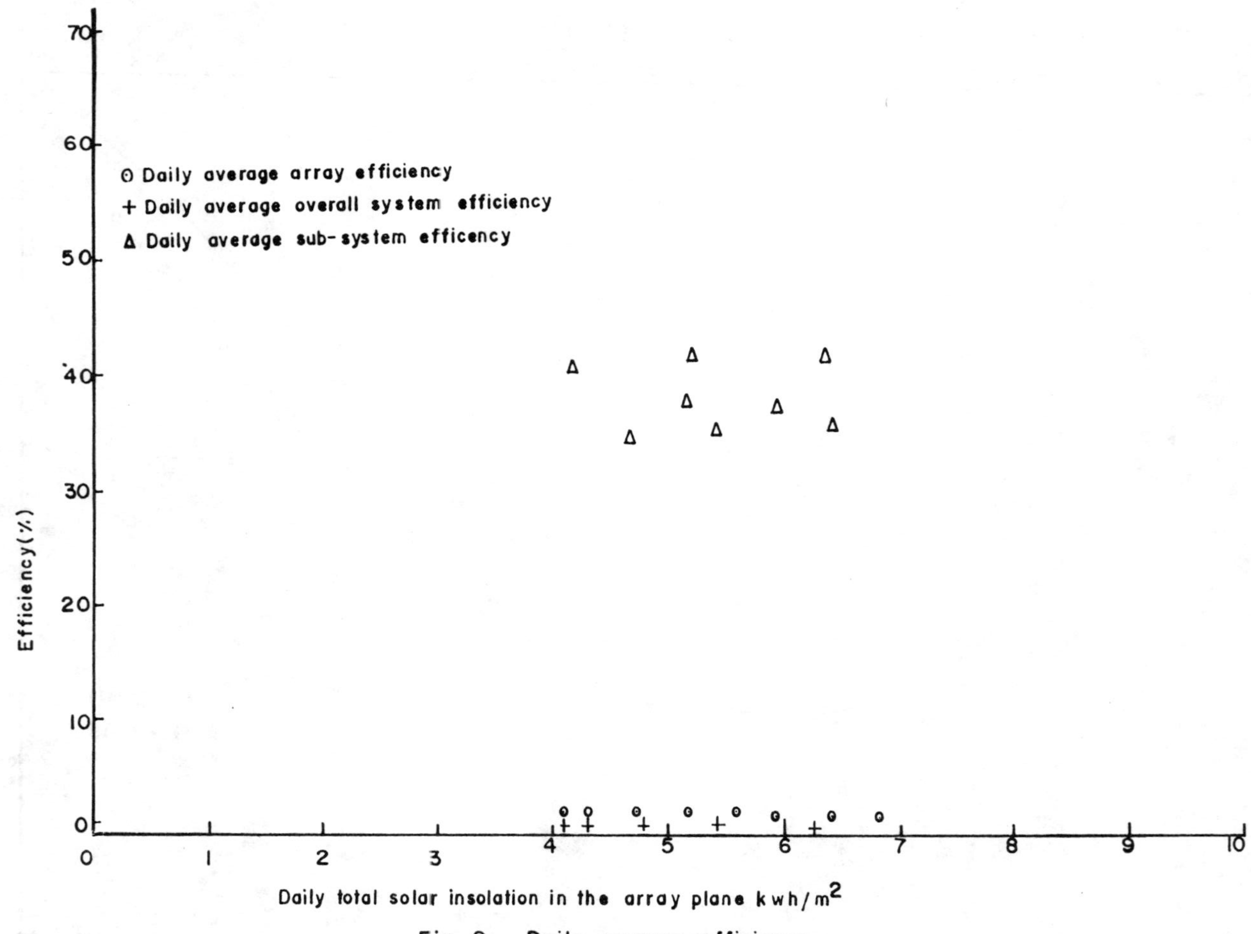

Fig. 9 . Daily average efficiency.

THE DESIGN OPTIMIZATION OF PHOTOVOLTAIC/THERMOELECTRIC REFRIGERATOR

D.Z. Chen and B.B. Wang
Dept. of Power & Energy Engineering
Xi'an Jiaotong University, Xi'an, Shaanxi Province, P.R. China

Abstract

The energy conversion processes and physical model of a photovoltaic thermoelectric refrigerator (PTR) are described in this paper. Using mathematical theory of optimization, an optimal design method of such a refrigerator are analyzed. And some results are presented. The proposed method may also be applied to other thermoelectric cooler with forced convection of air stream

INTRODUCTION

In developing countries and some remote areas where electrical supply are not available, then refrigeration may be made by combining solar power and thermoelectric cooling. Having the advantages of being small size, light weight and easy transportation, PTR is very suitable in areas for foods and medical drugs storage especially for some fieldworks

There are several refrigeration methods by solar power. The PTR disscussed here has some advantages over the others as follows: a) There are no moving-parts or fluids; b) A portable refrigerator being simple in structure can be made easily; c) photovoltaic cells produce DC current and TE requires DC current, so there is a natural opportunity for matching the output of an array of photovoltaic cells to a TE refrigerator.

The lower efficiency and the sensitivity to design variables lead to design of TE being more strict than other refrigeration. Moreover, PTR operate in field where temperature is much higher than indoor temperature, the cost of PV cells is higher and the efficiency is lower, all of these bring so much difficulties to the design of TE coolers. At present, most of TE design is still made by means of graphs and tables. Several graphs and curves are used to find a better solution. But as we know such method is not formulated. Only an approximate optimum solution may be made randomly and the work is also very complicated. Furthermore, it is very difficult to analyze the off-design performance.

The aim of this paper is to establish a complete mathematical model of the problem mentioned above, using mathematical theory of optimization, to present an optimal design method of such a refrigerator. A corresponding computer program is also presented here.

PHYSICAL AND MATHEMATICAL MODEL

1. Energy Convertion Process and Physical Model

As implied by Fig. 1, the system consists mainly of solar panels, batteries and thermoelectric refrigerator. Its thermal profile is illustrated on the left bottom of Fig. 1. At the right, the net of thermal-electrical simulation is given. In the graph, the heat flux is simulated to electrical current. The temperature difference across the TE pellets is:

$$\Delta T = \Delta T_h + \Delta T_{ho} + \Delta T_{oi} + \Delta T_{ic} + \Delta T_c = (T_h - T_h') + (T_h' - T_o) + (T_o - T_i) + (T_i - T_c') + (T_c' - T_c) \tag{1}$$

It is corresponding to the total voltage in the circuit. The relation of heat resistance and temperature difference can be decided according to that model.

1) Insulation Leak

To treat the problem by average method, thus:

$$(T_o - T_{Bo}) \cdot A_o \cdot h_o + I \cdot \alpha_s \cdot A_s +$$
$$+\alpha_o \cdot \sigma_b \cdot (T_o^4 - T_{Bo}^4) \cdot A_o = Q_o \tag{2}$$

$$(T_{Bo} - T_{Bi}) \cdot s \cdot k = Q_o \tag{3}$$

$$(T_{Bi} - T_i) \cdot h_i \cdot A_i = Q_o \tag{4}$$

The above equations can be combined to one:

$$Q_o = (T_o - T_i) / (R_o + R_k + R_i)$$
$$= (T_o - T_i) / (1 / A_o \cdot \overline{h}_o + 1 / s \cdot k + 1 / A_i \cdot h_i) \tag{5}$$

where

$$\overline{h}_o = h_o + (I \cdot \alpha_s) / (T_o - T_{Bo}) + \alpha_o \cdot \sigma_b (T_o^3 - T_{Bo}^3) \tag{6}$$

is called "effective heat transfer coefficient" on the outer wall. Because $R_k = 1 / s \cdot k$ is the main heat resistance from outer of box to inner, the R_o, R_i can be neglected, thus:

$$Q_o = (T_o - T_i) / R_k = k \cdot s \cdot (T_o - T_i) \tag{7}$$

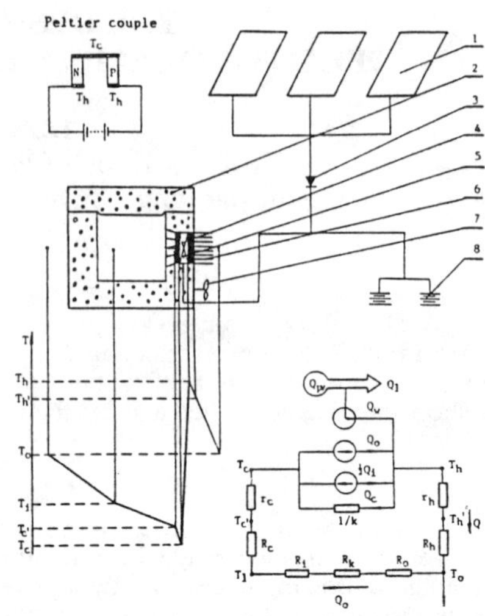

Fig. 1. A Physical Model of PTR

1. solar panels 2. cold box
3. diode 4. cold sink
5. TE modul 6. heat sink
7. fan 8. battery

2) Thermoelectric modules (TEM)

Three kinds of TEM operating condition are taken into account:

a) Single-stage Cascaded TE on the Maximum Coefficient of Performance (COP) Condition

$$\eta_{max} = \frac{T_c}{T_h - T_c} \cdot \frac{\left[1 + \frac{Z}{2}(T_h + T_c)\right]^{\frac{1}{2}} - T_h / T_c}{\left[1 + \frac{Z}{2}(T_h + T_c)\right]^{\frac{1}{2}+1}} \tag{8}$$

b) Tow-stage Cascaded TE on the Maximum COP Condition

$$T_m = \sqrt{T_h \cdot T_c} \tag{9}$$

$$\eta_{max.1} = \frac{T_c}{T_m - T_c} \cdot \frac{\left[1 + \frac{Z}{2}(T_m + T_c)\right]^{\frac{1}{2}} - T_m / T_c}{\left[1 + \frac{Z}{2}(T_m + T_c)\right]^{\frac{1}{2}+1}} \tag{10}$$

$$\eta_{mox.2} = \frac{T_m}{T_h - T_m} \cdot \frac{1 + \frac{Z}{2}(T_m + T_h) - T_h / T_m}{1 + (T_m + T_h) + 1} \tag{11}$$

$$\eta_{max} = \frac{1}{(1+1/\eta_{max\cdot1})\cdot(1+1/\eta_{max\cdot2})-1} \tag{12}$$

c) Single-stage Cascaded TE on Maximum Cooling Condition

$$\eta_Q = \frac{1}{2}(1-\Delta T/\Delta T_{max})\cdot(1-\Delta T/T_h) \tag{13}$$

3) Heat Sink

Considering the characteristics of a portable refrigerator which should be suitable for field-work, the heat sink employs forced air cooling heat exchanger with rectangular cross section fins. For simplification, the model may be treated as a one-dimensional problem, thus:

$$Q = h\cdot(F_f\cdot\eta_1 + F_b)\cdot(T_h - T) = h\cdot(F_f\cdot\eta_1 + F_b)\cdot\Delta T_m \tag{14}$$

$$W_i = G/\rho\cdot(\Delta P + u^2\rho/2) \tag{15}$$

here convection film coefficient h and pressure drop ΔP are the function of fin size and air flow velocity.

4) Cold Sink and Contact Heat Resistance

It is clear in Fig. 1., that

$$Q_o = (T_c - T_i)/R_c \tag{16}$$

2. The Quality Standard of Refrigerator

Usually, the quality of a refrigerant device can be evaluated from three aspects, that is, efficiency, size and price.

1) Device Efficiency

η_t is the efficiency of a whole device which is given by:

$$\eta_t = \eta_c\cdot\eta_{PV}\cdot\eta_e = \eta_{PV}\cdot\eta_e\cdot\frac{Q_o}{Q_o/\eta^{+W/\varepsilon}} \tag{17}$$

2) Device Size

The size of a refrigerant box depends on the capacity and is irrelative to the process of optimization. The size of solar panels depends on its area of a solar cell array which is:

$$F = Q/\eta_t\cdot E_P \tag{18}$$

Therefore, when the device efficiency is reached to a highest value, the area of solar panels required is smallest, e.g. the size of device is smallest.

3) Device Price

The device price is the price sum one of TE module, solar cells, cold box and other accessories. The price of box is held constant. The number of TE couples which determined its price depends roughly on its operating condition and the value l/s (couple height divided by module area).

It is known well that the price of solar panels depends on its area. Therefore, the smallest panels area means the lowest cost of device. In summary, both size and cost of PTR depend on device efficiency. In turn, η_e depend mainly on the quality of solar cells themselves, so the cooling efficiency of TE η_c can be considered as the operating quality standard of device. Maximizing η_c means optimizing the device efficiency, size and cost in a certain condition.

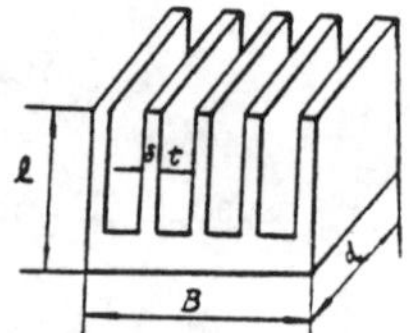

Fig. 2. Heat exchanger

MATHEMATICAL MODEL OF OPTIMIZATION

1. Object Function

The cooling efficiency of TE is the objective function in the problem since it is one of the most important parameters in the design procedure. Objective function is:

$$\eta_c = Q_o / (Q_o / \eta^{+W/\varepsilon} \tag{19}$$

2. Solution Variables

It is known by analyzing eq. (19) that Q_o is decided by the design required, and ϵ is fan efficiency, those two are all constants at optimizing process. Additionally, the work of air flow resistance $W = f(u,t,\delta,1)$, COP $\eta = f(T_h)$, so eq. 19 can be rewritten as follows:

$$\eta_c = f(T_h, u, t, \delta, 1) \tag{20}$$

where T_h,u,t,δ, and l are the variables in this problem. It is well known that the lower is T_h, the higher will be η and η_c. To reduce T_h, we must optimize the variables mentioned above when the heat load is kept constant. Moreover, except for η, η_c is also concerned with W, so the significance of optimization is maximizing η_c by changing T_h,u,t,δ, and l.

3. Constraints

Five solution variables mentioned above are not dependent. In fact,

$$Q = h \cdot (F_f \cdot \eta_c + F_b) \cdot \Delta T_m = h \cdot (F_f \cdot \eta_c + F_b) \cdot (\Delta T_1 - \Delta T_2) / \ln(\Delta T_1 / \Delta T_2) \tag{21}$$

$$\Delta T_1 = T_h - T_o \tag{22}$$

$$\Delta T_2 = T_1 - Q / (C_p \cdot G) \tag{23}$$

To eliminate Q from three equations above and after introducing the relation: $Q = Q_o(1 + 1/\eta)$, obtain:

$$T_h - T_o - \frac{Q_o(1+1/\eta)}{C_p \cdot G} \cdot \frac{\exp[h \cdot (F_f \cdot \eta_1 + F_b) / C_p \cdot G]}{\exp[h \cdot (F_f \cdot \eta_1 + F_b) / C_p \cdot G] - 1.} = 0. \tag{24}$$

That is the constraint equality of the problem. To make sure that heat transfer at surface with fins is much more effective than one without fins, the following inequality constraints are given by:

$$\lambda / (\delta \cdot h) \geq 2.5 \tag{25}$$

Considering fin efficiency, we have:

$$1 > m \cdot l > 0.5 \tag{26}$$

Processing and manufacture of fins gives the following constraints

$$t \geq t_{min} \tag{27}$$

$$\delta \geq \delta_{min} \tag{28}$$

$$l \leq l_{min} \tag{29}$$

4. Complete Mathematical Description

In summary, we can get the complete mathematical description of design optimization of TE with forced air cooling.

$$\min \eta_c = -Q_o / (Q_o / \eta^{+W/\varepsilon})$$

$$\text{s.t.}\, T_h - T_o - \frac{Q_o[1 + 1/(1 + 1/\eta)]}{C_p \cdot G} \frac{\exp[h \cdot (F_f \cdot \eta_l + F_b)/(G,.C_p)]}{\text{esp}(h \cdot (F_f \cdot \eta_l + F_b)/(G \cdot C_p)] - 1} = 0$$

$$\lambda / h \cdot \delta - 2.5 \geq 0$$

$$t - t_{min} \geq 0$$

$$\delta - \delta_{min} \geq 0$$

$$l_{max} - l \geq 0$$

$$m.l - 0.5 \geq 0$$

$$1 - m.l \geq 0$$

MATHEMATICAL METHOD AND RESULT ANALYSIS

1. Mathematical Method of Design Optimization

The "effective constraints set method" is applied for changing an inequality constrained problem into an inequality constrained one. Then, "generalized Lanrangian multipliers method" is used for changing equality constrained problem into unconstrained one. Finally, the solution is made by means of "BFGS" method (one of variable metric method). The result of calculation shows that the combined method mentioned above is very efficient. Its stability, convergence and precision is quite satisfactory. The block diagram of the computer program is shown in Fig. 3. The program is suitable for all TE device sub-assembled heat exchanger with rectangular cross section fins for air forced cooling.

2. Result of Calculation and Example

The result of optimizing calculation under various design requirements and the effect of solution variables deviation to optimal value can be found in Tab. 1 and Fig. 4., which indicated that the solution of unconstrained optimization is not practical because both fin spacing and thickness is too small, and fin height is not reasonable specified. Therefore the dimension of fins must be restrainted, and then the constrained optimization method should be used for this problem. Moreover, the temperature difference ΔT_{ho} between hot side temperature T_h and ambient temperature T_o are about 7°C when TEM operate at maximum efficiency condition. And ΔT_{ho} is larger at maximum cooling condition. The fan efficiency has little effect on the result.

The four curves represent the variation of η_{cop} with variables u,t,δ and l, respectively. It is shown that point * is the maximum efficiency subject to the constrains. So the results of calculation are reliable. In the region where constrains are not contented, the optimal solution of variables t,δ and l are not corresponding to the maximum efficiency. However, the value of η^*_{cop} is approximative to the maximum value, which proved that the selection of constrains concerned with either efficiency or other factors is reasonable.

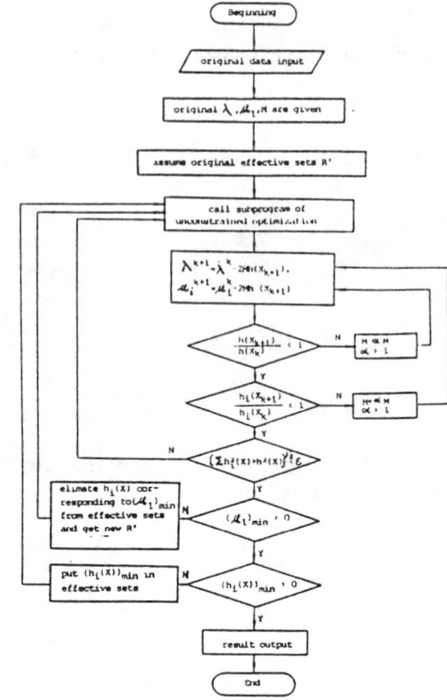

Fig. 3. Block Diagram

The effect of solution variables on optimum value is shown in Fig. 4.

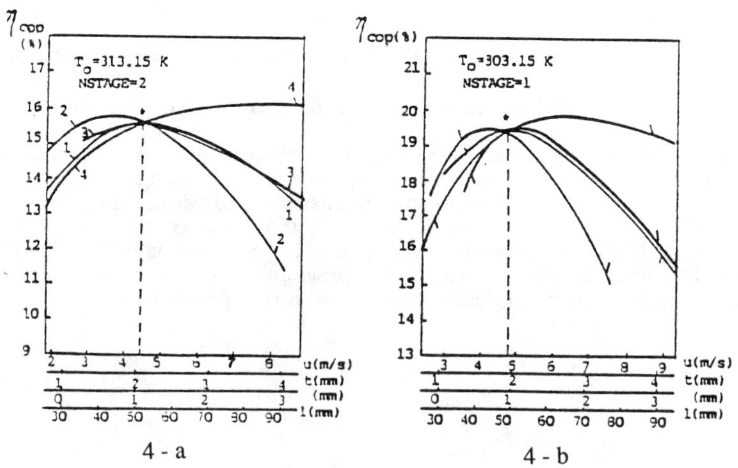

4 - a

4 - b

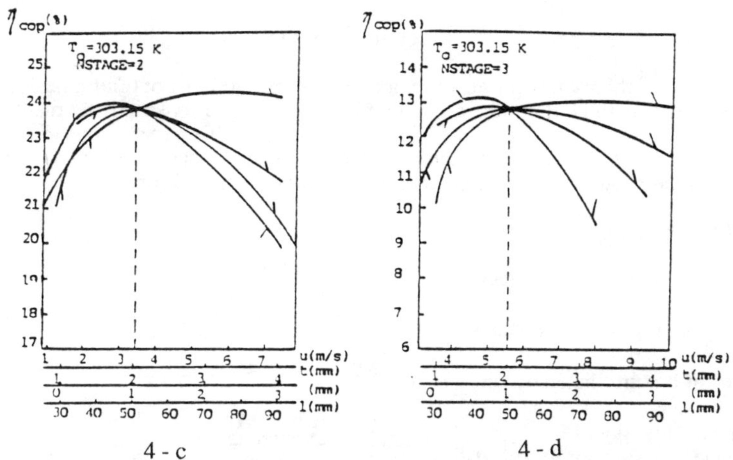

Fig. 4. Variation of η_{cop} with optimizing parameters

Table 1. The result of calculation

NSTAGE	u (m/s)	t (mm)	δ (mm)	ℓ (mm)	T_h (k)	η (%)	η_{cop} (%)	Constraint
1	0.6486	0.5000	0.3213	51.57	319.2	11.26	9.758	nothing
2	0.4047	0.6001	0.3470	55.19	318.8	18.01	17.21	
3	0.3546	0.3357	0.2048	108.1	319.2	9.145	8.182	

design requirement	NSTAGE 1				NSTAGE 2				NSTAGE 3				Constraint
	u (m/s)	T_h (k)	η (%)	η_{cop} (%)	u (m/s)	T_h (K)	η (%)	η_{cop} (%)	u (m/s)	T_h (K)	η (%)	η_{cop} (%)	
Q_o= 5.981(W)	12.09	322.3	8.472	6.398	4.382	321.2	16.53	15.59	16.07	323.4	6.571	4.774	$\frac{\lambda}{h\cdot\delta} \geqslant 2.5$
4.981	9.323	321.0	9.627	8.174	3.789	320.4	17.01	16.14	11.19	322.2	7.316	6.051	
3.981	7.012	320.0	10.58	9.40	3.191	319.6	17.51	16.71	8.087	321.0	8.087	7.133	$t \geqslant t_{min}$
T_o= 313.15(k)	12.09	322.3	8.472	6.398	4.382	321.2	16.53	15.59	16.07	323.4	6.571	4.774	$\delta \geqslant \delta_{min}$
308.15	6.834	315.5	14.95	13.45	3.867	315.5	20.36	19.27	8.100	317.0	10.53	9.435	$\ell \leqslant \ell_{max}$
303.15	4.888	309.6	22.18	20.59	3.429	309.9	25.16	23.88	5.794	311.2	14.29	13.32	
$\varepsilon\ell$ = 0.2	11.59	322.7	8.159	5.735	3.900	321.6	16.26	15.21	15.42	324.0	6.245	3.797	
0.3	12.09	322.3	8.472	6.398	4.382	321.2	16.53	15.59	15.72	323.7	6.421	3.899	
0.4	12.53	322.1	8.696	6.834	4.764	320.9	16.71	15.83	16.07	323.4	6.571	4.774	
Z= 0.0015(1/k)	9.982	324.9	6.568	5.669	7.490	318.1	9.281	8.332	5.939	311.7	12.72	11.65	
0.0020	4.382	321.2	16.53	15.59	3.867	315.5	20.36	19.27	3.429	309.9	25.16	23.88	
0.0025	3.088	320.3	26.08	24.97	2.826	314.6	30.92	29.64	2.591	309.3	36.94	35.42	

T_o	NSTAGE 1															Constraint
	u	T_h	δ	η	η_{cop}	u	T_h	δ	η	η_{cop}	u	T_h	δ	η	η_{cop}	$\frac{\lambda}{\delta\cdot h} \geqslant 2.5$, $t \geqslant t_{min}$, $\delta \geqslant \delta_{min}$, $\ell \leqslant \ell_{max}$
308.15	7.176	316.3	1.693	14.15	12.54	4.016	315.8	1.358	20.14	19.10	8.888	318.0	1.837	9.892	8.664	$m\ell \geqslant 0.5$
303.15	4.821	310.1	1.454	21.46	19.81	3.520	310.2	1.291	24.93	23.72	5.857	311.9	1.567	13.81	12.77	$m\ell \leqslant 1$

An example of PTR being designed by the method discussed here is given as follows.

The cooling unit employs two stage cascaded TE which is designed on the maximum COP condition. Assume the refrigerator is used in Yanan, China. The energy storage can be used for one cloudy day. The calculation of solar radiation and relative data can be found in Ref.2. The results of optimum design is shown in Table 2.

CONCLUSION

According to theory of optimization, a computer program of optimum design of PTR device has been made. The method of optimization (combined "effective constrains set method","generalized Lagrangian multipliers method" and "BFGS" method) used in this paper seems efficient. It is proved that the optimum design of TE device can be made by means of mathematical analysis method. The result of calculation shows that forced air cooling is suitable for a PTR device .

NOMENCLATURE

A_i: area of box's inner surface $[m^2]$
A_o: area of box's outer surface $[m^2]$
A_s: area of box irradiated by sun light $[m^2]$
G: rate of flow [kg/s]
h_i: convective film coefficient at box's inner side [w/m.K]
h_o: convective film coefficient at box outer side [w/m.K]
I: intensity of solar radiation [w/m]
Q: heat load of heat exchanger. [w]
r_c,r_h: contact resistance, [w/K]
R_c,R_h: resistance of cold and hot exchanger. [w/K]
s: shape factor, [m]
z: material figure of merit [l/K]
α_s: absorptivity of solar radiation,
α_o: absorptivity of environment
η: TE cooling eff.
η_c: TE device (included heat exchanger) eff.
e: eff. (due to battery loss etc.)
ε: fan eff.

Table 2. The results of calculation

No.	Parameter	Sign	Unit	Value
1.	heat load	Q_o	w	5.918
2.	ambient temp	T_o	K	313.15
3.	box's inner temp	T_i	K	273.15
4.	box's Vol	V	l	6.000
5.	box's heat cond.	k	w/m·k	0.025
6.	P.V. eff.	η_{PV}	/	0.100
7.	Seebeck coef.	α	v	200.0
TE module				
8.	hot side temp	T_h	K	321.2
9.	cold side temp	T_c	K	264.7
10.	current	I	A	3.500
11.	voltage	U	V	10.27
12.	efficiency	η	/	0.1653
heat exchanger				
13.	fin thickness	δ	mm	1.000
14.	fin height	l	mm	50.00
15.	specing	t	mm	2.000
16.	air velocity	u	m/s	4.382
17.	fan eff.	ε	/	0.400
18.	cooling eff.	η_c	/	0.1559
device				
19.	battery cap	Q_d	A.h	94.20
20.	array area	F	m^2	2.660
21.	total eff.	η_t	/	0.014

REFERENCE

1. B.B.Wang, "The Design Optimization of a Thermoelectric Refrigerator Using Photovoltaic Cells as a Power Source", M.D. Thesis of Xi'an Jiaotong Univ., 1987

2. R.Q. Cui, "The Calculation of Solar Radiation and the Design of Photovoltaic System" Science Report of Xi'an Jiaotong, Parv. 85-598.

SOLAR POWER

A SOLAR BOWL ELECTRIC POWER PLANT FOR NORMAL SOLAR FLUX

M.A.K. Lodhi*, M. Yusof Sulaiman and Shaharin Ibrahim
Department of Physics, Universiti Pertanian Malaysia
43400 UPM, Serdang, Selangor, Malaysia

Abstract

A solar bowl (SB) device is characterized by a fixed concentrator and a tracking linear receiver. The solar rays reflected from the concentractor are focused on a linear receiver moving across the bowl. An improved version of the first solar bowl is designed for operating at University Pertanian Malaysia on an experimental basis. The average daily solar flux has been measured for several places in the area. Thus the calculations and design have been made on the daily average solar flux of 850 Wm^{-2} for making a shallow bowl with a 32 m rim span and rim angle of 40^0 for a latitude of 3^0N. A 6 m deep pit of spherical curvature is to be excavated and a concrete surface is placed directly on the stabilized pit surface area. Small flat mirror glasses are glued directly onto the concrete much like tiling a floor. A hole at the bottom of the bowl is made to let the rain water run into the drainage system.

The cost of construction estimated for two solar bowls of the same size are compared. One is estimated for construction in the USA and the other in Malaysia. The cost of construction in Malaysia is about half of that in the USA.

A further improvement in the efficiency of the plant by about 1.5 to 2% is planned to be achieved by adding an iris on the rim of the bowl to catch the solar radiation when the sun is at small angle above the horizon. The iris is to slide on the rim of the bowl to track the sun.

1. INTRODUCTION

The solar bowl (SB) concept is characterized by a fixed concentrator and a tracking linear receiver (Fig. 1). The solar rays reflected from a concentrator are focused on a line receiver extending up from the surface a distance equal to one-half the radius of curvature. The line focus always points toward the sun and moves across the bowl each day. The solar bowl is primarily an on-site civil engineering technology in which 95%, by cost of the components, can be constructed at the site with conventional methods and materials in order to ensure minimum cost. Such a construction requirement is well suited for developing countries. The major components of the other solar thermal collector technologies, e.g. central receivers' heliostats, parabolic dishes and trough require factory type mass production facilities in order to obtain low costs. A developing nation must either build factories to manufacture those components or import them. Either approach will increase the cost of the initial solar system.

* To whom all correspondence be sent

A 60^0-rim angle spherical bowl was scuccessfully demonstrated on a small scale by Texas Tech University. A 20-m diameter spherical bowl consisting of 438 spherically-curved glass mirror panels, each of about 1 m square area, focusing the sun onto a movable 5.5 m-long receiver disigned, constructed and put in operation in 1982 [1]. Water was supplied into the receiver for thermal collection of solar energy. The water turned into superheated steam to about a temperature of 810 K working at a measured peak efficiency of 63%, thus providing a power of 853,000 Btu h^{-1} (250 kW_{th}). This was constructed as an exprimental plant to develop a 5-MW solar hybrid electric power plant to serve a small town, Crosbyton, near Texas Tech University. This was the first solar plant which generated electricity for local grid. This plant produces steam at high temperature which in turn runs a steam engine turbine and generates somewhat uneconomical electricity. However, the steam at high temperatures can, in principle, be put to use in a different manner more efficiently. For example, some of the thermochemical reactions up to that temperature can be brought about to produce hydrogen. A proper choice of intermediary substance to be used at various temperatures would lead to an economic production of hydrogen.

The solar bowl device requires a simple technical knowhow in construction and maintenance. It does not need many factory-manufactured sub assemblies. Most of the construction is done on site. In this system there is no or minimum motor moving parts and valves. The concept works very well for a plant in the range of 150 kW to 10 MW. It has a relatively short effective operational day, which can be overcome in principle using several bowls oriented at different angles.

2. MAIN COMPONENTS

A solar bowl plant has very few parts and most of them can be manufactured on the site or locally. The bowl is a civil engineering labor intensive work and will be done on the site. The mirrors can be obtained from a local glass manufacturing firm.

The receiver is a shell and tube type as schematically shown in Fig. 2. It is approximately equal to half of the radius of curvature of the bowl. For this design it can be manufactured at the UPM workshop. Depending on the total length of the receiver, the requisite numbers of segments, each three to four meters - long, are to join at common plenum to make the receiver. Spiral groves of 5/16" deep for water flow are cut on the outer surface of the inner tube. This tube is housed in an outer shell. The receiver can be counterbalanced by a weight and is pivoted on a two-axis tracking mount. The insulated pipes connect the receiver and the turbine (or load).

A computer is fed with a program for the receiver to track the sun. It can be obtained locally and can be used for other purposes.

Depending on how the steam available from the solar bowl receiver is used, other facilities will added to the plant. The sechematic diagram shown in Fig. 1 is one of the uses of the solar energy collected for generating electricity. Direct heat can be used for heating, drying, cogenerating, or for research purposes.

3. MALAYSIAN PARAMETERS

An improved version of that solar bowl is designed for operating at Universiti Pertanian Malaysia on an experimental basis. The average daily solar flux has been measured for several places in the area [2]. Thus the calculations and design have been made on the daily average solar flux of 850 Wm^{-2} for making a shallow bowl with a 16 m rim radius and rim angle of 40^0 (which need not to be tilted for latitude of 3^0 N). A 6 m deep pit of span of 32 m and 40^0 rim angle is excavated in the ground and a concrete surface is placed directly on the stabilized pit surface area. Small flat mirror glasses are glued directly onto the concrete much like tiling a floor. A hole at the bottom of the bowl is made to let the rain water run into the sewerage.

For a bowl of dimension given in TABLE I, heat energy in the form of steam at a temperature of 750^0F at 700 psi can be produced at a rate of about 0.7×10^6 kWh_{th}/yr. This is eatimated with an insolation of 850 W/m^2 normal to the surface for six hours. The mirror reflectivity is assumed to be 90% converting about 50% of the annual solar flux to useful steam. Should this heat be used in running a steam turbine to produce electricity, an electrical power plant of capacity 150 kW can be operated thus generating 0.27×10^6 kWh/yr. The generator is assumed to be working with an efficiocy of 40%.

4. COST COMPARISON

The cost estimate of construction of a solar bowl of same size as mentioned above is given in TABLE II. The cost of a similar bowl is estimated for construction in the USA and compared with one in Malaysia. The cost of construction in Malaysia is about half of that in the USA. This difference is mainly due to:

i. Cheap labor cost in Malaysia compared to the USA. The solar bowl is primarily an on-site civil engineering technology in which 95%, by cost of components, can be constructed at the site with conventional methods and materials in order to ensure minimum cost.

ii. It does not require much of the imported material.

iii. Locally manufactured material, primarily glass, is available much cheaper in Malaysia than in the USA.

iv. The location being at 3^0N parallel receives almost normal radiation during peak time in Kuala Lumpur, Malaysia. In the USA even the southern most part is much in the north of Tropic of Cancer.

v. There is not much dust to settle on the bowl in Malaysia and whatever settles on the bowl is washed away by rain. A cleaning device had to be installed particularly for Crosbyton Solar Bowl in the USA which costs extra and needs extra power supply to drive it.

vi. The year round temperature in Malaysia does not fluctuate much where as in the most of the USA including West Texas, where the solar bowl was constructed, the temperature variation is much wider. It goes below freezing during winter time. For that reason, either working fluid used should be other than water or water should be kept warmed running during frosty season in the receiver.

5. ADDITIONAL EFFICIENCY

A further improvement in the efficiency of the plant by about 1.5 to 2% is planned to be achieved by adding an iris on the rim of the bowl [3]. The iris is designed to catch the solar radiation when the sun is at small angle above the horizon in the morning and in the afternoon. The iris is to slide on the rim of the bowl to track the sun. The addition of the iris will increase the power output; and the cost per unit of energy is estimated to be reduced almost by a factor of two.

6. CONCLUSION

The development of the solar bowl technology in Malaysia (or for that matter in any of the developing countries with similar or better local parameters) can be perceived from several point of views. It would benefit the scientific and technical community directly in the first place, the cousumer and the country's economy by improving the energy supply. This economic feasibility study favors the solar bowl technology development in Malaysia (see Table II). A logical place to propose installing a solar bowl first is on some university campus within the existing electrical generation and transmission system. The bowl can be effective there in the fallowing ways:

i. Providing research and development in solar energy and through this vehicle offers opportunity to other research.

ii. Providing the training of personnel like engineers techicians, administrators and energy planners.

iii. Providing the electricity to the local grid to help overcome the peak hour load.

iv. Providing steam to steam turbine in existing system.

v. Providing preheated water to the boiler.

vi. Providing industrial process heat.

The solar bowl concept is a low level thermal technology. It can easily be adapted in a country like Malaysia where there is a moderate level of industrial capability and a low cost of labor.

Acknowledgment:

The authors are thankful to Dr. E.A.O'Hair for many useful discussions and advice during this study.

* Permanent Adress : Department of Physics, Texas Tech University, Lubbock Texas 79409 USA.

REFERENCES

1. M.A.K. Lodhi, Collection and Storage of Solar Energy, Int. J.Hydrogen Energy 14, 379 (1989).

2. Malaysian Meteorological services Data sheet 1988 - 89.

3. T.A. William et al., "Characterization of solar thermal Concept for Electricity Generation", Texas Tech Univ. Report No. PNL - 6129, USDOE (1988).

TABLE I : PARAMETERS OF A 32-m SPAN SOLAR BOWL 40^0-RIM ANGLE A LATITUDE OF 3^{0}N

Item	Dimension
Volume of the pit	2400 m^3
Cross sectional area	800 m^2
Curved area	1000 m^2
Depth	6 m
Receiver	12 m

TABLE II: COST OF A SOLAR BOWL OF 32 m SPAN, 40^0 RIM ANGLE IN MALAYSIA

Item	Cost in US $1000
Earth moving 2500 m^3 @ $3/$m^3$	7.5
2" concrete 1000 m^2 @ 5.5/m^2	5.5
2 mm mirrors 1000 m^2 @ 7.5/m^2	7.5
Receiver	15.5
Support structure	5.5
Equipment	4.5
Initial investment	45.0
Maintenance @ 1%/yr for 25 yrs	11.00
Total For Heat	56.00
Heat cost ¢/kWh_{th}	0.3
Turbine-generator: 150 kW @ 0.5/W	75
Initial investment for the electivity plant	120
Maintenance @ 1% /yr for 25 yrs	30
Total For Electricity	150
Electiricity cost ¢/kWh	2.2

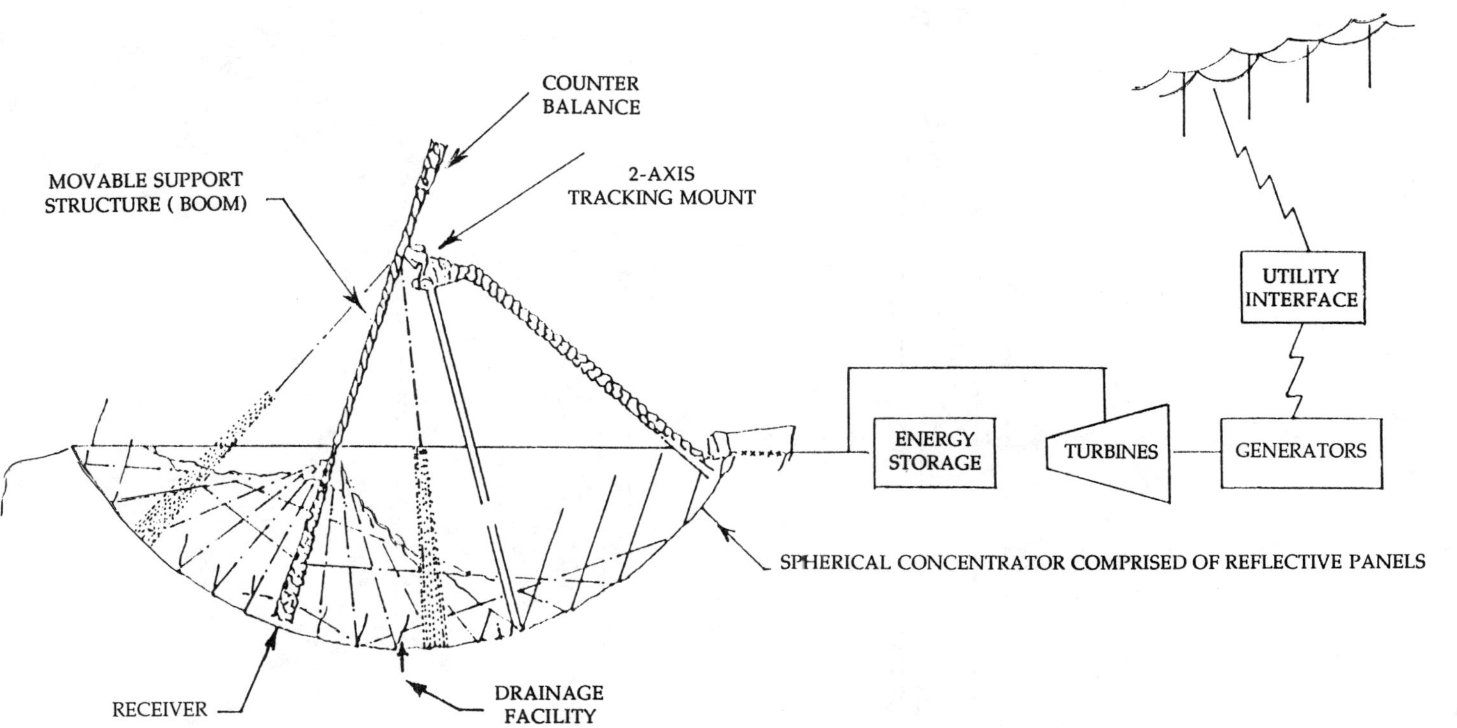

Fig. 1 A View of the Solar Bowl

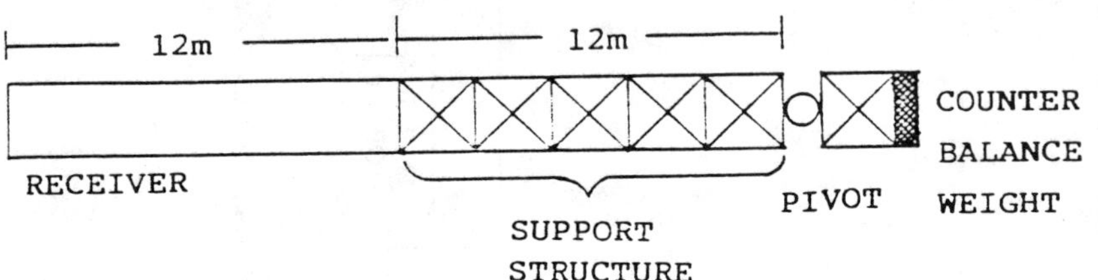

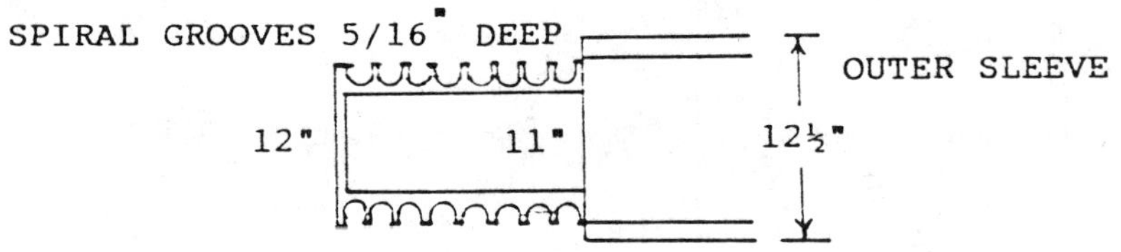

Fig.2 Schematic Diagram of Receiver.

THE CONVERSION OF SOLAR ENERGY INTO POWER IN SPACE

Jose L. Perez-Mira
University of Murcia, Chemical Engineering Dept.
c/. Norte, No 5, Valle de Escombreras, Cartagena, Murcia, Spain

In recognition of various terrestrial obstacles to large scale solar-generated power, therefore, a proposal has been advanced to locate a satellite solar power station (SSPS) in geosynchronous Earth orbit (GEO), where solar energy is available 24 hours a day during most of the year.

As originally conceived, an SSPS can utilize any of several current approaches to solar energy conversion-photovoltaic thermal-electric, thermonic-and others likely to be developed in the future. Among these conversion processes, photovoltaic conversion represents a useful starting point because solar cells are in wide use on satellites. An added incentive is the substantial progress being made in the development of low-cost, reliable photovoltaic systems and the increasing confidence in the capabilities to achieve the required production volumes. Since the photovoltaic process is a passive one, it could reduce maintenance requirements and achieve the desired 30-year operating lifetime for an SSPS. Several photovoltaic energy conversion configurations applicable to the SSPS concept are being considered, mainly differing in the specific choice of photovoltaic materials, structural arrangements, solar concentration factors and method of minimizing gravity gradients and disturbing torques due to microwave antenna recoil and solar radiation pressure. For these configurations, silicon, cadmium sulfide, and gallium arsenide solar cells are being evaluated.

Although these configurations are in an early stage of system analysis, the most promising could be determined during the next three years. But even at the this time, several of the design approaches indicate that the conversion technology on which future SSPS development can be based already exits.

SSPS Photovoltaic Design.

The photovoltaic design concept uses silicon solar cells in combination with solar reflectors to generate about 5 GW (from the end of the fifth year on) at the output of the receiving antenna. Two rectangular solar cell arrays, each about six kilometers long and five kilometers wide, are interconnected by a carry through structure of dielectric material.

A modified structure could be designed to eliminate the dielectric materials, if they would interfere with the passage of the microwave beam. A microwave transmitting antenna is located on the centerline between the two arrays and is supported by the central power transmission mast structure that extend the full length of the SSPS. The transmitting antenna is attached to the mast by a rotary joint system with unlimited freedom of East-West rotation and limited North-South elevation.

The silicon cell, carrying a system of electrical accumulation in the same cell, located between the concentrator mirrors (arranged in the form of a trough) operate at an efficiency of 30% (from the end of the fifth year on). The microwave subsystem operates at a frequency of 2,45 GHz with a dc-to-dc efficiency of 58%.

The solar cell degradation over 30 years due to radiation damage is assumed to be 20%. The rate of solar degradation will be influenced by the choice of materials. An increase in the thickness of the cover glass of the solar cell will reduce the effects of protons trapped in the Earth's magnetic field and of solar flares (occurring in II-year cycles) but increase the mass of the solar cell array. Gallium arsenide or cadmium sulfide solar cells would reduce the degradation because these materials are less affected by the radiation environment. Micrometeoroid impacts are projected to degrade 1% of the solar cell array area after a 30-year exposure. The mass of an SSPS delivering 5 GW is about 18×10^6 kg.

Orbital perturbations due to the gravitational potential of the Sun and Moon solar radiation pressure, microwave radiation recoil pressure, the ellipticity of the Earth's equatorial plane, rotary joint friction torques, magnetic field interactions, and aerodynamic drag can have a significant impact on overall SSPS performance unless appropriate maneuver corrections are applied by thrusters. Inclination effects (North-South drift) require correction annually. Solar radiation forces affecting orbital periods require continuous nulling. To achieve the desired thrust levels to control drift in altitude and inclination, thrusters will be required to burn about 50,000 kg. per year of propellant, e.g., Argon.

The placement and reliability of the propulsion units will be critical determinants of maintenance costs. Methods of achieving control during SSPS construction and maintaining operational control of the SSPS during eclipses will require further investigation.

Utility Power Pool Interface

The large power output potential of the SSPS requires careful design of the utility power pool interface to reduce the impact on the stability of a total utility system. Electrical power grids are designed to provide this stability of power supply to the user by incorporating redundant installations of reliable equipment. At present, it is technically and economically feasible to construct a utility system that will, over a period of 10 years, meet its demand on all days except one.

The SSPS could either reduce system reliability or, for the same reliability, increase the required amount of redundant equipment. Conventional power pool design and management methods rely on probability theory, based either on the availability or unavailability of each system component at full capacity. The installed capacity required to provide an LOLP of 0.1 day per year using terrestrial equipment to meet typical loads is usually 25% greater than the peak yearly demand in any one power pool. The LOLP of the SSPS will be influenced by unavoidable power losses during eclipses, failure of terrestrial equipment and failure of spaceborne equipment. Failure of the ground equipment can be treated in a conventional manner. The probability of catastrophic failure of the SSPS, for example, because of impact of a large meteorite or of an accident will be similar to that of a catastrophic failure on the ground.

Space Transportation

To be commercially competitive, the SSPS will require a space transportation system capable of placing large and massive payloads into synchronous orbit at low cost. The cost of transportation will have the most significant impact on the economic feasibility on the SSPS. The space transportation system which will be available during early phases of SSPS development for technology verification and component functional demonstration will be the space shuttle now well along in development and nearing its first horizontal flight. Compared to the previously used expendable launch vehicles, it will not only significantly reduce the cost of launching payloads, but also be a major step towards the development of space transportation system of greatly increased payload capability-the heavy-lift launch vehicle (HLLV).

The HLLV represents an advanced space transportation system with a planned capability to place payloads ranging from 200 to 500 metric tons into LEO.

The HLLV will be recoverable and repeatedly reusable. The fuel used in the lower stage will be liquid oxygen and a hydrocarbon; liquid oxygen and liquid hydrogen will be used in the upper stage. Both offshore and onshore launch facilities could be developed for the HLLV.

THE TECHNOLOGY OPTIONS FOR POWER

Transmission to Earth

To transmit the power generated in the SSPS to Earth there are two options
1º. - A microwave beam or
2º. - A laser beam.

The microwave method uses achievable technology to obtain high efficiency in generation, transmission, and conversion. Moreover, it promises to satisfy environmental requirements and safety considerations. Microwave generation, transmission and conversion technology is based on demonstrated results from commercial use. Present mass production of microwave devices is indicative of the commercialization of the technology.

The laser method is less preferable because of the low efficiencies associated with the conversion of electricity into laser power and the reconversion of laser into electricity, and because of optical limitations. In addition, the absorption of laser beams in the atmosphere, and by clouds would reduce the overall efficiency of power transmission to an unattractive level. Malfunction of the laser directional control system could result in unacceptable safety risks.

Microwave Power Transmission System

Figure A shows the functional blocks of the microwave power transmission system designed to transmit the electrical power generated by the solar energy conversion system to a receiving antenna on Earth, and the associated efficiencies goals. Although free-space transmission of power by microwaves is a relatively new technology, in recent years it has advanced rapidly and system efficiencies greater than 50%, including the interconversion between dc power and microwave power at both terminals of the system, are being obtained. The application of new technology will further raise this efficiency to almost 70%.

Microwave Power Generation

The device which is being considered for converting dc voltage to rf power at microwave frequencies in the SSPS is a crossfield amplifier (Amplitron). The Amplitron uses a cold platinum metal cathode operating on the principle of secondary emission to achieve a nearly infinite cathode life. The dc voltage required for the Amplitron is 20 kV.

Samarium Cobalt magnet material provides low mass compared to that obtainable from conventional permanent magnet material and makes it feasible to use such devices in the SSPS. Radiating surfaces are designed to reject waste heat (about 10% of the input power) to space.

Considerations of mass, cost, and efficiency at specific frequencies have led to the selection of frequency within the industrial microwave band of 2,40 to 2,50 GHz. The mass and cost of the Amplitron are near optimum at this frequency and at power output level of about 5 kW.

Microwave Beam Transmission

The transmitting antenna is designed as a circular, planar active phased array having a diameter of about one kilometer. Microwave power can be transferred at high efficiency when the transmitting antenna is illuminated with an amplitude distribution which is of the form (1-r2*)n and when the phase front of the beam is carefully controlled at the launch point to minimize scattering losses.

Space is an ideal medium for the transmission microwaves; a transmission efficiency of 99.6% is projected after the beam has been launched at the transmitting antenna and before it passes through the upper atmosphere. To achieve the desired high efficiency for the transmission system while minimizing the cost, the geometric relationships between the transmitting and receiving antenna indicate that the transmitting antenna should be about one kilometer in diameter while the receiving antenna should be ten kilometers in diameter. The power density at the receiving antenna will always be a maximum at the middle and will decrease with distance from the center of the receiver. The exact size of the receiving antenna will be determinated by the radius at which the collection and rectification of the power becomes marginally economical. To reduce the dimensions of the transmitting antenna, the illumination taper from the center to the edge can be reduced at the expense of losing more power in the side lobes.

Microwave Power Reception and Rectification

The receiving antenna is designed to intercept, collect and rectify the microwave beam into a dc output with high efficiency. The dc output can be designed to either interface with high-voltage dc transmission networks or be converted into 60 Hz alternating current. The receiving antenna consist of an array of elements which absorb and rectify the incident microwave beam. Each element consists of a half-wave dipole, and integral low-pass filter, diode rectifier and bypass capacitor. The dipoles are dc-insulated from the ground plane and appear as rf absorbers to the incoming microwave.

Because each dipole has its own rectifier the receiving array has the directivity of a single dipole. Thus the collection efficiency of the array is relatively insensitive to substantial changes in the direction of the incoming spatial variations in phase and power density of the incoming beam that could be caused by non-uniform atmospheric conditions.

The half-wave dipole is spaced about 0.6 wavelengths apart and is arranged in a triangular lattice at a distance from the ground plane of about one-fifth of the wavelength. This distance many be adjusted within limits for impedance and the incoming microwave beam. This match can approach 100%; refection losses of less than 1% have been experimentally achieved. The low pass filter is designed to minimize losses at the fundamental frequency and achieve effective rejection of the harmonic which are generated in proper phase to result in maximum rf-to dc conversion.

An rf-bypass capacitor acts as a smoothing filter to remove fundamental and harmonic ripple from circuit is created at the frequency of the incoming microwave power. Measured element has reached 90%. Field test have indicated that a combined collection and rectification efficiency of more than 82% can be achieved at the receiving antenna.

The author

Jose L. PEREZ-MIRA, Chemical Dr. of Chemical Engineering Dept. of University Of Murcia. Director of Research and Products Development for Cartagena Polytechnic University in Spain. (30350 Valle de Escombreras, Cartagena, Murcia, Spain. Apartado no I3). Has worked for Research Scientific Design Co. He holds an M.S. Degree from University of Murcia.
This article is based on the book "Vehicles Electrosun" which deals with rapid population growth in large cities and small towns, to be published in late I988.

References

1º. "The Technology and Application of Free-Space Power Transmission by Microwave Beam" W.C. Brown. Proc. IEEE, Vol. 62, No 1, January I974. pp. 11-25.

2º. "Transportation Options for Solar Power Satellites" G.R.Woodcock Proceeding, Vol. II AIChE, September I976 pp. 1400-1407.

3º. "Engineering a Space Manufacturing Center" Astronautics & Aeronautics. G.K. O'Neill. Vol. 14, No 10, 1976. pp. 20-28

4º. "Solar Power Stations Space" Committee on the Peaceful Uses of Outer Space. Background paper prepared by the Secretariat, United Nations, General Assembly (A/AC. 105) CXIX-CRP.I June 1976.

5º. "The Future of Power from the Sun" P.E. Glaser IECEC 1968, Record IEEE Publication 68C21-Energy. pp.98-103.

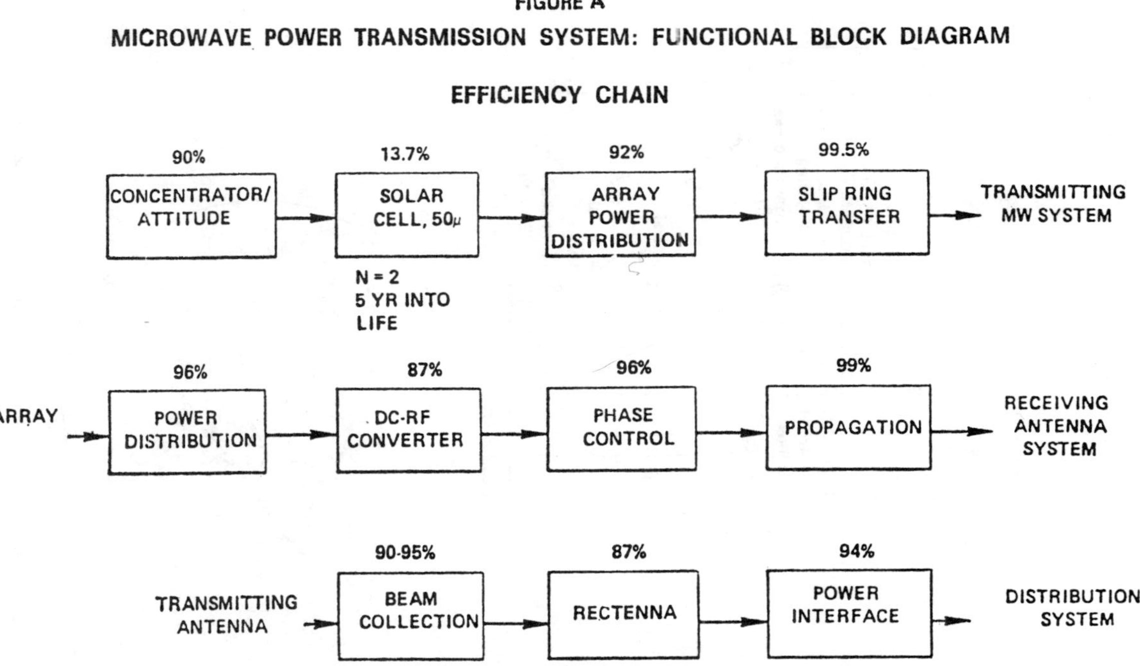
FIGURE A
MICROWAVE POWER TRANSMISSION SYSTEM: FUNCTIONAL BLOCK DIAGRAM
EFFICIENCY CHAIN
90%
CONCENTRATOR/ ATTITUDE
13.7%
SOLAR CELL, 50μ
N = 2
5 YR INTO LIFE
92%
ARRAY POWER DISTRIBUTION
99.5%
SLIP RING TRANSFER
TRANSMITTING MW SYSTEM
ARRAY
96%
POWER DISTRIBUTION
87%
DC-RF CONVERTER
96%
PHASE CONTROL
99%
PROPAGATION
RECEIVING ANTENNA SYSTEM
TRANSMITTING ANTENNA
90-95%
BEAM COLLECTION
87%
RECTENNA
94%
POWER INTERFACE
DISTRIBUTION SYSTEM

PHOTOVOLTAIC DESIGN

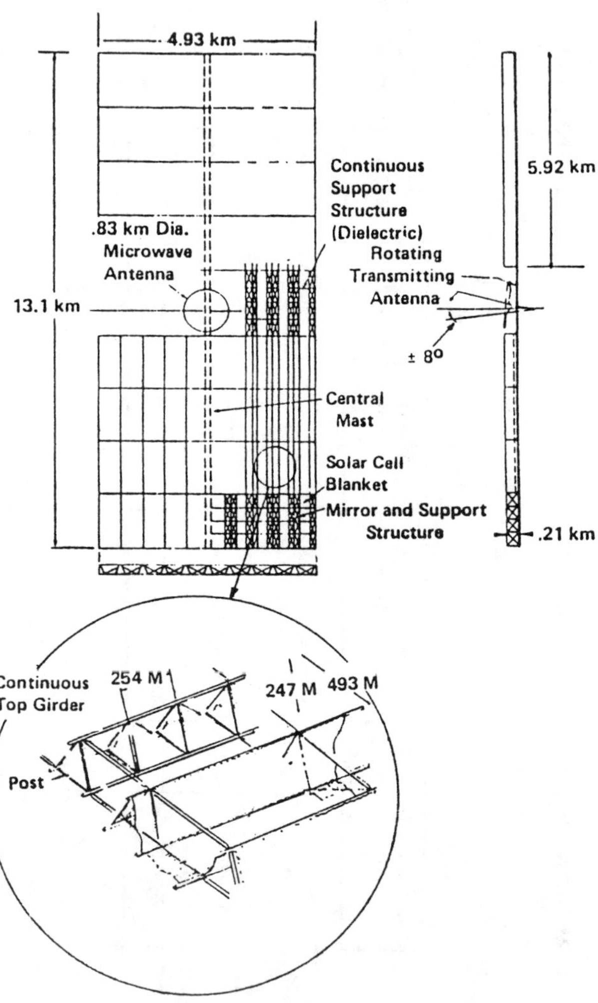

Source: Arthur D. Little, Inc., Grumman Aerospace Corporation, Raytheon Company, Spectrolab, Inc.

A PROPOSED SOLAR POND POWER PLANT USING OFFSHORE BRINE

M.A.K. Lodhi*, Shaharin Ibrahim, and M. Yusof Sulaiman
Department of Physics, Universiti Pertanian Malaysia
43400 UPM Serdang, Selangor, Malaysia

Abstract

Along with the offshore crude oil production in Malaysia a bulk of brine of nuisance value is produced. The amount and the concentration of sodium chloride in the brine produced from the oilfield varies from 2% to 30% by weight. This brine is proposed to be utilized to harness solar energy. This work involves a feasibility study of assessing the availability and cost of energy from the concept of salt gradient solar pond (SGSP) with special reference to Malaysia. It receives an abundant solar energy throughout the year, almost normal. The retail salt price in the country is about $75 - $80 per ton. The brine produced from the offshore oilfields which costs about 1.5¢ to 6.5¢ per ton-mile if transported by train or truck. For a pond of 1 acre x 5m, a total amount of salt needed would cost about $2000, obtained from a brine of 10% concentration by weight. A comparison of the cost of construction of this size SGSP in California, Texas and Malaysia is made. In estimating the unit energy cost, the solar radiation intensity is assumed to be the same ($1kW/m^2$) at the three places and SGSP efficiency to be 14 - 15%. There is a heavy precipitation in Malaysia during Monsoon season which last for about three to four months. The SGSP can be covered while it is raining. The heat energy is planned to be used for food processing and crop drying. The electrical energy may also be generated for remote areas. The feasibility of SGSP using the brine produced from the offshore oilfield in Malaysia looks promising.

1. INTRODUCTION

The principle and the application potential of solar ponds were first reported in the literature at the beginning of this century. The natural solar ponds have been existing in nature for millenia unnoticed by man until the turn of the last century when one of them was discovered with its amazing properties [1]. The Lake Madoe (Medie Lagoon) located at latitude 42°, 44′N and longitude 26°, 45′E in Transylvania was found to have temperatures up to 70^oC at a depth of around 1.32 m during the end of summer. These high temperatures were apparently due to the salinity gradient in the lake. Kalecsinsky reported this discovery and indicated its potential as a source of low grade heat.

Salt gradient solar ponds (SGSP) are simple and lowcost devices for collecting and storing solar energy employing the principles of gravitation and buoyancy. Solar pond technology has been demonstrated on the

* To whom all correspondence be sent

world-wide scene since 1960's. General conclusions are that solar ponds are technically and economically viable energy transducers that can and should be exploited in the applicable regions and markets where the potential is high.

The most important parameter affecting the economic feasibility of SGSP is the cost per unit surface area of the pond. This is primarily for the acquisition cost of salt which is critical to the near ferm feasibility of these ponds' application.

The storage is one of the most critical problems of almost all solar energy applications due to its intermittent nature. The low storage efficiencies and the consequential energy losses, the construction and maintenance of storage devices usually add up to a considerable fraction to the total cost of the solar energy conversion systems into some usable form of energy. A realistic solution to the storage problem, for low temperature applications, is provided by solar ponds. In a solar pond no distinction exists between the collecting and storage devices. The same bulk of water serves both purposes simultaneously. A one to two meter thick layer of water can provide a sufficient thermal insulation in order to let the water temperature rise to boiling point if desired [2].

2. WORKING OF A SOLAR POND

In an actual salt-gradient solar pond there are three distinc layers (as shown in Figs 1) namely (A) surface zone, (B) non-convective zone, (C) storage zone. The surface zone is in the topmost layer of about 0.25 m thickness. In this zone the salinity is minimum but uniform. The non-convecting or gradient zone is just below the surface zone which has a steep salinity and temperature gradient. In this zone the convection is suppresesed. This zone is about 0.8 to 1.5 m thick with the optimum thickness of about 1.0 m. The third zone has almost uniform temperature and salt concentration. This layer is typically 1.0 m deep for a workable pond. It can, however, be varied to several meters. This layer has the most salt concentration and thus the highest density. As the temperature of the fluid in this zone rises by direct solar heating, the hot fluid becomes light and tries to rise up due to buoyancy. However, this effect is reduced duc to high salt concentration until it is balanced and then overcomes. Consequently the hot fluid stays in the bottom zone where heat loses due to evaporation and temperature difference between the fluid and the surroundings are minimal. In this way the heat remairs trapped in the bottom zone.

3. IMPORTANT FACTOR FOR SOLAR PONDS

The cost and performance of a salt gradient solar pond depends on a number of items which may include the climatological conditions existing at the proposed sites and its geotechnical features, land in relation to environmental impact, earth moving cost and labor needed at different stages of the construction, cost of linear with respect to the soil and the size of the pond, water and salt. One should investigate well before starting the design of a solar pond at a particular location in order to make it cost effective.

Of course the most important parameter for SGSP is the insolation. The higher the insolation better it is for the performance of the pond. The ambient temperature is related to the insolation. The high ambient temperature and low humidity will cause the water to evaporate more thus causing the temperature of the remaining water in the pond go down. In all over Malaysia these factors favor the performance of the solar pond. It has high insolation and humidity but moderate ambient temperature through out the year. See TABLE I [3].

Water is neded for filling the pond initially and then for maintaining the required level of the SGSP. This can be obtained from a nearby stream or collected from the precipitation (rain). However, severe rainfall can adversely affect the performance of the pond. It may decrease the temperature, dilute the brine execessively, over flow the pond and accumulate water in its proximity. It is certainly desinable to have some rain but not to exceed evaporation. While the average rainfall in most of Malaysia is just about the same as the evaporation on the annual basis. Sometimes and in some parts the precipitation exceeds the evaporation. The location for the SGSP should be selected by taking this into account as one of the factors. Nevertheless, the excess rainfall can be handled by constructing the pond on a slopy terrain and covering it during sever rain. The water is abundantly in Malaysia either from rivers or from collected from the rain.

A strong wind can create waves on the pond's surface causing some problems if they are not suppressed. The technique to supress the gusts of wind blowing ever at 100 m/s is available to nullity its adverse effect. However the wind around Kuala Lumpur is not found strong to worry about this effect. The annual average wind specd measured for 1986 at University Pertanian Malaysia is 0.45 m/s.

The amount of salt needed for a SGSP is huge. In the suface layer where the density is minimum the salt needed there, for maintaining a specific gravity of 1.018 in the solution, is 25 kg/m^3. In the storage layer for maintaining a specific gravity of 1.170 the salt needed is 260 kg/m^3. The salt concentration varies uniformaly between these fwo figures in the salt-gradient zone. For a SGSP of 1 acre x 5.25 m the salt needed is about 4.6×10^6 kg. Unless the salt available is at a very low cost the economic feasibility of a SGSP is not favorable. These is a lot of salt available from the brine produced as a byproduct with the crude oil in Malaysia. Unless this salt is used for SGSP in Malaysia the SGSP will cost very high and can not compete with the existing available fuel inspite of all other good fratures favoring the SGSP.

The land cost, soil characteristics with respect to insulation, seepage and infiltration of brine are particularly important in determining the cost construction and operation of a SGSP. The land used for SGSP should not be of prime value unless the need is dire for a particular location. A pit of appropriate dimension may be selected for this purpose and modified to the required measurement. This will reduce the cost of land and labor for excvation. The thick smectitic and other expanding clays would permit minimal infiltration of brine into the underlying strata.

These fratures warrant for a cost estimation of a SGSP in order to make its feasibility study for Malaysia.

4. COST COMPARISON

The retail salt price in the country is about $75 - $80 per ton. If the cost of the salt is more than $15 - $20 per ton, the economic feasibility of SGSP goes against the solar pond. There is a potentially plentiful source of salt in Malaysia which is not being utilized presently. It is the brine produced from the offshore oilfields which can be utilized for constructing a SGSP, see Fig. 2. The brine transportation costs about 1.5¢ to 6.5¢ per ton-mile if transported by train or truck. For a pond of 1 acre x 5 m, a total amount of salt needed would cost about US $2000, obtained from a brine of 10% concentration by weight. A comparison of the cost of construction of this size SGSP in California [4], Texas [5] and Malaysia is given in TABLE II.

In estimating the unit energy cost, the solar radiation intensity is assumed to be the same $1 kW/m^2$. at the three places and SGSP efficiency to be 14 - 15%. The Malaysian pond construction cost in given in TABLE III.

There is a heavy precipitation in Malaysia during Monsoon season which last for about three to four months. The SGSP can be covered while it is raining heavily and continuously to keep the SGSP from mixing the zones to a uniform density. The cost of the cover has not been includd in TABLE III. During cloudly days the pond will retain heat and keep the water warm for weeks to months.

5. CONCLUDING REMARKS

There are several uses of the low grade heat available from the SGSP in Malaysia. The agricultural and related economy of the country can be benefited significantly with the application of the solar pond technology. The heat energy can be used for crop drying and food processing. Low temperature industrial process heating, or preheating industrial processes in oil seeds can consume a lot of SGSP heat. In principle this heat can be used for direct heating and cooling of space. It can be utilized for draft power for water holes for grazing livestock and farm irrigation operations. For biomass processing and thus methane generation this energy can be well utilized. Integrating the SGSP with biomaes plant will give a further boost to energy generation for domestic, agricultural commercial and industrial use. The electrical energy may also be generated for remote areas and for peak hours for urban areas. An experimental SGSP system is designed for its construction at Universiti Pertanian Malaysia. In that system, the brine is collected, skimmed and concentrated for the final use for SGSP. The feasibility of SGSP using the brine produced from the offshore oilfield in Malaysia looks promising.

*
Permanent Address: Department of Physics
Texas Tech University
Lubbock Texas 79409 USA.

REFERENCES

1. a. Kalaecsinsky, Ann. D. Physics 1, 708 (1902).

2. H.C. Bryant, New Scientists 17, 734 (1981); D.D. Weeks, S.M. Long, R.E. Emery and H.C. Bryant, Proc. SW Conf. on Astropysics VII (1981).

3. Department of Environmental Science, University Pertanian Malaysia Annual Weather Report 1986.

4. M. Edesses, Techn. Rev. 85, 59 (1982).

5. M.A.K. Lodhi, Int. J. Hydrogen Energy 14, 379 (1989).

TABLE I : AVERAGE ANNUAL CLIMATOLOGICAL DATA FOR A $3^{o}N$ PARALLEL IN MALAYSIA BASED ON 1986

Insolation	4.5 kWh/m^2
Ambient temperature	27.7°C
Relative humidity	82.8%
Rain fall (total for 1988)	2.28 m
Evaporation (total for 1988)	2.36 m
Wind speed	0.45 m/s

TABLE II : COST COMPARISON OF A SGSP OF 1 ACRE X 5 m SIZE IN US $1000

	California	Texs	Malaysia
Pond Construction	305	54	38
Unit pond cost ($/m)	75	13	8.5
Unit cost of energy (¢/kWh)	19	3	2

TABLE III. COST ESTIMATE OF A 1 acre x 5m SOLAR POND IN MALAYSIA

Item	Cost in $1000.00
Remodelling the pit	10
Liner	2
Salt	1 - 2
Equipment and instrumentation	25
Total Pond Construction	38 - 39

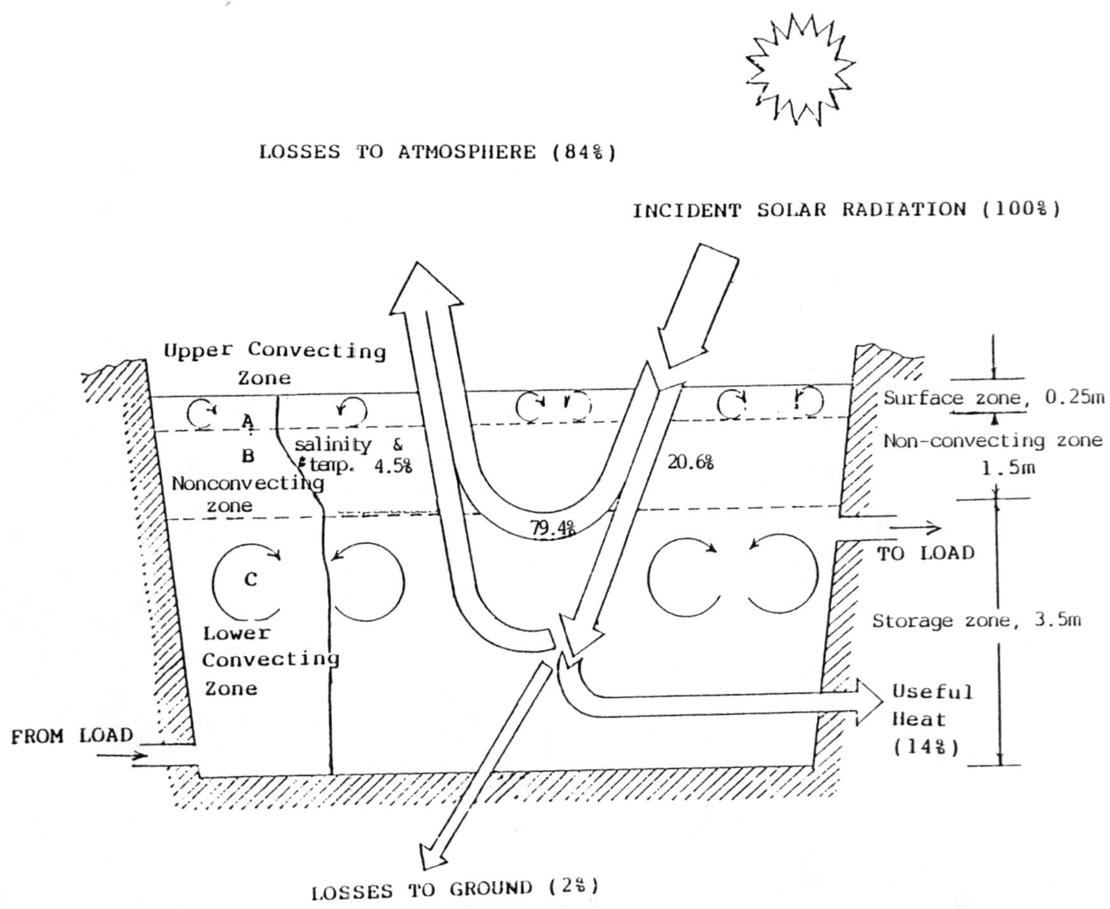

Fig. 1 Energy Balance for Solar Pond at Various Layers.

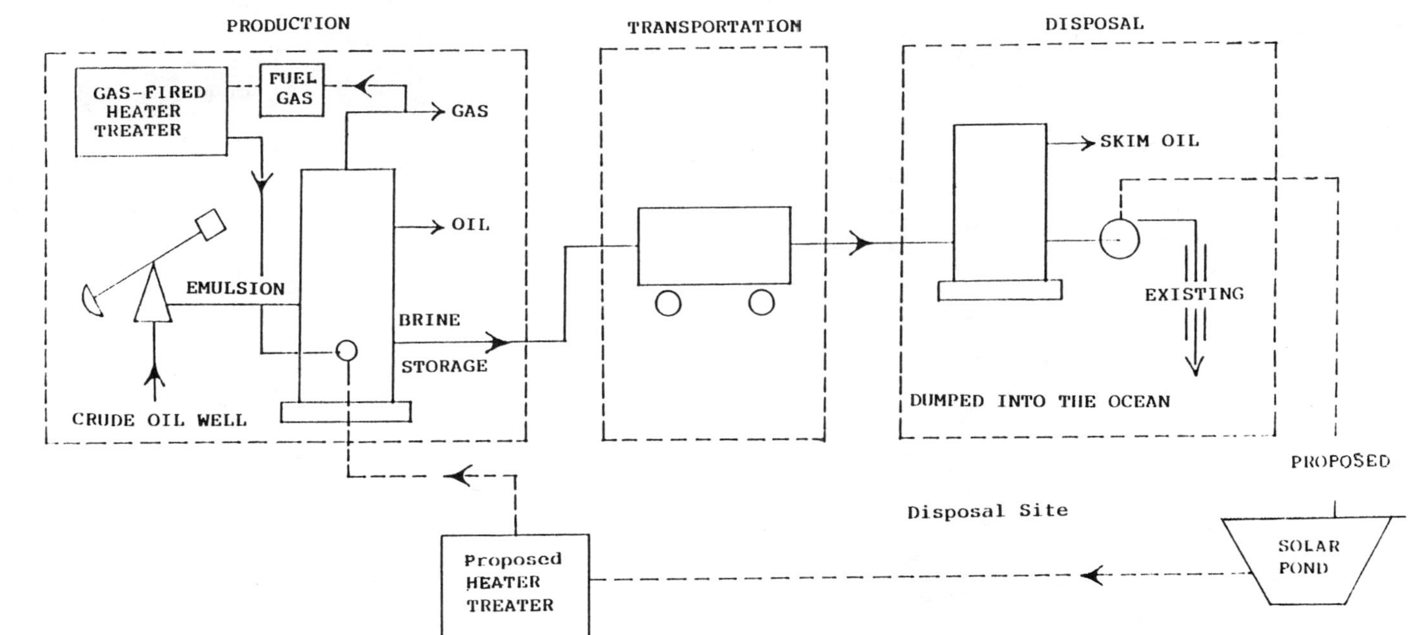

Fig. 2 Existing and Proposed Waste Brine Disposal Process.

THE SOLAR ENERGY UTILIZATION SYSTEM OF A THERMAL POWER PLANT

Yongkang Xi, Guangduo Liu, and Wanchao Lin
Energy and Power Engineering Department
Xi'an Jiaotong University, Xi'an, Shaanxi Province, P.R. China

Abstract

The fuel-solar energy power plant use solar energy to heat the exit condensate of turbine condensor for saving part of turbine extraction steam and the extra electric energy can be done by saved extraction steam. The rated output of the generating unit is kept by fuel burning, thus the expensive heat-storage device of solar energy is no longer needed to be installed. This idea can easily be broadly applied only with limited investment. If 67.5 - 159.5 °C is exit condensate temperature span of the solar energy collector, then relatively the percentage 4.95 - 8.71% is the span of the solar energy thermal efficiency, which of the series connection input system is higher than that of the parallel connection input system.

1. INTRODUCTION

The fuel-solar energy power plant by means of heating condensate with solar energy in the power plant thermal system, in comparison with the solar energy thermal power plant, is more competitive in the respects of reliability and economy, for the needed electric load can be assured by using fuel and the solar energy can be fully utilized without installing the expensive heat-storage device. Since the exit condensate temperature of a common turbine condenser is about 33 °C, various solar energy collectors can be used to heat the condensate so that part of turbine extraction steam, which is used to heat condensate in the deaerator and low pressure feed water heaters, might be reject to keep on working in the turbine and producing more electric power. The extra electric power is gained through the utilization of solar energy. If there is no change of electric power in the generating unit, the energy-saving is possible.However, the part of extraction steam being rejected by the utilization of solar energy increases condensing steam heat loss due to its being through the turbine condenser, hence the reduction of thermal economy in the power plant. Thus, the counting analysis on the solar energy thermal efficiency, which is utilized in such a way, and the choice of solar energy input system, becomes necessary.

2. SOLAR ENERGY INPUT SYSTEM

(1) The Series Connection System (shown by the actual line in Figure 1) The solar energy collector is in series connection with the low pressure feed water heater. When valve A closes, all the condensate gets to the solar energy collector through valve G. After being heated by the solar energy, it enters the entrance of the low pressure feed water heater No.1 through valve F. Any change of the solar energy radiation intensity will in turn cause the change of exit condensate temperature of the solar energy collector and that of the extraction steam quantity in

each low pressure feed water heater. If the exit condensate temperature of the solar energy collector is higher than that of the point C in the thermal system, then all the extraction steam of the low pressure feed water heaters No.1 and No.2 as well as part of the extraction steam of No.3 can be replaced by the input solar energy. Since the condensate temperature at point D only depends on the turbine extraction steam parameter, so the low pressure feed water heater No.4 is no longer influenced. When the exit condensate temperature of the solar energy collector equals to the entrance temperature, close valve F and valve G, and open valve A totally, so that the normal operation of original thermal system will be recovered. If this system is applied, the flow resistance from the solar energy collector as well as pipework should be overcome by the condensate pump, so there is a higher demand for the fall head of the pump.

(2) Parallel Connection System (shown by the dotted line in Figure 1) The solar energy collector is in parallel connection with the low pressure feed water heater. The exit condensate of the solar energy collector can be led to each exit of the low pressure feed water heaters such as the point B, point C, and point D in the Figure 1. Since the part of condensate having entered the solar energy collector bypasses at least one low pressure feed water heater, the increase of condensate flow resistance in both the solar energy collector and the connecting pipework can be compensated by the flow resistance reduction of it. This is benefitial to the working power plant where solar energy is utilized.

3. THE THERMAL EFFICIENCY OF THE SOLAR ENERGY REJECTING EXTRACTION STEAM

The solar energy, which is inputted to the thermal system to heat the condensate, by means of rejecting. The extraction steam of the low pressure feed water heater, makes the rejected extraction steam go on working in the turbine. This part of work is gained by the utilization of solar energy. The ratio between the work and the input solar energy in the thermal system is called the thermal efficiency of the solar energy rejecting the extraction steam.

(1) The Type of The Low Pressure Feed Water Heater

a. The series connection drain feed water heater is surface-type heater whose drain is of return-flow step by step, such as the low pressure feed water heaters No.1, No.2, and No.4 in Figure 1. Its counting feature is:rejecting extraction steam 1 kg of low pressure feed water heater No.j

Input Solar Energy $$q_j = i_j - i_{sj} \quad (1)$$

Input Drain Heat $$r_j = i_{s(j+1)} - i_{sj} \quad (2)$$

Condensate Heat Absorption $$Z_j = i_{cj} - i_{c(j-1)} \quad (3)$$

In the formulas :

i —— steam enthalpy kJ.kg

i_s ——drain enthalpy kJ/kg

i_c—— condensate enthalpy kJ/kg

Lower mark j —— number of the low pressure feed water heater

b. The mixing type feed water heater is a heater in which the drain is pumped up the main condensate by the drain pump, as shown by the low pressure feed water heater No.3 in Figure 1. Its counting feature is: the rejecting extraction steam 1 kg of the low pressure feed water heater No.j

Input Solar Energy $q_j = i_j - i_{c(j-1)}$ (4)

Input Drain Heat $r_j = i_{s(j+1)} - i_{c(j-1)}$ (5)

Condensate Heat Absorption Z_j = formula (3)

(2) The Thermal Efficiency of Solar Energy Rejecting No.J Block Extraction Steam

The extraction steam of the low pressure feed water heater No.j can be simply called No.j block extraction steam. The thermal efficiency of solar energy rejecting No.j block extraction steam is

$$N_j = W_j / q_j \quad (6)$$

In the formula: W_j —— work from rejecting No.j block 1 kg extraction steam kJ/kg

q_j —— Solar energy needed in the thermal system in rejecting No.j block 1 kg extraction steam, counting by formula (1) or (4) kJ/kg

W_j is counted on the condition of the stability both in the new steam flow and the quantity of the fuel supply, when the change of the turbine electric power is only related with the solar energy heating. If the input of solar energy q_1 in the low pressure feed water heater No.1 just makes it reduce its extraction steam by 1 kg, the work done by the extraction steam in the turbine

$$W_1 = i_1 - i_k$$

In the formula i_k is the exhaust steam enthalpy of the turbine.

If the input of solar energy q_2 in the low pressure feed water heater No.2 makes the latter lose 1 kg extraction steam, then the entering drain will be reduced in the No. 1 heater by 1 kg. In order to compensate this part of drain heat, the extraction steam quantity of the low pressure feed water heater No.1 can be increased to

$$a_{1.2} = r_1 / q_1$$

With the $a_{1.2}$ being drawn out, the working is reduced by $a_{1.2} \times (i_1 - i_k)$, so the work done by the input solar energy in rejecting extraction steam of block No.2 is

$$W_2 = i_2 - i_k - a_{1.2} \times (i_1 - i_k) = i_2 - i_k - (Y_1/q_1) \times W_1$$

If the extraction steam of the low pressure feed water heater No.3, after the input of the solar energy, is reduced by 1 kg, part of the rejecting extraction steam will work on the condenser while another part will work till the low pressure feed water heaters No.2 and No.3 are drawn out to heat the condensate, which is increased owing to the rejecting of the extraction steam. The partial rating is

$$a_{2.3} = Z_2/q_2$$

$$a_{1.3} = (Z_1 - a_{2.3} \times r_1)/\ q_1$$

$$= Z_1/q_1 - (Z_2 \times r_1)/(q_2 \times q_1)$$

The work done by the input of solar energy in rejecting extraction steam of block No.3

$$W_3 = i_3 - i_k - a_{2.3} \times (i_2 - i_k) - a_{1.3} \times (i_1 - i_k)$$

$$= i_3 - i_k - (Z_1/q_1) \times W_1 - (Z_2/q_2) \times W_2$$

From the above-mentioned counting, it is known that the W_j formulas are regular —— they all minus some fixed part from the enthalpy drop by rejecting 1 kg extraction steam, hence the general formula as follows:

$$W_j = i_j - i_k - \sum_{r=1}^{j-1} (A_r/q_r) \times W_r \qquad (7)$$

In the formula A_r depends on the type of the low pressure feed water heater. If No.j is a mixing type feed water heater, then A_r should all be replaced by Z_r ; and if No.j is series connection drain feed water heater, then from No.j downward till the mixing type feed water heater, A_r should be replaced by r_r while from the mixing type heater downward, A_r by Z_r.

The thermal efficiency of the extraction steam of block being rejected by the input thermal system solar energy in the 100 MW generating unit is shown in Table 1.

(3) The Thermal Efficiency of Solar Energy in Rejecting Multi-block Extraction Steam

$$N_{1-j} = W_{1-j}/Q \qquad (8)$$

$$Q = D \times (\ i'' - i'\ ) \qquad \text{kJ/hr} \qquad (9)$$

In the formulas:

N_{1-j} —— thermal efficiency of solar energy rejecting extraction steam from block No.1 to block No.j

W_{1-j} —— work done by solar energy rejecting extraction steam from block No.1 to block No.j

Q —— the input solar energy in the thermal system

D —— the entering feed water flow quantity of the solar energy collector

i' —— the feed water enthalpy from the thermal system to the solar energy collector

i'' —— the condensate enthalpy from the solar energy collector to the thermal system

If the exit condensate of the solar energy collector is combined with the exit point B of the low pressure feed water device No.1 in Figure 1, and the condensate enthalpy $i'' >$ that of i_{cl} at the exit of the low pressure feed water heater No.1, then the entering condensate temperature of the low pressure feed water heater No.2 can be raised so that both No.1 and No.2 blocks' extraction steam will be rejected at the same time by the input solar energy. The work done by the solar energy in rejecting block No.1 extraction steam

$$W_1 = D \times (i_{cl} - i_{ck}) \times N_1 = D \times Z_1 \times N_1$$

In the formula i_{ck} is the entrance condensate enthalpy of the No.1 low pressure feed water heater. The work done by the solar energy in rejecting block No.2 extraction steam

$$W_2 = D \times (i'' - i_{cl}) \times N_2$$

The work done by solar energy in pushing both No.1 and No.2 extraction steam $W_{1-2} = W_1 + W_2$, and if it is put into formula (8), then

$$N_{1-2} = \frac{D \times Z_1 \times N_1 + D \times (i'' - i_{cl}) \times N_2}{D \times (i'' - i')} = \frac{Z_1 \times N_1 + (i'' - i_{cl}) \times N_2}{i'' - i'}$$

If the exit condensate of the solar energy collector is combined with the exit point C of the No.2 low pressure feed water heater, and the exit condensate enthalpy $i'' > i_{c2}$, the exit condensate enthalpy of No.2 low pressure feed water heater, then the extraction steam from No.1 to No.3 block will be rejected at the same time by the input solar energy.

$$W_1 = D \times Z_1 \times N_1$$

$$W_2 = D \times Z_2 \times N_2$$

$$W_3 = D \times (i'' - i_{c2}) \; N_3$$

The work done by solar energy in rejecting extraction steam from block

No. 1 to block No.3 $W_{1-3} = W_1 + W_2 + W_3$, and put it into formula (8)

$$N_{1-3} = \frac{Z_1 \times N_1 + Z_2 \times N_2 + (i'' - i_{c2}) \times N_3}{i'' - i'}$$

It is known from above that, if extraction steam from block No.1 to block Noj is rejected at the same time by the input solar energy of the thermal system, then the general thermal efficiency formula of rejecting multi-block extraction steam by solar energy will be

$$N_{1-j} = \frac{\sum_{x=1}^{j-1} Z_x \times N_x + (i'' - i_{c(j-1)}) \times N_j}{i'' - i'} \qquad (10)$$

4. SOLAR ENERGY THERMAL EFFICIENCY

The solar energy inputting to the thermal system is collected by the solar energy collector, so the solar energy thermal efficiency N_s by means of the solar energy to heat condensate of the thermal system is

$$N_s = N_{1-j} \times N_c \times N_d \qquad (11)$$

$$N_c = Qu/(3600 \times A \times I) \qquad (12)$$

$$Q = Qu \times N_d = D \times (i'' - i') \quad \text{kJ/hr} \qquad (13)$$

In the formulas:

N_{1-j} —— solar energy thermal efficiency rejecting of the extraction steam from block No.1 to block No.j

N_c —— efficiency of solar energy collector

N_d —— pipework efficiency

Qu —— output heat quantity of solar energy collector

A —— area of solar energy collector

I —— solar energy radiation intensity

Q —— input solar energy in the thermal system

D —— the entering feed water flow quantity of the solar energy collector

i' —— the feed water enthalpy of the thermal system to the solar energy collector

i'' —— the condensate enthalpy of the solar energy collector to the thermal system

As the actual counting example of the solar energy thermal efficiency, the CPC vacuum tube solar energy collector of row arrangement, can be expected to gain the working temperature of 200-300 ^oC, and Figure 2 is the relationship curve of the thermal efficiency and temperature difference of the solar energy collector divided by solar energy radiation intensity. If the solar energy radiation intensity $I = 860 \ W/m^2$, the environment temperature $T_a = 20\ ^oC$, and piping efficiency $N_d = 0.97$, then solar energy thermal efficiency of the 100 MW generating unit series connection system gained by counting according to formula (10) and formula (11) can be shown in Table II.

5. SOLAR ENERGY INPUT SYSTEM

Since the entering water temperature of the solar energy collector is low, various types of solar energy collector can be used and the solar energy can be fully absorbed, so the feed water whose temperature is about 33 ^oC at the exit of the condenser is supplied to the solar energy collector. Since the multiplying of solar energy collector efficiency and piping efficiency is the same when compared to solar energy input system, so only the thermal efficiency of rejecting the extraction steam of the solar energy should be counted when the series connection system is compared with the parallel connection system.

(1) Parallel Connection System

When the condensate enthalpy, inputted from solar energy collector to the thermal system, $i'' > i_{c(j-1)}$, then the temperature of the condensate entering the No.j low pressure feed water heater is high, and part of No.j block extraction steam can be reduced so as to work in the turbine. This can be shown in the formula (10):

$$i'' - i_{c(j-1)} \times N_j > 0$$

Otherwise the extraction steam of No.j block should be increased and the part of shortage of work ruled out. This item shows negative value. When the numerator in formula (10) is of negative value, the appearance of negative value in the thermal efficiency of the solar energy rejecting multi-blocks of extraction steam shows that the work done by the input solar energy in the thermal system is less than the work lost by increasing higher level steam extraction. In this situation, the utilization of solar energy in the respect of thermal economy can not be paid off.

(2) Series Connection System

The item $(i'' - i_{c(j-1)}) \times N_j$ in the formula (10) can not result in negative value for when $i'' < i_{c(j-1)}$, the extraction steam quantity can be automatically increased by the No.(j-1) low pressure feed water heater (s); and if there is a stable entering water temperature in the No.j low pressure feed water heater, the extraction steam at No.j block will not be affected. Therefore, in the series connection system, there is a comparatively stable change of thermal efficiency of the solar energy rejecting the extraction steam, following the solar energy collector exit water temperature. This is shown by the curve 1 in Figure 3.

In Figure 3 is the changing curve of the solar energy thermal efficiency of the rejecting steam extraction according to formula (10) for the 100 MW generating unit, following the condensate temperature of input thermal system in the solar energy collector. The curve 3 is of parallel connection system, whose solar energy collector exit condensate is drawn to the exit point B of the No.1 low pressure feed water heater, and the entering water flow quantity of the solar energy collector is smaller than 32% of the overall condensate quantity. In order to get rid of the negative value in the system in the respect of thermal economy, there should be a water temperature $t'' \geq 46.5$ °C of the input thermal system in the solar energy collector. If the solar energy collector exit condensate is drawn to the exit point C of the No.2 low pressure feed water heater, namely, the curve 2, then value t which does not have a negative value in the respect of thermal economy will be raised to $t'' \geq 54.3$ °C.

It is known from Figure 3 that the solar energy series connection system has equally the same curve as the parallel connection system within some temperature range, but the former is better than the latter in the respect of thermal economy. When the solar energy series connection input system is used, all the condensate should pass the solar energy collector, so the scope of solar energy utilization is larger, which results in great area of solar energy collector in command. If the limited solar energy collector is used, the condensate temperature at the solar energy exit by means of series connection system will be low, which gives rise to the low solar energy thermal efficiency. If an automatically-controlled mixing system is used, according to the counting in the formula (10), the thermal efficiency of solar energy rejecting extraction steam, which is equally the same as the series connection system, can be obtained when the following conditions are satisfied.

a. When part of proper condensate is fed by the solar energy collector and the condensate temperature of the input thermal system $t'' \leq 67.5$ °C, the flow throttle can be done by means of valve A and the exit condensate of solar energy collector can be drawn to the entrance of No.1 low pressure feed water heater. This is now of the series connection system.

b. When $67.5\ ^{o}C < t \leq 98.2\ ^{o}C$, the exit condensate of solar energy collector is automatically cut and changed to the exit of No.2 low pressure feed water heater, which is now of the parallel connection system.

c. When t'' is in turn bigger than 98.2 °C, 113.2 °C, and 138.6 °C respectively, the exit condensate of the solar energy collector will in turn be automatically cut and changed to the entrances of No.3 and No.4, and the exit of the No.4 low pressure feed water heaters.

5. CONCLUSION

(1) The fuel-solar energy power plant by means of heating condensate with solar energy in the power plant thermal system, in comparison with the solar energy thermal power plant, is easier to be broadly applied and more competitive in the respects of reliability and economy, for it can both assure of the rated electric load by the utilization of fuel, and make full use of solar energy without installing the expensive heat-storage device.

(2) The solar energy thermal efficiency mainly depends on that of the solar energy collector as well as the extraction steam being rejected by the solar energy.

(3) The solar energy thermal efficiency both in series connection and in

mixed-type connection system is higher than that of the parallel connection system, and, compared with the parallel connection system, is more stable in change following the exit water temperature of the solar energy collector.

REFERENCES

(1) Wanchao Lin, Wang Zhai, " On Analysis of Equivalent Heat Falling Approach of The Thermal System", Xi'an Jiaotong University Journal, No.1, 1979, pp.9-12. (in Chinese)

Table I The Thermal Efficiency of Solar Energy Rejecting Each Block Extraction Steam

Items	Rejecting the Extraction Steam Block				
	No.1	No.2	No.3	No.4	No.5
P MPa	0.0377	0.123	0.202	0.424	0.929
t' °C	32.6	67.5	98.2	113.2	138.6
t'' °C	$\leq$ 67.5	$\leq$ 98.2	$\leq$ 113.2	$\leq$ 138.6	$\leq$ 159.5
N_j	0.08943	0.1492	0.1751	0.2208	0.2668

In the table: P -- extraction pressure

t' -- feed water temperature from thermal system to solar energy collector

t'' -- the condensate temperature of the thermal system from solar energy collector

N_j -- the thermal efficiency of solar energy rejecting No.j block extraction steam

Table II Solar Energy Thermal Efficiency (Series Connection System)

Input thermal system condensate temperature t °C				
67.5	98.2	113.2	138.6	159.5
Solar Energy Efficiency of Rejecting Multi-Blocks Extraction Steam N_{1-j}				
N_{1-1}	N_{1-2}	N_{1-3}	N_{1-4}	N_{1-5}
8.94	11.74	12.81	15.03	16.95
Solar Energy Thermal Efficiency N_s				
4.95	6.38	6.83	7.87	8.71

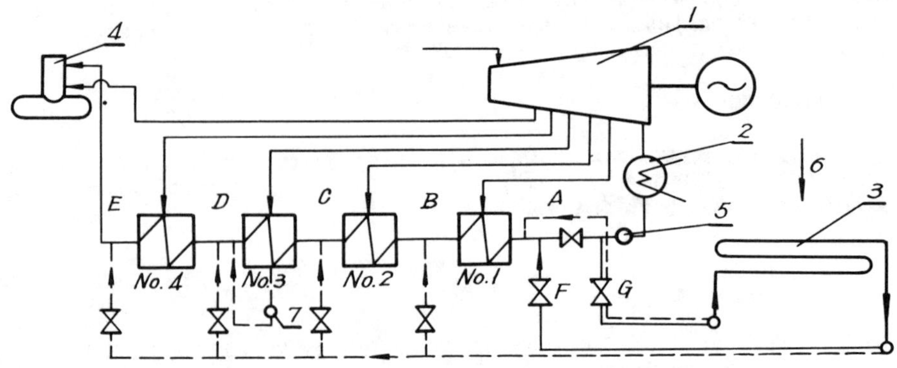

1 —— turbine 2 —— condenser 3 —— solar energy collector

4 —— deaerator 5 —— condensate pump 6 —— solar energy

7 —— drain pump

No. 1, No.2, No.3 and No.4 low pressure feed water heaters

Figure 1 Solar Energy Input System

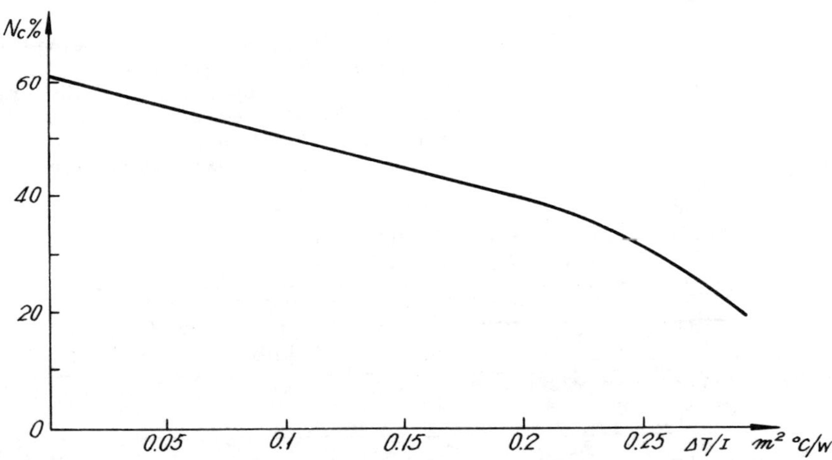

Figure 2 CPC Vacuum Tube Solar Energy Collector thermal efficiency

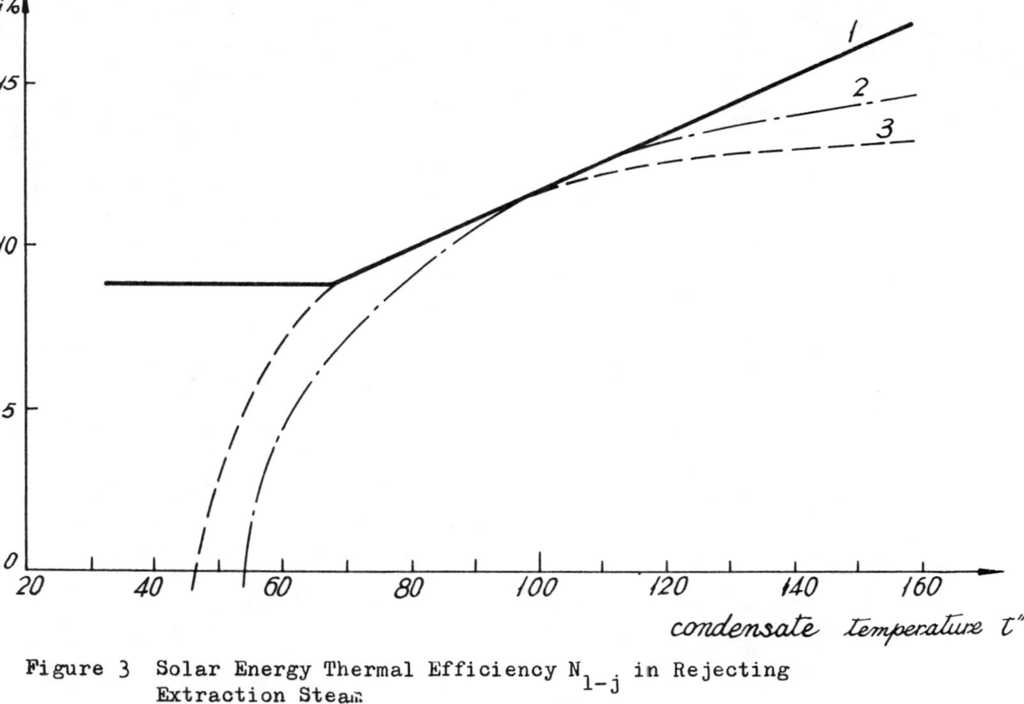

Figure 3 Solar Energy Thermal Efficiency N_{1-j} in Rejecting Extraction Steam

EFFICIENCY OF THERMOCOMPRESSORS WITH RESTORABLE POWER SOURCES

R. Zakhidov, R.Mukhamedov, B. Inogamov, M. Savochkin, R. Abdurakhmanov, and K. Tokhtakhunov
Power and Automation Institute of Academy of Sciences of the UzSSR
700143, UzSSR, Tashkent, Akademgorodok, USSR

Abstract

The present article deals with the processes of heat and cold generation with thermocompressors when solar energy, low - mineralized thermal water, biologic heat are low-potential heat sources. The article gives general characteristic of restorable power sources in Uzbekistan and presents various diagrams of combined thermocompressors with restorable power sources.

1. INTRODUCTION

Development of home and world economy is characterized by increased rate of production and utilization of fuel-energy resources consumption of which will be increased by 1990 (by preliminary data) nearly by a factor of two. XIII Congress of WEC [1] defined three peculiarities in production and utilization of fuel-energy resources:

1. Mismatch between modern structure of consumption and the structure of provision with resources.
2. Depletion of energy supply sources in most of the countries having the highest level of energy consumption.
3. Inevitable rise in the cost of energy production, its transportation to consumers.

All this requires improvement of fuel-energy balance structure, wide utilization of restorable power sources, energy saving.

At present restorable power sources utilized with the help of thermocompressors allow to save energy and protect environment as they convert low-potential heat flows into heat which can be used for heat supply (due to additional energy, for example mechanical or electric energy). In the USSR heat supplying systems are the main consumers of power sources which consume up to 750 million tons of equivalent fuei annualy. At the same time heat sources with low efficiency are still widely in use (small boiler rooms, individual heat generators, stoves). The principle of direct incineration of fuel in order to obtain low-potential heat is by itself a wasteful one as the portion of fuel energy which could have been used at combined heat generation is lost.

The sources of low-potential heat for thermalcompressors can be restorable heat sources: solar radiation, thermal water, soil, secondary power sources, namely exhaust gases of gas-motor engines operating on biogas.

2. SOLAR RADIATION. THERMOCOMPRESSOR PLANT

According to the reported data one fifth of all energy utilized in the world is provided by solar resources, wind, water, biomass and directly by sun rays. [2] About 47% of solar energy radiated into outer part of atmosphere reaches the Earth. This energy can be converted into mechanical, thermal, electrical, chemical and other types of energy utilized by mankind.

Latitude is the main factor determining intesity and duration of solar radiation. In Central Asian regions and in Uzbekistan in particular annual total radiation amounts to 4-5 GJ per square meter of level surface, and average annual sunshine period for Tashkent city is 2500 hours. The most intensive sunshine is in summer months when the sun shines for 60-65% of solar day, and it makes up 30-35% in December.

The main difficulty in solar radiation utilization is connected with wide change in radiation intensity during the day. In this context, accumulation of heat as during the day so by seasons is an acute problem.

Utilization of solar energy requires special devices absorbing this energy (solar collectors, solar power plants, "solar pools", etc). Efficiency of solar power plants is substantially increased when thermocompressor plants are used in heat supply systems. Joint use of thermocompressor plants (TCP) and solar power plants allowed the operation hours to be increased. This resulted from a possibility to change over solar-heat collector to solar energy lower temperature potentials in autumn-spring period with their following increase in TCP up to suitable levels. The experiments carried out in Power Institute of Georgian Academy of Sciences showed that efficiency of solar power plant when it operated with TCP was by 10-15% higher than without TCP. During these experiments, in heating season TCP conversion coefficient varied from 3.5 to 9, average value being 5. Such high conversion coefficients as well as the use of TCP in summer period for air conditioning are good grounds to consider it as being a promising one. [3]

Our contry experience shows that in the field of TCP design and use the problem of decentralization of heat supply, i.e. heat supply to individual consumers using TCP on the basis of combined utilization of restorable power sources is not settled.

Since 1987 Power and Automation Institute of Academy of Sciences of UzSSR have been conducting the research work on developing of solar-thermocompressor plants (STCP) for air conditioning in apartment houses, agricultural products drying (summer operation mode), and for heating and hot water supply (winter operation mode).

STCP shown in Fig. 1 operates as follows:

Summer operation mode: valves 2, 5, 7, 10 are open while valves 3, 6, 9, 14 are closed. During this period STCP operates in air conditioning mode. The heat extracted from the room is used for fruits drying. Large amount of solar radiation (up to 850 MJ/sq.m in July) allows water to be heated for everyday necessities by solar collector. Heat excess can be accumulated in seasonal tank-accumulator or in the soil with the following use in winter period.

Winter operation mode: valves 2, 3, 5, 9, 14 are open while valves 6, 7, 10 are closed. For heating and hot water supply for everyday necessities low-potential heat of solar collector 20 and tank-accumulator 12 converted in thermocompressor are used. During the days of solar activity low-potential heat source (water) is fed from tank-accumulator by thermocompressor 13 through evaporator 17 to solar collector and again to tank-accumulator. During nights and dull days when solar collector does not function, water is circulating by evaporator 17-tank-accumulator 12 loop. Hig-potential heat obtained in condenser 4 (water temperature is T=70°C) passes to hot water tank and that obtained in condenser 8 (air temperature is T=45°C) passes to a heating system. In this mode of operation, average temperature of low-potential heat source being 15-20°, theoretical conversion coefficient is estimated to be 5-5.5. In the end of heating season tank-accumulator temperature drops to less than 10°C and this cold can be used at the beginning of air conditioning season.

Theoretical calculations showed that combined use of thermocompressor and solar-heat collector increases conversion coefficient of thermocompressor by a factor of 1.5-2.0 and efficiency of solar radiation utilization by a factor of 2.0-2.5 as well as decreases consumption of organic fuel and improves environmental conditions.

3. GEOTHERMAL RESOURCES. GENERAL CHARACTERISTIC AND PECULIARITIES OF GEOTHERMAL RESOURCES

Central Asia is the region of the USSR where the use of thermal water in the national economy is promising.

At present in the USSR geothermal resources are not widely used (less than 2%), possible reserves of thermal water being 19750 thousand cubic meters per day or 230 cubic meters per second. [4]

The Earth heat exists in three forms:
- the heat accumulated by underground water (hydrogeothermal resources);
- the heat in practically "dry" rocks (petrothermal resources);
- heat energy of volcanic focuses.

Within each form the resources are classified depending on temperature, type of heat carrier, degree of mineralization, occurance depth and discharge. Thermal waters in Uzbekistan can be divided into three groups : low-potential (up to 50°C), medium-potential (from 50°C to 80°C) and high-potential (more than 80°C) 4

Thermal water in Uzbekistan can be used for heating, hot water supply, balneology, minerals(Sr , J , B , etc) extraction. From this point of view thermal water of artesian basins may be considered as commercial mineral what is very important when developing low-potential thermal water.

In spite of wide-spread occurence of thermal water and its great reserves not all regions are equal for utilization of deep-seated heat. This is because of wide spread of aquifers over thousand square kilometers and their non-uniformity which affects thermal water discharge and temperature.

Unfortunately, the data on thermal waters in Uzbekistan are so scarce that the figures on reserves differ from author to author. We were the first who proposed optimum flow diagrams for utilization of the Earth heat.

Thus, our group carried its own investigations based on initial data obtained during geological and hydrogeological field studies rather than on existing general conclusions. As the number of wells drilled for thermal waters prospecting is very small we had to collect data on aquifers temperature using oil and gas wells.

In the course of our study we have investigated about 500 wells of different depth (from 50 to 3500 m)distributed over the whole republic.

The main parameters considered were occurence depth, temperature, mineralization of underground water. Later on we shall take into account piezometric levels as well.

The territory of the republic was conventionally divided into 6 sites:

1. Western territory near the Aral Sea (Ustjurt plateau)
2. Southern territory near the Aral Sea (Southern Kara-Kalpac Autonomous republic, Khorezm region)
3. Group of artesian basins of the Kizil-Kum (Samarkand region)
4. Bukhara-Karshi-Surkhadarja artesian basin.
5. Artesian basin in Tashkent region (Samarkand, Sirdarja regions)
6. Ferganski artesian basin (Fergana, Andijan, Namangan regions).

We have throughly studied territory of Kara-Kalpac Autonomous Republic. On the basis of the data obtained from about 150 wells (total number of wells being 1400) it is seen that this artesian basin is inexhaustible storage of energy. 4 The cost of 1 Gcal of supplied heat is in this region is 2 roubles while the same within the republic when using traditional sources is from 50 to 60 roubles. It should be added that water content in this region is rather high and amounts in Southern region to, according to our data, 740 thousand cubic meters per day. At average temperature of 40°C saving in equivalent fuel will be 4200 thosand tons per day or 1.5 million tons per year.

Ustjurt plateau is also very promising. Here annual saving will be 1.0-1.2 million tons of equivalent fuel.

As the main component of thermal energy production cost is the cost of well sinking, having such a number of already drilled wells, utilization of thermal energy in the region near the Aral Sea is very promising.

If we take into account that 20% of wells are flowing ones and no expensive pumps are required, electric power consumption for their use will be minimum it becomes obvious that the use of geothrmal sources is very efficient.

Opposed to the wide use of this artesian basin is high mineralization of water (up to 400 gm/l). But using intermediate heat exchangers, water softening plants and ultrasound cleaning of heat exchange surfaces this becomes possible.

Central Kizil-Kum zone of uplift is also of interest and has some areas with anomalous thermal gradients (up to 3.7°C/100m). Discharge of this region is from 3 to 70 l/sec, mineralization being within 10 gm/l. This water can be used for agriculture (greenhouses heating in combination with biogas plants, thermocompressors, solar collectors). Here there is no acute problem with waste thermal water drainage as some part of local waters is drinkable and can be drained to the surface basins, 6

Prospective water storage makes up 450 milllion cubic meters per year at temperature of 40°C what is equal to 2.6 million tons of equivalent fuel per year.

Bukhara-Karshi-Surkhandarja region has more uniform temperature pattern and more depth of occurence of thermal water. We have studied about 170 wells. Mineralization of water varies from 5 gm/l to 120 gm/l. [4, 7]

As the depth of occurence is as large as 2500 m, water extraction with the temperature of 60-70°C and cooling in branched heat exchange surfaces with the following pumping into production level seems to be promising. The water can be also used in combination with alternative power sources. Water storage is estimated to be 300-320 million cubic meters per year what saves about 1.7 million tons of equivalent fuel, the temperature being 38°C on the average.

Thermal waters in Tashkent region artesian basin occur in three levels. The first aquifer contains drinkable water. It is not reasonable to use this water for technical needs mainly because of low temperature (21-24°C), and low discharge. At present it is successfully used in balneology and water supply system. [4, 8]

Two lower level are actively used for fields irrigation and other agricultural needs. With some capital investments water from these aquifers could be used for heating and after chemical treatment for hot water supply of livestock farms and industrial complexes. Under conditions of high level of integration and population density development of optimum patterns of geothermal sources use is very important.

Fergana artesian basin occupies rather small territory but here Adrasmano-Chutskaja anomaly occur. This anomaly was discovered in 1968 and stretches on the territory of Uzbekistan and Tadjikistan. Its length is 180 km from north-west to north-east along Fergana valley margin. Anomalous temperatures were recorded in 28 wells at the depth of 300-5400 m. At 2000 m rocks are heated to the temperature of 100-160°C, at 5000 m in oil prospecting well the temperature is recorded to be more than 200°C.

Density of heat flow within Adrasmano-Chutskaja anomaly varies from 63 to 207 **mWt** per square meters, at normal background it is 62 **mWt** per square meter in Fergana depression. Maximum density was recorded in Djarkamar area, Such a high density of heat flow is specific to Earth heat deposits in Italy, Hungary, Mexico and the USA. Inflow was equal to 15 l/sec with the temperature of up to 50°C, mouth pressure was 21-23 atm.[8]

Constant initial temperature during the whole period of use is specific to thermal water. This feature is very important for all-the-year-round utilization of thermocompressor plants where thermal water is used as the source of low-potential heat. If the temperature of heat carrier in solar thermocompressor plant changes with climatic conditions, the temperature of thermal water is constant as it has accumulated the heat of rocks. This provides stable operation of geothermal thermocompressor plant (GTTCP).

The diagram of geothermal thermocompressor plant used for heating and hot water supply (winter operation mode) and conditioning as well as initial thermal treatment of juice and milk (summer operation mode).

The plant operates as follows:

Summer operation mode: Low-mineralized thermal water (T = 318 K) is fed through valves 3 and 4 to evaporator 5 and condenser 7 respectively. In evaporator 5 thermal water is cooled to T = 288K, after this it is fed to heat exchanger 14 which is used as air cooler. In condenser 7 thermal water is heated to T=353K by working body fumes condensation and fed through valve 11 for the initial thermal treatment of juices and milk.

Winter operation mode: GTTCP is used for heating and hot water supply. Thermal water at temperature of T=353K is fed to heat exchanger 14 through condenser 7. In this operation mode heat exchanger is air heating unit. Cooled thermal water is fed from evaporator 5 through valve 15 for technical needs.

In this type of GTTCP such working substances as isobutane, butane, dimethylamine which are ecologically clean substanses are supposed to be used as cooling agents.

4. BIOGAS PLANT IN COMBINATION WITH TCP

Bioenergetics is one of the most promising directions in nontraditional energetics in Uzbekistan.

The main advantages of alternative source of heat and energy under consideration in our opinion are:

a) simplicity in implementation of this proccess;

b) biogas can be converted into thermal and electric power, to use in engines and gas engines in cooling and thermocompressor plants.

In other words bigas production solves at a time three problems: ecological, agrochemical (fertilizers production), energy problems.

In Central Asian region annual farm waste amounts to 13.4 million tons of dry substance, out of this 50% accounts for our republic (7.02 mln t). Processing of this waste can give 5.4 billion cubic meters of biogas annualy on the whole and 2.4 billlion cubic meters in UzSSR what allows to save 2.1 million tons of equivalent fuel.

About 80% of farm waste in Uzbekistan accounts for cattle (5.4 mln t), the rest 20% for sheep-breeding (0.92 mln t), pig-breeding (0.14 mln t) and poultry (0.56 mln t) [9]

Among other types of promising waste is guza-paja (cotton stem). Cotton in the republic is grown on the territory of 20 mln ha, the waste being more than 8 mln tons. Utilization of waste is necessary in order to prevent the fields from pathogenic fungi. Processing of this waste can give the republic annualy 2 billion cubic meters of biogas and will save 1.18 million tons of equivalent fuel.

Biogas can be obtained from plants grown on land and in water. Estuary area of the Aral Sea is suitable for biomass plantations. Economists from Lomonosov University counted that 20 gm plant yield per square meter of water surface will give 24 tons of biomass from 1 hectare during summer period of vegetation. Its processing in reactor of biopower plant will yield 12 thousand cubic meters of gas what is equal to 10-11 tons of equivalent fuel.

It should be noted that up to 60% of generated biogas is consumed for maintaining suitable temperature. The use of low potential solar heat, the heat of thermal water and waste gas of biogas-operated engines as well as waste heat of fermented substrate in combination with TCP decrease amount of biogas concumed for process maintenance.

The diagram of biogas plant combined with TCP is shown in Fig. 3.

Raw material is fed to preparatory tank 2 where it is heated up to suitable temperature by solar collector 1 and TCP 6. Heated raw material passes to methane tank 3 from which after being fermented it passes through TCP where some part of heat potential is removed. Biogas obtained in methane tank is burned in engine 4 which is the drive of electric generator. The heat of waste gas is utilized in heat exchanger 5 and also for heating. To maintain process temperature electric heater is supposed to be used.

At present we are developing mathematical model of this process which will take into account solar radiation in the region, fermentation temperature, plant capacity and will optimize parameters of thermal converter on the basis of these data.

REFERENCES

1. World energetics. Documents of XIII Congress of **WEC** , M: Energoizdatomizdat, 1989, p. 430.
2. McVeig D. Utilization of solar energy. - M: Energoatomizdat, 1981, p. 216.
3. Darchia G.I. Ratiani G.V. et al. Solar thermocompressor system of heat and cold supply. - Holodilnaja teknika, 1982, No 6, p. 43-46.
4. Mavritski B.F. Thermal water in folded and platform regions of the USSR. - M: "Nauka", 1971, p. 242.
5. Grebenshikova T.B. Thermal water in eastern regions of Turanskaja plate in Sir-Darja-Amu-Darja interfluve. - M: "Nedra", 1976, p. 238.
6. Burak M.T. Kizil-Kum underground water - Tashkent, "Fan", 1968, p. 120.
7. Balashov L.S. Surkhandarja artesian basin. Tashkent, "Fan", AN SSSR. M: 1960, p. 280.
8. Sultankhodjaev A.N. Fergana artesian basin. Tashkent, "Fan", 1972, p. 245.
9. Wiestur U.E., Dubrovski V.S. Methane fermentation of agricultural waste. Riga,"Zipatne", *1988, pp. 29-31.

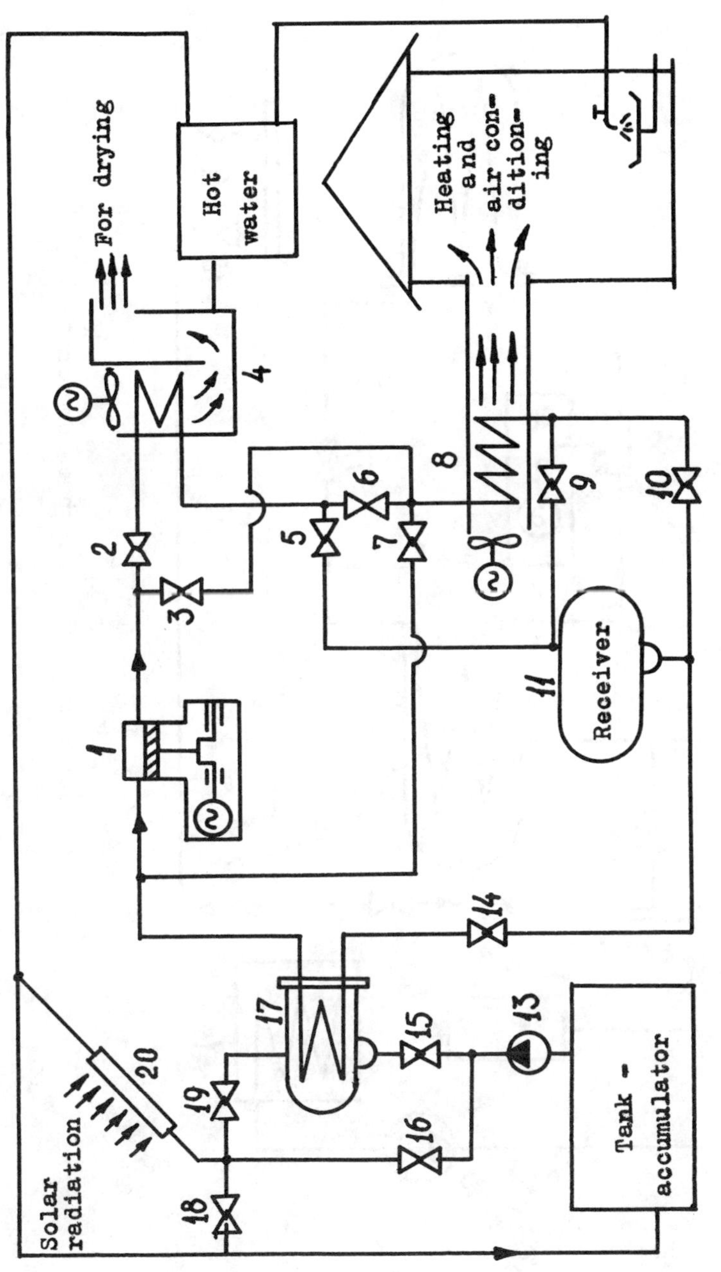

Fig. 1 - Solar thermocompressor plant

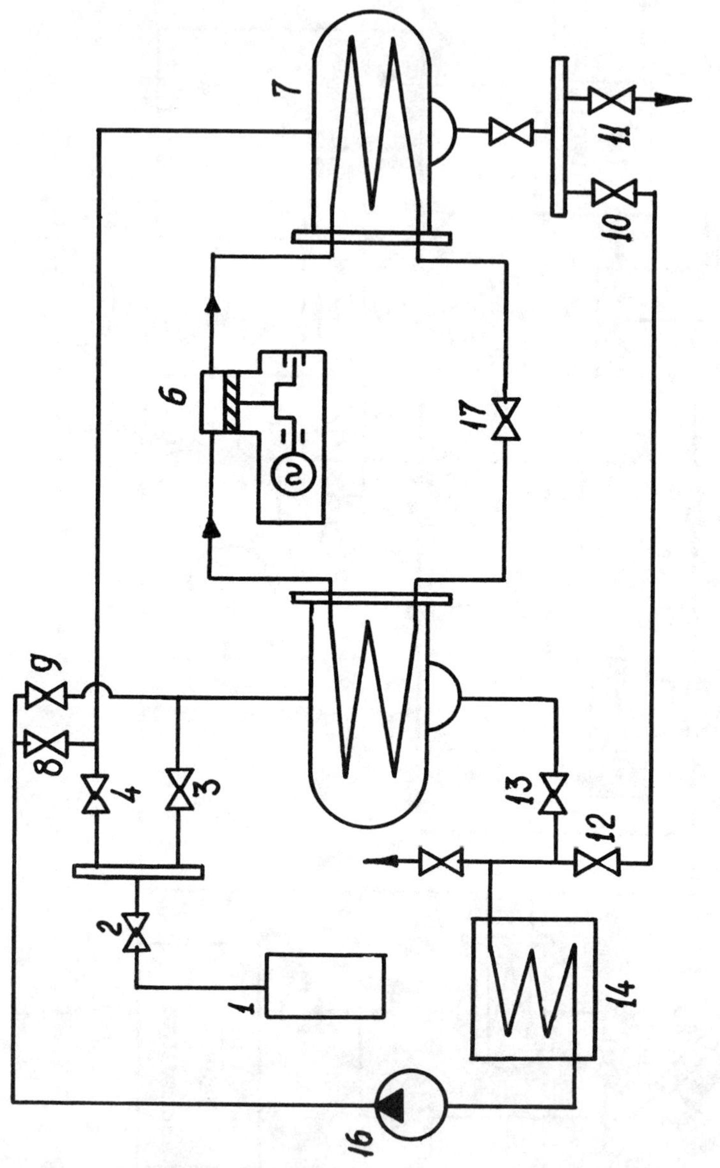

Fig. 2 - Geothermal thermocompressor plant

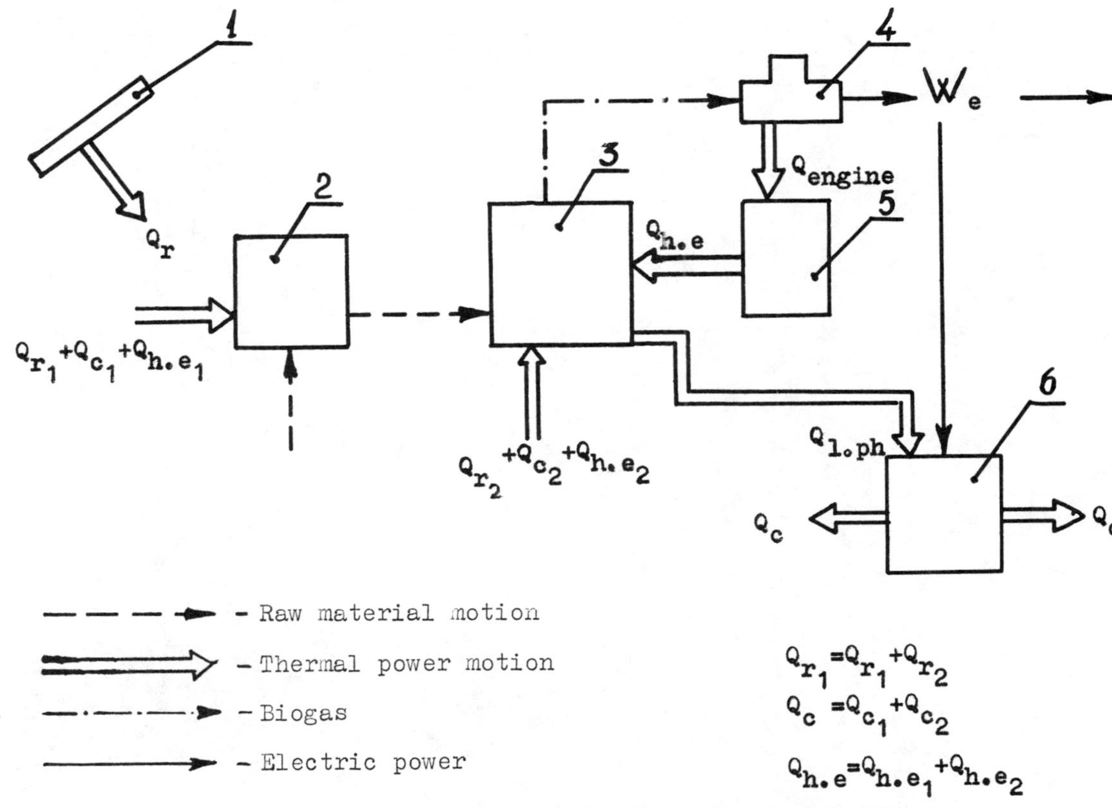

Fig. 3 - Biogas plant combined with TCP

INDUSTRIAL APPLICATIONS

LARGE SCALE INDUSTRIAL USE OF SOLAR ENERGY: THE ARAB POTASH COMPANY

James E. Jonish
International Center for Arid and Semiarid Land Studies
Texas Tech. University, Lubbock, Texas, U.S.A. 79409

ABSTRACT

The Arab Potash Company produces potash from the Dead Sea brine utilizing solar evaporation pans to condense and crystallize the brine. This industrial project, one of the largest in Jordan (investment cost over $460 million) produces for the export market which exhibits considerable year to year fluctuations in the world price. The efficiency and cost effectiveness of the solar evaporation process is critical to the success of the project.

The original design and subsequent modifications of the facilities are described. The initial years of operating experience were at a loss, but a positive return was achieved in 1988. Using a simple life cycle model and a company cost of capital of 10 percent, the levelized cost per ton of output at design production levels falls within the range of world potash prices experienced in the last five years. Thus, the project economics appear favorable.

1. INTRODUCTION

Jordan is a country with very few natural resources. In recent years, the Jordan balance of trade has been in deficit with remittances from Jordanians abroad a major item of foreign exchange earnings. The balance of payments has been accomplished through foreign assistance from neighboring countries and loans from western allies.

A major area of export potential is the chemical character of the Dead Sea brine. It contains a unique assemblage of salts, calcium, magnesium, potassium and bromine. It is expected that the Dead Sea will play an increasingly important role in the industrialization of Jordan. Mineral reserves are estimated at 43 billion tons.

The Arab Potash Company plant on the Dead Sea is a multi-Arab joint venture which was completed in 1982. It utilizes two of Jordan's abundant natural resources: solar energy and the mineral brine of the Dead Sea. Production of potash by the Arab Potash Company relies heavily on solar energy in the concentration and precipitation of the brine. The resulting potash product, which meets all international agricultural grade potash standards, has become a major export item in the Jordan balance of trade since 1985.

The purpose of the paper is twofold: 1) to describe the project design and operations of the Arab Potash Company with emphasis on the solar processes, and 2) to describe the current project economics, given the export promotion of the potash product. The company, in operation since 1983, had to incur several initial modifications to the plant design and processes to overcome initial design and production bottlenecks. Production in 1988 totaled 1.3 million tons.

Information on the Arab Potash Company has come from external sources, company records, interviews and on-site visits to the company during visits to Jordan in June and December, 1988.

2. THE ARAB POTASH COMPANY

More than four-fifths of Jordan is arid or semiarid in nature, and the country has two high plateaus, one in the east and one in the west rising to heights of 600-1000 meters. Between these mountain ranges lies a north-south valley that lies 200-400 meters below sea level. Through this valley flows the Jordan River into the Dead Sea.

The local climate in the Dead Sea region is extremely arid and hot with annual rainfall of less than 50 millimeters. Water evaporation is low due to dissolved salts but the Dead Sea level has been decreasing slowly for the last 30 years as much of the fresh water flow has been diverted to agricultural and industrial uses by Jordan and Israel.

The chemical composition of the Dead Sea brine is typically as follows:

Sodium Chloride	7.65%
Potassium Chloride	1.20%
Magnesium Chloride	13.76%
Magnesium Bromide	.48%
Calcium Chloride	3.80%
Water	73.11%

The average salinity of the Dead Sea is 300 grams of salt per kilogram as compared to an average of 30-35 grams per kilogram in the oceans. Estimates vary, but the total amount of dissolved salts is about 43 billion tons [1].

The major commercial product being exploited from the Dead Sea brine today is potash. Both Israel and Jordan have production facilities along the Dead Sea. More than 90 percent of the potash produced in the world is used for fertilizers. Canada is the world free market leader in potash reserves, production and exports, although the Soviet Union annually produces more. Other large producers are East and West Germany, France and Israel [2].

The international potash industry has significant reserves and current production capacity relative to the demand for the product. This has led to frequent price wars in the past and rationalization (i.e., cartelization) of production and markets [3]. Year to year variations in world prices of fertilizers are affected by not only variations in agricultural demands for fertilizers but the natural gas market (especially nitrogen) as well. From the perspective of Jordan, the potash industry is an export based industry with significant variability in export prices of the potash products.

The Arab Potash Company plant was completed in 1982. It is an Arab joint venture with the following shareholders (as of November, 1988):

The Government of Jordan	56.7
The Arab Mining Company	22.8
The Islamic Bank	5.5
The Iraq Government	5.2
The Libyan Government	4.3
The Kuwaiti Government	4.3
The Saudi Government	.4
Others	.8
	100.0

The initial capital cost to complete the plant was $460 million, of which 55% was international loans and 45% was provided by the equity shareholders above. The original plant design was for 1.2 million tons of potash per year [4].

Internationally, mining of potash varies considerably by location. In some cases open pit mining techniques are used. In Saskatchewan, the room and pillar method is used at 3100-3500 feet to mine seams that are 12 feet deep. Electrical mining equipment with 3-4 man teams produce more than 400 tons per hour [5].

At least three major producing facilities utilize solar evaporation in the production process. In the U.S., Bonneville Limited uses solar evaporation on the western edge of the Salt Lake basin in Utah, Israel's production is in the lower Dead Sea and Jordan's Arab Potash Company's plant is also in the southern end of the Dead Sea.

An overview of the production process for the Arab Potash Company is illustrated in Figure 1. The input source is the Dead Sea brine with four intake pumps drawing brine through a gravity flow brine canal to be concentrated in a series of eight solar evaporation pans. Here the brine is increasingly concentrated through the solar evaporation process [6].

The evaporation process results in precipitated carnallite which reaches a thickness of 40-60 centimeters. The bed is harvested as a slurry by four floating harvesters. Computer operated, the harvesters deliver the slurry to booster pumps on the shore via a floting line.

The carnallite is then pumped to the potash refinery plant through steel pipes for further processing. The refinery has a design capacity of 1.2 million tons per year of potash. The first refinery step is to dewater and decompose the carnallite slurry. The resulting solids from decomposition are called sylvinite (potassium chloride and sodium chloride).

The sylvinite is leached in a two stage process using heated brine from the crystallization process. The resulting thickener overflow is cooled successively in a five stage crystallizer system (from 98°C to 49°C). The potassium chloride decreases in solubility and crystallizes. Product dewatering and drying of the crystallized cake is accomplished by an oil fired rotary dryer. The product goes to a screening section where it is segregated into four grades: granular, coarse, standard and fine. Fine or standard materials can be sent to a compaction plant to be converted into granular product.

The finished product is sent to the plant storage warehouse (60,000 tons capacity) or loaded into storage bins to load one of forty trucks with a capacity of 50 tons each. Daily transit from the refinery to the port is via the Safi-Aqaba road, approximately 200 kilometers south.

Aqaba is Jordan's only commercial access to the world shipping lanes. The Arab Potash Company has a port terminal with storage capacity for 160,000 tons of potash. Loading facilities are modern and capable of loading vessels at rates up to 2000 tons/hour. The major export markets for Jordan's potash is India (1), China (2), Brazil (3), France (4), Korea (5), Malaysia (6), and Italy (7).

The major points of this overview are the following:

1) The project cost of the Arab Potash Company plant facilities are significant ($460 million), and particularly so for a developing country,

2) The potash product is directed at the export market, where overcapacity in the industry has led to extremely competitive conditions and fluctuations in product price,

3) The major source of energy used in the potash production process is the solar evaporation process used in the concentration stage which yields the precipitated carnallite. Conventional energy sources are minor and are used in the pumping of slurry, product dewatering and drying, and shipment to the port.

Therefore, at the Arab Potash Company, the solar energy component is a very large part of the total energy utilized and the product produced competes in a very competitive market. The cost effectiveness of the solar process is thus critical to the Arab Potash Company's economic viability.

3. SOLAR EVAPORATION PROCESS

The Arab Potash Project is located near the southern end of the Dead Sea, approximately 180 kilometers south from Amman, the capital and 200 kilometers north from Aqaba, the port city from which the potash product is shipped throughout the world.

Figure 2 illustrates the main components of the solar evaporation process [7]. Brine intake from the Dead Sea is pumped via four intake pumps (annual capacity 250 million tons of brine) to a gravity flow brine canal from a 1 kilometer pipeline. The brine flows about 10 kilometers where it reaches the salt pan (70.2 km^2), the largest of the evaporation pans with the highest solar evaporation rate in the system. When the brine reaches its optimum operating specific gravity (1.290 GMS/CC), it is pumped into the first and second precarnallite pans (PC-1 and PC-2) for further concentration and precipitation of sodium chloride.

The final five pans are the carnallite pans (C_1, C_2, C_3, C_5 and C_6) which are now operated in series fashion. Carnallite ($M_gCL_2.KCl.6H_2O$) is precipitated in the pans by solar evaporation until the average thickness of the carnallite deposited is 40-60 cm. The carnallite product is then harvested by floating harvesters that are computer controlled. The slurry is delivered to booster pumps and then to the refinery by a floating line attached to the harvesters.

Brine from pan C-6 which now contains very little potassium chloride is discharged into the return flow channel back to the Dead Sea.

The system depicted in Figure 2 differs from the initial design as a result of early operational experience in 1981-1983 [8]. Early results showed:

1) Evaporation in Pan C-1 was only 70-80 percent of that obtained in experimental trials. This would reduce the capacity of the system to less than 1 million tons per year from a design capacity of 1.2 million tons,

2) The brine transfer pumping capacity was inadequate, and

3) Brine in the salt pan was not of a uniform consistency.

Modifications in the solar pans included:

1) Converting carnallite pans from parallel flow to series flow improving solar evaporation efficiency,

2) Converting PC-1 to PC-2 brine transfer to gravity flow made additional pumping capacity available elsewhere,

3) Builling of a diversion dike in the salt pan improved brine consistency.

Coupled with modifications in the refinery recovery rate, production in 1989 is now estimated at 1.4 million tons of potash, or 16 percent over the original design capacity. Expansion to 1.8 million tons per year is anticipated by the mid 1990's [9].

4. PROJECT ECONOMICS

The size of the Arab Potash Company in Safi makes it one of the largest single investments in Jordan. There are both macroeconomic and microeconomic aspects of the project that are critical to the Jordanian economy and the economic viability of the solar evaporation process in potash production.

As an industrial concern, the Arab Potash Company is second only to the Jordan Phosphate Mine Company in terms of employment and industrial value added. At the beginning of 1988, the plant facilities employed 1335 persons. Of this number, 595 were engineers, chemists and technicians and 410 were truck drivers and day laborers. As Jordan has had significant unemployment among skilled people, including engineers, the large proportion of professional and technical people employed in the company helps to alleviate that problem. Formal and informal training programs are an integral part of the Arab Potash Company personnel policies. This will develop a core of trained people as Jordan decides to develop other products from the Dead Sea brine.

Infrastructure in the region around Safi was also developed in conjunction with the potash plant and solar pans. A town for nearly 3000 plant employees and dependents was constructed as well as necessary roads, electricity and water facilities.

The foreign exchange earnings potential of the Arab Potash Company is a major social benefit to Jordan. For many years, phosphates, fruits and vegetables dominated exports and phosphate exports represented 30 to 50 percent of all exports during the 1970's and early 1980's. Imports had been two to five times as large as exports, so Jordan has constantly had an unfavorable balance of trade since the 1948 war and the influx of Palestinians [10].

Given the gap between imports and exports, the balance of international payments for Jordan has been obtained through worker remittances of Jordanians abroad, government transfers or aid from other Arab countries, and soft loans from western countries [11]. With the prolonged slump in oil prices since 1984, aid from Arab countries is declining and Jordanian workers in nearby countries are being sent home. Thus, potash exports offer an opportunity to develop alternative sources of foreign currency earnings.

Since 1983, sales of potash have risen from 211,211 tons valued at 5,677,000 Jordanian Dinars (JD) to 1,222,373 tons valued at 27,417,000 JD in 1987. Since sales are directed to the export market, potash exports now rank second to phosphate exports. In 1988, potash exports were 20.7 percent of total exports of Jordan [12].

In addition to the above considerations, the operating results and experience gained in the production of potash via solar evaporation has led the Arab Potash Company to engage in technical and economic feasibility studies to diversify and expand their range of products and export potential. Such studies include potassium sulfate, sodium carbonate, magnesium oxide, bromine and compound fertilizers NPK.

The economic feasibility of the Arab Potash Company will depend heavily on the efficiency of the solar evaporation system and its impact on cost competitiveness since world market conditions will determine the revenue side and Jordan is not the price leader in this market.

As Table 1 illustrates, the initial years of start up and operations have resulted in net losses each year from 1983-87 [13]. In 1988, due to higher potash prices in the world market, and the APC production reaching and even exceeding design capacity at 1.309 million tons, a net profit of 6,800,000 JD was obtained. In the early years, subsidies from the Jordan government covered the operating losses.

Figure 3 illustrates the historical behavior of average cost, production and average revenues for recent years. It appears that as production has reached or exceeded design capacity (1.2 million tons), average cost is falling within the band or range of average price fluctuations over the same time span. Thus, the project appears capable of yielding a return on investment. The average cost depicted in Figure 1 does not include a return to shareholder equity (some $260 million), but it does include charges for debt or loan repayment.

A simple life cycle levelized cost approach might be more instructive in this regard. Let:

(1) $LEC = \frac{K \cdot CRF + MO}{CAP \cdot CU}$

(2) $CRF = \frac{R(1+R)^N}{(1+R)^{N-1}}$

where LEC = Levelized cost/ton

K = Investment cost of plant (183,000,000 JD)

CRF = Capital recovery factor

MO = Annual maintenance and operations cost (16,476,000 JD)

CAP = Annual capacity of plant (1,400,000 tons)

CU = Capacity utilization (.86)

R = Company cost of capital

N = Life of project (30 years)

The levelized cost of production varies directly with the company cost of capital and thus the capital recovery factor (CRF). It is this levelized cost which must be covered by the world price or average revenue. With an R=.20 (and CRF=.201), the levelized cost per ton (covering all costs including debt service and a return on equity) is 44.38 JD. With an R=.10 (and CRF=.106), the levelized cost is 28.90 JD per ton. The latter cost is within the range of average revenues or prices received for the potash in the last few years. The former is not.

In summary, the Arab Potash Company with its use of solar evaporation pans is making a number of significant contributions to the Jordan economy. It is a large employer of professional and technical personnel, its exports provide needed foreign currency and a vehicle to develop further products from the Dead Sea brine.

Early start up operations operated at a loss, with subsidies provided by the government. As production reached and then exceeded design capacity, the project economics improved and resulted in an initial year profit in 1988. Using a simple life cycle model with a CRF=.10, a levelized cost of 28.90 JD per ton falls in the interval of average revenues received over the past 5 years, suggesting that the project economics will continue to be favorable while recognizing the extreme year to year variability in prices.

Since world prices are not determined by Jordan's production, efforts to improve the economic performance of the company will focus on increasing productive capacity or modifying techniques in production reducing costs per unit.

REFERENCES

1. The Arab Potash Company Ltd. Booklet. N.D.

2. U.S. Department of Interior Minerals Yearbook, Vol. I. Metals and Minerals, 1986. pp. 755-766.

3. Louis Kurrelmeyer, The Potash Industry: Analysis of Recent Developments (University of New Mexico Press: Albuquerque, 1951).

4. The Arab Potash Company Ltd. Research and Development Department. Potassium Chloride 1/RD/86.

5. Richard Shaffner, New Risks in Resource Development: The Potash Case. MRI Observation Series No. 12 (C. D. Howe Research Institute: Montreal, 1976).

6. On site visit to plant site December, 1988 and interviews with the Plant Manager and company personnel.

7. The Arab Potash Company Ltd. Booklet. N.D.

8. The Arab Potash Company Ltd. Research and Development Department. Potassium Chloride 1/RD/86.

9. Personal Communications with Mr. Nasser Al-Sadoun, Plant Manager of Arab Potash Company. Dated March 8, 1989.

10. Darrel Eglin, "The Economy," in Richard Nyrop (ed.), Jordan: A Country Study. Foreign Area Studies Handbook Series (The American University: Washington, 1979).

11. International Monetary Fund. Balance of Payments Yearbook. Various Issues.

12. Central Bank of Jordan, Monthly Statistical Bulletin. June, 1989.

13. The Arab Potash Company Ltd. Annual Report 1987.

TABLE I: REVENUES AND COSTS: ARAB POTASH COMPANY
(JD 000's)

Total	1983	1984	1985	1986	1987
Sales (tons)	211,211	449,608	932,668	1,123,096	1,222,373
Revenues	5,677	14,316	28,769	29,821	31,278
Variable Costs	8,145	10,134	15,534	16,414	16,476
Capital Costs	11,372	13,898	19,760	18,382	18,691
Net Profit (Loss)	(13,840)	(9,716)	(6,525)	(4,975)	(3,889)
Unit Revenue/Cost					
Average Revenue	26.88	31.84	30.85	26.55	25.59
Average Cost	92.40	53.45	37.84	31.01	28.77
Unit Profit (Loss)	(65.52)	(21.61)	(6.99)	(4.46)	(3.18)

Source: Derived from The Arab Potash Company, Limited, Annual Report 1987, pp. 24-25.

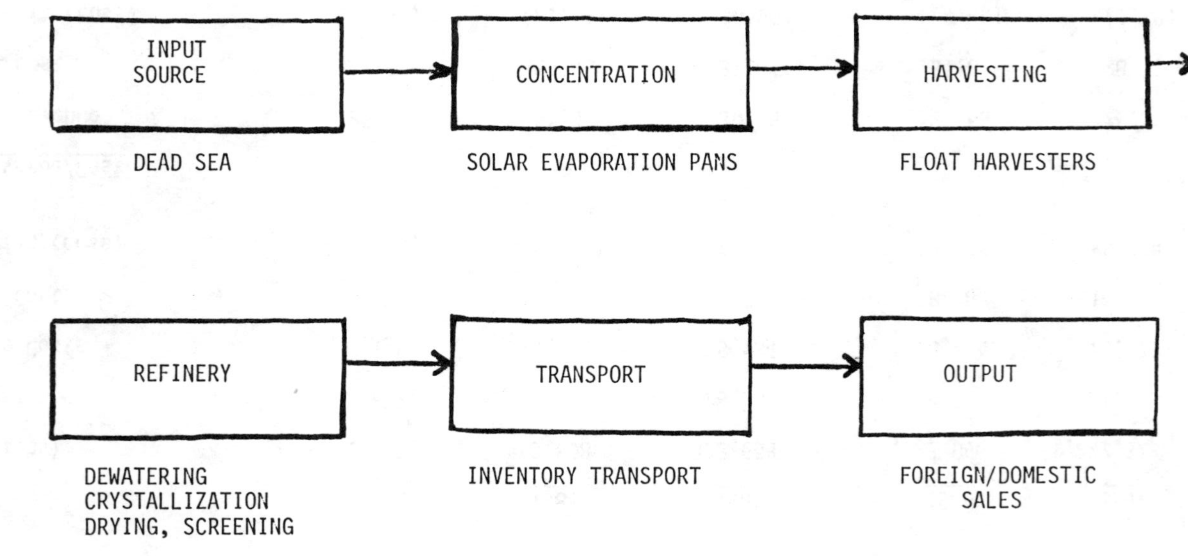

FIGURE 1: PRODUCTION PROCESS: ARAB POTASH COMPANY

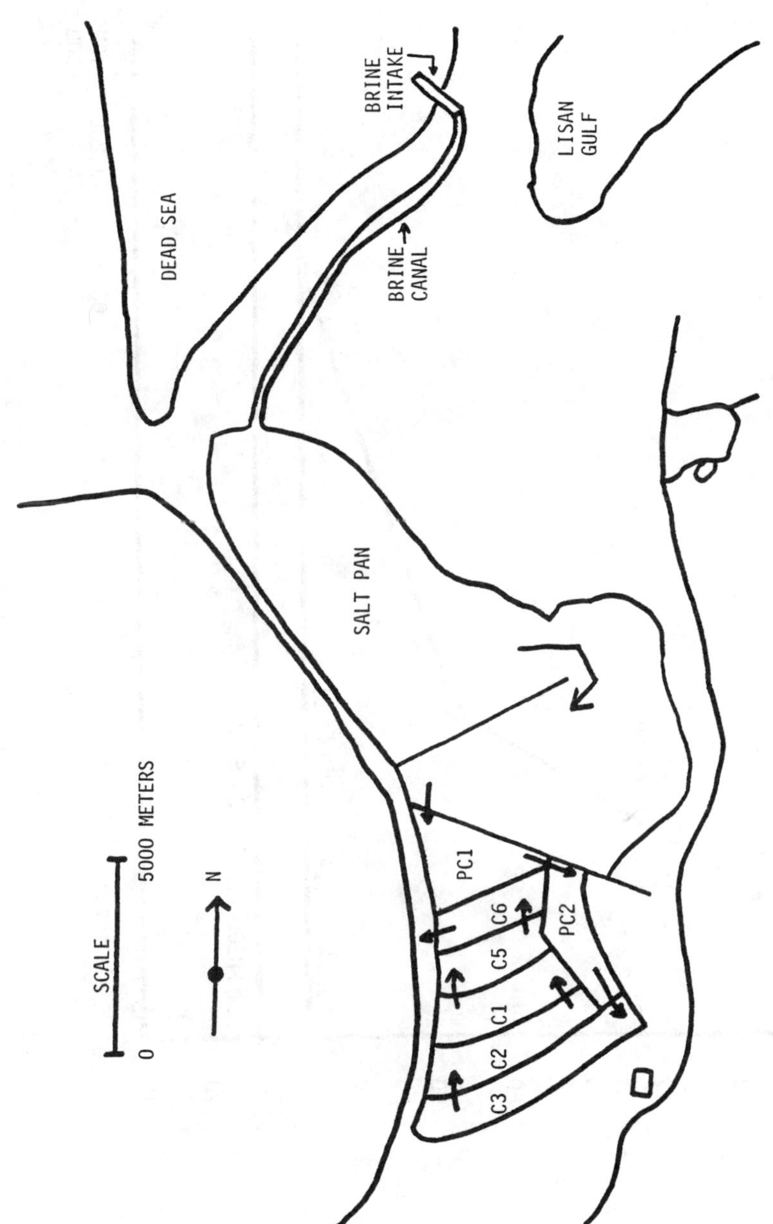

FIGURE 2: SOLAR EVAPORATION SYSTEM

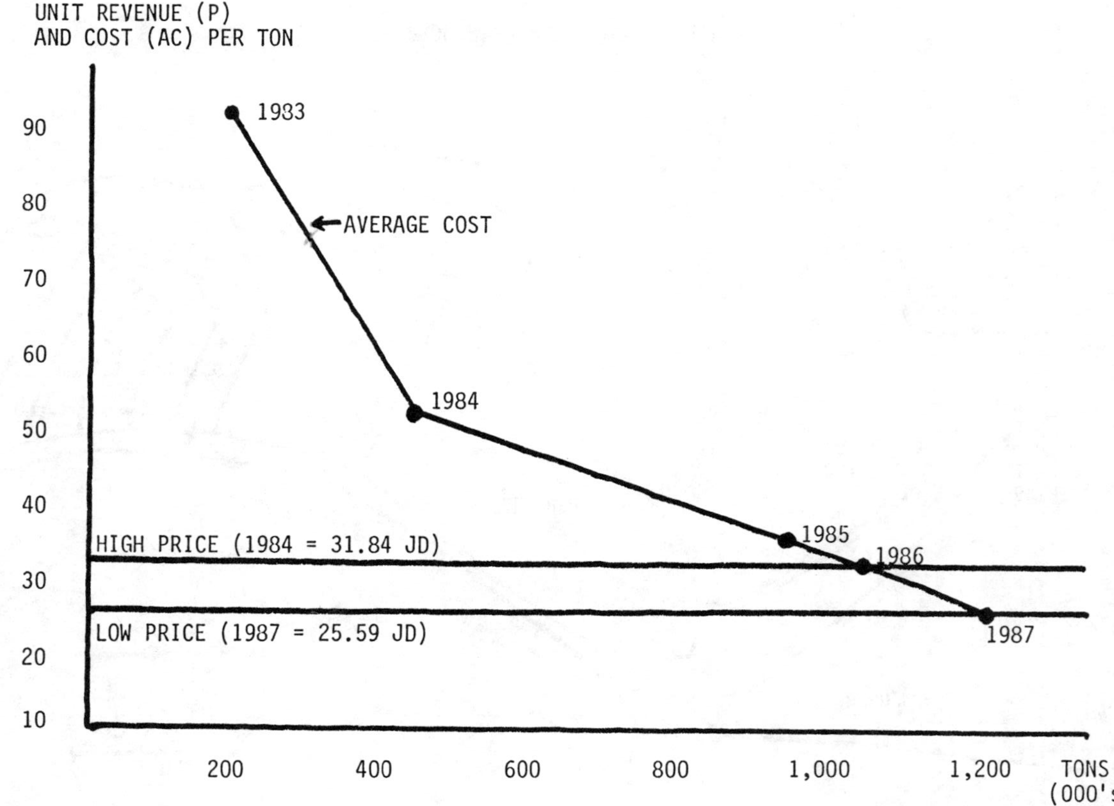

FIGURE 3: UNIT REVENUE AND COST

AN EMPIRICAL STUDY OF ENERGY MANAGEMENT CONTROL SYSTEMS: OBSERVATIONS, RECOMMENDATIONS, ADVICE

H. Norman Clark,
Bonneville Power Administration, P.O. Box 3621, Portland, Oregon 97208
and
Susan Majewski
Branch, Richards, Anderson & Co., Seattle, Washington, U.S.A.

ABSTRACT

This two-phase study establishes guidelines for installing computer-based Energy Management Control Systems (EMCS's) in commercial buildings. These guidelines are derived from studying the effectiveness of a variety of existing EMCS installations. The guidelines will be used in current and future Bonneville Power Administration (Bonneville) conservation programs.

This first phase reports preliminary recommendations on the suitability of EMCS's installed in buildings in the Pacific Northwest. The recommendations are based on an assessment of the effectiveness of twenty-two EMCS installations that are currently in operation.

A preliminary evaluation showed that not all EMCS's are obtaining projected energy savings. A detailed technical evaluation was initiated to determine criteria for their future use. This study evaluates EMCS's in field operations.

The project's first phase is complete. It gathered information on existing EMCS installations in Washington, Oregon, and Idaho. The work consisted of two parts, the first reviewed and summarized available files for each installation.

In the second part, each site was visited and building operating personnel were interviewed. Each interview determined the institution's expectations and satisfaction level regarding the performance of the EMCS. Additionally, the interviews were used to verify file information and to record changes in the facility subsequent to EMCS installation.

The combined results were used to develop a list of observations, recommendations, and advice on EMCS use. This list covers an array of factors associated with EMCS's, from the planning stages through installation, as well as operation and maintenance. These observations and recommendations provide information on the appropriateness of installing EMCS's in both these specific building situations and in general.

Phase II will involve field verifications of EMCS performance to determine the effectiveness of the installations. Results from Phase II are expected in December 1990.

Background and Objectives

The purpose of this two-phase study is to establish guidelines for installing Energy Management Control Systems (EMCS's) in commercial buildings. These guidelines are derived from studying the effectiveness of a variety of existing EMCS installations. The guidelines will be used in current and future Bonneville Power Administration (Bonneville) conservation programs.

This first phase consists of preliminary recommendations on the suitability of EMCS's installed in buildings in the Pacific Northwest. These buildings are located throughout the region and had EMCS's installed under Bonneville's Institutional Buildings Program (IBP). The recommendations are based on an assessment of the effectiveness of twenty-two EMCS installations that are currently in operation. The assessment was performed for Bonneville by Branch, Richards, Anderson and Co., a Seattle, Washington, engineer firm.

The Pacific Northwest Electric Power Planning and Conservation Act, signed into law December 5, 1980, mandated energy conservation practices. Under the Act, Bonneville is required to identify and implement cost effective electric energy options and programs for the Pacific Northwest. To support Bonneville's efforts, the Commercial Technology Section of the Division of Resource Management designs and carries-out various research and development projects, of which this study is one.

The IBP was a six-year program (October 1982-September 1988) sponsored by Bonneville to achieve energy savings in institutional buildings. The program was administered by the state energy offices in Washington, Oregon, Idaho, and Montana. The energy offices arranged for the installation of energy conservation measures in buildings owned and operated by hospitals, schools, governments, and Indian Tribes. Energy Management Systems were eligible for fundings under the IBP and more than 100 were installed under the program. Essentially, an EMCS is a computer-based system that controls a building's energy-using equipment to optimize energy consumption.

A preliminary evaluation of the program by Bonneville showed that not all EMCS's are obtaining projected energy savings. A detailed technical evaluation was initiated to determine criteria for their use under current and future Bonneville conservation programs. This two-phase study is designed to evaluate EMCS in-field operations.

Project Approach and Findings

The project's first phase is complete. It focussed on gathering information on existing EMCS installations in Washington, Oregon, and Idaho. Twenty-two sites were chosen for evaluation. The actual work consisted of two parts, the first of which was reviewing and summarizing available files for each installation.

The material collected was used to develop information "packets" for each installation in the study. The packets include:

- File summary information (building type, age, area, mechanical system description, lighting, and conservation measures in use);
- Photographs of the building;
- Local maps;
- EMCS manufacturer information; and
- When available
  the operating schedule at time of site visit,
  a listing of active control points and zones,
  control diagrams, and
  past energy consumption.

These packets are summarized in the Bonneville Report of April 1989, titled: Phase I Energy Management System Study. These packets will be used for the Phase II work.

The second part of Phase I involved visiting each of the installations in the study and interviewing personnel at each site, usually an administrator and the person most familiar with the operation of the EMCS. Each interview was conducted using a standardized form to determine the institution's expectations and satisfaction level regarding the use and performance of the EMCS. Additionally, the interviews were used to verify file information and record and relevant changes in the facility subsequent to installation of the EMCS.

The combined results of the file reviews and the site interviews were used to develop a list of observations, recommendations, and advice on EMCS use. This list covers an array of factors and considerations associated with EMCS's, beginning with the planning stages through installation, as well as operation and maintenance. These observations and recommendations provide preliminary information on the appropriateness of installing EMCS's in both specific building situations and in general.

Phase I generalizations have been formed using information found in the IBP project files and from on-site interviews. Also, Phase I gathered background and support information for the Phase II work.

Phase II will involve field verification and analysis of EMCS performance to determine the effectiveness of the installations.

Building Selection

In April 1988, twenty institutions with EMCS's installed during the IBP were selected by Bonneville for this study. Two institutions were added later. Ten of the selected EMCS's are in Oregon, ten in Washington, and two in Idaho.

Two substitutions were made to the original Washington selections. A social services facility was substituted for a court house. The court house has been completely remodeled and the original court house EMCS has been replaced. (The original unit has been reinstalled in another location.) The file review found that a middle school did not have and EMCS installed. A high school was selected as a substitute.

Process

The work was separated into two parts designated below as A and B. B.

Part A

Part A consisted of gathering background information on the selected IBP projects. This required a review of the project files for each installation.

Information for each project was recorded on a summary form. These forms, with supplemental pages from the records, have been collected as "packets" for each installation. This information includes the building type, age, floor areas, mechanical system descriptions, lighting, and the conservation measures installed under IBP with their estimated savings and costs. When available, project specifications have been included in the packets.

Part B

Part B involved a visit to each institution and an interview with their personnel. Typically there were two interviews, one with an administrator and another with the maintenance person most familiar with the EMCS operation. These visits also were used to verify the information found in the files, and to note conditions that may have changed since the EMCS was installed. An "on-site" survey form was developed for Part B responses.

The form was designed to help determine:

- How the institution feels about conservation.
- How they like the EMCS.
- How they use the EMCS at present.
- Any changes to the original EMCS installation.
- How the building occupants respond to the system.
- What kind of problems have occurred with the EMCS.

The opinions and feedback gathered during "on-site" visits were used to develop the observations, recommendations, and advice portions of this report.

Special Note

One system selected for the study is not an EMCS. The facility has warm-up and night set-back controls installed as separate controllers. Although not an EMCS, it is a control system that is reliable and satisfies the occupants. This installation can be contrasted to the multi-thousand dollar systems.

General Observations

- Building owners and operators with simple mechanical systems such as baseboard resistance and fixed operating schedules are happy with simple EMCS equipment, for example, central time clock functions.

- Most institutions are satisfied with a one year warranty on the EMCS. They do not recommend a service agreement and see other ways to spend their limited funds.

- To administrators, it is important that the EMCS is going to save them money.

- "Overrides" with fixed time settings give the occupants some additional flexibility and improve their opinions of their system's reliability.

- Perceptions of how well the system functions are based on how well the HVAC system performs.

- An EMCS will not cure inherent control malfunctions or shortcomings.

- Success of the EMCS is based on the quality of the system to which it is retrofitted.

- Most operators can troubleshoot the HVAC system. Several operators use the EMCS to help identify HVAC problems. However, if something goes wrong with the EMCS hardware, they call the installer or the manufacturer to fix it.

- Many institutions have only one person who is comfortable working with the EMCS.

- Most facilities employ a person who is familiar with the operations of their EMCS. At the locations where this is not the case, modifications to the operating schedules are limited.

- The operator can make a big difference in the effectiveness of the EMCS installation. The most successful operators understand their mechanical systems and use their EMCS aggressively. They constantly modify and check their operating schedules and they use the EMCS to monitor their HVAC equipment.

- Building operators know that the EMCS is reducing the amount of energy used while maintaining comfort in the building.

- Some building operators understand how the EMCS is saving energy.

- Most operators believe they know their HVAC systems and EMCS. However, only those who understand their EMCS, and work with it, achieve more than minimal savings.

- In schools it is common to find a custodian in charge of day-to-day programming of the EMCS. Usually there is a maintenance supervisor with a thorough knowledge of the mechanical system and EMCS. This supervisor can review the programming and correct entries with which he disagrees.

- The more the operator understands the mechanical system and the EMCS, the more likely the EMCS installation will be successful.

- When the operator is not sure of how to operate the EMCS, very few changes are made to the programming.

- At places where the administrators are not confident of their operators' abilities to use the EMCS, few programming changes are allowed.

- The institutions achieving significant savings usually have at least one person aggressively using the EMCS and the information they extract from it. Often these employees like their job and enjoy the additional control the EMCS gives them.

- Where the institution hired an outside consultant to come-in and modify EMCS programming, scheduling changes are infrequent and frequent uses of the overrides occur.

- Institutions with more than one EMCS recommend using one brand of EMCS throughout the installation. These institutions also find a central control computer to be helpful.

- Many of the institution's personnel mentioned that the IBP let them fund projects that the institution would never have been able to do on its own. In many cases all conservation efforts were developed because of the IBP program.

Recommendations

Prior to EMCS Selection:

- Select and EMCS that will be installed by a company which is familiar with that particular brand.

- Select one manufacturer to supply all controls for the building. This simplifies both maintenance and service.

- Match the EMCS selection to the HVAC system complexity and size.

- Require a one-year warranty for all systems.

- Provide timed manual controls for occupants' spontaneous use.

- Consider a service agreement for complex systems.

Prior to Installation:

- Get the mechanical system in top condition.

- Tell the occupants of the building what the system will be doing and their role in its effective operation.

- Develop a procedure by which the building occupants will request schedule or temperature changes.

After Installation:

- Conduct a "fine tuning" period when the EMCS is first operating.
- Familiarize at least two people with the proper operation of the EMCS. Buildings with only one operator may have difficulties if the operator leaves; a new operator may take a long time to learn how to monitor, adjust, and use the system.
- Educate occupants on what the EMCS is doing and how to arrange schedule modifications.
- Encourage conservation and technical training for maintenance personnel.
- Demonstrate energy accounting methods to the building's maintenance staff and business administrators.

Advice

The selection of the EMCS designer and specifier should be carefully considered. This person should be a specialist with good experience and reputation. The designer should become familiar with the mechanical system, building use, and the future system operator before the EMCS is selected.

Factors to consider when recommending an EMCS installation:

- Energy savings potential.
- Building occupancy patterns.
- Installation contractor reliability.
- Improvement in consistency of energy savings.
- Technical expertise of the maintenance staff.
- Complexity of systems to be controlled.
- Number of systems to be controlled.

Considerations for central control of EMCS's:

- Number of EMCS's managed by one maintenance group.
- Distance between EMCS installations.
- Complexity of EMCS installations.
- Technical expertise of maintenance staff.
- Administrative support for maintenance work.
- Lead time for scheduling changes.

The designer of the EMCS should be very familiar with:

- The brand and model of EMCS installed.
- The mechanical system to be controlled.
- The institution's needs.

The installer of the EMCS should:

- Have a good record.
- Be based near the institution.
- Train the maintenance staff.
- Provide follow-up service.
- Allow for a fine-tuning period after installation.
- Be very familiar with the brand of EMCS installed.

<u>Observations on Conservation</u>

- Institutions are concerned about the amount of energy they use; however, the focus of their concern is usually first cost.
- Most institutions in this survey do not have a formal conservation/energy management program.
- Maintenance and operation departments usually have constrained budgets. Reducing funds for maintaining equipment and facilities is typically one of the first steps in a budget cut. There seems to be a perceived difference between paying to maintain vs. paying to replace equipment. This same perception seems to apply to investing money in conservation now, or paying increasing utility charges. In both cases it seems that paying a large amount of money for either a necessary replacement or an operating expense is preferred to investing additional funds now to reduce expenses later.
- Most institutions are not aware that continuing efforts can consistently reduce their energy costs. Often, maintenance is viewed as an expense rather than an investment. This same perception applies to conservation: only the initial expense is seen; the long term savings are not tracked and not recognized.
- An absence of energy conservation education was observed throughout the study. Examples include:
  - Poorly maintained mechanical systems.
  - Lack of funding for efficiency improvements.
  - Little or no training for maintenance staff.
  - Reductions in maintenance staff.

- The low priority given to conservation presents a problem for all efforts including EMCS installations. In many cases more can be done with the EMCS installed. Many operators and maintenance personnel don't know how to achieve additional savings with their EMCS. They are not confident using the systems they have. Most operators want additional training.

- In general, institutions do not invest in conservation and technical training. Their operating budgets do not include funds for these functions. They do not see that conservation and proper maintenance can reduce operating expenses. This perception affects not only the use of their EMCS, but the performance of all energy using equipment.

- When conservation investments are made, little effort is made to track and record the results. Most of the maintenance staff are aware of the need for improved training and equipment, but the public, and those in charge of budgets generally are not.

Additional Recommendations

In response to the conservation attitudes observed, it is recommended that a conservation education program be designed for institution administrators, budget advisors, and the governing boards of institutions.

This program should emphasize:

- How conservation can reduce operating expenses.

- How maintenance can:

  Produce additional energy cost reduction.

  Maintain these reductions.

  Lower operating budgets.

  Prolong life of major equipment.

  Improve environmental conditions.

- Conservation and maintenance programs as investments.

- Some basic evaluation techniques.

Future Efforts

Phase II of this study will consist of a detailed technical evaluation of the same twenty-two EMCS installations. The work will include in-field verification and analysis of the performance of each EMCS. This second phase of the study is designed to determine whether each EMCS, as currently installed, is

- suitable for the operating characteristics of the building and;

- presently achieving or capable of achieving projected energy savings.

The results of Phase II are scheduled for Spring 1991.

(VS5-RMC-2451c)

DESIGN OF INDUSTRIAL PUMPING STATIONS: A COMPUTERIZED APPRORACH

M. Coluccia and D. Miconi
Department of Energetics - University of Pisa
Via Diotisalvi, 2 - 56100 Pisa - Italy

Abstract

The present study deals with a computer simulation code engineered at the Department of Energetics (University of Pisa) for the design of pumping stations. The problem is to define the type, size, number (including auxiliary), the linking of each pumping unit and above all to define the flow rate regulation system of the pumping station in order to minimize the sum of the investment and operating costs. In particular, the problem of the flow rate regulation is very important to date when we consider the energy saving aspect. The code is characterized by its high degree of flexibility with respect to the nature of the fluid being pumped, the type of plant configuration adopted, and the variability in the supplied flow rate. A numerical application of the code to a particular case, reveals its principal advantages.

1. INTRODUCTION

The ever increasing complexity of industrial plants and the need to maintain their low costs has stimulated the use of a computer design to meet this requirement.

The use of simulation models permits a rapid search for an optimal solution. This is effected by means of a fast determination of numerous alternative solutions. This technique seems to be well suited for the present study which is aimed at optimizing a particular type of pumping station. Due to the high power inherent within the pumping station, a possible error in its design could cause large energy losses.

The pumping station to be optimized has to satisfy the utility requirements by means of a connecting hydraulic network with well known hydraulic characteristics.

The absence of a collecting reservoir located downstream from the pumpimg station makes it difficult to investigate the characteristics of this type of station. The station should be capable of adapting to different consumption requirements and this can be attained by modifying its functioning characteristics.

Obviously, an optimization study cannot be carried out if the regulation problem is not taken into consideration. Consequently, we will examine the more frequently used regulation systems.

2. THE REGULATION SYSTEMS

The operation of the pumping systems (composed of N pumping units in parallel, of which one is the reserve) could be adjusted in different ways, /1,2/. Limiting our study to the widely used regulation systems, we will consider the following plant configurations:

-pumping station which feeds into an elevated collecting reservoir;
-pumping station which directly feeds the hydraulic network.

Pumping station which feeds into an elevated collecting reservoir

It deals with a widely used regulation system that allows for the complete separation between the pumping station and the hydraulic network. The collecting reservoir is situated at a height so as to guarantee an adequate supply of flow to the most unfavourably located user.

The collecting reservoir is located downstream from the pumping units and regulates, by means of a feedback mechanism, the switching on-off of the pumping system (activated at a low water level and deactivated at a high water level) (see fig.1-A).

The uncoupling of the pumping station and hydraulic network allows for the optimal functioning of the pumping unit, assuming we have a constant suction head /1/.

The advantage of working under optimal conditions is minimized if we consider the high cost of the collecting reservoir. In addition to having an on-off switch, the reservoir offers two added functions that determine its size: the first is the storage of water, and the second is the certainty of having a continuous water supply.

Pumping station which directly feeds the hydraulic network

In the absence of a collecting reservoir, the pumping station should be capable of modifying the water supply (by means of a specific regulation system) so as to compensate for the ensuing fluctuation in the demand /2/. During its operation, the pumping station could function in non-optimal conditions.

In order to reduce to a minimum the energy loss, a series of pumps are employed (with a fixed speed) which supply a basal flow rate, while the additional fluctuations in the demand are compensated by elementary regulation systems. The switching on-off of each pump could be controlled by pressure or flow rate.

In the first case, it is necessary to add (in parallel of network) a water pressure tank which serves the dual function of regulation (also for a very small demand) and protection against water hammer. The water pressure tank is equipped with a feedback mechanism that regulates the on-off switch of each pumping unit. The pressure variations within the tank are generated by the difference in the supply and demand (see fig.1-B). The main advantage that accompanies this system is its simplicity, but this is counterbalanced by the high cost of the water pressure tank and by the non-optimal operating conditions.

In the second case, a flow meter (placed downstream from the pumping station) controls the on-off switch of each pumping unit in relation to the demand in flow rate (see fig.1-C). If the flow meter's precision remains constant throughout a wider range, it is possible to control a larger number of pumps with respect to the first case (limited by the necessity of separating the intervening threshold of the regulation system) and therefore cover a wider range of flow rate. For this plant configuration type the problem of small demands is frequently encountered as well as the necessity of maintaining a predetermined pressure in zero flow rate. This is resolved by activating a pumping unit by means of a pressure controlled mechanism (small water pressure tank).

3. ELEMENTARY REGULATION SYSTEMS

Apart from subdividing the pumping station into some pumping units, we will investigate the principal mechanisms that permit the adaptation of the supply of one pumping unit to the demand variation. With reference to fig.2, we will take into consideration the following elementary regulation systems: employment of a check (in series to the pumping unit) valve or a bypass check valve, variation in rpm of the pumping unit.

System equipped with a check valve (in series)

The supply is adjusted to the demand variation through the activation of a check valve located downstream to the pumping units /3/. This type of regulation gives rise to an energy loss caused by two factors: the loss inside the valve and the non-optimal operation of the pumping unit (see fig.3-a).

System equipped with bypass check valve

A check valve, mounted in parallel to the pumping unit, controls the backflow of water into the aspirating reservoir, in order to supply the demand (see fig.3-b). This system also prevents the occurrence of anomalous conditions of the pumping unit when subjected to small demands /4/.

System equipped with a variable speed pumping unit

The speed of pumping unit is adjusted step by step in order to furnish to the fluid the required energy to satisfy the demand /5, 6/. The functional range of the pumping unit is outlined by three curves: $H_p(Q,N_{min})$, $H_p(Q,N_{max})$ and η_{min}. N_{min} and N_{max} represent the acceptable limit values for the speed, while η_{min} is the acceptable minimum value for the efficiency of the pumping unit. In general, this regulation system allows the pumping station to save a considerable amount of energy.

4. THE SIMULATION CODE

The code is characterized by its open structure and modularity. It permits the rapid simulation of many plant configurations, and furnishes the relative technical and economic informations that can be utilized in decision making.
These are the basic input data: the demanded flow rate as a function of time, Qr(t); the hydraulic characteristics of the network; the functional characteristics of one or more series of commercial pumps (pressure to flow rate characteristics; power to flow rate characteristics; recommended functioning range); the hydraulic characteristics of each accessory; the necessary data cost for the evaluation of plant and operating costs of the pumping stations. All of the above mentioned inputs, including those concerned with the simulation process (convergence criteria, simulation steps, ect.), are grouped in the appropriate files so as to permit their easy handling (see fig.4). For each regulation system mentioned above, the simulation code is capable of generating different pumping system configurations, to evaluate for each of them the number of necessary pumping units, and is also capable of simulating their operation. Table I illustrates the conditions of acceptability of the pumping unit. These conditions are valid only if the pumping station is composed of one active pumping unit and employs an elementary regulation system (see fig.3).

The subsequent completion of the simulation of the plant operation allows the code to calculate the absorbed energy from the pumping station at the investigated time.

The final operation involves the calculation of the total cost (plant and operating cost) of each simulated plant configuration /8/. The total cost is determined by a specific routine which takes into consideration the estimated life span of the pumping station. Fig.5 illustrates a schematic flow chart which shows the entire simulation process. At the present stage the simulation code is composed of a main program plus 30 routines (written in fortran 77 for IBM or IBM compatible personal computer). The program also works interactively with the user.

The main program controls the input and output files, the passage between the subroutines (effected by common or labelled common instructions) employed for the specific simulation of the plant under study, and the immediate updating of the parameters on which each subroutine operates.

The different routines could be grouped according to their specific functions as outlined below:

- simulation routines: these are capable of generating different plant configurations and simulate their operation while representing a crucial part of the code;
- mathematical problem solving routines: include several algorithms /7/, finalized at polynomial and exponential interpolation, solving the root of the equation, sorting of vectors and matrices, integration, etc.;
- general calculation routines: permit the calculation of the cost and performance indexes which characterize each plant configuration and allows a comparison between alternative solutions.

The code outputs are stored in two seperate files. The first file contains the alternative generated plant configurations with its corresponding type of pumping units employed. The second file holds information pertaining to the energetic performance of each configuration plant (instantaneous and integrated during simulation) and their respective cost.

The developed code allows the storage of the results in an appropriate file that can be used as input for other programs that are aimed at a more indepth analysis of the plant configuration under investigation (see fig.4).

5. AN APPLICATION OF THE CODE

The code is currently being tested and perfected. An extended trial program was created aimed at the application of the code to the design of pumping stations with an ever increasing complex configuration.

An application of the code is carried out with the inputs plotted in fig.6 (typical daily and monthly demand, network characteristic) and in fig.7 (a series of commercial pumping units).

The three configurations shown in fig.3 are take into consideration. Pumping station is characterized by two pumping units in parallel (one auxiliary) equipped with an automated regulation system.

Fig.8 shows the minimal energy E_m needed to satisfy the demand (curve m) and the energy $E_{(a,b,c)}$ consumed in the three optimal pumping stations (curves a, b, c).

Fig.9 portrays the efficiency of three optimal pumping stations calculated on a monthly basis:

$$\eta_{(a,b,c)} = \frac{E_m}{E_{(a,b,c)}} \qquad (1)$$

Fig.10 and fig.11 dictate the cumulative progressive energy consumption (E^*_m and $E^*_{(a,b,c)}$) and the progressive energetic efficiency ($\eta^*_{(a,b,c)}$) of the optimal pumping stations. All the curves refer to a generic year of the plant's life.

In fig.12 is shown the present value of the total cost of the three optimal pumping stations, assuming a ten year life span.

Table II shows the most significant values obtained through the application of the simulation code. The pumping station equipped with the variable speed pumping units seems to be the most economically convenient solution.

6. CONCLUSION

The first applications of the code, which are only a part of a large program of trials aimed at its verification, proved the flexibility, rapidity and efficiency of the code. For each possible regulation system, the code defines a certain number of pumping stations which all proved valid to satisfy the demand. The code also simulates for each pumping station its operation (in its life span), and also evaluates its respective energy consumption and performance indexes (economic and energetic). In conclusion, the code simplifies the designer's task and enables him make better decisions. This major decision making capacity enables him to avoid the possibility of uselessly oversizing the pumping station caused by the designer's fear concern of its insufficiency to meet the demand.

7. NOMENCLATURE

- C Present value of the total cost, Lit;
- E energy, kWh;
- H hydraulic head, m;
- N speed, rpm;
- P electric power, kW;

- Q flow rate, m^3/h;
- t time, h;
- η efficiency.

Subscripts
- d Right;
- p pumping unit;
- r hydraulic network;
- s left.

Superscript
- * Cumulated.

REFERENCES

/1/ Association Générale des Hygiénistes et Techniciens Municipaux; "Les Stations de pompage d'eau"; Technique et Documentation, Paris, 1981.

/2/ J.Jost; "Alimentation en eau potable d'un réseau de distribution par pompage a la demande, sans réservoir en charge sur la réseau"; Travaux, n.12, 1969.

/3/ G.Supino; "Idraulica generale"; Ed. R.Patron, Bologna,1965.

/4/ R.Segù; "Pompe acqua circolazione e risparmio energetico"; Impianti, n.2, 1987.

/5/ I.Karassik, W.Krutzsch, W.Fraser, J.Messina; "Pump Handbook"; Ed.McGraw Hill, 1987.

/6/ I.Karassik; "Pompe centrifughe e sistemi idraulici"; Ingegneria e fluidi, n.1, 1984.

/7/ W.T.Vetterling, S.A. Teukolsky, W.H. Press, B.P. Flannery; "Numerical Recipes Example Book", Cambridge University Press, 1985.

/8/ D.Miconi; "Criteri nell'analisi di convenienza economica dei progetti di investimento industriale"; Bollettino degli Ingegneri, n.7/8, 1980.

/9/ D.Miconi, G.Caratti; "Ottimizzazione di impianti di distribuzione di liquidi: il programma CLIO"; Fluid, n.243/244, 1984.

/10/ G.Caratti, F.Martelli, D.Miconi; "Optimal Design of Water Distribution Networks for District Heating"; 5th World Energy Engineering Congress, Atlanta, September 14 - 17, 1982.

TABLE I: ACCEPTANCE CONDITIONS FOR EACH PUMPING UNIT
(a, b, c: see fig.3)

(a)	$Q_{p,min} \leqslant Q_{r,min}$ $Q_{p,max} \geqslant Q_{r,max}$ $H_p(Q_{r,max}) \geqslant H_r(Q_{r,max})$
(b)	$H_p(Q_{p,max}) \leqslant H_r(Q_{r,min})$ $H_p(Q_{p,min}) \geqslant H_r(Q_{r,max})$ $H_p(Q_{r,max}) \geqslant H_r(Q_{r,max})$
(c)	$H_r(Q_{r,min}) \geqslant H_p(Q_{r,min}, N_{min})$ $H_r(Q_{r,min}) \leqslant H_{p,s}(Q_{r,min})$ $H_r(Q_{r,min}) \geqslant H_{p,d}(Q_{r,min})$ $H_r(Q_{r,max}) \leqslant H_p(Q_{r,max}, N_{max})$ $H_r(Q_{r,max}) \leqslant H_{p,s}(Q_{r,max})$ $H_r(Q_{r,max}) \geqslant H_{p,d}(Q_{r,max})$

TABLE II: THE MOST SIGNIFICANT FINAL RESULTS
(Life span=10 years; interest rate=0,05;
a, b, c,: see fig.3)

Outputs	Optimal solutions		
	(a)	(b)	(c)
Pumping unit (see fig.7)	Type: A P = 56 kW	Type: B P = 54 kW	Type: C P = 37 kW
Energy consumption (at the final time)	3.69×10^6 (kwh)	4.70×10^6 (kWh)	2.89×10^6 (kWh)
Global efficiency (at the final time)	0.65	0.51	0,83
Present value of total cost (at the final time)	473×10^6 (Lit)	585×10^6 (Lit)	383×10^6 (Lit)

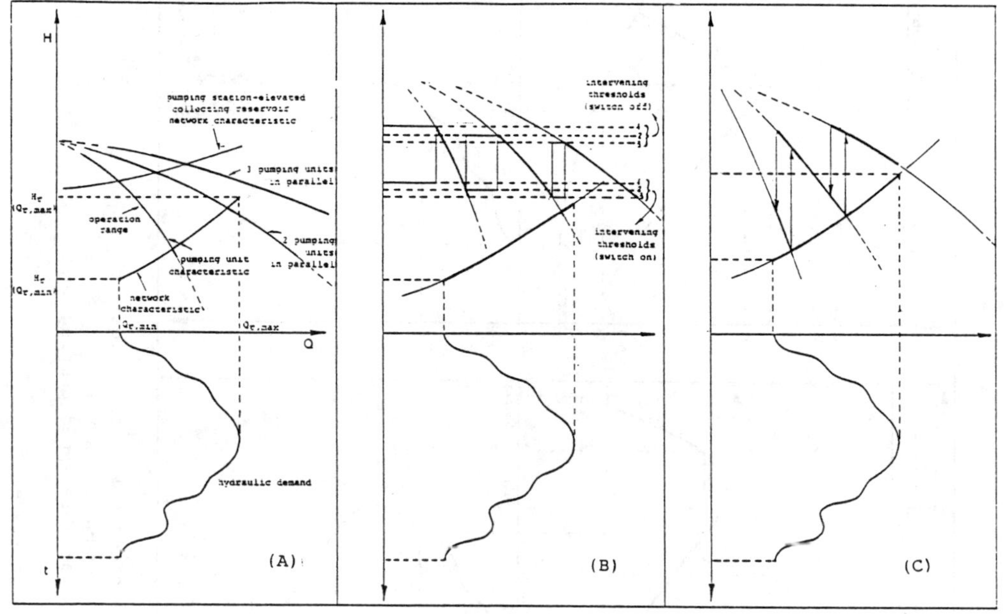

Fig.1 - Regulation systems

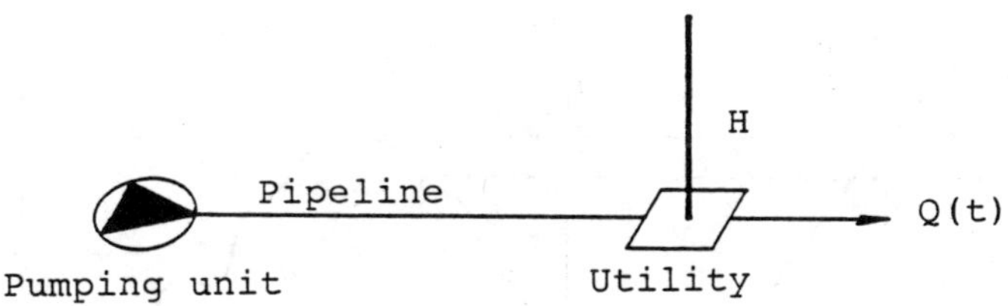

Fig.2 - Schematic plant configuration

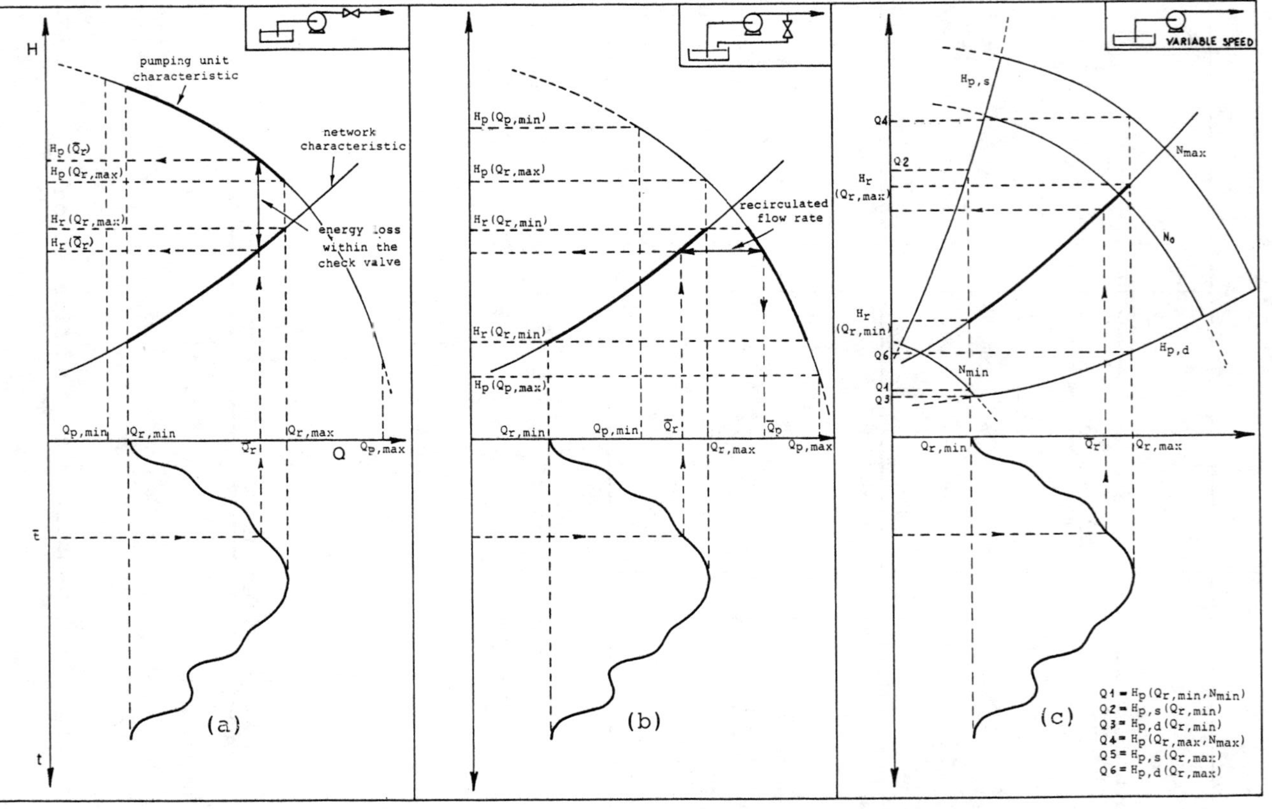

a - system equipped with a check valve (in series)
b - system equipped with a bypass check valve
c - system equipped with a variable speed pumping unit

Fig.3 - Elementary regulation systems

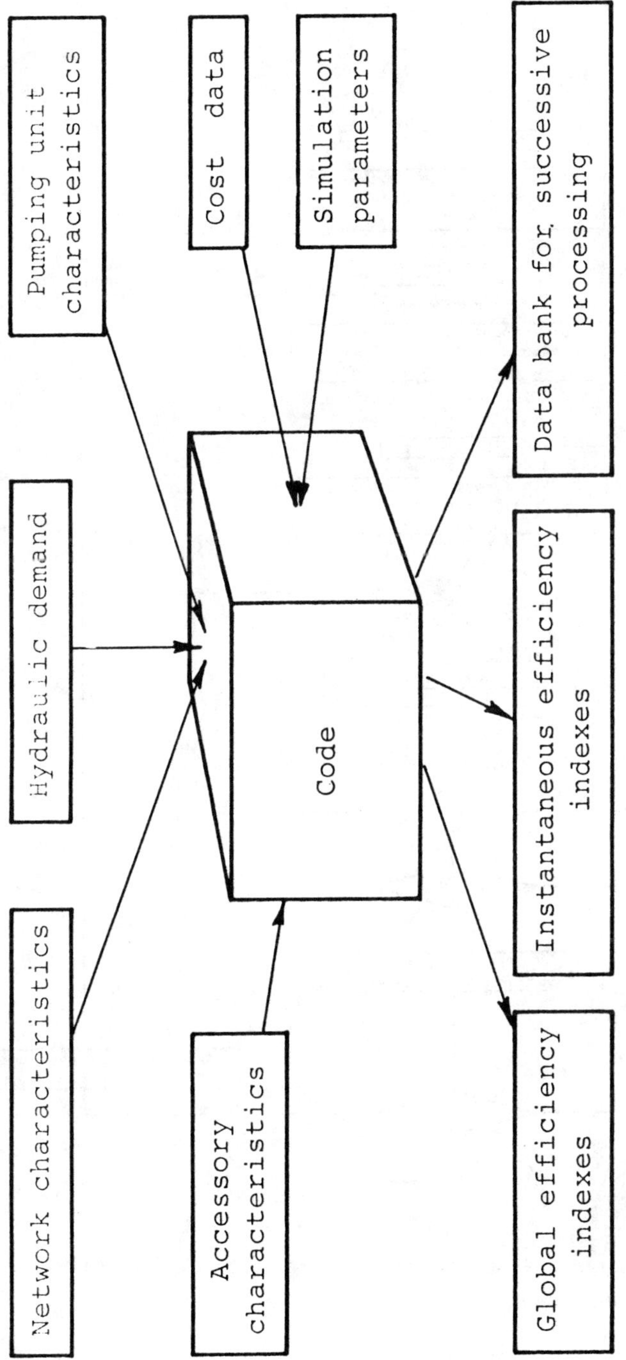

Fig.4 - Inputs and outputs of the code.

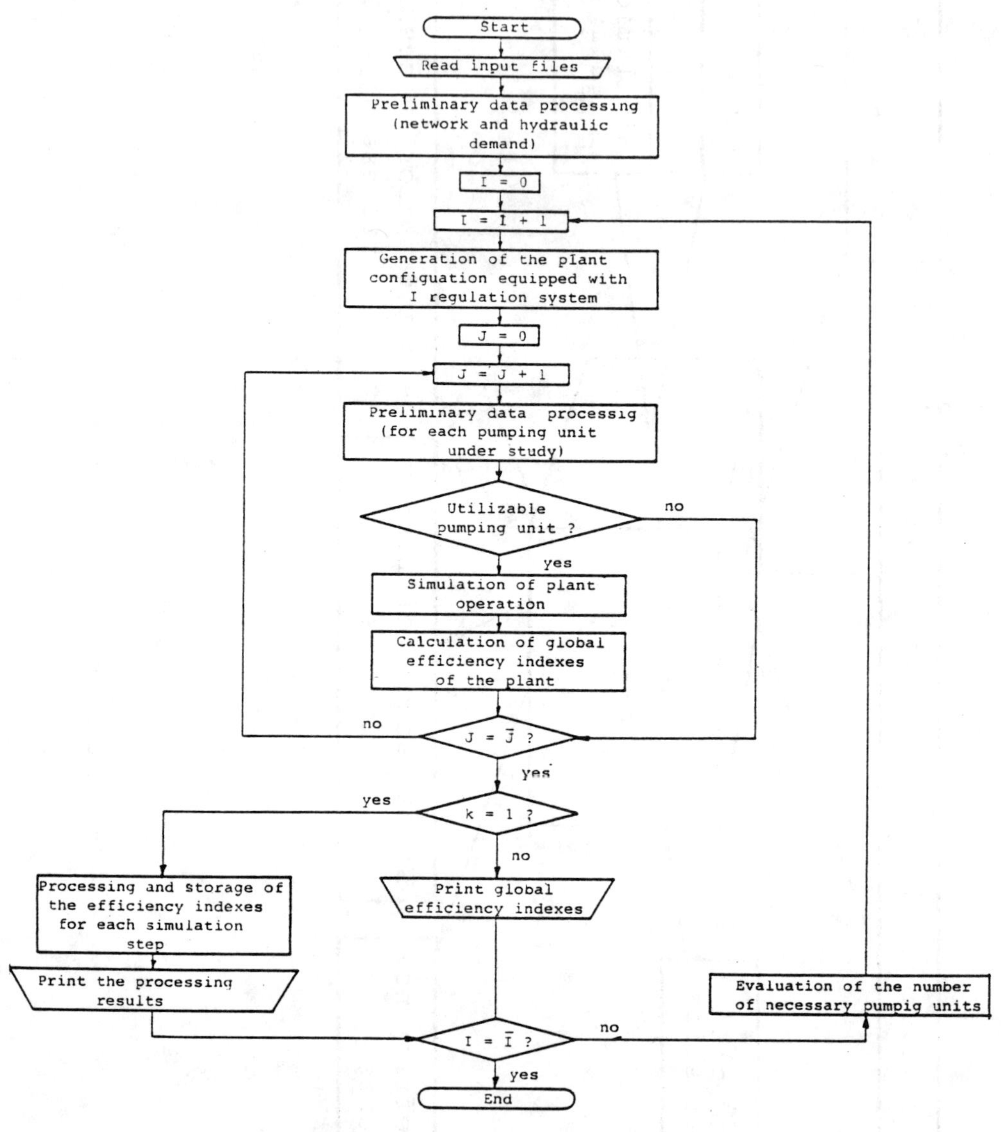

Fig.5 - Schematic flow-chart of the entire simulation process

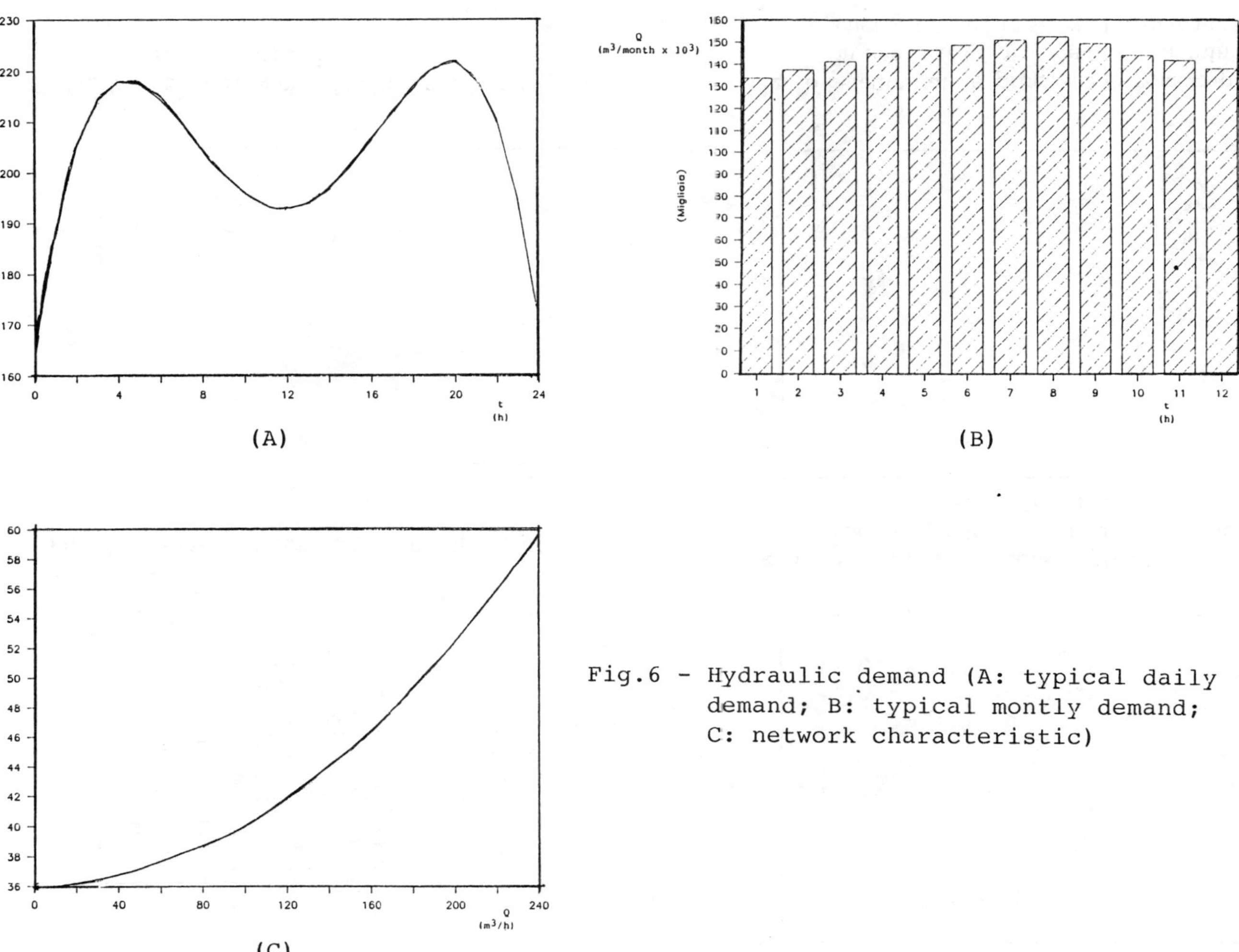

Fig.6 - Hydraulic demand (A: typical daily demand; B: typical montly demand; C: network characteristic)

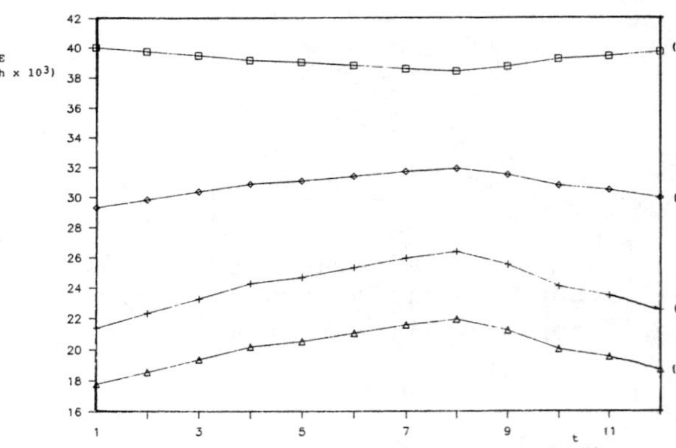

Fig. 7 - Series of a commercial pumping units

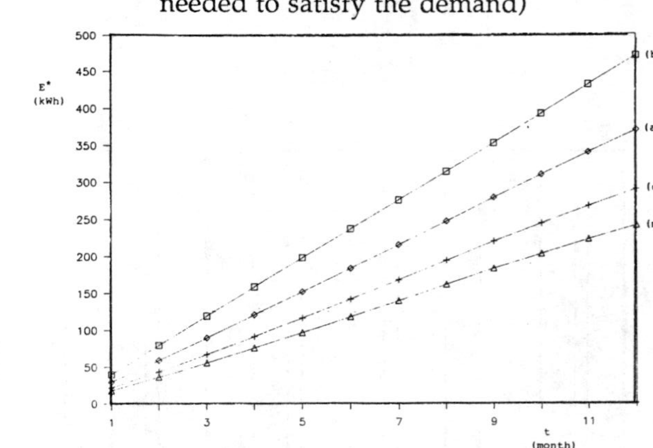

Fig. 8 - Monthly energy consumption in one year (a, b, c: see fig. 3; m: minimal energy needed to satisfy the demand)

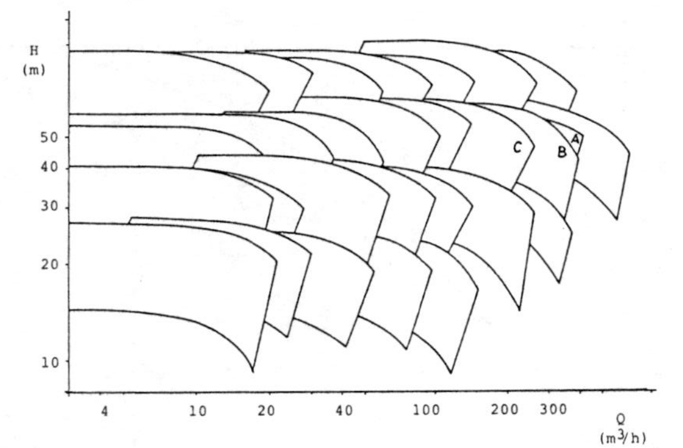

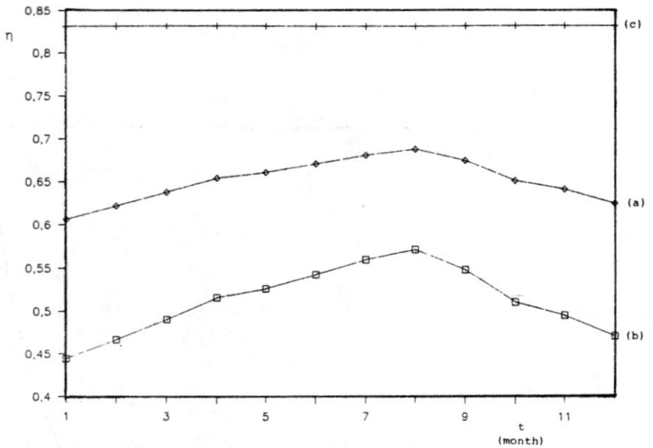

Fig. 9 - Monthly energetic efficiency in one year 9a, b, c: see fig. 30

Fig. 10 - cumulative progressive consumption in one year (a, b, c: see fig. 3; m: minimal energy needed to satisfy the demand)

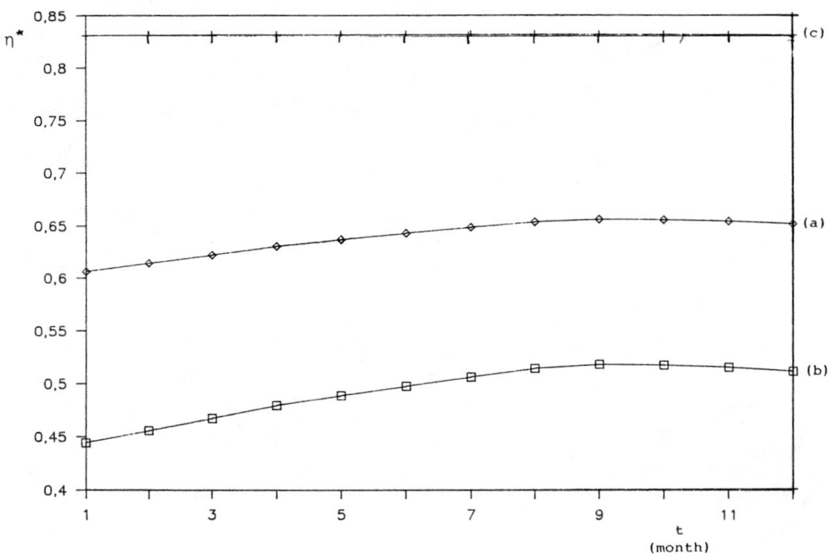

Fig.11 - Progressive energetic efficiency in one year
(a, b, c: see fig.3)

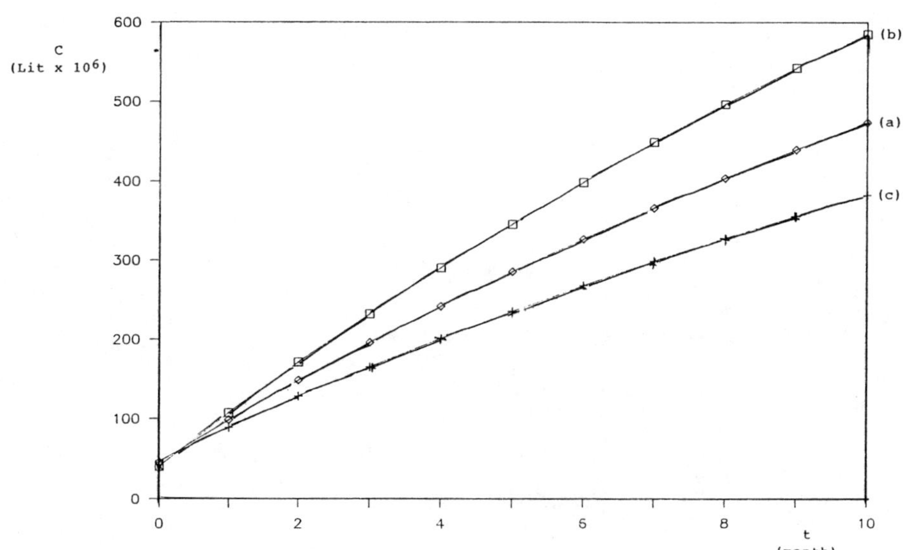

Fig.12 - Present value of total cost (a, b, c: see fig.3)

THE EFFECT OF SENSITISER ON THE PHOTOCATALYTICAL OXIDATION OF PROPAN-2-OL BY Pt/TiO_2 AND OTHER CATALYSTS

F.H. Hussein, S.W. Radi and Naman*
Department of Chemistry,
College of Education, University of Salahadin, Arbil, Iraq
*Solar Energy Research Center, Jadiriyh, Baghdad, P.O. Box 13026, Iraq

ABSTRACT :

Photoconversion of pure propan-2-01 to propanone has been done in the presences of TiO2 (0.7 g/l) and Rhodamine-B($2.5x10^{-6}$M)inside the cylinderical reaction vessel of 35 ml using 250 W Xenone lamp,with constant flow of 30 ml/min air . The photoproduction of the propanone has been measured by GC. with time of irradiation for temperature range of(280-350 K) . The effect of dopping of the TiO2 by 1% Pt , Pd and Rh on the photo-conversion have been studied,also other sensitisers such as thionine and eosine were examined

Arrhenius activation energies have been calculated for this photocatalytical conversion and suitable mechanism proposed for this photoreactions.

* To whom all correspondence be addressed.

INTRODUCTION:

There have been world-wide efforts to develop photochemical systems based on semiconductor materials for the utilization of solar energy to chemical potential (fuels). Bard [1] Suggest "Integrated Chemical System" (ICS)approach to the constraction of photochemical conversion systems. An ICS is a heterogeneous,multiphase containing different component ,designed and arranged for specific functions or to carry out specific processes . The different components such semiconductor particles will often have a particular structure and organization to improves the performance of the system , Indeed biological photosynthesis generally follow ICS approach.

The simplest semiconductor ICS are involved irradiation of particle dispersion such us TiO_2 and CdS [2] for photoproduction of hydrogen from water [3] or photofixation of CO_2 to fuel or nitrogen fixation [4-5] and photosplitting of H_2S [6-7].

Then later TiO_2 is doppted by Pt and other catalysts to assist in the reactions of many organics molecules (CH_3COOH, $Ph_2C=CH_2$ ...) in the presence of oxygen [8]. Photocatalytic dehydrogenation of liquid propan-2-01 by suspension of Pt and other metals supported on anatase has been investigated to produce propanone [9]. Pt/TiO_2 Photocatalyses various gases and liquid-phase reactions,Pt assists in the separation of photogenerated electrons and holes, which may otherwise recombine within the TiO_2, this is in addition to participating in conventional surface reaction steps.

When suspensions of Pt /TiO2 in liquid- phase reactants or reactant mixtures,photocatalyses the dissociation of water [10], the decarboxylation of carboxylic acid [11]the oxidation of hydrocarbons,[12] the raidical polymerization of methyl methacrylate,[13] the dehydrogenation of alcohols,[14] and synthesis of amino acids from CH_4 , H_2O and NH_3 [15]. These reactions proceeds according to ICS approach to convert light energy to chemical potential with widely different efficiencies [1-3] .

This paper will stress the effect of sensitiser on the photocaalytic oxidation of propan-2-01 by Pt and other metal supported on anatase , using a double-layer cylinderical pyrex reactor with plane window 2 cm dia of 35 ml and 250 W Xenone lamp for irradiation reaction was followed by GC. monitoring propanone formation.

EXPERIMENTAL:

Materials: TiO_2,Propan-2-01 , proponone,thionine ,rhodamine-B and eosine were purchased from Fluka AG. Switzerland,1%Pt/TiO_2, 1%Pd/TiO_2 and 1%Rh/TiO_2 were prepared by using Degussa P25 anatase of surface area 50 $m^2 g^{-1}$,the method has been described . [9] .
Apparatus: The apparatus used to determine photocatalytic reaction has been described [16],a known mass of catalyst was suspended in 35 ml of propan-2-01 in a cylinderical double layer pyrex reaction vessel equipped with 2 Cm pyrex window for irradiation by 250 W Xenone lamp thermostat (Haake FE2) has been used to maintain the reaction vessel

temperature , propanone was measured by taking the sample through septum on the top of the vessel by gas syrings to Pye-Unicam 304 Gas Chromatography using column (10 PEG 20 M) 0.4 mm x 1.5 m at 368 C, flow rate 20 ml/min,injection Temp. 368C,the reaction vessel is supplied with inlet and outlet tubes for saturating the solation with oxygen (air). A magnetic stirrer is usually used to agitate the solution mixture during the photolysis.

The solar 118 has been used for measuring the intensity of light ,ultra-centrifuge MSE minor 35 was used for separation TiO2 form the reaction mixture,UV-Visible-Spectrophotometer Cecil- 599 was used for measurments of absoption spectra.

RESULTS:-

The following blank experiments have been done:- first photolysis of propan-2-01 in the reaction vessel without the TiO2 with and without nitrogen the results shows that there is no propanone , and no hydrogen formation that was at room temperature and 45 ^{o}C, but the photolysis in the presence of oxygen there was propanone formation.

Series of experiments have been performed with different concentration of TiO2 (0.005-o.46 g/l) in 35 ml propan-2-01 in reaction vessel, at 300K,with continuous flow of air inside the reaction vessel and stirring . The concentration of propanone was measured by G C. during the time of irradiation, the results are in figure(1a). The rate of photoproduction of propanone against concentration of TiO2 are plotted in figure(1b) at 300K .

Three sets of expriments have been performed using 1% Pt,Pd and Rh metal as dopping material to the TiO_2,(0.8 gm/l of each).Photoproduction of propanone with time of irradiation have been recorded at 300 K, with continuous flow of air to the reaction mixture, the results are in figure (2) .

The effect of Rhodamine-B in different concentration on the photoproduction of propanone using TiO_2 at 300 K with continuous flow of air to the reaction mixture figure(3a) and figure(3b)give the best concentration of Rhodamine-B in the 0.8 gm/l of TiO_2. The effect of sensitisers such as thionine,eosin, Rhodamine-B in 2.6×10^{-6} mol/l to 0.8 g/l of TiO_2/at 300 K on the rate of photoproduction of propanone using same conditions. Figure (4) shows that Rhodamine-B gives higher-value than other sensitiser .

The influmence of the light intensity on the rate production of propanone using 0.8 g/l TiO_2 at 300 K using continuous flow of air the results are in the figure (5). Series of experiments have been performed using 0.8 g/l TiO_2,air,and light intensity 90 MW/CM . the concentration of propanone with time of irradiation were measured for different temperature (280-328K) same experiments have been repeated after addition of Rhodamin-B as sentiser figure(6). Arrhenius plot for photocatalytical production of propanone in the case 0.8 g/l TiO_2 only and for 0.8 g/l TiO_2 with 2.6×10^{-6} mole/l of Rhodamine-B figure(7).

DISCUSSION:-

The overall reaction in this paper is the conversion of propan-2-01 to propanone, using necked TiO_2, metalized TiO_2 and TiO_2 with Sensitiser. Table 1 shows the rate of photoproduction of propanone . The difference in the rates of production are probably due to the different components of ICS . The incident light is absorbed in the semiconductor figure(5) to form an electron(e^-) hole (h^+) pair. The support of semiconductor TiO2 with metal Pt,Pd and Rh and associated with sensitizer Rhodamin-B,eosine and thionine can play an important role for separation of the e^- and h^+ pair, minimizing losses due to recombination. Thus the TiO2 material plays as transducer, and converting the radiant energy to an internal chemical potential (photocatalysis of propan-2-01).

In the presence of oxygen at constant flow rate of air, oxygen will adsorbed on the surface of TiO2 and traps electrons from conduction band form an ionic oxygen (O_2^-) or will form ionic oxygen radicals such us (O3,O2,O) on the surface of TiO2, then OH will trapped by the holes on the valance band and form hydroxyl radical ($OH^\bullet$) on the surface [17]

$$TiO_2 + h\nu \longrightarrow e^- + h^+ \quad \ldots\ldots\ldots\ldots(1)$$

$$e^- + O_{2\,ads.} \longrightarrow O^-_{2\,ads.} \quad \ldots\ldots\ldots\ldots(2)$$

$$h^+ + OH^-_s \longrightarrow \dot{O}H_s \quad \ldots\ldots\ldots\ldots(3)$$

The rate of reaction 2 and 3 is a strong function of the types of TiO2 in case of metalized TiO2, Pt/TiO2 gives propanone more than Pd/TiO2 and Rh/TiO2 (Fig.2). After e^- h^+ separation, considerations generally involve in the secondary reactions of adsorbed oxygen radicals or ions on the surface of the catalyst.

$$OH + (CH_3)_2\,CHOH \longrightarrow (CH_3)_2\,\dot{C}OH + H_2O \quad \ldots\ldots\ldots\ldots(4)$$

$$O^-_{2\,ads.} + H_2O \longrightarrow OH^-_s + H\dot{O}_2 \quad \ldots\ldots\ldots\ldots(5)$$

Then $\dot{O}H$ and $H\dot{O}_2$ are very active species for formation of propanone from $(CH_3)_2\ \dot{C}OH$ [18].

$$(CH_3)_2\ \ddot{C}OH + H\dot{O}_2 \text{ ---------> } (CH_3)_2\ CO + H_2O_2 \quad ....(6)$$

$$(CH_3)_2\ \dot{C}OH + \dot{O}H \text{ ---------> } (CH_3)_2\ CO + H_2O \quad ....(7)$$

$$2(CH_3)_2\dot{C}OH \text{ ------> } (CH_3)_2\ CO + (CH_3)_2\ CHOH \quad ....(8)$$

Beside reaction (6) peroxide H_2O_2 can be formed by reaction between two $H\dot{O}_2$

$$2H\dot{O}_2 \text{ ------> } H_2O_2 + O_2 \quad ....(9)$$

and the peroxide molecules and hydroxyl ion on the surface will react with photogenerated electrons and holes.

$$e^- + h^+ + H_2O_2 + OH^- \text{ ------> } 2\dot{O}H + OH^- \quad ....(10)$$

then OH^- will react with h^+ through reaction (3) to form $\dot{O}H$ on the surface of the TiO_2 .

The addition of sensitiser such us Rhodamine-B to the TiO_2 despersion system increases the rate of the photogeneration of propanone (Table 1), that may be due to the fact that more light absorbed by Rhodamine-B between 460 nm to 580 nm Fig.8, then energy transfer from Rhodamine to TiO_2 or to any other active species and hence promote the reactions to increases the formation of propanone ,but in case of eosine and thionine the photoproduction decreases even less than necked TiO_2 that may be due the covering the surface of TiO_2 by these dyes Table(1) , olso these sensitisers have a narrow absorption bands in the visible region Fig.(8). Efficient sensitisation of TiO_2 electrode using transition metal charge transfer complexe has been done by Gratzel [19] .

The effect of temperature have been studied in this system between (280 - 328 K) for TiO_2 and TiO_2/Rhodamine-B . Figure(6) shows that the rate of photogeneration of propanone is slightly higher than necked TiO_2,that may be due to the longer wavelength absorption,and the molecules of Rhodamin-B does not cover the surface of TiO_2 but it absorb the different photon then it promote the energy to enhanced the photoproduction of propanone.

The Arrhenius plot of the rate of photoproduction of propanone Fig.(7) give value of 22.59 kj mol^{-1} for TiO_2 and 21.94 kj mol^{-1} for TiO_2/Rhodamine-B . These results are similar to our result for propanone formation by photocatalytic of TiO_2 after O_2 treatment for same temperature range (282-303K), while Arrhenius energy for same reaction at dark is 120 kj mol^{-1} [9].

In conclusion,propan-2-01 can be converted to propanone, by TiO_2/Rhodamine-B, in photocatalytical reaction, but both eosine and thionine decreases the photoproduction of propanone. The conversion value is about the same as in the Pt/TiO_2 system in the range of (282-303K).

Table 1

Conversion of propane 2-01 to propanone

at 300K in 35 ml reaction vessel.

Semiconductor* 0.8 g/l	Rate of production of propanone mol/h x 10^{2}
TiO_2	0.62
Pt/TiO_2	0.92
Pd/TiO_2	0.81
Rh/TiO_2	0.70
TiO_2/Rhodamine-B	0.72
TiO_2/eosin	0.41
TiO_2/thionine	0.30

* Concentration of sensitizer (Rhodamin-B,eosin and thionine were 2.6×10^{-6} mol/l .

REFERENCE:

1- A.J.Bard,Ber.Bunsenges.Phys.Chem.92,1187-1194 (1988).

2- S.N.Frank and A.J.Bard.T.Phys.Chem. 81,1484 (1977).

3- T.Inone,A.Fujishima,S.Konoshi and K.Honda Nature 277,637.(1979).

4- M.Halmann and B.Ourian-Blajeni Nature 275,115,(1978).

5- G.N.Schranzer and T.D.Guth.J.Am.Chem.Soc 99,7189,(1977).

6- E.Borgarello,K.Kalyanasundaram and M.Gratzel Helv.Chim Acta 65,243 (1982).

7- S.A.Naman S.M.Aliwi and K.Al-Emara Int.J.Hydrogen Energy.11,33-38, (1986).

8- T.Kawai and T.Sakata Chem. Phys.Cett 80,391 (1981).

9- F.H.Hussein and R.Rudham,J.Chem.Soc.Faraday tran.I. 80,2817-2825, (1984);ibid 83 , 1631-1639 (1987).

10- A.Mills and G.Porter J.Chem.Soc.Faradsy Tran 1 78,3659,(1982).

11- A.J.Bard J.Phys.Chem.87,1417,(1983).

12- I.Izumi,W.W.Dunn,K.O.Wilbourn,F.F.Fan and A.J.Bard J.Phys.Chem. 84,3207,(1980),.

13- B.Kraentler,H.Reiche,A.J.Bard and R.G.Hocker.J.Polym.Sci.Lett.17, 535 (1979).

14- P.Pichat,M-N.Mozzanega,J.Disdier and J.M-Merrmann Nouv.J.Chim. 6,559, (1982).

15- W.W.Dunn,Y.Aikawa and A.J.Bard J.Am.Chem.Soc.103 ,6893, (1981).

16- S.A.Naman.K.Al-Emara Int.J.of Hydrogen Energy . 12,629,1987.

17- A.P.Griva,V.V.Nikisha,B.N.Shelimone and S.J.Teichney,Kinetics and Catalysis 15,86,(1974).

18- R.I.Bickley,G.Munuera and F.S.Stone J.Cata.31,398,(1973).

19- N.Vlachopoulos and M.Gratzel.ISES Solar World Congress Hamburg. abstract. Vol.2,7.2-07,(1987).

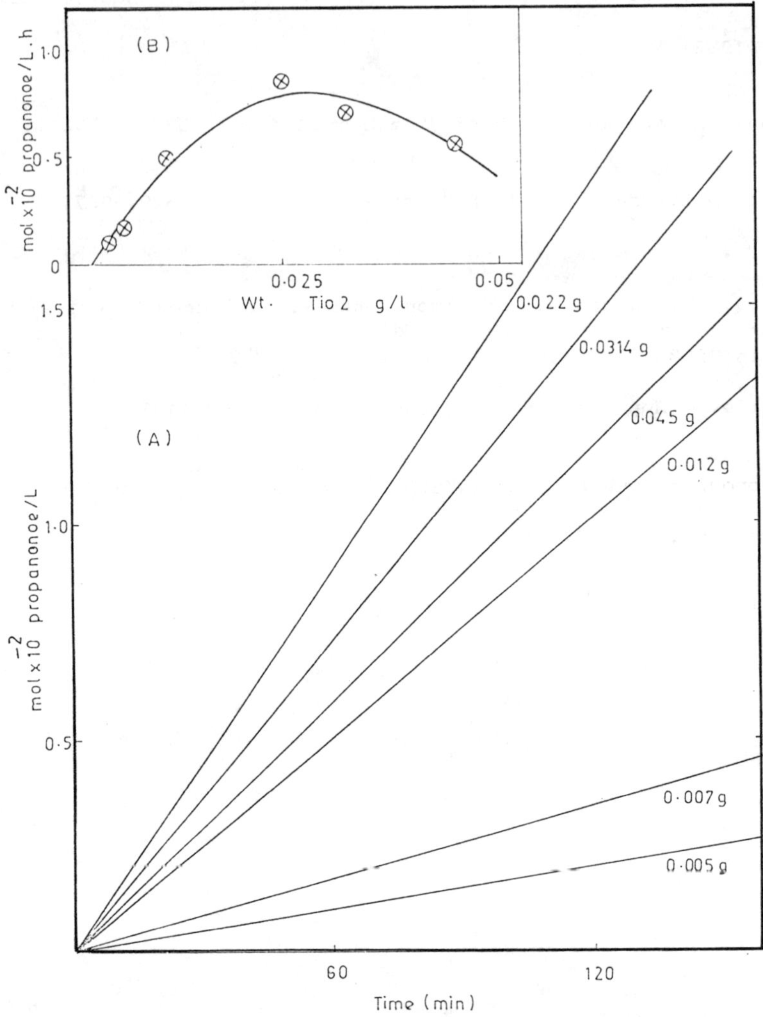

Fig. (1): (a) Propanone yield as a a function of time of irradiation in photoconversion of propan-2-Ol over different concentration of TiO2 in 35ml, at 300K. (b) Rate production of propanone mol/h as a function of concentration of TiO2 at 300K.

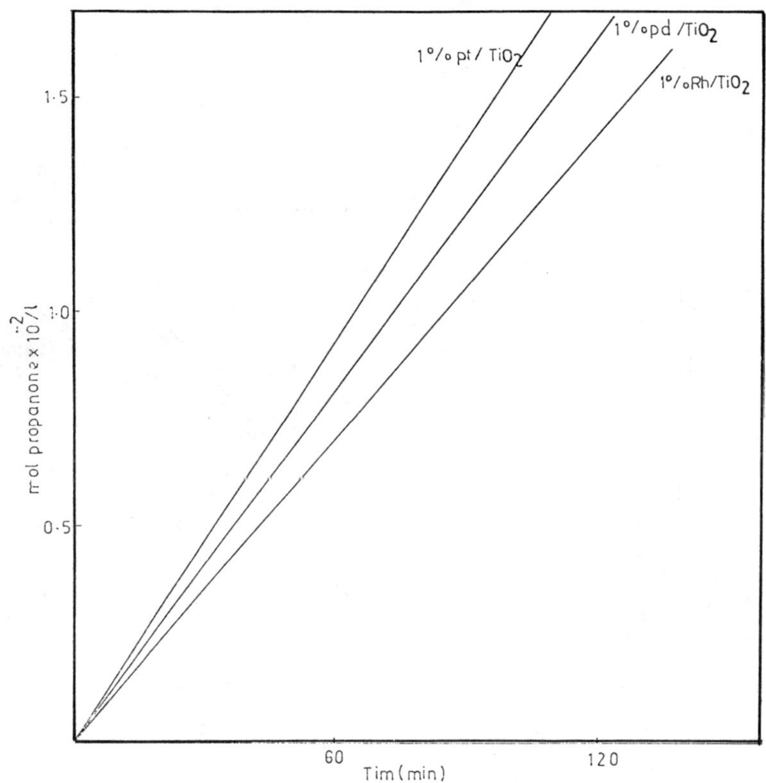

Fig. (2): Propanone yield as a function of time of irradiation for 1% pt, 1% pd and 1% Rh.

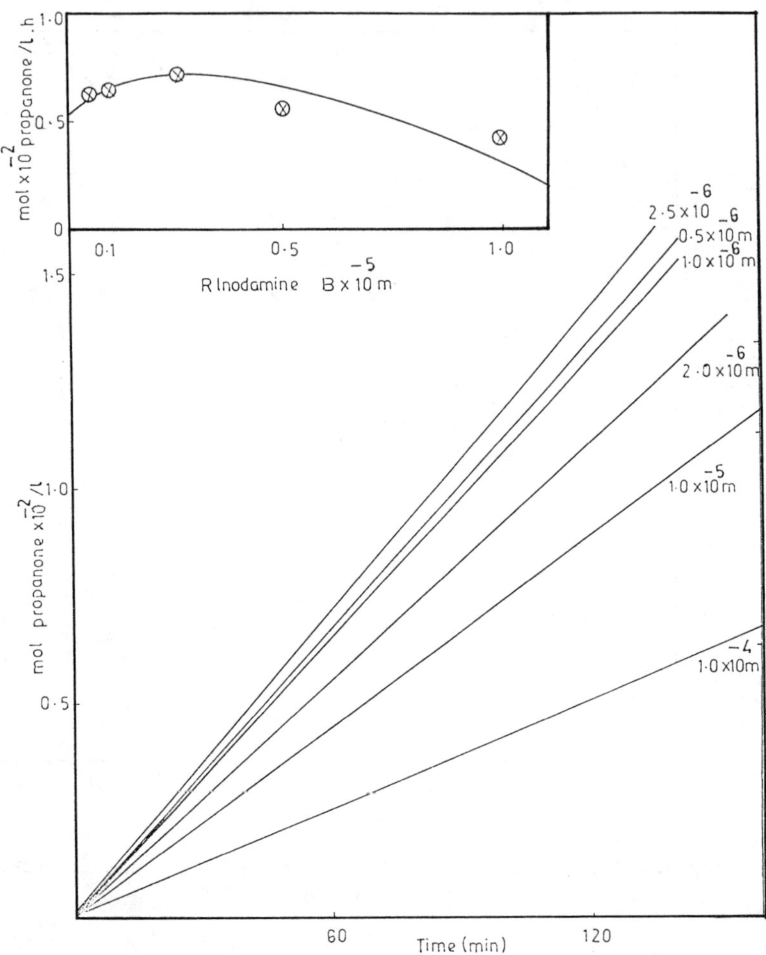

Fig.(3): (a) Propane yield as a function of time of irradiation for 0.8 g/l TiO2, at 200K at different concentration Rhodamine-b (b) rate production of propane as a function of concentration of Rhodamine-B mol/l at 0.89/l TiO2.

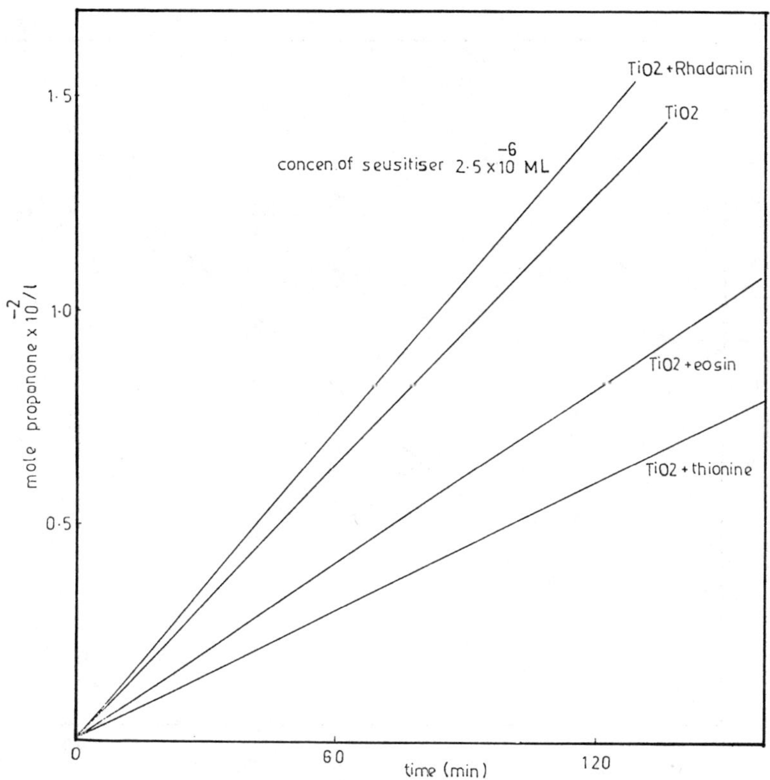

Fig.(4): Concentration of propane as a function time of irradiation for Rhodamine-B, eosine and thionine 2.6 x 10^{-6} mol/l in 0.8 g/l TiO2 at 300K.

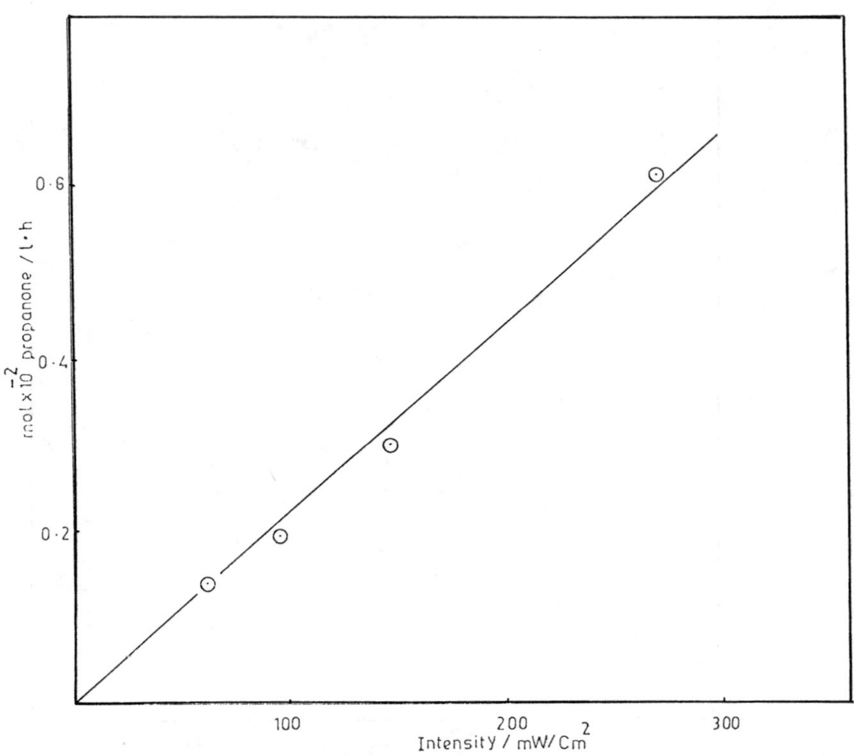

Fig.(5): Rate of photoproduction of propane with light intensity for TiO2 0.8 g/l at 300K.

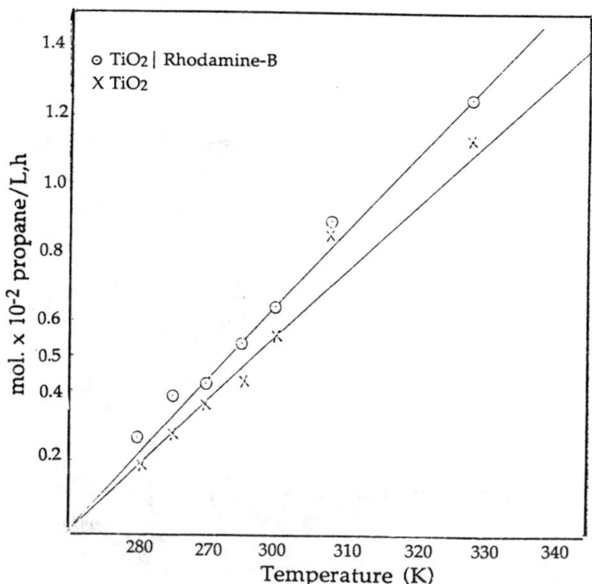

Fig.(6): Rate of photoproduction of propane with temperature from TiO2/Rhodamine-b and TiO2, 0.8 g/l in 35 ml vessel.

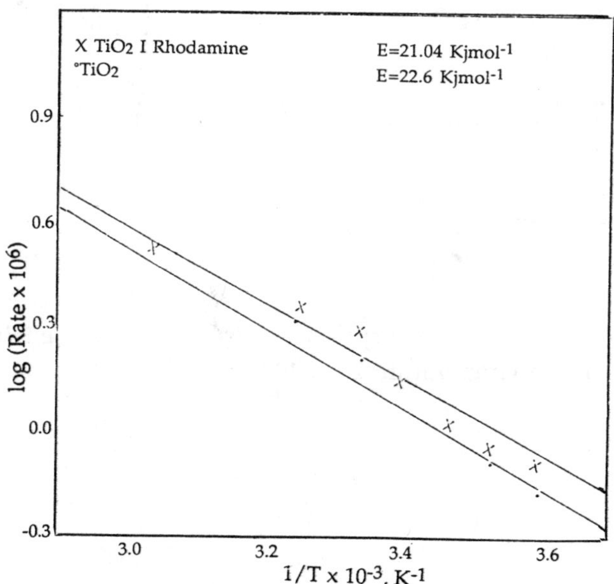

Fig.(7): Arrhenius plot for log rate of photoproduction of propane with 1/T for temperature range (280-303K).

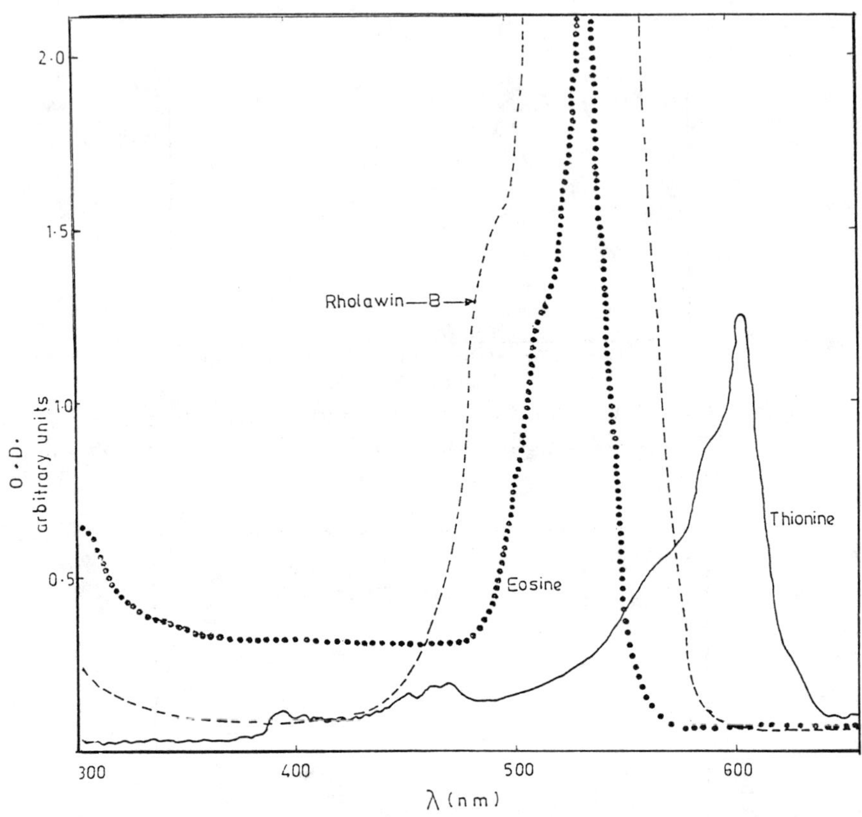

Fig.(8): UV-Visible absorption spectra of three sensitizer at room temperature 2.6 x 10^{-6} mol/l.

EFFECT OF PCM INCORPORATION ON THE PROPERTIES OF GYPSUM WALLBOARDS

S. Cadoux[1], J.F. Houle[2], and J. Paris[1]

[1] Chemical Engineering Department, Ecole Polytechnique de Montréal C.P. 6079, Succ. A, Montréal, QC, Canada, H3C 3A7

[2] Parinova Inc., 2975 ave. Douglas, Montréal, QC, Canada, H3R 2E2

Abstract

Incorporation of a phase change material (PCM) into gypsum wallboards has been shown to be an effective way to distribute passive heat storage in buildings. The purpose of this work was to assess the impact of PCM incorporation (by encapsulation and by impregnation) on relevant properties of the boards. Four PCM were used: polyethylene glycol (PEG), paraffins (PAR), butyl stearate (BS) and a mixture of capric and lauric acids (CLA). Incorporation of a PCM reduces sligthly the thermal conductivity and the thermal diffusivity of boards. Heat capacity has been influenced by aging of boards with PEG capsules. Under same experimental conditions, boards with paraffins absorbed more heat than the other boards. All boards have marginal mechanical properties. Those experiments have lead to the exploration of new incorporation techniques.

1. INTRODUCTION

Previous studies, based on simulations, by Rudoy and al. [1] and by Goodwing and al. [2] have shown that an increase in building thermal mass leads to a substantial reduction of cooling energy requirements. Work performed by Burch and al. [3], using field tests houses, confirms that heavy weight buildings consume less heating energy than light weight buildings and are more comfortable because of a reduction of temperature swing. The effect on heating loads was greater when wall mass was positioned inside as opposed to outside. The same conclusions were also observed for summer space cooling loads [4]. Kammerud and al. [5] has shown that ventilation has the potential to substantially reduce cooling energy and that an increased thermal mass enhance the effectiveness of ventilation and reduces peak energy use.

The traditional method to increase the thermal capacity of interior partitions is to increase sensible heat by using heavy construction materials such as concrete and bricks. An alternate way to increase the thermal inertia of inside surfaces is to make use of the latent heat or Phase Change Materials (PCM). PCM have several advantages: a large amount of heat is absorbed within a small temperature variation and for a small volume. Incorporating PCM to gypsum wallboards is a practicle way to realize increased thermal inertia. For this applications, the PCM must have its transition zone within the allowable temperature swing. Heat is transfered to the PCM during the solid-liquid phase transition, and restored during the reverse process.

A current concern is to develop and assess methods for the incorporation of PCM to the boards. Recently, real size gypsum wallboards have been evaluated under realistic conditions and over a long period; it was concluded that the concept is viable [6,7,8]. The purpose of this work was to investigate further the effect of several methods for PCM incorporation on the most important properties of the boards. Assessment of thermal and mechanical properties of six different designs of gypsum wallboards containing a PCM was performed; the description of the experimental boards, the experimental procedures and the results of the tests are presented here.

2. EXPERIMENTAL BOARDS

The specifications of the gypsum wallboards used in the experiments are given in Table 1. The capsules of polyethylene glycol and paraffins were placed in grooves dug out from the back side of the boards and subsequently covered by a standard board-finishing paper. The capsules arrangements are shown in Figure 1. The impregnated boards were prepared by dipping the boards into the melted PCM for a predetermined duration.

The PCM used are CARBOWAX-600 from Union Carbide, a polyethylene glycol with a mean molecular weight of 600, a mix of technical grade paraffins from Humphrey Chemical (hexadecane at 25 W% and octadecane 75 W%), butyl stearate and a mix of capric acid (45 w%) and lauric acid (55 w%) from Emery Industries. According to manufacturers specifications, all these PCM melt within the range 18-30 °C. The thermal storage capacities of the experimental boards for this temperature interval, based on manufacters supplied thermal properties, are given in Table 2; the values include latent and sensible heat.

3. METHODOLOGY

The thermal conductivity under steady state conditions was evaluated. The thermal diffusivity of boards without PCM and of boards with impregnated PCM (uniform dispersion) was evaluated. An experimental procedure was developed to determine the actual thermal capacity of the boards. A current concern is the impact of PCM incorporation on mechanical properties of the boards. The flexural strength and the humid deflection were evaluated. Also to verify if impregnated boards with organic PCM could be considered as moisture resistant boards, water absorption tests were performed with impregnated samples.

The thermal conductivity of boards was measured at the Center for Building Studeis of Concordia University, in Montréal. The thermal diffusivity and heat capacity were measured at Ecole Polytechnique de Montréal. All the measurements of mechanical properties were performed at the Research Centre of Westroc Industries Ltd in Mississauga, Ontario.

Thermal Conductivity

Tests have been performed using a conductivity meter Dynatech R-Matic following the ASTM C-518 procedure. During the two hours of data acquisition, after steady-state conditions were reached, the average temperature of the boards was 22 °C and the differential from one side to the other was 14°C. Those experimental conditions were selected so that the thermal conductivity is evaluated with a partially melted PCM. Three different boards were tested to determine the impact of PCM incorporation method on thermal conductivity: standard boards without PCM (SB), boards containing capsules (PAR) and impregnated boards (BS).

Thermal Diffusivity

The experimental apparatus for the determination of thermal diffusivity within boards is illustrated on Figure 2. The apparatus including the sample is placed in a cold chamber at -10 °C for five hours before the beginning of the test. The test is performed at ambiant temperature and consists in measuring the temperature at the interface between the hot plate and the sample (T_S), at the interface between the sample and the cold plate (T_Y) and into the cold plate (T_0). A typical graph showing the evolution of those temperatures with time is given on Figure 3. Thermal diffusivity was determined after two different time intervals: 275 and 550 seconds, using the following equation applicable to a semi-infinite surface:

$$(T_S - T_Y)/(T_S - T_0) = \mathrm{erf}[y/2(\alpha t)^{0.5}] \qquad (1)$$

Where: erf = Gauss error function.
t = Elapsed time since the beginning of the test (s).
y = Sample thickness (m).
α = Thermal diffusivity (m^2/s).

Actual Thermal capacity of boards

The experimental apparatus, illustrated in Figure 4, consists of a cube with sides of 600 mm made of plywood and insulated with 50 mm of extruded polystyrene. An electric element is fixed to the base of the cube. The five other internal faces of the cube are lined with square samples of a board to be evaluated. The cube is placed in a cooled chamber (15 °C). The experimental procedure is as follows: the apparatus is left unheated during 48 hours to reach thermal equilibrium with its 15 °C environment. Then, a constant heat input of 42 W is provided to the cube until its inside air reaches at least 30 °C. During the heating period, the temperatures of the air in the cube and in the chamber and of the experimental board samples are measured and recorded every six minutes. It is possible to evaluate the rate of heat absorbed by the board samples ($Q_{AB}(t)$), at any given time, by the equation:

$$Q_{AB}(t) = Q_P - Q_{AA}(t) - Q_{AS}(t) - Q_{LOST}(t) \qquad (2)$$

Where: Q_P = Rate of heat provided (42 W)
$Q_{AA}(t)$ = Average rate of heat absorbed by air in the cube (W)
$Q_{AS}(t)$ = Average rate of heat absorbed by the cube assembly (W)
$Q_{LOST}(t)$ = Rate of heat lost to outside air (W)

The values of $Q_{AA}(t)$ and $Q_{AS}(t)$ are evaluated using the specific heat and the mass of the air and materials in the cube assembly and the average temperature variation rate at the particular measuring time. $Q_{LOST}(t)$ is evaluated using the temperature difference between inside and outside air and the UA value for the complete cube assembly. This value was determined in a preliminary experiment for each board and for steady-state conditions with a board temperature of approximatively 24 °C. Considering that the rate of temperature increase in the apparatus is approximatively 2 °C/h, this method should provide a sufficiently precise evaluation of the coefficient UA even if the cube assembly is not rigorously at steady-state during the measurement.

The total energy absorbed by the board samples during the experiment is computed by stepwise numerical integration, the rate of energy absorption $Q_{AS}(t)$ being assumed constant during a data taking interval (6 min). A typical graph for one experiment is shown in Figure 5.

Flexural Strength

The principle of the flexural strength measurement, following the ASTM C-473 procedure, is illustrated on Figure 6; the applied force F increases at a constant rate and the value of F at the instant when the board ruptures is recorded. For each tested board, the experiment is repeated four times: board face up (U) and face down (D), larger dimension of the sample parallel to the length of the board (L) and parallel to the width of board (X). Prior to the measurements, the samples are conditioned, in a horizontal position, at constant humidity (48 to 52 % RH) and temperature (20 to 24 °C) during 24 hours.

Humidified Deflection

Humidified deflection, as defined by the ASTM C-473 procedure, consists in mesuring the difference of deflection of a board under its own weight, before and after it has been conditioned for 48 hours at constant temperature (30 to 34 °C) and at constant humidity (87 to 93 % RH). Prior to the first measurement, the samples are pre-conditioned at the same conditions as for the flexural strength test. Samples are 300 x 600 mm and are supported at the extremities of the larger dimension which is parallel to the width of the board.

Surface Water Absorption

For the surface water absorption test a column of water (25 mm high, 21 ± 1 °C in temperature) is placed for two hours over an horizontal board sample (125 mm x 125 mm) and directly in contact with the finishing paper. Prior to the test, the samples are conditioned as for the flexural strength test. The test is repeated for both faces of the board (front, F and back B).

4. RESULTS AND DISCUSSION

Thermal Conductivity

The global thermal conductivity of the boards is measured with a PCM partially melted. Since, the average temperature of the board was of 22 °C and the difference between each side of the sample was of 14 °C. Results of thermal conductivity determinations, under steady-state conditions, are given in Table 3. The impregnated boards (BS) and the boards containing PCM capsules (PAR) have similar thermal conductivities both lower than the conductivity of the standard board (SB); the reduction is relatively small: about 15 %. Gypsum wallboards are not used for their insulation properties, and this reduction should not affect the global thermal resistance of a room finished with PCM wallboards.

In unsteady-state conditions, for example when the inside temperature is changing, the heat transfer through the boards is more affected by PCM during its solid-liquid transition due to the increased heat accumulated into the PCM. In this case, the PCM acts as a thermal barier and then the thermal conductivity is no longer the determinant factor.

Thermal Diffusivity

The results of thermal diffusivity determinations for the two elapted times at which measurements were made are presented in Table 4. There is a major difference between the standard and the PCM boards. It can be assumed that the reduction in diffusivity is directly related to the increase of the apparent specific heat of the boards during the phase change since the apparent specific heat of a board is then the sum of the latent heat of fusion and the specific heat of the materials (including the gypsum matrix). As illustrated by equation (3), the diffusivity is inversely proportional to the apparent specific heat.

$$\alpha = k/\rho C \tag{3}$$

Where: α = Thermal diffusivity (m^2/s).
k = Steady-state thermal conductivity ($W/m.^{o}C$).
ρ = Density (kg/m^3).
C = Apparent specific heat ($kJ/kg.^{o}C$).

Thermal Capacity of Boards

The results of the thermal capacity tests are given in Table 5. The value obtained for the standard boards is in good agreement with the nominal value given in Table 2, the difference being only 5.5 %. The values obtained for all the PCM boards are quite lower than those of Table 2. Differential Scanning Calorimeter (DSC) tests (Figures 7 and 8) performed after the experiments, have shown that polyethylene glycol melts within a wide temperature range (from -27 to 30 °C) and that only 38 J/g (latent plus sensible heat) is really available for heat storage between 18 and 30 °C, while the nominal value based on the quantity of PEG incorporated was 152 J/g. A DSC test for the paraffin mixture showed an actual capacity of 98 J/g (262 J/g was assumed). No DSC measurement was made for the butyl stearate and for the fatty acid mix. It was assumed that the entire board temperature was 30 °C when the middle point was at that temperature. However since the probe was placed within the gypsum, it is possible that the temperature of the melting PCM was quite lower, espacially for boards where the PCM is segregated from the gypsum (capsules).

Boards with paraffin capsules have absorbed more heat than other boards under identical experimental conditions. It was also observed that there is a significant difference in thermal capacity between boards with new and used PEG capsules. The thermal properties of the PEG may have been affected during the field tests because of water absorption by the hygroscopic PCM following capsules leakage.

Flexural Strength

As shown in Table 6, all samples tested failed to meet the minimum specifications for flexural strength required by the Canadian Standards Association. For the boards with capsules, the grooves have affected the internal bonds and the resistance of the back paper. For the impregnated boards one can suppose that the PCM affects the bond between the gypsum and the paper by reacting with or dissolving the binding compound (starch).

Humidified Deflection

Results of humidified deflection tests are given in Table 7. Only one set of samples (PEG) failed to meet the CSA standard. It is accepted in the trade that a humidified deflection greater than 5 mm would tranlsate into an unacceptable sagging in installed boards. By this more rigorous criteria only the BS boards would be acceptable.

Surface Water Absorption

This test is not a mandatory test for standard board products. It was included to show the effect of PCM impregnation to the wallboard water resistance. Although, the fatty acid mix and butyl stearate did reduce the amount of water absorbed as shown in Table 8, they did not meet the criteria for a moisture resistant wallboard which is 1.6 g maximum.

5. CONCLUSIONS

Incorporation of a PCM to gypsum wallboards significantly affects the physical and mechanical properties of the boards, whether the PCM is incorporated by impregnation or as macro-capsules. Macro-capsules which occupy large cavities reduce the binding strength of the boards. Impregnated PCM seems to reduce the adherence between the gypsum and the finishing paper by reaction with the binding element. Hygroscopic PCM may also affect the properties of the boards unless they are rigorously segregated from ambient moistures.

Incorporation of a PCM to gypsum wallboards does not appear to have an important effect on their thermal properties except of course on their heat storage capacity in the PCM transition zone which is greatly increased. However, considering the lack of reliable data on the latent heat of such materials, precise laboratory measurements are a prerequisite to any product design. These measurements, should include aging and humidity effects. For non segregating methods of incorporation, potential interaction between the PCM and components of the gypsum paste should also be considered.

Work is presently under way on incorporation methods which alleviate those drawbacks. Results will be presented elsewhere.

6. NOMENCLATURE

C Apparent specific heat (kJ/kg.°C).
erf Gauss error function.
k Thermal conductivity (W/m.°C).

$Q_{AB}(t)$ Average rate of heat absorbed by air in the cube (W).
$Q_{AS}(t)$ Average rate of heat absorbed by the cube assembly (W).
$Q_{LOST}(t)$ Rate of heat lost to outside air (W).
Q_P Rate of heat provided (42 W).
t Time (s).
THC Total heat capacity of the boards including the latent and sensible heat.
T_O Temperature of the cold plate used in the thermal diffusion measurement (°C).
T_S Temperature at the interface between the hot plate and the board sample used in the thermal diffusion measurement (°C).
T_Y Temperature at the interface between the cold plate and the board sample used in the thermal diffusion measurement (°C).
UA Overall heat transfert coefficient of the cube assembly (W/°C)
y Sample thickness (m).
α Thermal diffusivity (m^2/s).
ρ Density (kg/m^3).

Acknowledgements

This work was performed under contract with Energy, Mines and Resources Canada. Westroc Industries Limited provided additional financial support and experimental facilities for some tests. The National Research Council of Canada also provided financial support for the work (IRAP L and H grants). Some tests were performed at the Center for Building Studies of Concordia University. The contributions of the following persons is gratefully acknowledged: Anne-Marie Jodoin and John Christopher Paris who carried out most of the experimental work at Polytechnique, P. Purins from Westroc Ind. Ltd, N. Low and D. Feldman from Concordia University and M. Shapiro, consultant.

REFERENCES

1- RUDOY, W. et DOUGALL, R.S., "Effects of Thermal Mass on Heating and Cooling Loads in Residences", ASHRAE Transactions, 85, (I), 903-17, 1979.

2- GOODWING, S.E. et CATANI, M.J., "The Effects of Mass on Heating and Cooling Loads and on Insulation Requirements of Buildings in Different Climates", ASHRAE Transactions, 85, (I), 869-84, 1979.

3- BURCH, D.M., KRINTZ, D.F. et SPAIN, R.S., "The Effect of Wall Mass on Winter Loads and Indoor Comfort - An Experimental Study", ASHRAE Transactions, 90, (I), 94-120, 1984.

4- BURCH, D.M., MAZCOLM, S.A. et DAVIS, K.L., "The Effect of Wall Mass on the Summer Space Cooling of Six Test Buildings", ASHRAE Transactions, 90, (2B), 5-21, 1984.

5- KAMMERUD, R., CEBALLOS, E., CURTIS, B., PLACE, W. and ANDERSSON, B. "Ventilation Cooling of Residential Buildings", ASHRAE Trans., 90, (1B), 226-251, 1984.

6- HOULE, J.F. et PARIS, J., "Développement de plaques de plâtre avec emmagasinage thermique par changement de phase", Contrat 54SZ.23216-6-61-34, EMR Canada, Projet PAR-115, Parinova Inc., Rapport final, 1989.

7- PARIS, J. et HOULE, J.F., "Tests en vrais grandeur de plaques de gypse à grande inertie thermique", Energy Solutions for Today, Comptes Rendus, Edit. SESCI, Ottawa, 212-6, 1988.

8- CADOUX, S., HOULE, J.F. et PARIS, J., "Réduction de la surchauffe par plaques de gypse à haute capacité thermique", Renewable a Clean Energy Solution, 15ième Conf. Annuelle SECSI, Penticton, Comptes Rendus, 400-4, 1989.

TABLES

TABLE 1: EXPERIMENTAL GYPSUM WALLBOARDS

Symbol	MCP Incorporation
SB[a]	None
PEGU[b]	Polyethylene glycol in elongated capsules (300 x 20 x 10 mm)
PEG[c]	Polyethylene glycol in elongated capsules (300 x 20 x 10 mm)
PAR[d]	C_{16}-C_{18} paraffins (25/75 W%) in square capsules (75 x 20 x 10 mm)
PAR+[d]	C_{16}-C_{18} paraffins (25/75 W%) in square capsules (75 x 20 x 10 mm) containing steel wool
BS[d]	Butyl stearate impregnation
CLA[d]	Capric and lauric acids (45/55 W%) impregnation

Notes:

a: Standard boards used as reference in experiments.
b: Boards used during one year in field tests prior to measurements.
c: Boards made at the same time as PEGU but not used in field tests.
d: All other boards newly prepared.

TABLE 2: NOMINAL PROPERTIES OF BOARDS

Board	SB	PEGU	PEG	PAR	PAR+	BS	CLA
Thickness[a] (mm)	15.9	15.9	15.9	15.9	15.9	12.7	12.7
PCM concentration							
(kg/kg)	-	0.32	0.32	0.19	0.19	0.39	0.45
(kg/m^2)[b]	-	3.6	3.6	2.2	2.2	3.3	3.8
Heat capacity							
(kJ/kg°C)	1.09	1.42	1.42	1.10	1.10	1.46	1.43
($kJ/m^2 °C$)[b]	12.5	16.3	16.3	12.6	12.6	12.4	12.2
Latent heat							
(kJ/kg)	-	37.5	37.5	47.7	47.7	54.3	56.4
(kJ/m^2)[b]	-	429	429	528	528	462	480
THC_{12}[c]							
(kJ/kg)	13.1	54.5	54.5	59.3	59.3	71.8	73.6
(kJ/m^2)[b]	150	625	625	679	679	611	626

Notes:

a: The density of the 12.7 mm gypsum wallboards is: 670 kg/m^3 and the density of the 15.9 mm gypsum wallboards is: 720 kg/m^3.
b: For practicle reasons, we define the effective value based on surface unit for PCM concentration, heat capacity, latent heat and total heat capacity of the boards.
c: THC_{12} is the total heat capacity of the boards (including sensible and latent heat) for a 12°C variation between 18 and 30 °C (i.e. under measuring conditions).

TABLE 3: STEADY-STATE THERMAL CONDUCTIVITY

Board	Conductivity, k (W/m°C)
SB	0.131
PAR	0.108
BS	0.113

TABLE 4: THERMAL DIFFUSIVITY

Boards	Elapsed time t (s)	Temperature			Thickness	Thermal diffusivity
		To (°C)	Ts (°C)	Ty (°C)	y (mm)	α ($10^{-6} m^2/s$)
SB	275	-1.4	41.4	4.6	15.9	0.32
	570	-1.4	41.4	8.9		0.20
BS 1[(a)]	275	-0.6	43.8	6.9	12.7	0.18
1[(a)]	570	-0.6	44.0	11.3		0.14
2[(a)]	275	4.0	41.2	10.7		0.19
2[(a)]	570	4.0	41.4	14.1		0.14
CLA	275	3.0	41.2	9.5	12.7	0.17
	570	3.0	41.8	14.4		0.14

Notes:
a: number of determinations made.

TABLE 5: EFFECTIVE THERMAL CAPACITY

Board	Duration of test $t_1 - t_0$ (h)	Heat Absorbed by Boards, Ea_{12}		Total Heat Capacity of the Boards, THC_{12} (kJ/kg°C)
		(kJ/m²)	(kJ/kg)	
PS	4.9	133	11.7	0.98
PEGU	5.3	147	12.9	1.08
PEG	5.8	169	14.9	1.24
PAR	6.6	253	22.2	1.85
PAR+	6.8	275	24.2	2.02
BS	6.1	194	22.7	1.89
CLA	6.6	225	26.4	2.20

TABLE 6: FLEXURAL STRENGHT

Board	Sample	Position	Breaking Strenght F (N)
PEG	U	L	801
	D	L	748
	U	X	151
	D	X	-
PAR	U	L	196
	D	L	365
	U	X	312
	D	X	329
BS	U	L	685
	D	L	632
	U	X	294
	D	X	329
CLA	U	L	614
	D	L	587
	U	X	267
	D	X	258

CSA Specifications: F=667 N minimum for the L position.
F=222 N minimum for the X position.

TABLE 7: HUMIDIFIED DEFLECTION

Board	Deflection (mm)
PEG[a]	31.0
[b]	14.5
PAR[a]	8.0
[b]	5.0
BS[a]	4.5
[b]	1.0
ACL[a]	8.5
[b]	1.5

CSA Specification: 16 mm maximum

Notes:

a: First measurement.

b: Second measurement.

TABLE 8: SURFACE WATER ABSORPTION

Board	Weight Increase (g)
PEG[a]	63
[b]	65
PAR[a]	67
[b]	74
BS[a]	47
[b]	46
ACL[a]	50
[b]	46

Notes:

a: First measurement.

b: Second measurement.

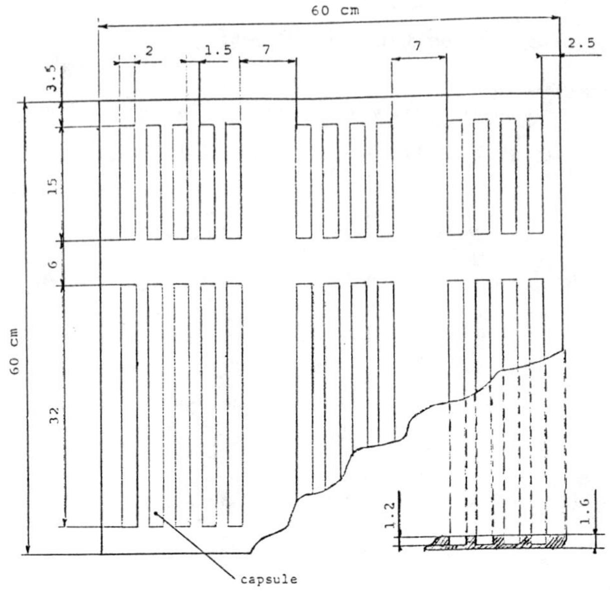

1a.-Elongated Capsules (PEG and PEGU)

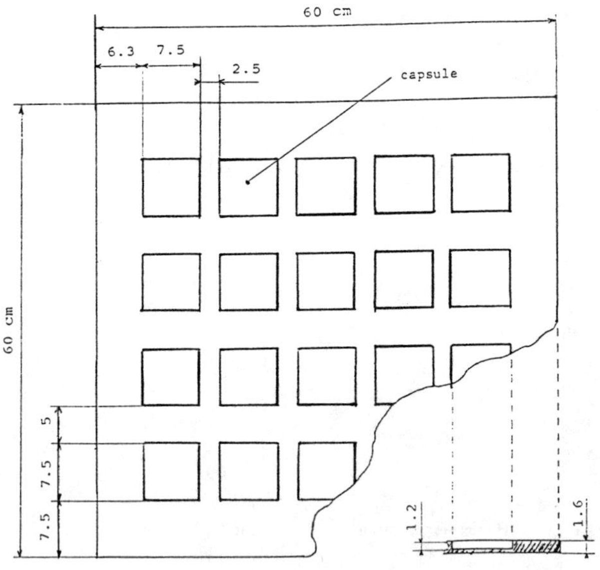

1b.-Square Capsules (PAR and PAR+)

Fig.1.-Arrangement of Capsules in Boards

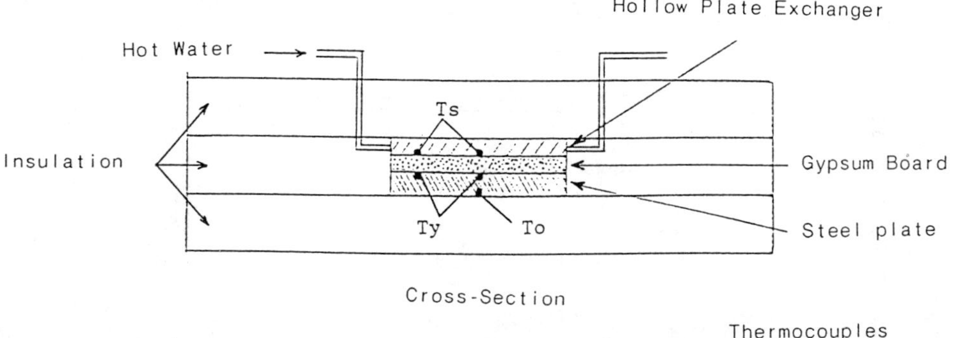

Fig.2.-Experimental Apparatus for Thermal Diffusivity

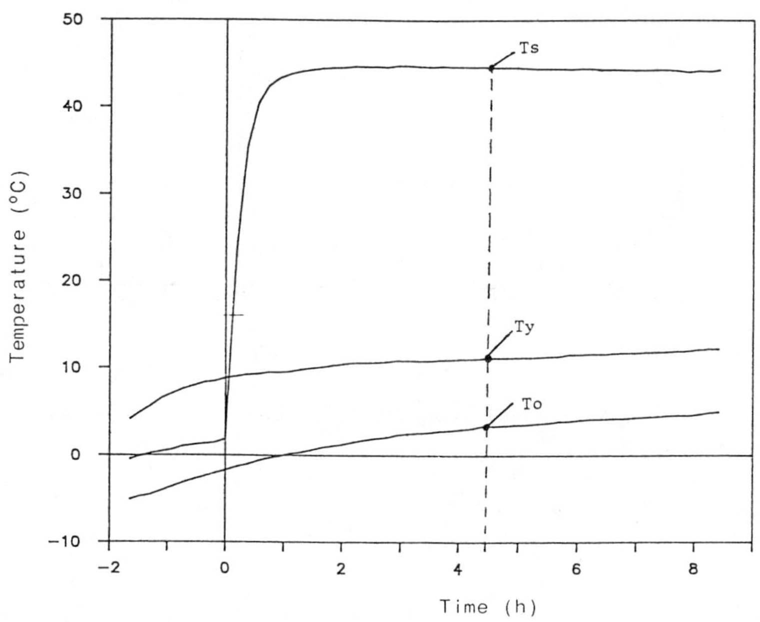

Fig.3.-Temperatures Curves for Diffusivity Assessment

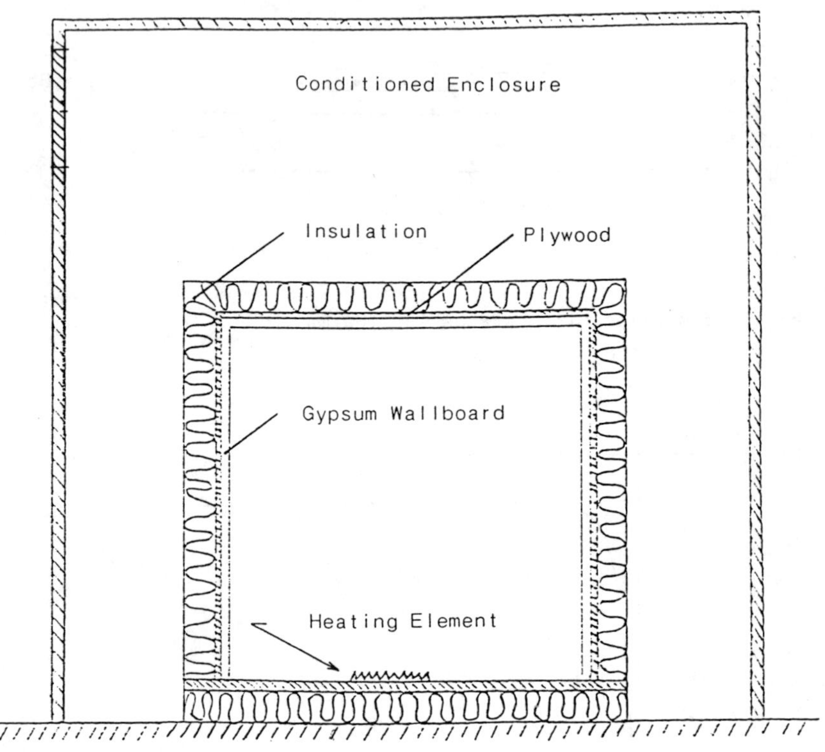

Fig.4.-Thermal Capacity Apparatus

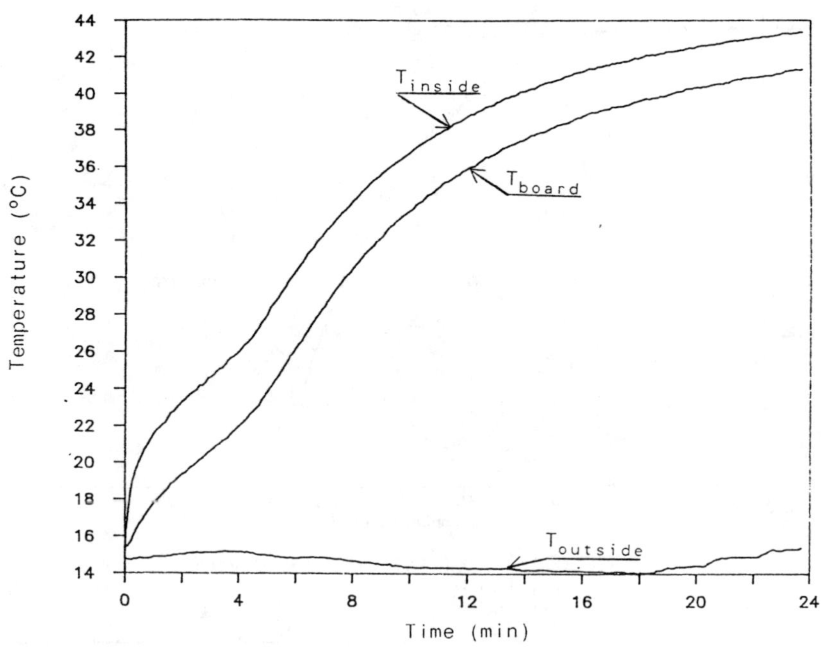

Fig.5.-Temperatures Curves for Thermal capacity Assessment

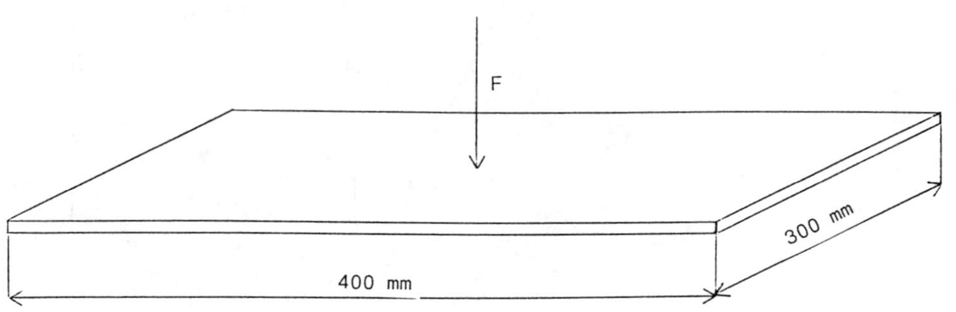

Fig.6.-Principal of flexural strength measurement

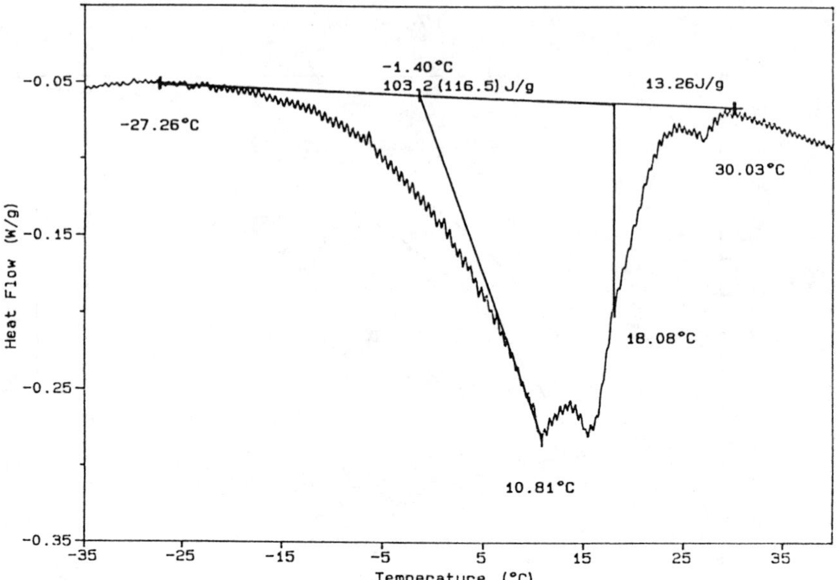

Fig.7.-DSC Graph of Polyethylene Glycol
Heating Rate: 2°C/min

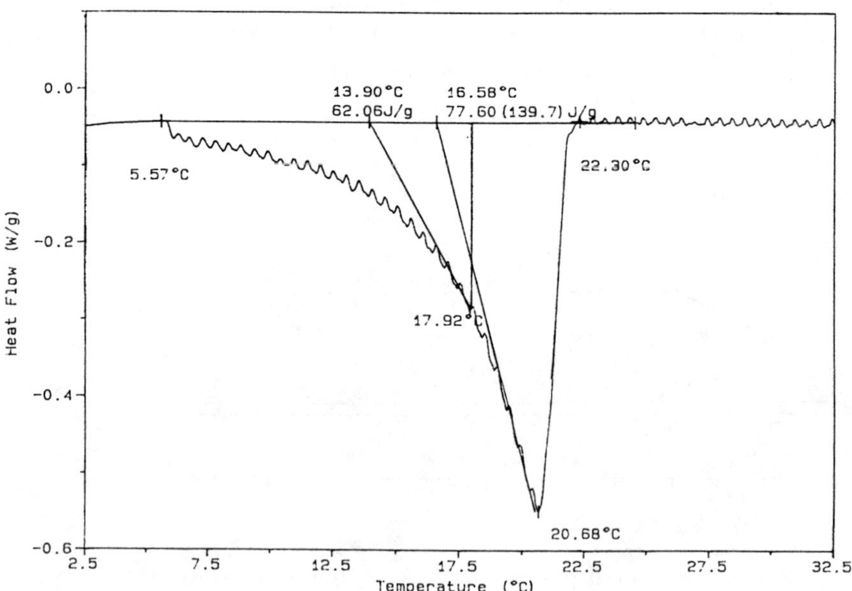

Fig.8.-DSC Graph of C_{16}-C_{18} Paraffin Mix
Heating Rate: 1°C/min

IMPROVEMENT OF TRAFFIC: SOLAR ENERGY ON THE ROAD

Clauwaert Frans
Constr. Engineer, Belgium, Switzerland

1. INTRODUCTION

Mobility is one of our legitimate needs, as attested by the steadily growing number of motor vehicles, worldwide soon 500 millions. On the other hand, the same citizen who exercizes his right to mobility must also be concerned about the consequences of his actions.
The following report presents a solution which makes use of the latest technology to utilize solar energy in a rational manner.

To a great extent, the sun determines the rhythm of man's life. It forms the basis of the nutrition and growth of living nature consisting of man, animals and plants.

The amount of energy that radiates from the sun exceeds our powers of imagination. It is thought that the sun has been fulfilling its role as a source of light and heat for about 4 - 5 billion years now, and we can certainly work from the fact that this process will continue for again as long.

Even though only an infinitesimal part of the sun's energy, one-half a billionth of it, reaches the earth in the form of electromagnetic radiation, it is nevertheless considerable. This amount alone is 10 000 times greater than today's primary energy requirements of the entire earth.

When we speak of solar energy (SE) we first think of its mass-bound influence on the climate and - in the case of photosynthesis - its effect on the growth of living nature.

Among the familiar examples of the technical application of solar energy are its passive and active utilization in the preparation of hot water and for heating purposes.A relatively new phenomenon is the use of electromagnetic radiation by the photovoltaic effect for the production of electrical energy to run vehicles. At least since the Tour de Sol 85 in Switzerland even the most diehard doubters must now be convinced that this phenomenon is a technical reality. This event, which took place during the un-sunny month of June 1985, far exceeded the expectations of its organizer, the Swiss Solar Energy Society. From the fully electronic Solarmobile with every technical refinement all the way to a bicycle supplemented by muscle power, proof was given that this idea must be pursued further.

Recently the specialist press has raised the question as to wether the solar-powered vehicle may one day actually displace the conventional motor car.

It may be that this question is based on a certain skepticism. In any case, the question should not be left hanging, now that the first experiences and evaluations are available.

Due to the fact that these solarmobiles required very little energy, chiefly thanks to their low weight, that they caused no exhaust or pollution and that they drive without making any noise, the attempt will be made in the following to provide a basis for their possible use in everyday traffic.

2. URGENT REASONS FOR ACTION

Based on the existing situation, the following arguments are proposed which from various points of view, require us to forge a new relationship between man, technology and nature.

Energy:

From a global standpoint, our present energy situation is rather uncertain to many people and in some parts of the world has already led to tension on economic, political and military levels. Based on the latest reports, it can be seen that the phase now seems to have been reached which was already predicted by many people. This is self-evident, but it requires the proper insight. What is meant is the finding that in the oil sector fewer new wells are being found than its direct consumption requires, despite the massive drop in sales. Oil will remain the world's major source of energy. But most of the easily accessible reserves have been found already, what means that its price may rise to a higher level. On the other hand, the struggle for black gold is being carried on more and more openly, which leads us to the conviction that this precious liquid must be more sparingly used. It may be assumed that the amount of traffic and the consumption of fuel will continue to rise.

Unheard-of amounts of energy and fuel are thus being consumed by motor vehicles, and this with an overall degree of efficiency of at best 4,4 %.
The so-called catalyst car does not improve the situation at all; just the opposite! But even such improvements as better aerodynamics, ceramic motors, smart fluids, 5-speed gearboxes, fuel injection, turbo additions, car pooling and more roads will not achieve that which has always been desired. And again, not only the power but also the manufacture of vehicles requires a great deal of energy and vast amounts of raw materials.

Traffic:

According to one study, in the Federal Republic of Germany 60 % of all trips made with a car are less than 5 km. Even if the conditions in other industrialised countries are not quite the same, based on the standard of living, the number of cars and the level of petrol consumption, to mention only a few factors, it may be assumed that this figure is at least as high in these countries. It is precisely in this short trips with their stop-and-go conditions, traffic signals, etc. that use of an internal combustion motor is very inefficient. The well-known results are a very high level of fuel consumption and poor combustion with a corresponding production of exhaust gases. In addition, parking space in city centres requires huge investments for expensive buildings and high property costs.

Air:

For some considerable time now, homo sapiens has concerned himself more intensively with the consequences and secondary effects of the combustion of non-replaceable fuels. At the beginning of the present decade, the Max Planck Institute reported on the effects of acid rain in Sweden: 20 000 lakes became over-acidified. Every form of life in them died off. And this is not even taking account of the condition of the forests, which may be mentioned here only in passing.

The harmful effect of photooxidants on plants has been known in the USA for years, and it is considered to be very serious. It is calculated that the losses in crops in that country are in some cases over 15 % because of increased ozone contamination. These toxic photooxidants are formed from nitrous oxides and hydrocarbons between spring and autumn under the effect of solar radiation. It is just during this period that a reduction in air pollution is essential.

In heavily polluted parts of Switzerland it has been reported that the range of visibility has become appreciably lower during the summer months. In 1950 the number of hours per day in the summer with more than 10 km visibility was 9,7. In 1985 this figure tended to be around 7,5 hours. The Swiss Federal Department of the Interior has reported that this reduction by 2 hours can only be explained by the increasing air pollution. It is therefore logical to conclude that in addition to the quality of the environment, safety and economy are also seriously involved.

No individual citizen can be expected to know how much, when and which substances are still bearable for his health. The days when it was only the CO-content which represented the degree of air pollution are long gone; since we regularly read about CO_x, NO_x, C_xH_y, lead halogenides, peroxyacetylnitrate and other substances; we need assistance.

Here the tasks of public authorities have reached a level which borders on the impossible, especially when one considers the synergistic effects of these pollutants.

Noise:

"Some day man is going to fight noise just as relentlessly as cholera or the plague." This statement was made more than 70 years ago by Robert Koch, a well-known physician. It can be affirmed that a great deal has already been undertaken to reduce the effects of traffic noise, which is directly perceived to be the greatest noise pollution we face.

A considerable part of our efforts are only based on decreasing noise immissions. It is far more important to reduce noise emmissions at their source, i.e. the vehicle itself or driving habits. Here again our cities are the most affected due to daily commuter traffic.

The West German Automobile Club has calculated, for example, that on Saturdays when people go shopping, up to 75 % of the entire traffic consists of people just looking for a place to park!

And did you know that the noise level of a motorcycle going 50 km an hour is more than that of 10 cars put together?

Economy:

In the early 19th century, David Ricardo, the outstanding representative of the political economy system, thaught that there are 3 production factors in our society: labour, capital and land. Today, we know that these factors no longer suffice; in this writer's opinion, at least the factors of the limited supply of raw materials and the qualitiy of the environment belong in cost calculations. It is the task of our society to quantify these factors. If people are to be responsible for their own future, then they must also be informed as to the consequences of their actions.

For example, it can no longer be a matter of no concern to us when we hear that the North Sea looks like the Gulf of Mexico; 100 different companies are drilling for oil and gas (more than 2000 wells are being drilled in British waters alone!). We may be admire such technical achievements, but we should be full of concern, as those people who are making efforts to eliminate the dangers which may result from drilling into the earth at the bottom of the sea.

The question is whether it is right to burden all peoples with air, land and water pollution. Are we ethically justified in creating ecological conditions in which the weaker peoples and future generations neither participated nor desired?

The extend to which the biosphere can be stressed has become a bottle neck for technical development. As compared to 1860 the present atmosphere must, among other things, deal with 40 times more man made carbon dioxide (CO_2).

One way to counter further increases in CO_2 would be for every human being to plant one tree every day of his life!

A mature tree produces enough oxygen for ten people, whereas a medium sized automobile uses the same amount of oxygen in one second as an adult needs in ten minutes! It is not hard to see why many trees are doing so poorly.

A lot can be done to improve traffic, especially in urban and suburban areas. We should continue to take advantage of the many advantages of the internal combustion engine. In my opinion, however, there are very serious problems associated with its overuse.

The state of the art continues to be refined, but these changes constitute only small improvements which are neutralized by such things as increase in traffic and changes in driving behavior. Even if all gasoline powered engines were fitted with catalytic converters, noise, pollution and consumption of energy would remain, indeed, more energy would be needed. It is clear that additional kinds of motivation are necessary to effect real improvements.

The present consumption of energy and raw materials for our legitimate transportation needs far exceeds our real capabilities. Technology is neither good nor bad, however, its use must be an integral part of an ethical value system and a natural order. Do we need a book of ethics? Perhaps it will be easier in the future to assign a value to work, time, raw materials and quality of life now that our system of values is changing and affluence is no longer considered to be synonymous with well being (owning is not being).

Duerrenmatt claimed that "we are living in an explosives factory in which smoking is allowed", which might be considered courageous in light of the precarious state of affaires.

Time is running out, especially since the amount of time it takes to reverse ecological damage is ever increasing.

3. DEVELOPMENT OF NEW TECHNOLOGIES

The period of time in which it takes a new technology to establish itself has always been one during which the inventor was considered rather eccentric or odd. Let us take a look at the length of time it has taken technologies to become well established.

- Cugnot's invention in 1769 is considered to be the predecessor of both the locomotive and the automobile. His steam engine only attained a maximum speed of 6 km/h and the fuel lasted for just one mile. At the time of its patent in 1877 the internal combustion engine was of little importance in Switzerland due to its poor efficiency. The low average speed of 18 km/h during the first automobile race in 1891 posed no obstacle to the eventual acceptance of the automobile. Engines soon had capacities of 50 to 100 horsepower even though the first ones were much less powerful.

- The long history of the computer started with the abacus which the Chinese developed about 1100 B.C. and continued through the use of punch cards developed by Hollerit in 1891, and the famous ENIAC built in 1946. The real breakthrough took place in the second half of this century and development of the computer is certainly not yet finished.

  It remains to be seen whether in the coming century robots will be constructed by robots and computers by computers but most people are still rather skeptical. Predictions are not to be believed so much as simply taken note of.

- Production of high quality electricity began modestly in 1884 at Berlin's Friedrichsdorf in the first coal-powered plant which only supplied 1500 light bulbs and 18 street lamps with current. Today photovoltaic cells are producing electricity by a simple method which produces neither smoke nor noise. The past few years have demonstrated that solar energy has great potential.

  The long period of development between Helios' mythical chariot and the solarmobile has been nurtured by fantasy, inventiveness, and technical know-how.

4. WHAT WAS LEARNED

The first Tour de Sol in 1985 showed that it is possible to build a Solarmobile, that can run at 40 km/hour or even 70 km/hour. This can be done with an energy consumption of less than 5 kwh per 100 kilometer. A normal car uses up to 20 times more energy for the same distance.

This minimal use of energy is only possible by using light weight parts, electronics, computers, space technology and other special areas of 'high tech' technology.

We know that space technology uses energy very conservatively, because only then is man in the position to convert Jules Verne's ideas into reality. Why shouldn't these ideas be used on earth too?

The new Solarmobile should neither be for high speed nor for long trips. We especially need this vehicle to reduce the summer ozone pollution and the general air pollution in large cities.

For many it may be a disappointment that only 4 % of the sun's energy is available to drive the vehicle. As already noted, the efficiency of the energy chain (raw oil - refining - burning in the Otto-Motor) is also only about 4 %.

If we look at the ratio payload/net weight for both a Solarmobile (1 person) and an automobile (1,3 people, this being the average number of passengers in a car), then the picture changes in favour to the Solarmobile. The future Solarmobile has the following properties:

- Speed	30 - 50 km/h
- Weight	max. 1/10 that of a car
- payload	max. 120 kg
- number of passengers	1 (max. 2)
- parking space	max. half or a third of a car
- high safety	
- easy handling	
- pleasant form	

This data is not only optimum for the Solarmobile; they conform with the goals of some auto producers for a pollution conscience future vehicle.

5. FUTURE EXPECTATIONS

With a yearly influx of solar beams in Switzerland of about 1000 kwh per m^2 and with corresponding concentrators (e.g. lenses of mirrors) it is possible to obtain enough energy for a Solarmobile for example from the roof of a garage.

By use of concentrators it is possible to reduce the area of expensive solar panels. The winning of energy would be decentral; at home, at public parking places or in electro service stations.

Several important components are already available which would help to realize these ideas. For example, we mention the use of Kevlar in light weight construction, and the storage of electricity in sodium-sulfur batteries. The profis will be using better regulation switches so called maximal power tracking (MPT) units , to optimize fuel consumption a.o.m.

The life of the photo cells is longer than the motor vehicle, but it shouldn't be necessary to transport the cells during every trip. A Solarmobile is a vehicle that will be used for short traveling times and long waiting times, it would be efficient to accumulate and store this energy and in some cases change the batteries while it is parked.

Let's take the case where half of the potential users in winter are not able to get enough electricity from the sun to drive the motor. This would mean an absolute increase in electricity use from the electrical network. This would be a true substitution of oil with electricity that would be accepted by the majority of the population, because it brings among other things the following advantages:

- Reduction of raw material, oil, noise, photochemical smog etc.
- New meaningful jobs that would increase balance of payments.
- Larger city areas would become more attractive without further building of streets.

The scientific and technological knowledge that is already available must not be forgotten. From our experience we must only draw the right conclusions.

The advantages and disadvantages of cars and motorcycles are known, and public transportation unfortunately does not come to our door step. This market gap must be closed.

According to the newest scientific news, I will allow myself to write down the following assumptions. Thorough team work between companies from the branches of automobile production, electronics, computers and space technology ideas could develop through innovation the reality of a Super Solarmobile. The end product would allow us to accept that "Global Thinking" and "Local Doing" is possible.

With the new growth wave there is a signal for an new life style.

Therefore it is worthwhile to consider this new transportation system, especially in case of cities, work places, and infrastructures.

Progress must not be paid for with noise and exhaust. After extensive examination of its long term effects with a thorough technology assessment this new technology will prove itself.

It is said, that solar energy is of low intensitiy and only available at intermittant intervals.

A comparison in this matter with the hydropower - as a part of the harnessed solar energy - shows the following:

> The availability of hydro-electricity is linked with the precipitation of rain or snow.
> Physicaly considered, the energy from falling or flowing water in function of surface, altitude and time would be to neglect and nevertheless it supplies about 7 per cent of the worldwide energyamount. In Switzerland this part is near to 12 per cent.
>
> We know, that the potential of the sunradiation is many times higher than those of the hydropower. It is therefore evident, that her contribution may be very significant and a practical alternative to conventional energyforms.

Cleverly conceived technologies will play an important role to help our energy-hungry world. So we may say yes to a thoughtful mobility in according with man and his environment.

Following the Kondratjeff-cycle, we are now at the end of the 3th phase, that means that we have "to stop the wrong technologies and to choose the right ones."

HANDLING AND TRANSPORTATION EQUIPMENT IN RURAL RAJASTHAN-INDIA

A.K. Sharma and N.K. Mehta
College Tech. Agricultural Engineering,
Rajasthan Agricultural University, Udaipur 313001, Rajasthan, India

ABSTRACT

Much of the farmer's time is spent in handling and transporting the crops he grows, the material and machines he uses to cultivate them as well as other materials-bulky or fluid-produced, needed on the farm-stead and finally transporting the ready material for marketing.

This paper deals with the various manual, animal and engine powered equipment used by the farmers of the State for handling and transportation at different levels. Manually powered equipments include forks, binding materials, various containers, carrying loads on shoulders and carts. Animal powered equipments are the carts powered by bullocks, camels and donkeys. Tractors, buses and trucks fall under engine powered equipments commonly used for transportation.

1. INTRODUCTION:

Rajasthan is located between 23 to 30 latitude and 70 to 78 longitude. Its total geographical area is 342,239 square kilometers. About 47 percent area of the State is under cultivation out of which 25 per cent area is under irrigation. Rural population is about 79 percent of the total population of the State.

Much of the farmer's time is spent in handling and transporting the crops he grows, the material and machines he uses to cultivate them as well as other materials - bulky or fluid- produced, needed on the farm-stead and finally transporting the ready material for marketing.

Various tools, machines, vehicles, and other equipments facilitate these operations and are being adopted by the farmers of the State and can be broadly classified in three categories.

- Manually operated
- Animal operated
- Engine operated

2. MANUALLY OPERATED EQUIPMENTS:

Human power is the main source of power in rural Rajasthan. Various handling and transporting operations are performed by the manually powered equipments.

Forks

Forks of many shapes and sizes are used to handle farm produce of various weights and consistencies (Fig.1). Forks are locally made of wood, but the demand for stronger types with steel tynes is increasing. The tynes are straight or curved, round or oval in cross section. Wooden handles are used on the tynes for easy operation. Length of the handle varies according to the requirement of individual farmer and operation.

Binding Materials

The handling of bulky material can often be facilitated by binding it into bundles. Wires, ropes (made of jute, straw, iron, etc.) are common materials used for binding the crop while handling and transporting (Fig. 2).

Containers

Liquids, loose grain, fertilizer, tubers, fruits, and many other such things are carried and stored in buckets, pitches, cases, baskets, sacks, drums, etc. made of material such as jute, plastic, bamboo sticks, pottery, and metal sheets. Sacks, buckets, drums/bins made of jute, iron/plastic and mild steel sheets, respectively, (Fig. 3) are most popular containers used for storage.

Transportation by Carrying

Transportation by carrying is mainly determined by the weight of the load in relation to the carrying capacity of the person (Fig. 4).

Carrying Aids for Man

A load should be carried in such a way that it's centre of gravity is vertically above the carrier's centre of gravity. This means that the load should be placed on the head which is a common practice. Method of load carrying can be improved by carrying the load on the shoulders with supports (Fig. 5). This practice is also well adopted by the farmers of the State.

Transportation on Wheels

Manually pulled four wheel carts (Fig. 6) are used in the State. Use of the above increases the load transporting capacity of the individual. Many people of the State earn their living by use of this cart. It is made of wood and has pneumatic rubber wheels. Carts are fabricated by local artisians at village levels.

3. ANIMAL OPERATED EQUIPMENTS

Animals, such as bullocks, camels, and donkeys, are the main source of power in the State. Carts powered by these animals are most commonly used for transportation.

Bullock Cart

Bullock cart (Fig. 7) is the most common mode of transportation in rural Rajasthan. It is used for transporting seeds, fertilizer, implements, harvested crops to the threshing yard, processed crops to the market, family members for visiting and many other purposes. Usually a pair of bullocks and sometimes a single bullock is used as a power source. It is made of wood and mild steel and is fabricated by local village artisians.

Camel Cart

Camels are the most popular source of power in deserts of Rajasthan. Camel carts (Fig. 8), two and four wheels, are most commonly used by the farmers of Central and Western Rajasthan. Camels have good draftibility and carry heavy loads for long distances on sandy soils. Research work on camels for their maximum utilization is in progress at Camel Research Institute (ICAR) in the State.

Donkey Cart

The donkey is also a major draft animal in Northern parts of the State. The donkey is used to carry loads on its back. Donkey carts (Fig. 9) are the most common transportation equipment in Northern Rajasthan.

4. ENGINE POWERED EQUIPMENTS:

Tractor Powered

Use of tractor powered equipments such as tractor trolley (Fig. 10) is a common transporting equipment in most of the parts of the State, specially in Northern and Eastern Rajasthan. Hiring system for tractor powered equipments is also well adopted by the farmers of the State.

Trucks and Buses

Trucks and buses through rural parts are major transport medias of the State. They transport people, processed crops, raw material for various uses etc. from one part of the State to other parts. They are either run by Government or private agencies.

5. CONCLUSIONS

Handling and transportation equipments can be human, animal and engine powered. As the economic condition of the farmer in the State is poor, human and animal powered equipments are more popular. In certain parts of the State where irrigation facilities are good because of better economic conditions of the farmer, engine powered-tractor operated equipments are commonly used.

Farmers are inclined to go for mechanized handling and transporting equipment but it is their economic feasibility which restricts them. Emphasis on research work to improve the commonly used equipments is the immediate need of the State.

REFERENCES

1. Depande, S.D. and Alam, A. (1982). Performance evaluation of bullock cart of Bhopal region. Paper presented at XIX, Annual convention of ISAE, Udaipur.

2. Despande, S.D. and Ojha, T.P. (1984). Design of animal drawn vechicles. Technical bulletion, C.I.A.E., Bhopal.

3. Despande, S.D. and Ojha, T.P. (1985) Development and evaluation of an improved ox cart for Central India. Journal of Terramechanics, Vol. XXII, No. 3, pp. 135-146.

4. F.A.O. (1972). The employment of draught animals in agriculture, F.A.O., Rome.

5. Khande, M.D. (1982). The modern bullock cart and rural transportation, draught animal power. A report prepared by Dunlop India Ltd.

6. Mehta, N.K. (1988) Evaluation of draftability of camel. An unpublished M.E., Thesis, CTAE,l Udaipur.

Table 1 Carts used in rural Rajasthan

S.No.	Type of cart	Material of construction	No.of wheels	Type of wheels	Capacity		Cost of operation Rs/100kg/ km
					Wt. kg	Floor area sq.m.	
1.	Manually operated cart	wood,mild, steel	4	Spoked, Solid/ pneumatic rubber	500	1.35	0.35
2.	Bullock cart	Wood mild steel	2	Wood spoked with mild steel on outer periphery/ Pneumatic rubber	800	1.23	0.20
3.	Donkey cart	Wood,mild steel	2	Pneumatic rubber	1500	3.36	0.12
4.	Camel cart	Wood mild steel	2,4	Pneumatic rubber	2000	4.11	0.08, 0.06

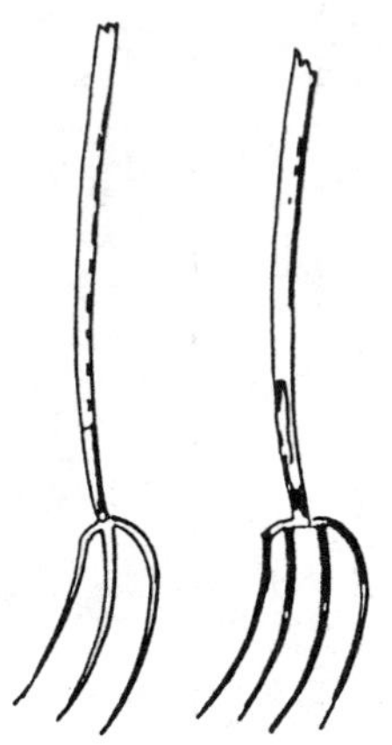

Fig. 1. Forks

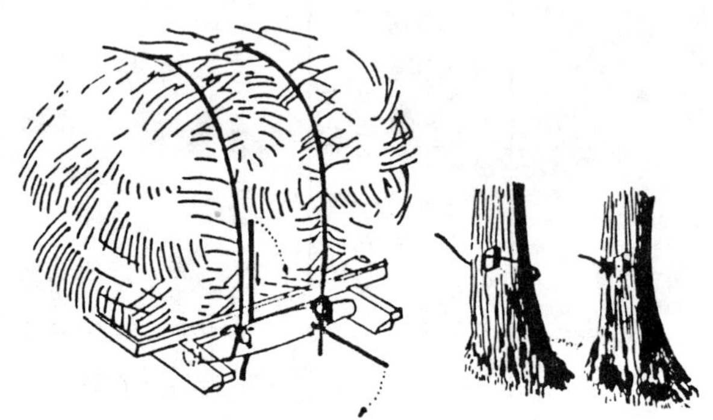

Fig. 2. Binding materials

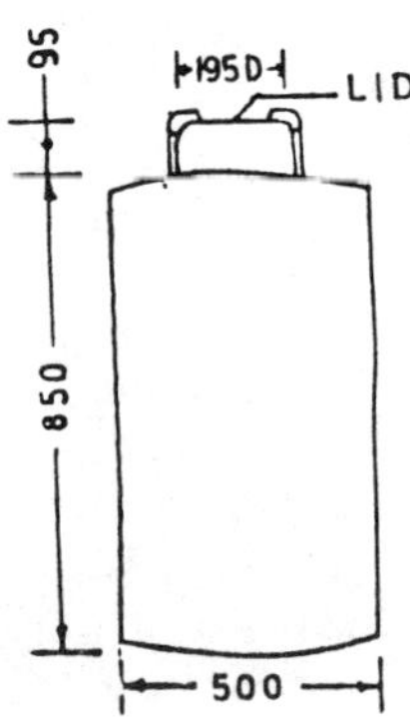

ALL DIMENSIONS ARE IN MM
SCALE= 1:20

Fig. 3. Bin

Fig. 4. Transportation by carrying

Fig. 5. Carrying aid for man

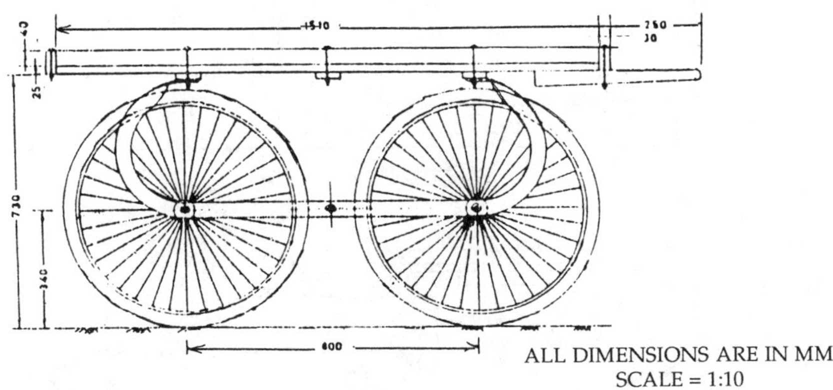

Fig. 6. Manually pulled cart

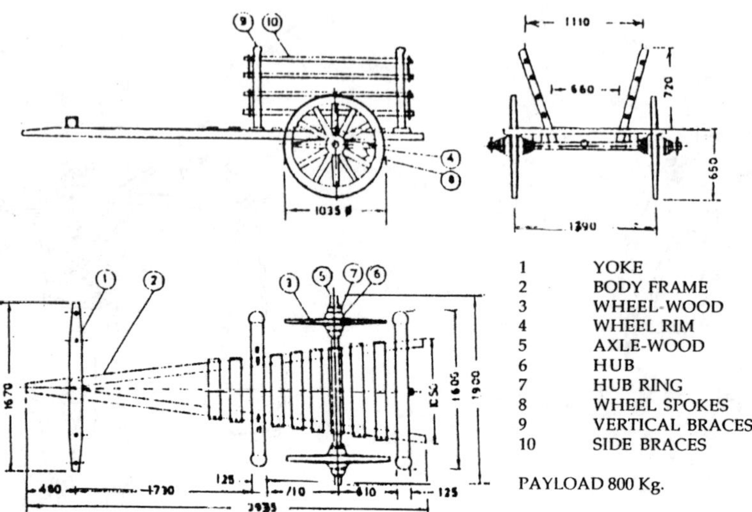

Fig. 7. Bullock cart

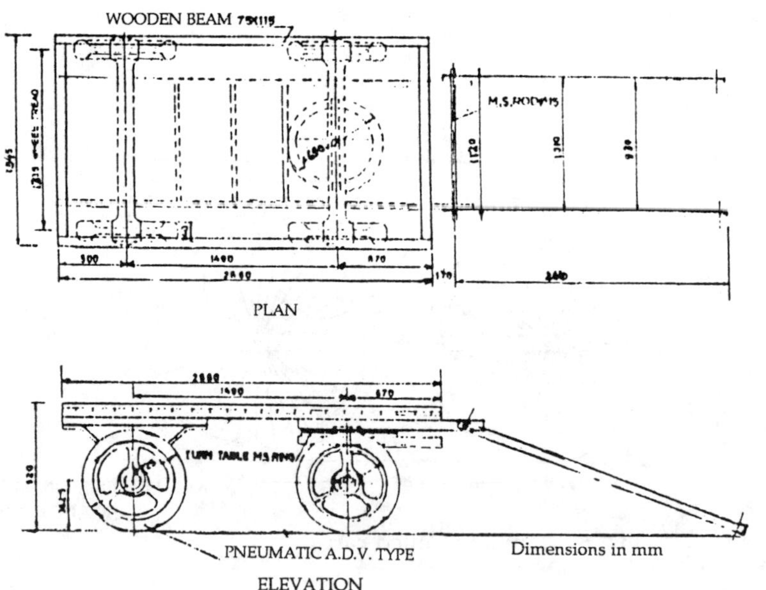

Fig. 8. Camel cart

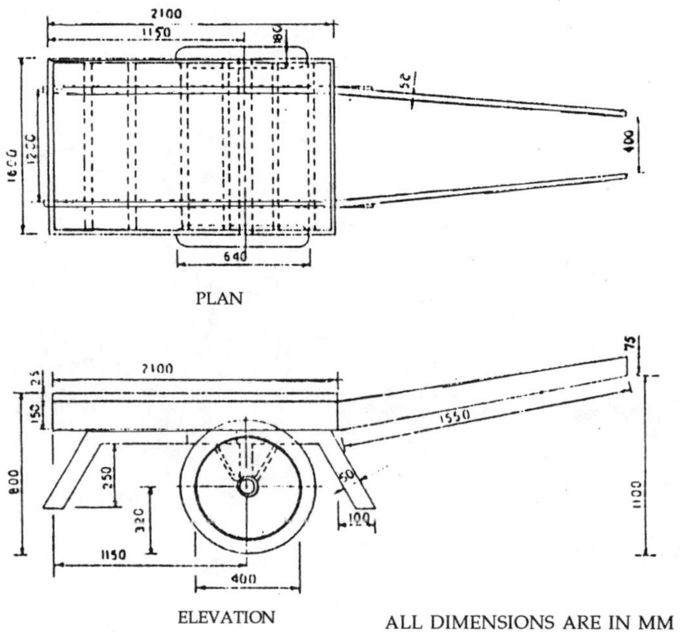

Fig. 9. Donkey cart

Fig. 10. Tractor trolley

APPLICATION OF SELF-RECUPERATIVE BURNERS TO LOW CALORIFIC VALUE GAS

Waldir A. Bizzo and Raymond B. Peel
UNICAMP - FEM/DETF, 13081 - Campinas, SP - Brazil

Abstract

A conventional self recuperative burner has been tested burning low calorific value gas supplied at 200 or 400°C from a wood burning producer with the objectives of using low C.V. gas in higher temperature processes without enrichment with oxygen or oil, and of economising fuel. During tests out puts up to 140.000 Kcal/hr, furnace temperature of 1400°C was reached, with air preheat to 600°C, and wood gas entering at 400°C. In parallel with the experimental tests a computer simulation was made. This showed good a agreement between model and practice and demonstrating that heat transfer between the walls of the recuperator section was largely by radiation, showed that no advantage would be gained by thermal insulation between the hot inlet gas and the cold incoming air.

1. INTRODUCTION

In the past the use of low calorific value gas such as producer gas in high temperature furnaces was common, for example, metallurgical or ceramic furnaces. These furnaces were mostly large and the combustion air was pre-heated by regenerators. For most applications the use of producer or other low C.V. gas has been superseded by oil or natural gas.

In Brazil there has been since the mid 1970s a renewal of interest in the gasification of indiginous fuels such as wood and vegetable residues to provide low C.V. gas which is used in various industries, notable ceramics to heat small or medium size furnaces [1]. The maximum temperature that could be reliably attained in small furnaces using this low C.V. gas, rea-ching the furnace from the producer at 700-900 K and air without preheat was c.1370 K. Where higher temperature is needed the most common recourses have been to enrich the combustion air with oxygen from liquid storage or to burn oil jointly with the gas. Combustion air preheat has not been used probably because of the cumbersomeness of the conventional recuperator or regenerator when applied to a small furnace.

It was thought that an autorecuperative burner similar to those used with natural gas might be adapted for burning low C.V. gas (in this instance producer gas from wood; wood gas).

The major differences between wood gas and natural gas affecting their use in autorecuperative burners may be sumarised:

	Natural Gas	Wood Gas
C.V. Kcal/kg	11000	1000
Stoiquiometric ratio	10:01	1:1
kg products/1000 Kcal	1	2
Theoretical flame temp.	3070 K	1950 K
Temperature of fuel gas	ambient	300-870K

The mass of air available to carry sensible heat into the furnace is only half that of the combustion products carrying heat out; in compensation the fuel gas is often available hot, and is thermally stable. A typical composition of wood gas is: CO: 18%, H_2: 15%, CH_4: 1%, H_2O: 12%, CO_2: 10%, N_2: 44%.

2. EQUIPMENT AND TEST PROCEEDURE

The autorecuperative burner, based on an existing type [2,3] was initially designed for an output of 115 KW.

The dimensions of the gas passages and of the heat transfer surfaces were calculated using available published heat transfer data (4-10) to give 600 K of preheat to the air when the exhaust products were exhausted at 900 K. The burner and integrated recuperator (Fig. 1) were constructed from 2 mm stainless steel (AISI 310) with a 10 mm mild steel mounting flange. The outside of the recuperator was lagged with mineral wool; to avoid overheating and possible failure of the flange and furnace wall these were left unlagged. During the test programme the air injector holes in the burner were enlarged from 9 mm to 12 mm so that the output could be increases to 160 KW cal/hr without incurring excessive pressure drop on the air side.

Burner temperature measurements were made using type K chromel/alumel thermocouples. The temperature of the furnace gas entering the recuperator was taken as the mean between the readings of two thermocouples in the gas flow, one at the entry to the annulus wich had a large angle of view of furnace radiation, and another 20 mm inside the annulus seeing little furnace radiation. The difference between the two readings did not exceed 60^oC.

The test furnace (Fig. 2) was a mild steel box 1 m^3 with internal refractory blanket insulation. The chimney was fitted with a flap valve so that the furnace could operate at a small pressure.

Fuel gas was provided by a Codegas CD250 parallel flow producer burning eucalyptus wood; this gasifier produces largely tar free gas at temperatures up to 600^oC. Gas reached the burner at 470 K; after relagging the rather long connecting pipe this temperature was increased to 710 K.

Flow(F), temperature(T), and pressure(P) measuring points were provided as indicated in Figures 1 and 2. Sample points were gas samples could be withdrawn for analysis are also marked on the diagram. Samples were taken from each charge of eucalyptus to the gasifier for humidity measurement. The humidity of the wood which largely controls the C.V. of the gas (Fig.3) was maintained between 11% and 19% giving a gas of C.V. between 4180 and 4473 KJ/kg.

The furnace contained a simulated load of firebricks whose temperature was measured by a total radiation pyrometer (Magdeburg G SP1871 07). Temperatures inside the furnace were also measured using pyrometric cones and , below 1520 K a type K thermocouple.

Testes were with the furnace stabilised at energy inputs between 80 and 160 KW.

The proportion of the total combustion products passing through the recuperator was varied:

- Total: The furnace chimney was closed so that all exhaust gas had to pass through the recuperator, the air ejector was used to reduce the rise in furnace pressure and consequent leakage. Care was taken to ensure that the pressure inside the furnace did not fall below atmospheric.

- Partial (between 40 and 60% of the combustion products passing through the recuperator): The chimney was open but the ejector was drawing furnace gas through the recuperator. The injector had insufficient capacity to allow good flow control.

- Zero: All the combustion products leave by the chimney; the burner functions as a simple nonrecuperative burner.

Most tests were made with wood gas entering the burner at 470 K the gas pipe was then lagged and the final tests were made with gas entering the burner at 710 K.

3. RESULTS AND DISCUSSION

Test results are shown in Figs. 4,5 and 6. It may be seen that with full extraction and fuel gas entering at 470 K the air preheat was 740-820 K, only reaching the design value of 870 K when the fuel gas was supplied at 710 K, however sensible heat gained by the gas is included in an "equivalent air preheat" $[(mC_p\ T)_{gas} + (mC_p\ T)_{air}]/(mC_p)_{air}$ 870 K is exceeded. It may also be seen that the equivalent preheat is reduced when the fuel enters at a higher temperature. Halving the furnace gas flow through the recuperator only reduced the preheat by c.25% this in conjunction with the calculated temperature profiles suggests that the surface/air heat transfer coefficient could usefully be increased.

The maximum furnace temperature reached, which would be influence both by the burner and by the furnace, without air preheat, but with fuel at 710 K was 1570 K; With full air preheat this was only increased to 1670 K. However with total extraction the energy input needed to reach 1570 K was reduced by 40% or by 20% with partial extractions; it may be observed that this energy saving is greater than the proportion of energy returned in the preheated air: 16-19%, presumably because of the advantage of higher flame temperature in improving heat transfer.

As the burner output was increases the air preheat temperature decreased for a given furnace temperature.

Computer modelling of the recuperator was based on a method described by Harrison et al [3]. The three concentric tubes were divided longitudinaly into sections; the properties of air, wood gas and combustion products (ref. 4,5,6 and 7) varying with their respective temperatures in each section.

Corrections were made for end effects at gas entries and exits.

The main resistance to heat flow was convection from surfaces 3 and 4 (Fig. 10) to the air: the air entered tangentially in an attempt to increase its velocity and therefore the heat transfer by convection. A problem arose in estimating (from ref. 9 and 10 wich gave widely differing values for two different surface conditions) the emissivity of AISI 310 steel.

An intermediate value was used in initial calculations; a more recent (11) provided justification of values chosen to give good agreement with experiment, Figs. 7, 8 and 11.

From (Fig. 10) it may be seen that the wall between the hot fuel gas and air is always at a temperature above that of the fuel gas. There is therefore no heat transfer from the hot incoming fuel gas and the air which might reduce the total effective preheat.

4. CONCLUSIONS

An autorecuperative burner can be used to increase the operating temperature of a furnace fired by wood gas or other low C.V. gas. The maximum temperature of 1670 K reached in our test furnace could probably be increased in a better lagged furnace.

The use of the autorecuperative burner gave fuel economy of 40% as compared with a non recuperative burner at temperatures where this could be used: When only partial furnace gas extraction was used a 20% fuel saving was still possible.

A model is available capable of predicting the performance of concentric tube recuperative burners.

In a previous report it was suggested that performance might be improved if insulation were provided between the incoming hot gas and cool air eliminating heat transfer between the two before the air had been preheated by the furnace exhaust gas to the some temperature of as the fuel gas.

However the model shows that the wall is heated by radiation to a temperature above that of either gas stream.

It was also suggested that the use of extended surface on the air side of the recuperator would improve its performance, or the recuperator could be made longer to give the same effect. Is this case the temperature of the exhaust gas would be reduced, the radiation to surface 4 also reduced, and the possible advantages of using thermal insulation between the coal air and the hot fuel gas increased.

Acknowledgements

FAPESP - Fundação de Amparo à Pesquisa do Estado de São Paulo for financial support.

TERMOQUIP Energia ALternativa Ltda. for construction of the burner and for use of a gas producer and test furnace.

REFERENCES

[1] Codetec - Cia. deDesenvolvimento Tecnológico,"O Gás de Madeira como Combustível", internal publication.

[2] Masters, J. et al., "The Use of Modeling Techniques in the Design and Application of Recuperative Burners", J. Inst. Energy, 1979, 12, p. 196.

[3] Harrison, W.P., "The Design of Self-Recuperative Burners", J. Inst. Gas Engineering, 1970, 10(8) p. 538.

[4] Raznjevic, K., "Tables et Diagrammes Termodynamiques" Edition Eyrolles, Paris, 1970.

[5] Silva, R.B., "Manual de Termodinâmica e Transmissão de Calor", Gremio Politécnico, São Paulo, 1979.

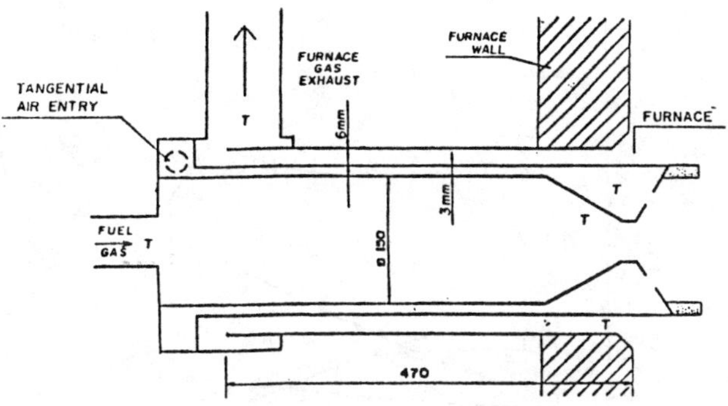

Fig. 1. - Autorecuperative Burner.

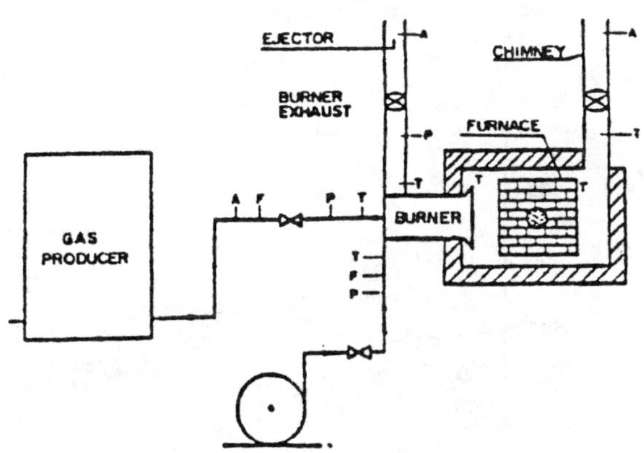

Fig. 2. - Layout of Test Furnace.

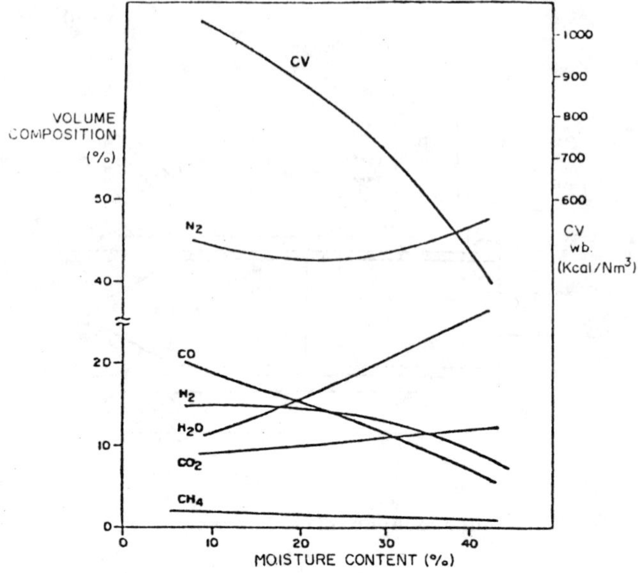

Fig. 3. - Gas Composition and C.V. as Function of Humidity of Wood. (From Codetec, 1980).

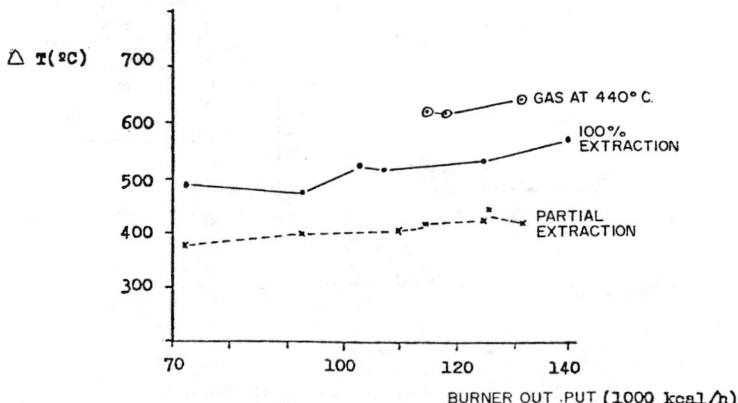

Fig. 4. - Air Preheat.

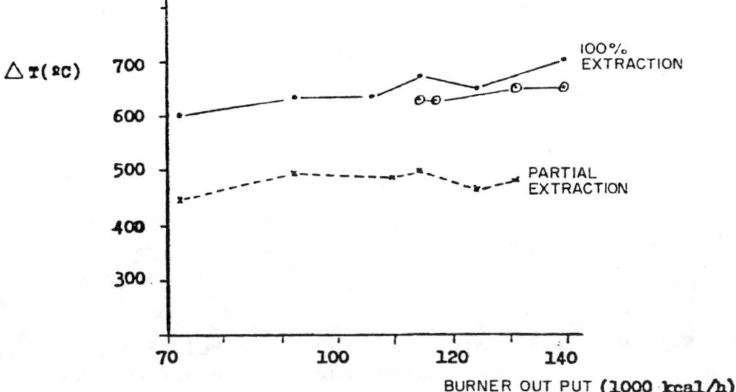

Fig. 5. - Equivalent Air Preheat.

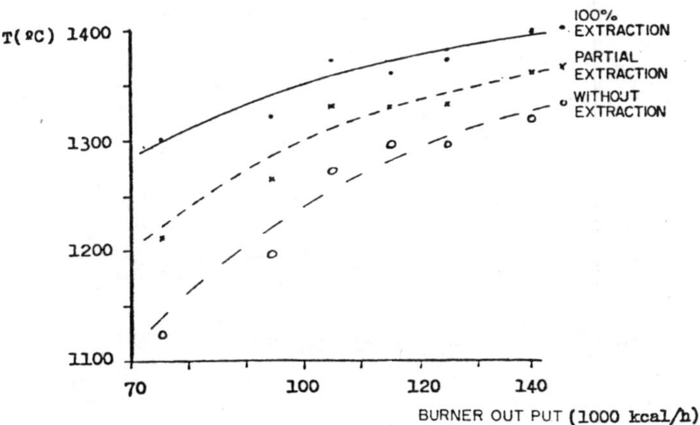

Fig. 6. - Furnace Temperature Extraction Rate.

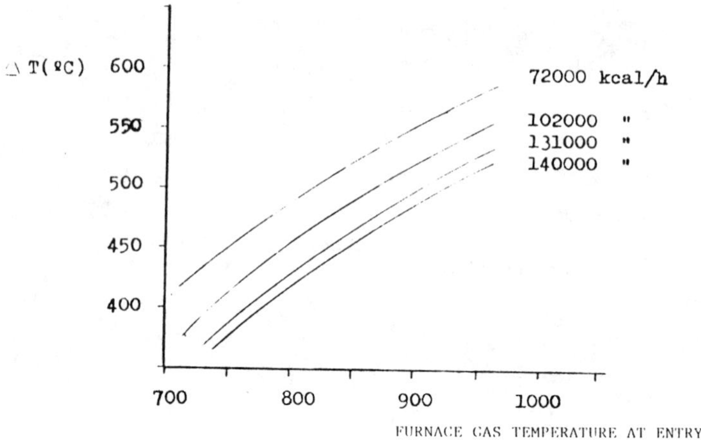

Fig. 9. - Calculated Air Preheat in Relation to Furnace Gas Temperature and Burner Out Put.

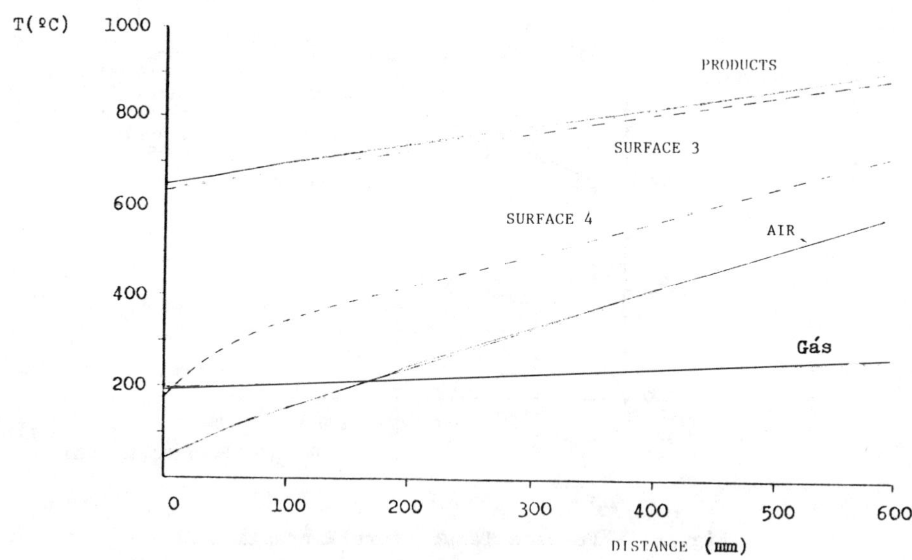

Fig. 10. - Temperature Profiles Along the Recuperator.

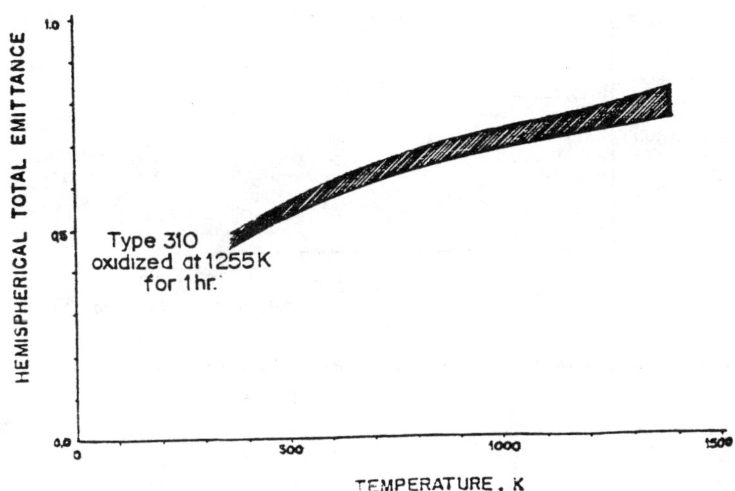

Fig. 11. - Emittance of Stainless Steel (310) (from Touloukian and Dewitt, 1970).

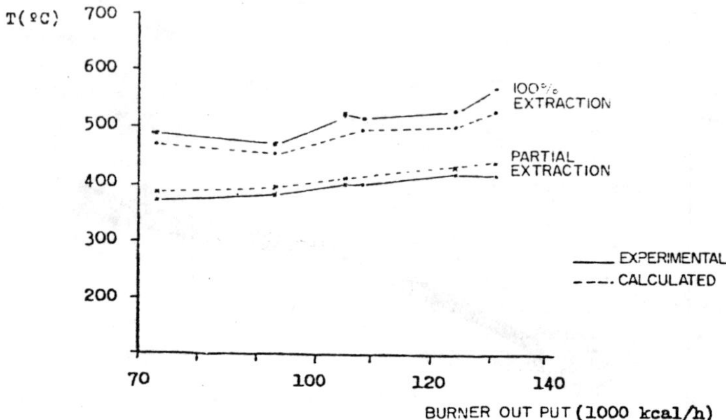

Fig. 7. - Experimental and Calculated Air Preheat.

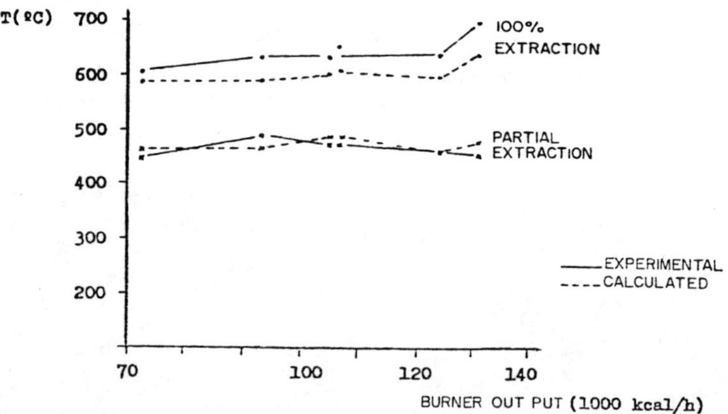

Fig. 8. - Experimental and Calculated Equivalent Air Preheat.

A CONCEPT FOR USING SOLAR ENERGY IN A COAL GASIFICATION PROCESS

Vijay K. Gupta
Chemistry Department, Central State University
Wilberforce, Ohio 45384, U.S.A.

ABSTRACT

A fluidized bed system called CO_2 acceptor process to convert lignite or sub-bituminous coal into synthetic gas or pipeline gas has been developed by Conoco Coal Development Company under contract with Department of Energy at a pilot plant stage of 40 tons per day. The background of the process, and its advantages over the Lurgi coal gasification process have been discussed. The potential methods how solar energy can be stored in lime and recovered at night time have been presented. The possible strategies how solar energy can be utilized in the CO_2 acceptor process to gasify the coal have been also been suggested. The preliminary analysis indicates that the product gas contains 50% more energy per kilogram of coal as compared to standard CO_2 acceptor process, and the gasification can be carried out during night time using the stored solar energy in a solid material. Some of the advantages and disadvantages of CO_2 acceptor process using solar energy as compared to solar coal gasification process have been pointed out.

I. INTRODUCTION

During the last few years, coal has again become a desiarble fuel, but only in applications where its handling cost, and particularly its pollution-control cost, are not prohibitive. At the same time coal is also an important raw material that can furnish the feedstock for synthetic oil and gas. The hydrogen to carbon atomic ratio in coal varies from 0.6 to 1.0, and in methane the ratio is 4.0, therefore coal gasification essentially means addition of hydrogen to coal. The routes for coal gasification, indirect hydrogenation or simply gasification, direct hydrogenation or hydrogasification, and pyrolysis, are described in Figure 1. These schemes can result from low heating value fuel gas to high heating value substitute natural gas. The types of gasifier beds used are: moving or fixed; fluid; entrained or dilute phase; and molten bath (salt or iron) reactor. Various gasification processes and their status have been summarized elsewhere (1-2).

Gregg (3-7) and coworkers have applied focussed solar energy for the gasification of coal, charcoal, activated carbon, wood, paper, biomass mixtures and oil shale. Focussed solar energy can be used to drive the endothermic reactions that gasify coal. The economics are attractive, the precess helps conserve coal and the coal reactors can easily be adapted to solar energy facilities. The major advantages of the solar coal gasification processes are summarized below:

1. Synhtesis gas (a nitrogen free product) can be produced without using pure oxygen which is expensive.

2. For a given amount of product only half as much coal is required because no coal is burnt to provide process heat and the product gas contains energy from both coal and the sun.

3. The sysytem has very low thermal inertia and is insensitive to thermal shock, making it very adaptable to rapidly changing solar conditions such as passing clouds.

4. The coal is heated directly, which allows one the option of operating in a mode where the reactor wall and other components are at a lower temperature than the coal, and helps minimize the material problems with the reactor.

The major disadvantage is that the solar energy is available only about 8 hours per day and thus hinders the continous operation of coal gasification. However, this problem can be resolved by applying the solar energy to CO_2 acceptor process for coal gasification. The description and the current status of the CO_2 acceptor process are discussed in the following section. In this paper the potential methods of storing solar energy in regenerated lime that can be used to gasify coal on through the night are discussed. The performance parameters and the possible problems with the system have also been pointed out. The possible layouts and the operating strategies for solar coal gasification using the CO_2 acceptor process are also discussed.

II. CO_2 ACCEPTOR PROCESS

The CO_2 acceptor process has been developed by Conoco Coal Development Company, with a funding from Department of Energy through the 40 tons per day pilot plant stage to a point of commercial decision. The progress of the development of the process was presented at several Pipeline Gas Symposia (8-12), and was finally reported (13) to DOE. The objectives of this 40 tons per day pilot plant process were achieved as desired and the process was deactivated in 1977.

The CO_2 acceptor process is a fluidized bed system to convert lignite or subbituminous coal to pipeline gas. The schematic diagram of the process is shown in Figure 2. There are two fluidized bed reactors, a gasifier and a regenerator both operate at a pressure of 10.2 atm. Raw dry lignite or subbituminous coal (sized to nominally 8X100 mesh) is fed at the bottom of the gasifier whereafetr rapid hydrodevolatilization, the gasification of the bed occurs. The gasifier operates at an average temperature of $1089^{o}K$. The heat required for the gasification is provided by the exothermic CO_2 acceptor reaction. The acceptor which can be derived from either limestone or dolomite (sized nominally to 6X10 Tyler mesh) enters above the fluidized char bed, showers through the bed and collects in the gasifier boot. Steam flow at the bottom of the boot is adjusted to strip cleanly the char from the acceptor so that a sharp stable interface exists between the acceptor and the char-acceptor mixture above it. The remaining steam is used for gasification of fixed carbon. The reaction of acceptor with CO_2 is reversed in the regenerator that operates at an average temperature of $1294^{o}K$ where the heat is supplied by burning the char from the gasifier with air. The ash is removed from the regenerator by elutriation and and collected via an external cyclone and lockhopper system. Since the acceptor loses reactivity during recirculation thus about 2% of the acceptor material is purposely withdrawn from the gasifier and replaced with fresh limestone. The make up acceptor amount is added to the acceptor returning to the regenerator.

The reactions taking place in the acceptor process are given in Table 1. The enthalpy diagram in Figure 3 shows the distribution of sensible heat and the heat of reaction as supplied by 1 g.mole of dolomite acceptor with 40% activity and $1294^{o}/1089^{o}K$ temperature differential between the gasifier and the regenerator. Note that only 24% of the heat supplied is sensible heat, the remainder 76% energy is the heat of reaction. The process satisfied the criteria set by AGA/ERDA operating committee for establishing the process feasibility. The details including the successful operation and status of the process as compared to a Lurgi coal gasifier have been reported elsewhere (13). The advantages of the acceptor process are listed below:

1. The acceptor is heated and calcined with air in a separate reactor thus keeping air away from the synthesis gas and eliminating the need for an oxygen plant.

2. About 76% of the total heat supplied by the acceptor comes from the CO_2 acceptor reaction.

3. The acceptor reacts with both S and CO_2, thus minimizing the gas clean up requirements.

4. The raw gasifier product is hydrogen rich compared with all other gasification processes which use steam. No water gas shifting is required prior to methanation since the raw gas contains enough hydrogen gas to methanate all of the CO and part of CO_2.

5. The cost of environmental clean up is reduced, because no tars, oils, or phenols appear in the gaseous effluents.

6. The process with its relatively low gasifier bed temperature, is the only known fluidized bed process which can cope with relatively high sodium content lignites (8 wt. percent Na_2O in the SO_3 free ash).

7. Ability to produce synthesis gas or high heating value pipeline gas.

8. The absence of hydrocarbons other than CH_4 in the product gas. Trace amounts of heavier hydrocarbons are removed by adsorption on the char fires

recovered in the quench system and are ultimately used as fuel in the regenerator.

9. Substantial total carbon utilization as shown by the fact that the carbon content in the ash removed from the regenerator is less than one percent of the carbon in the feed.

10. It is a relativelw low pressure process.

11. The capital and operating costs for the CO_2 acceptor process for high BTU gas are slightly better than for the Lurgi process. If full by-product value of the by-products from the Lurgi process can be achieved, then the cost of gas delivered to the pipeline is essentially equal (13).

The composition of the unquenched product gas, the composition of the regenerator flue gas, and the composition of the product gas from the solar coal gasification plant are given in Table 2. The product gas from the gasifier is very low in particulates, ammonia, sulfur, and organics, hence fewer environmental problems. The product gas from acceptor process is higher in hydrogen and methane and lower in carbon monoxide and carbon dioxide concentration as compared to the solar coal gasification process, thus a higher BTU by product gas is produced.

III. METHODS FOR STORING SOLAR ENERGY

The energy required to gasify coal can be provided by lime (CaO) or dolomite (CaO.MgO) reaction with CO_2. In the acceptor process the energy needed to regenerate lime is obtained by burning char from the gasifier that contains 30% of the heating value of the coal going into the gasifier. In the proposed methods (14-15) lime can be regenerated using focussed solar energy and the regenerated lime can be stored to be used during the night. Three different illustrations of heat recovery have been investigated. To carry out the analysis following assumptions have been made:

1. Temperature of the gasifier be $1100^{\circ}K$.

2. Temperature of the regenerator be $1300^{\circ}K$.

3. Heat capacity data for CaO and $CaCO_3$ at 1 atm. and $298.15^{\circ}K$ is applicable within the temperature range of $298.15^{\circ}K$ to $1300^{\circ}K$. Heat capacity data (16) as a function of temperature are:

$$C_p \text{ (CaO in cal./deg./mole)} = 10.00 + 0.00484\ T - 108000/T^2$$

$$C_p \text{ (CaCO}_3\text{ in cal./deg./mole)} = 19.68 + 0.01189\ T - 307600/T^2$$

4. Only 40% of the lime is converted to $CaCO_3$ in the gasifier and 60% remains as CaO based on the temperature enthalpy diagram.

5. All the energy analysis has been carried out for 1 gram-mole of lime in kilo joules.

In the first illustration, it has been assumed that the used lime from the gasifier is cooled down to $300^{o}K$ (the ambient temperature) and the lime after regeneration is also cooled to $300^{o}K$. The energy requiremnets from one step to another are given in Figure 4. Solar energy required to regenerate lime without any heat loss is 152.783 KJ and the energy that is required to heat the used lime from $300^{o}K$ to $1300^{o}K$ (81.864 KJ), and to regenerate (133.588 KJ) with a total of 215.272 KJ. The energy obtained in the gasifier from the acceptor reaction is 127.559 KJ and only 83.833 KJ of this energy can be recovered at night since 43.727 KJ are required to heat up regenerated lime from $300^{o}K$ to $1100^{o}K$. The energy of 83.833 KJ out of 215.272 KJ results in only 38.9% of the stored solar energy recovered at night time.

For the second illustration the data is given in Figure 5 and it has been assumed that part of the heat can be exchanged to heat the material from a lower temperature to higher temperature such as 90% of 43.727 KJ can be obtained from 62.489 KJ leaving an energy requirement of only 4.373 KJ to heat regenerated lime from $300^{o}K$ to $1100^{o}K$, similarly 56.895 KJ of energy at $1300^{o}K$ requiring only 24.783 KJ of energy in addition to regeneration energy of lime (133.588 KJ), thus a total of 158.377 KJ of solar energy are required and only 123.186 KJ of it are recovered at night, giving the energy recovery efficiency of 77.8%.

The data for the third illustration is given in Figure 6 where it has been assumed that with reasonable amount of insulation for the lime storage system the temperature drop can be reduced to 10% of the difference of the reactor temperature and the ambient temperature i.e., the gasifier may cool to $1020^{o}K$ and the regenerator to $1200^{o}K$ and no heat is recovered through heat exchangers. In this case the energy requirements are: to heat used lime from $1020^{o}K$ to $1300^{o}K$ (26.501 KJ), to regenerate (133.588 KJ), with a total of 160.089 KJ of solar energy to regenerate a mole of lime. Of this 134.039 KJ are recovered which includes sensible heat of 6.480 KJ, giving the energy recovery efficiency of 83.3%. The data for the three illustrations is summarized in Table 3. Since the third method of storing solar energy appears to be the most efficient one, so it has been utilized to demonstrate solar gasification of coal using CO_2 acceptor process.

IV. SCHEME FOR SOLAR COAL GASIFICATION

In the standard CO_2 acceptor process thirty percent of the heating value of coal comes out as char from the gasifier which is burnt with air to provide energy for the regeneration of lime, whereas in the solar coal gasification scheme the char will be gasified in the regenerator alongwith the regeneration of lime. The regenerated lime will be stored and used to gasify the coal during the night. In this scheme instead of air going into the regenerator as in the standard CO_2 acceptor process, excess steam goes into the regenerator alongwith the spent lime, char, and the focussed solar energy. For the purposes of analysis in this scheme it has been assumed that the ratio of CO to CO_2 production from char is one. An outline of such a scheme is shown in Figure 7 and the energy data for two cases, one solar-char gasification during the day (without any heat loss) and the other solar-char gasification during the night (with heat loss) are compared with the standard CO_2 acceptor process in Table 4 for North Dakota lignite coal.

Based upon the above analysis, it is clear that in case of solar system the product gas contains 50% more product energy per kilogram of coal as compared to the standard CO_2 acceptor process, and the gasification can be carried out during the night also using solar energy stored in a solid material. The scheme will however provide all the usual advantages of the standard solar coal gasification process, and the standard CO_2 acceptor process as described earlier in sections I and II. Other possible strategies for solar coal gasification using CO_2 acceptor process could also be explored. It may be better to design a scheme in which coal can be gasified completely or near to completion in the gasifier itself so that the need to gasify the coal in the regenerator does not arise.

V. CONCLUSIONS

Based upon the information presented in the previous sections, following are the conclusions:

1. This is a scheme which involves storage of solar energy in a solid material and allows operation of the gasification process both during day and night.

2. The product gas contains 50% more energy per kilogram of coal than the standard CO_2 acceptor process.

3. Since most of the sulfur present in the coal reacts with the acceptor material and is removed from the process stream, hence the product gas does not contain sulfur as an impurity.

4. The scheme has all the usual advantages of the standard CO_2 acceptor process and solar coal gasification process.

VI. ACKNOWLEDGEMENTS

The author would like to thank Lawrence Livermore National Laboratory, Livermore, California for providing him with an opportunity and financial support for the project. The author also expresses his thanks and appreciation to Dr. David W. Gregg and Dr. William R. Aiman for providing helpful comments and suggestions in completing this project.

Table 1. Chemical Reactions in the CO_2 Acceptor Process.

Reaction	Enthalpy
$C + H_2O\ (g) = CO\ (g) + H_2\ (g)$	$\Delta H = +131.42$ KJ/mole at 298.15^oK
$CO\ (g) + H_2O\ (g) = CO_2\ (g) + H_2\ (g)$	$\Delta H = -41.173$ KJ/mole at 298.15^oK
Acceptor Reaction: $CaO + CO_2\ (g) = CaCO_3$	$\Delta H = -177.24$ KJ/mole at 298.15^oK
Regenerator Reaction: $CaCO3 = CaO + CO_2\ (g)$	$\Delta H = +177.24$ KJ/mole at 298.15^oK
$2C + H_2\ (g) + H_2O\ (g) = CH_4\ (g) + CO\ (g)$	$\Delta H = +56.52$ KJ/mole at 298.15^oK

Table 2. Chemical Composition of Gas From Gasification Processes.

Component	Gas Composition (%)		
	CO_2 Acceptor Process		Solar Process
	Gasifier	Regenerator	
Hydrogen	60	2	57
Carbon Monoxide	15	2	19
Methane	12	-	3.51
Carbon Dioxide	10	24	20.51
Nitrogen	2	72	-
Ammonia	0.8 ppm	-	-
Sulfur	0.15 ppm	-	-
Particulates	7 g/scf	1 g/scf	-
Organics (Benzene & Naphthalene)	6 ppm	Traces	-
Cyanide	20 ppm	0.2	-

Table 3. Energy Storage Data in Regenerated Lime in KJ/mole.

Energy Step	Illustration 1 Figure 4	Illustration 2 Figure 5	Illustration 3 Figure 6
Solar Energy Required Without Heat Loss	152.783	152.783	152.783
Solar Energy Required With Heat Loss	215.272	168.377	160.089
Stored Energy Delivered To The Gasifer	83.833	123.186	134.039
Percentage of Sotred Solar Energy Recovered	38.9	77.8	83.3

Table 4. Comparison of Energy Data in KJ/mole for Coal Gasification Processes.

Energy Quantity	Energy in Different Processes		
	CO_2 Acceptor	Solar-Char Without heat Loss	Solar-Char With With Heat Loss
Coal (HHV) N.D. Lignite	484	484	484
Product Gas			
Gasifier	342	342	342
Regenerator	-	170.8	170.8
Solar Energy	-	87.5	96.2
Product Energy/Coal Energy	0.71	1.06	1.06
Product Energy/Solar Energy	-	5.86	5.33
Product Energy/(Coal+Solar)	-	0.90	0.88

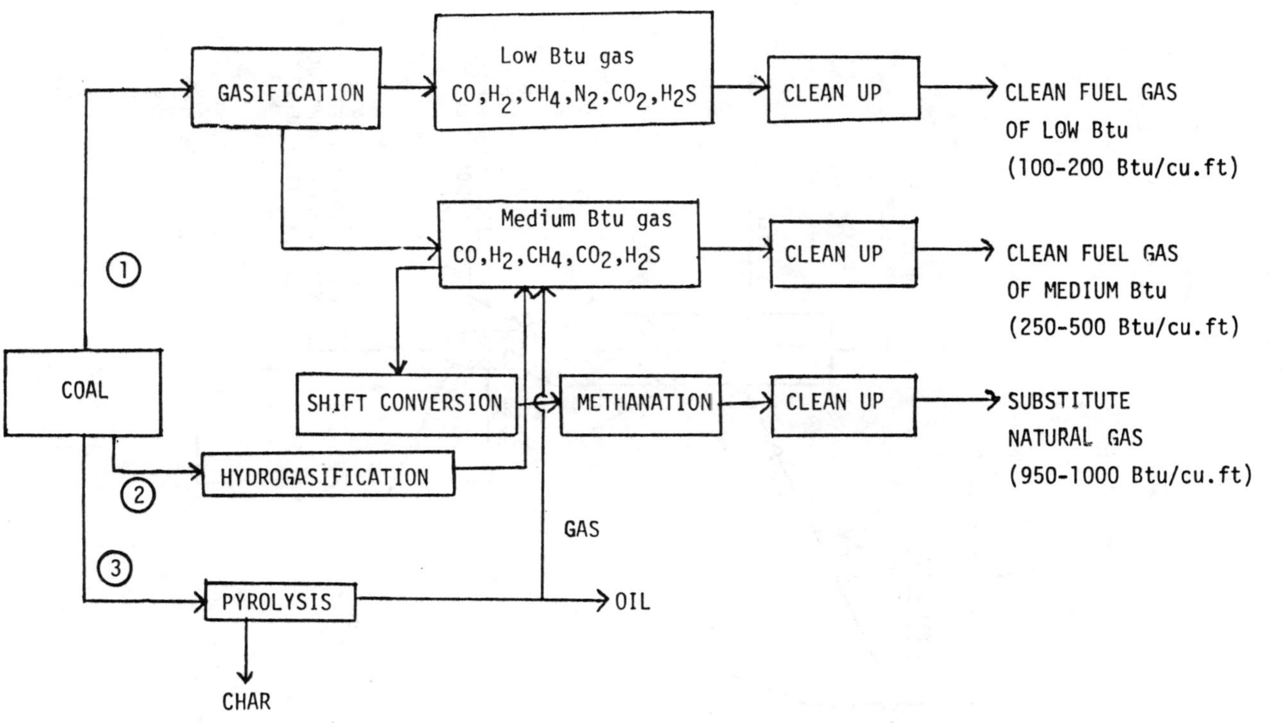

Figure 1. Routes to Substitute Fuel Gas From Coal.

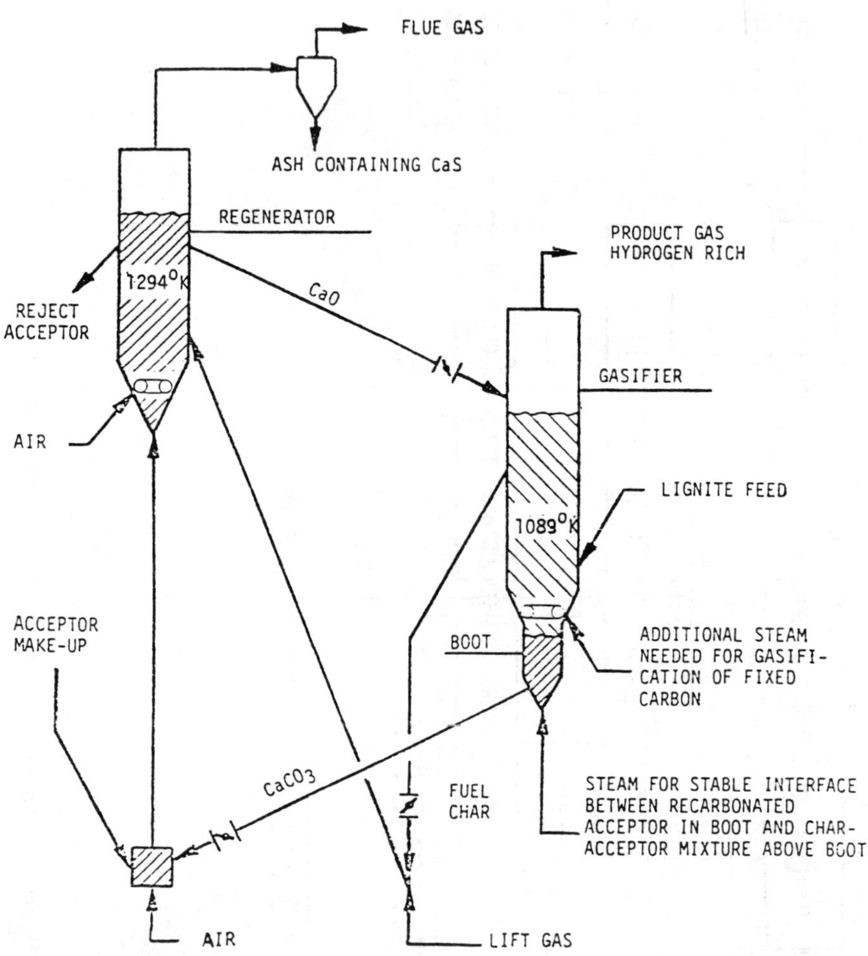

Figure 2. CO_2 Acceptor Gasification Process.

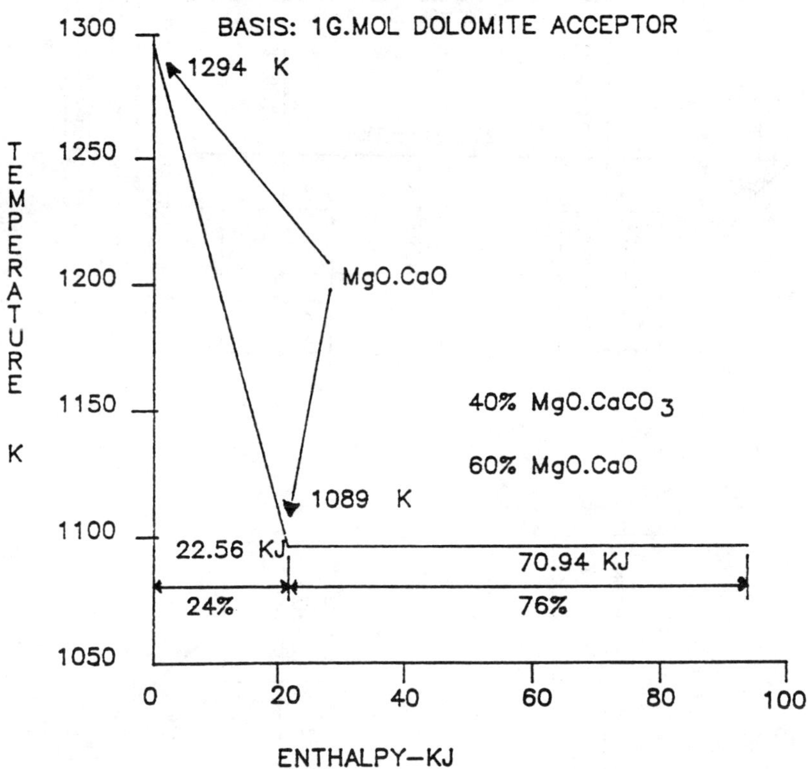

Figure 3. Heat Supply of Acceptor at 40% Activity.

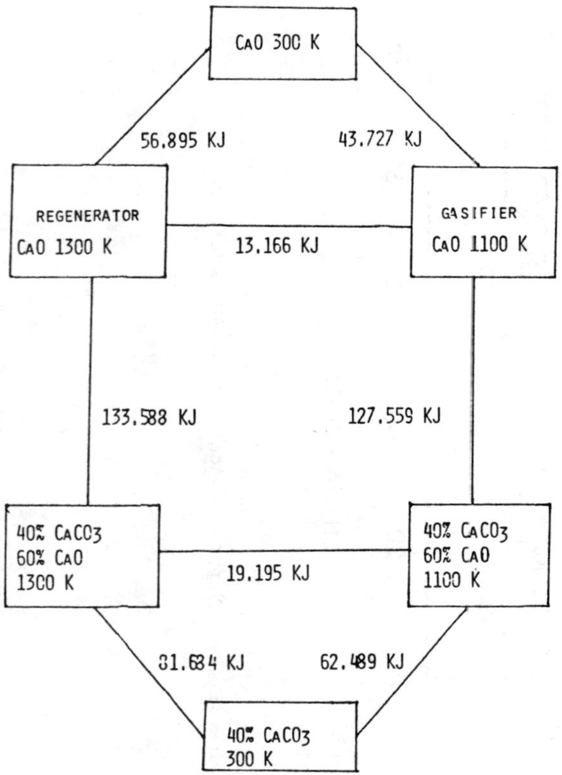

CaO 300 K
56.895 KJ
43.727 KJ
REGENERATOR
CaO 1300 K
13.166 KJ
GASIFIER
CaO 1100 K
133.588 KJ
127.559 KJ
40% $CaCO_3$
60% CaO
1300 K
19.195 KJ
40% $CaCO_3$
60% CaO
1100 K
81.684 KJ
62.489 KJ
40% $CaCO_3$
300 K

Figure 4. First Illustration of Heat Recovery.

CaO
300 K
56.895 KJ
43.727 KJ
REGENERATOR
CaO 1300 K
13.166 KJ
GASIFIER
CaO 1100 K
USED TO HEAT UP
PROVIDES 90% OF HEAT UP
133.588 KJ
127.559 KJ
40% $CaCO_3$
60% CaO
1300 K
19.195 KJ
40% $CaCO_3$
60% CaO
1100 K
81.684 KJ
62.439 KJ
40% $CaCO_3$
60% CaO 300 K

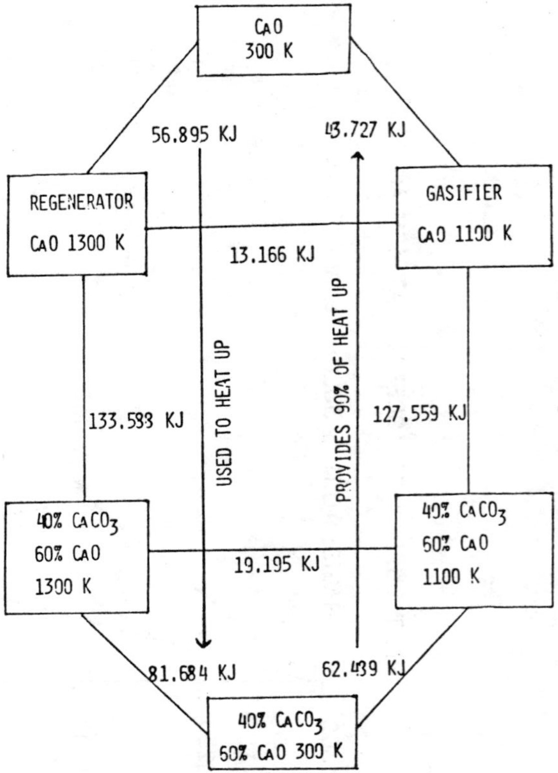

Figure 5. Second Illustration of Heat Recovery.

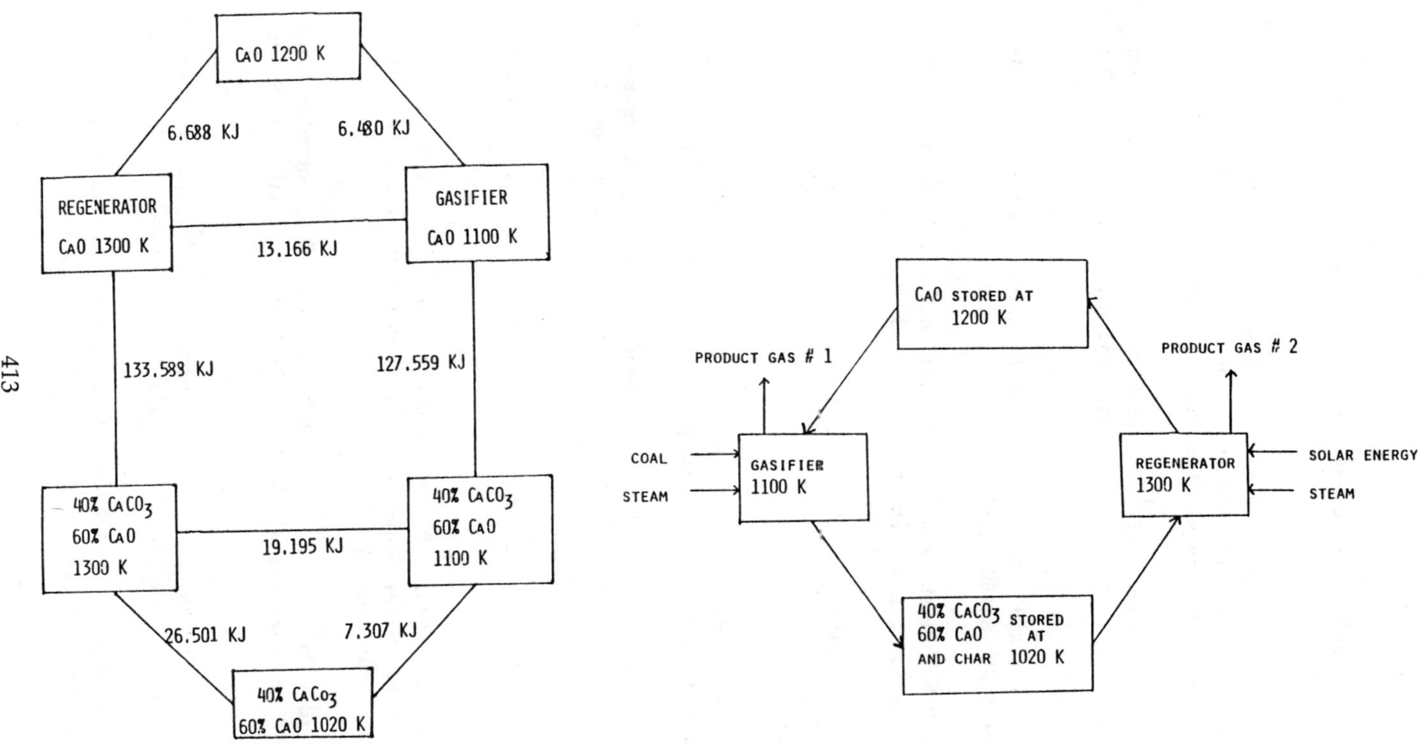

Figure 6. Third Illustration of Heat Recovery.

Figure 7. Illustration of Solar Coal Gasification Using CO_2 Acceptor Process.

REFERENCES

1. Verma, A., "From Coal to Gas", - Part I: Chem. Tech., 372, 1978; Part II: Chem. Tech., 626, 1978.

2. Gupta, V.K., "Synthetic Fuels From Coal - The Energy in Transition", Advances in Energy Utilization Technologies, Proceedings of the 4th World Energy Engineering Congress, Atlanta, GA, 487-98, 1981.

3. Gregg, D.W., Grenz, J.Z., Taylor, R.W., and Aiman, W.R., "Solar Retorting of Oil Shale", Lawrence Livermore National Laboratory (LLNL), Report UCRL, 84142, 1980.

4. Gregg, D.W., and Aiman, W.R., "Gasification of Coal With Solar Energy", LLNL Report UCRL, 84458, 1980.

5. Gregg, D.W., Aiman, W.R., and Otsuki, H.H., and Thorsness, C.B., "Solar Coal Gasification", LLNL Report UCRL, 81853, 1979.

6. Gregg, D.W., Taylor, R.W., Campbell, J.H., Taylor, J.R., and Cotton, A., "Solar Gasification of Coal, Activated Carbon, Coke, Charcoal, and Biomass", LLNL Report UCRL, 83440, 1979.

7. Taylor, R.W., Berjoan, R., and Coutures, J.P., "Solar Gasification of Charcoal, Wood, and Paper", LLNL Report UCRL, 84411, 1980.

8. Sudbury, J.D., Curran, G.P., and Heunisch, G.W., "Environmental Program Used to Characterize the CO_2 Acceptor Process", Proceedings of the Tenth Pipeline Gas Symposium, 223, 1978.

9. Sudbury, J.D., "The CO_2 Acceptor Process", Proceedings of the Ninth Pipeline Gas Symposium, 55, 1977.

10. Fink, C., Curran, G.P., and Sudbury, J.D., "CO_2 Acceptor Process Pilot Plant-1975 Rapid City, South Dakota", Proceedings of the Seventh Synthetic Pipeline Gas Symposium, 7, 1975.

11. Fink, C., Curran, G.P., and Sudbury, J.D., "CO_2 Acceptor Process Pilot Plant-1973", Proceedings of the Fifth Synthetic Pipeline Gas Symposium, 91, 1973.

12. Fink, C., Curran, G.P., and Sudbury, J.D., "CO_2 Acceptor Process Pilot Plant-1974, Rapid City, South Dakota", I.G.T. Symposium, 243, 1975.

13. Sudbury, J.D., "CO_2 Acceptor Process Gasification Pilot Plant", Final Report, 13, DOE Contract No. EX-76-C-01-1734, 1978.

14. Gupta, V.K., "Solar Gasification of Coal Using CO_2 Acceptor Process", Final Report, LLNL, CA, 1980.

15. Gupta, V.K., "Application of Solar Energy to CO_2 Acceptor Process of Coal Gasification", Joint Central and Great Lakes Regional Meeting of The American Chemical Society, Dayton, OH, Abst.# 126, 1981.

16. Chemical Engineers Handbook, McGraw Hill Series, 1973.

BIOCONVERSION

BIOGAS: AN ASSESSMENT OF POTENTIALS, TECHNOLOGIES AND UTILIZATION IN ASIA

Dr. M.A. Aziz, Associate Professor
Department of Civil Engineering,
Hydraulic and Environmental Engineering Division
National University of Singapore, Kent Ridge, Singapore

ABSTRACT

Biogas technology has reached a fairly advanced level in research and development in Asia and has been demonstrated to be economically viable when applied on an industrial scale .

The purpose of this paper is to critically review the potentials, production technologies and utilization of biogas for various beneficial uses in Asian countries. The paper also highlights the research findings of the author and other researchers on cost-effective biogas digester systems to be commonly applied in the Asian region.

1. INTRODUCTION

Energy is a fundamental need for all people . It has an impact on quality of life, on world politics, on international economics and on national development strategies . Energy is a prerequisite to trigger and sustain development. Therefore , there is a real need for a rapid increase of energy sugplies in the developing countries, especially thDse related to the development of new and renewable forms of energy. The spiraling price of fossil fuels and their limited mature of supplies are the primary concern of many non-oil producing developing countries. Many of these countries have resorted to crash programmes on research and development on non-conventional The renewable sources of energy [1-6]. Important new and renewable sources of energy that have been identified include solar energy, geothermal energy, wind power, hydro-power , biomass energy and ocean energy (wave power and tidal power). Of all these sources of renewably energy, the biogas, a by-product of anaerobic decomposition of organic matter has been considered as an alternative source of energy for cooking, heating, lighting and production of electricity in Asian countries. This region has abundant raw materials like human excreta, animal manure, sewage sludge, crop residues, trees, plants, vegetations, refuse and fruit-food-meat processing wastes which coupled with favourable climatic conditions, provide a great potential for biogas production.

Biogas consists mainly of methane and carbon dinxide at a ratio approximately 3:2 aud trace amounts of ammonia,hydrogen sulphide and hydrogen. The energy from biogas originates mainly from methane gas whose calorific value is about 8150 to 9960 kcal/m^3 (900 to 1100 BTU/ft^3) while the, approximate value for the biogas is 4500 to 6300 kcal/m^3 (500 to 700 BTU/ft^3). Some properties of biogas, natural gas and liquified petroleum gas (LPG) is given in Table 1 (7,8].

TABLE 1: SOME PROPERTIES OF BIOGAS, NATURAL GAS AND LIQUIFIED PETROLEUM GAS (LPG)

Properties	Biogas	Natural gas	LPG
Gross calorific value, kcal/m^3	4500-6300	9070	24000
Specific gravity	0.874-1.002	0.584	1.5
Inflammable, percent	7.5-21	5-15	2-10
Ignition temperature, °C	650-750	650-750	450-500
Combustion air requirement, m^3/m^3	0.874-1.002	9.6	13.8

Production of methane gas from organic matter through anaerobic decomposition is not a new technology and it has been known since Volta discovered methane in marsh gas in the 18th century. Although its interest waned in the middle of the century when the availability of cbeap fossil fuels minimised the need for methane gas, the use of biogas has become popular currently especially in Asian countries [8]. The Camerouns, Ethiopia, Lesothe and Upper Volta are using biogas in rural development schemes. Indonesia and Singapore are making use of biogas for experimental and training purposes, while in Brazil, China, Iran, Japan, Korea and Taiwan are producing biogas for heating, lighting , cooking and generating electricity for improvement of public health, pollution control and feed production. Pakistan and Papua New Guinea, on the otherhand, are harnessing biogas primarily for fertilizer production. In Algeria , Bangladesh, Congo , Costa Rica and Zaire , production of biogas is still at the experimental stage, but Australia, Barbados, Canada, El Salvador, Fijii, France, Guatemala, Guyana, Honduras, India, Jamaica, Kenya, Malaysia, Mauritius, Nepal, the Netherlands, New Zealand, Nicaragua, Philippines, Rwanda, Senegal, Sri Lanka, Thailand, Tanzania, United Kingdom, United States and West Germany are all utilizing biogas for domestic uses as well as for agriculture.

2. BIOGAS PRODUCTION TECWNOLOGIES

Anaerobic decomposition process of converting organic materials into biogas is carried out in a closed container known as digester (Fig. 1). Anaerobic process in which the biogas is formed consists of three phases of decomposition: hydrolysis or liquefaction, followed by acid formation and then methane formation. The basis for the terminology is the sequence of events that typically takes place in a digester. In the first phase, saprophytic bacteria attack complex organic substances such as proteins, carbohydrates and fats, then convert them into organic compounds by external enzymes. These simple organic compounds normally dissolve in the surrounding water solution. Secondly, acid forming bacteria which can be anaerobic or facultative degrade the while organic compounds to volatile fatty acids and some ammonia. Most of the volatile fatty acids produced are acetic acid and a less portion of propionic and butyric acid. These acid-forming bacteria can grow at a relatively rapid rate because of their ability to buffer against environmental changes. In the third phase, methane forming bacteria utilise these acids as their food to form methane. The anaerobic decomposition process is schematically illustrated in Fig. 2. There are many species of acid forming bacteria and methane forming bacteria. The stabilization of waste-organics requires a balance between acid formation and methane formation among these bacteria. The establishment and maintenance of this balance could be indicated by the concentration of volatile acids that should not be more than 2,000 mg/L in the digester [5]. The simple overall biochemistry of anaerobic decomposition of organic substances to biogas (methane and carbon dioxide), neglecting cell formation is as follows (9):

$$\underset{\text{(organic substance)}}{C_aH_bO_cN_d} + \frac{4a-b-2c+3d}{4} \underset{\text{(water)}}{H_2O} \longrightarrow \frac{4a+b-2c-3d}{8} \underset{\text{(methane)}}{CH_4} + \frac{4a-b+2c+3d}{8} \underset{\text{(carbon dioxide)}}{CO_2} + \underset{\text{(ammonia)}}{dNH_3}$$

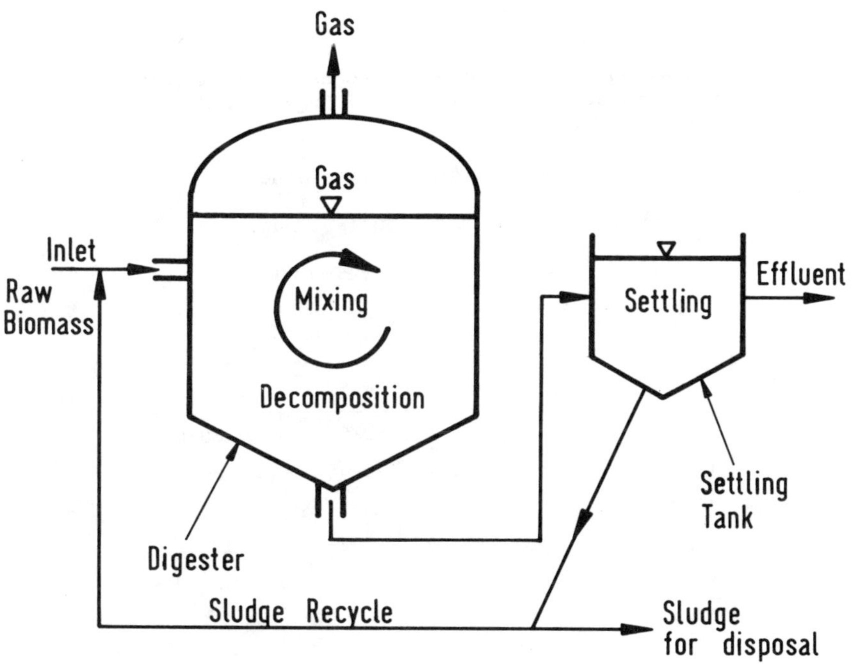

Fig. 1 - Anaerobic Decomposition Process of Biogas Production

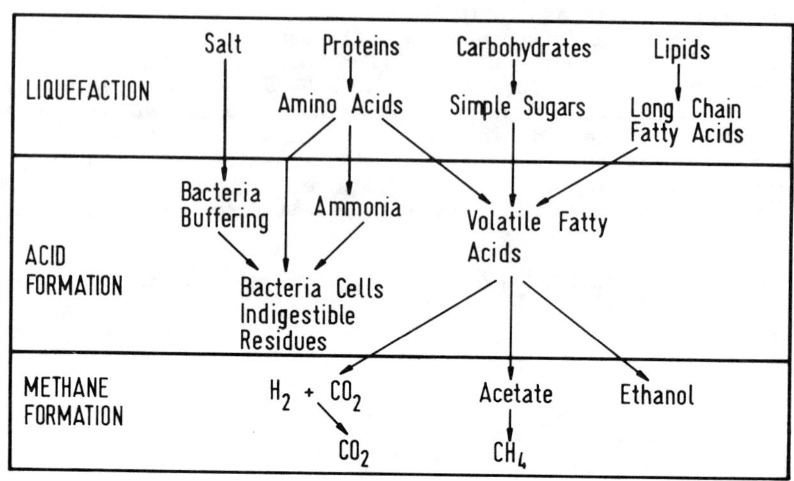

Fig. 2 - Diagram Showing Biochemical Process of Biogas Production from Anaerobic Decomposition of Organic Materials (after Meynell, 1976)

In general, the biochemical process is mildly exothermic. The composition of biogas from two different types of raw materials is shown in Table II. Methane is a colourless, tasteless and odourless gas.

TABLE II: BIOGAS COMPOSITION

Constituent		Gas, percent	
		From Dung	From Feces
Methane	CH_4	50-55	60-65
Carbon dioxide	CO_2	40-45	30-32
Nitrogen	N_2	2-5	2-5
Hydrogen	H_2	1-2	0.5-1
Oxygen	O_2	0.1-0.4	0.1-0.4
Hydrogen sulphide	H_2S	<1	<1

2.1 Raw Materials for Biogas Production

Raw materials for biogas production are generally naturally occurring organic matters generally cellulosic in mature. Wastes generated from human activities and agro-based industrial processes are also included in the list of raw materials. Various types of raw materials are shown in Table III. Important characteristics of raw materials that influence the biogas production rate are the carbon to nitrogen (C/N) ratio and the percent volatile solid contents.

TABLE III : RAW MATERIALS FOR BIOGAS PRODUCTION

Origin	Types
Agricultural wastes	Crop related stuhbles, straws, spoiled fodders and weeds.
Human wastes	Excreta, sewage sludge, refuse.
Animal wastes	Cattle-shed wastes, piggery wastes sheep and goat wastes, slaughterhouse wastes, tannery wastes, fishery wastes and wood wastes.
Agro-based industrial wastes	Palm oil and rubber mill wastes, sugarcane bagasse, rice bran, tobacco wastes, brewery and distillery wastes, food-fruit-vegetable processing wastes, sugar and tapioca mill wastes, tea and coffee wastes, textile and jute mill wastes.
Forestry wastes	Twigs, barks, branches, leaves, dead trees, plants and vegetation.
Aquatic wastes	Algae, weeds, waterhyacinth and other aquatic plants

2.2 Biogas Plant Design

Biogas plant has two main components: (a) digestion chamber where fermentation of organic matter occurs, and (b) gas storage chamber where the gas is collected and stored. The digester can be designed either batch or continuous operation. In the batch digester system, the organic waste materials are replaced only when the gas production has ceased. In this system, several digesters are employed at different stages of digestion process in order to obtain regular gas supply. In the continuous digester system, fresh slurry is fed everyday. This system is more efficient and has a higher gas production rate per unit digester volume than the batch operation [6-9]. Figure 3 shows the cross-section of a typical digester.

Various design options of biogas plants are available [10-19]. Major factors governing the selection of a particular type include availability of construction materials and technical know-how, material and labour costs, raw materials availability, trained personnel for construction, operation and maintenance. Factors influencing the selection of the size of the plant are the quantity of raw materials available, the quantity of gas required, the money available and the proposed retention time. Following factors need due

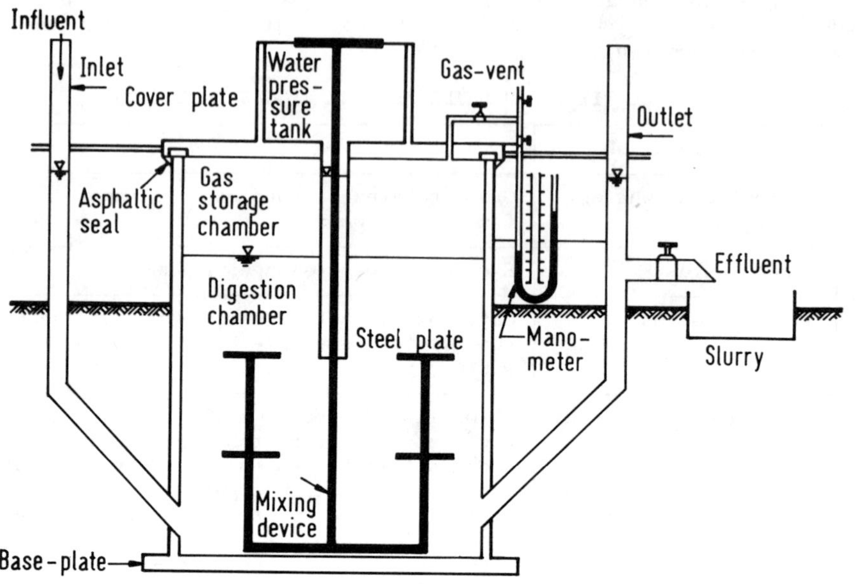

Fig. 3 - Cross-Section of a Typical Digester

consideration while selecting a biogas plant site: nearness of source of raw materials, nearness of gas consumption, far from drinking water sources, enough space for storage of raw materials and digested sludge not affected by flood.

2.3 Factors Affecting Biogas Production

Various factors affecting biogas production include carbon to nitrogen (C/N) ratio, volatile solids, loading rate, temperature, pH, toxicity, dilution and mixing.

Carbon to Nitrogen (C/N) Ratio: Carbon and nitrogen contents are two important criteria for selecting raw materials. Biogas production can be increased by supplementing raw materials that have high carbon content with other raw materials containing nitrogen. Table IV shows the approximate nitrogen (N) content and C/N ratios (by weight) of various types of raw materials.

It has been reported [7] that only 6 gm of nitrogen per gm of raw materials (0.6 percent) is needed to sustain anaerobic decomposition. A C/N ratio between 25 and 30 is needed to achieve optimal rate of decomposition. However, if the C/N ratio is too high, the process is limited be the nitrogen availability and if it is too low, ammonia may be found in quantities large enough to be toxic to the bacterial population.

TABLE IV: APPROXIMATE C/N RATIOS BY WEIGHT OF VARIOUS RAW MATERIALS (9)

Raw materials	Total N	C/N
Horse and cow manure	1.7-2.3	18-25
Farmyard manure	1.02-2.15	12-16
Nightsoil	5.5-6.5	6-10
Plant wastes	3.0-4.0	10-14
Raw garbages	1.8-2.2	22-27
Wheat straw	0.20-0.40	125-132
Rotted sawdust	0.20-0.30	200-216
Raw sawdust	0.08-0.12	500-520

Volatile Solids: Volatile solids content is another criteria in the selection of biogas raw materials because it represents the portion of the organic matters being anaerobically decomposed. It has been reported [7] that the gas production ranges from 0.25 to 0.70 m^3/kg of volatile solids added. It is also a common practice to determine the required volume of biogas digested based on volatile solids loading rate. The amount of methane produced depends mainly on the nature of the raw materials and on the efficiency of the fermentation process. Tables V and VI give the rates of gas production for various experimental digesters fed with different types of raw materials [8].

Loading Rate: Quantity of gas produced in a digester per day depends upon the amount of wastes materials per unit volume of the digester capacity. The loading can be expressed in terms of organic loading (kg of BOD or kg of COD/m^3.d or kg of volatile solids/m^3.d) and hydraulic loading. A loading rate between 1.0 and 1.5 kg of volatile solids per m^3 of the digester capacity per day was found to be satisfactory [10] although the recommended loading rate for standard sewage treatment plant digesters varies from 0.48 to 1.6 kg VS/m^3.d. Of course, an optimum loading range has been reported to be within 1.0 to 4.0 kg of COD/m^3.d for pig and poulty manure wastes (2,3,6). The hydraulic loading or detention time has an equally significant effect to the digester performance. An optimum detention time for a digester is dependent on types of raw materials and other operational parameters. A detention time ranging between 30 to 35 days was found satisfactory although it can vary from 25 to 90 days.

Temperature: Temperature is a key factor in gas production from organic waste materials. In general, decomposition starts at about 15°C and increases rapidly with increasing temperature up to 35°C. At high temperature (between 50 and 55°C), thermophilic pacteria take over and the gas production rate is higher. Relative rates of 1:1.8 at 30 to 35°C and to 50 to 55°C have been reported [10]. In this investigation, experiments were conducted at room temperatures varying between 28 to 32°C which was found satisfactory.

TABLE V: GAS PRODUCED THROUGH ANAEROBIC DECOMPOSITION OF VARIOUS TYPES OF ORGANIC WASTE MATERIALS [10]

Waste type	Gas production m^3/kg	Percentage distribution	
		Carbon dioxide	Methane
Pig waste	0.40-0.50	28-32	55-60
Poultry waste	0.45-0.60	27-34	57-62
Cow-dung	0.30-0.45	25-30	60-65
Food wastes	0.45-0.65	26-30	60-65
Palm oil waste	0.25-0.40	30-35	45-55
Pineapple waste	0.35-0.50	30-35	55-60
Sewage sludge	0.50-0.70	30-33	60-65
Raw garbage	0.45-0.65	30-35	55-60

TABLE VI: GAS PRODUCED THROUGH ANAEROBIC DECOMPOSITION OF MIXED ORGANIC WASTE MATERIALS [10]

Mixed Waste	Proportion	Gas Production m^3/kg	Percentage of methane
Poultry sweeps Cow-dung	1 : 1	0.40-0.55	55-60
Poultry sweep Food waste	1 : 1	0.45-0.65	57-62
Food waste Palm oil waste	1 : 1	0.30-0.50	50-60
Pig waste Cow-dung	1 : 1	0.35-0.55	58-65
Cow-dung Raw garbage	1 : 1	0.33-0.60	60-65
Sewage sludge Pineapple waste	1 : 1	0.42-0.62	60-65
Sewage sludge Pig waste	1 : 1	0.43-0.64	60-65
Sewage sludge Raw garbage	1 : 1	0.44-0.66	60-65

pH: Methane bacteria are sensitive to changes in pH. The optimum pH range for methane production is between 7.0 to 7.4. When the pH drops below 6.6, there is a significant inhibition of the methanogenic bacteria and the acid conditions of a pH of 6.2 are toxic to these group of bacteria. In most of the experimental runs, the pH was observed to be in the proper range. However, in some cases, imbalance was observed amd buffering with lime was found necessary.

Toxicity: Some substances, both organic and inorganic , are toxic or inhibitory to the anaerobic process. Common toxins are ammlnia (> 1500 mg/L of total ammonia nitrogen at pH 7.4), ammonia ion (> 3000 mg/L of total ammonia nitrogen at any pH), soluble sulphides (> 50 mg/L) and soluble salts of metals such as copper, zinc and nickel. However, generally, concentrations of heavy matal ion upto 100 mg/L cause no harm. Other materials can be toxic to anaerobes, if present in high concentration include common alkali and alkali-earth cations such as sodium (3500-5500) , potassium (2500-4500), calcium (2500-4500) and magnesium (1000-3000). Some organic materials can also inhibit the anaerobic decomposition process. Crop residues that have been heavily treated with organic pesticides (such as DDT) can inhibit anaerobic process in the digesters. Another source of toxic substance can be excreta of animals that have been receiving large dosages of antibodies or metals as growth factors in their feed. These substances can inhibit bacterial growth and their activities in the digester.

Dilution: Dilution of concentrated wastes was found necessary because the gas yield decreased as the solid content inside the digester increased. This may be attributed to the fact that the gases produced may accumulate at the bottom and some of them, such as hydrogen sulfide and ammonia, have an inhibitory effect on the bacterial population. The viscous liquid also prevents the digested material with its high bacterial population from dispersing uniformly throughout the fermentative liquid to accelerate decomposition. Water requirement for dilution depends on the concentration of input materials and the hydraulic retention time. Typical dilution ratio was found to be 1:1 (waste materials : water) but with some raw materials (poultry and pig manure), high dilution (1:1.2 to 1:1.5) was found necessary).

Mixing: Mixing was found necessary for most of the raw materials fed into the digester in order to obtain an intimate contact between the materials and the bacteria activating bioconversion process. Mixing was also found helpful to prevent stagnant corners, to release gases produced and also to reduce accumulation of hydrogen sulfide and ammonia which may be toxic to the system.

2.4 Biogas Production Rate

It has been reported [1-9] that the biogas production per kg of volatile solids depends on types of raw materials fed as shown in Tables V and VI. Gas production is greater than that from a single source as shown in Table VII [7]. The quantity of gas produced per day depends on the raw materials fed per unit volume of the digester capacity. As the loading increases, the gas production unit volume of the digester also increases while the gas production per unit of volatile solids added is reduced. On the other hand, with a long retention time, the gas production per volume of digester decreases, though gas production per unit of volatile solids added increases. For the optimum operation of a digester, the main variables to be considered are temperature, loading rate and dilution together with gas yield and the

TABLE II: INCREASE IN BIOGAS PRODUCTION RATE FROM MIXTURES OF RAW MATERIALS

Type of raw materials		Total gas production (m^3/kg.VS)	Increased rate by mixing materials
Pig		0.569-0.628	-
Chicken		0.617-0.643	-
Nightsoil		0.265-0.304	-
Weeds		0.277-0.316	-
Cow + pig	(1:1)	0.510-0.544	102-107
Cow + chicken	(1:1)	0.528-0.554	103-106
Cow + nightsoil	(1:1)	0.407-0.474	116-128
Cow + weeds	(1:1)	0.363-0.421	105-112
Pig + chicken	(1:1)	0.634-0.663	104-106
Pig + chicken + cow	(2:1:1)	0.585-0.601	103-111
Chicken + nightsoil	(1:1)	0.413-0.478	101-103
Chicken + weeds	(1:1)	0.495-0.590	101-123
Nightsoil + weeds	(1:1)	0.387-0.437	121-139

rate of degradation of organic matters of the feed. For a temperature range of 30 to 40°C, the optimum retention time would range from 20 to 25 days. Tables VIII amd IX shows the average gas production based on head count and on waste amount respectively under the above operating conditions [7,10].

TABLE VIII: AVERAGE DAILY GAS PRODUCTION BASED ON END COUNT

Waste source	Waste production (kg/d)	Gas production m^3/d
1 Buffalo	14-16	0.50-0.74
1 Cow	9-11	0.25-0.40
1 Calf	4-6	0:15-0.25
1 Pig	2-3	0.05-0.10

Multi-Purpose Approach

Major costs associated with the biogas plant system are for the digester, the gas collection and the storage systems, and for the sludge disposal systems. The whole system may become economically attractive when a multi-purpose project shown in Fig.4 is adopted. In addition to biogas, other products are sludge and spent liquor. Sludge is very rich in nutrient and can be used

TABLE IX: AVERAGE GAS PRODUCTION BASED ON WASTE AMOUNT

Waste source	Gas production m^3/1000 kg wastes
Cattle (cows & buffaloes)	30-50
Pig	40-60
Poultry	60-100
Crop wastes	40-50
Water hyacinth	40-50

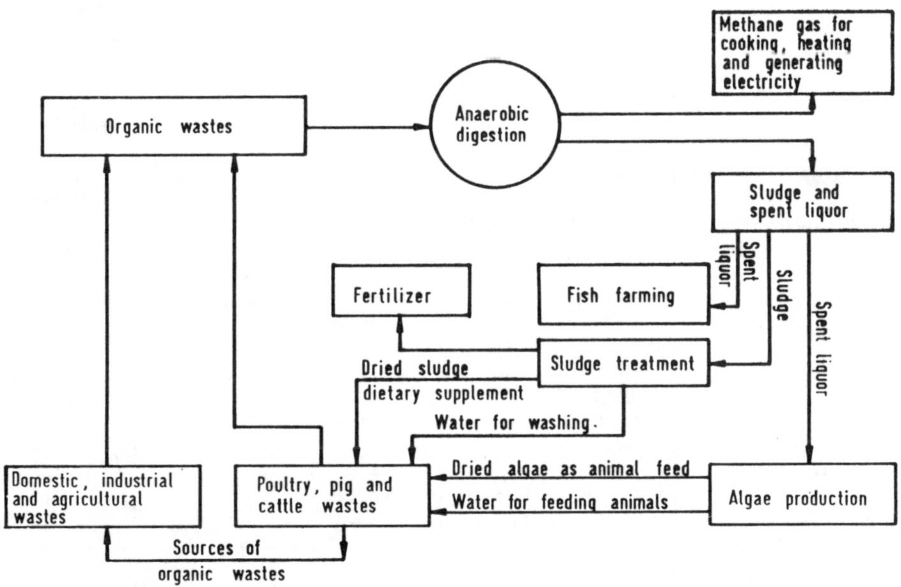

Fig. 4 - Schematic Diagram Showing Production of Methane Gas and Useful By-products from Anaerobic Digestion of Organic Wastes

as the dietary supplement to cattle and poultry feed and as fertiliser. The spent liquor can be disposed off in a tank to be reused for fish culture, irrigation, watering plants and crops, and for producing algae for cattle feed. With some preliminary treatment, the spent liquor can also be used for feeding cattle and poultry. Other significant advantages are abatement of environmental pollution from wastes and control of fatal diseases by destroying most of the pathogens (disease producing microorganisms) in the wastes through anaerobic decomposition/digestion.

3. BIOGAS UTILIZATION FOR VARIOUS USEFUL PURPOSES

Biogas can be used for heating, cooking, lighting and for motive power. One m^3 of biogas can generate 5200 to 5900 kcal of heat energy which is sufficiant to bring 130 kg of water from $20^{\circ}C$ to a boil or can light a biogas lamp with a brightness equivalent to 60 to 100 W for 5 to 6 hours. Figures 5 and 6 and Table X shows the possible applications of biogas and the amount to be used in each application respectively [7,20]. Biogas can also be used for running oil engines. One hp internal combustion engine can be kept running for 2 hours by 1 m^3 of biogas which is equivalent to about 0.65 kg of petrol.

TABLE X: VOLUMETRIC REQUIREMENTS FOR VARIOUS APPLICATIONS OF METHANE GAS

Application	Rate of use	Amount (m^3)
Cooking	5 cm dia. burner/hour	0.3256
	10 cm dia. burner/hour	0.4672
	15 cm dia. burner/hour	0.6371
	per person/day	0.339-0.4247
	boiling 4.545 liters of water	0.283
Lighting	1 mantle lamps/hour	0.071-0.085
	2 mantle lamps/hour	0.140
	3 mantle lamps/hour	0.167
Refrigerator	Flame operated, 0.028 m^3 per hour per 0.028 m^3 of refrigerator space (1 cft/hour/1cft refrigerator space)	0.034
Incubator	0.028 m^3 per 0.028 m^3 of incubator space (1 cft/hour/1 cft incubator space)	0.014-0.02
Gasoline Engine*		
CH4	per brake horse power/hour	0.312
Biogas	per brake horse power/hour	0.453
Equivalent to		
(a) Gasoline		
CH4	per 4.546 liters (gallon)	3.82-4.53
Biogas	per 4.546 liters (gallon)	5.09-7.07
(b) Diesel oil		
CH4	per 4.546 liters (gallon)	4.25-5.32
Biogas	per 4.546 liters (gallon)	5.66-7.87

* at 25% efficiency

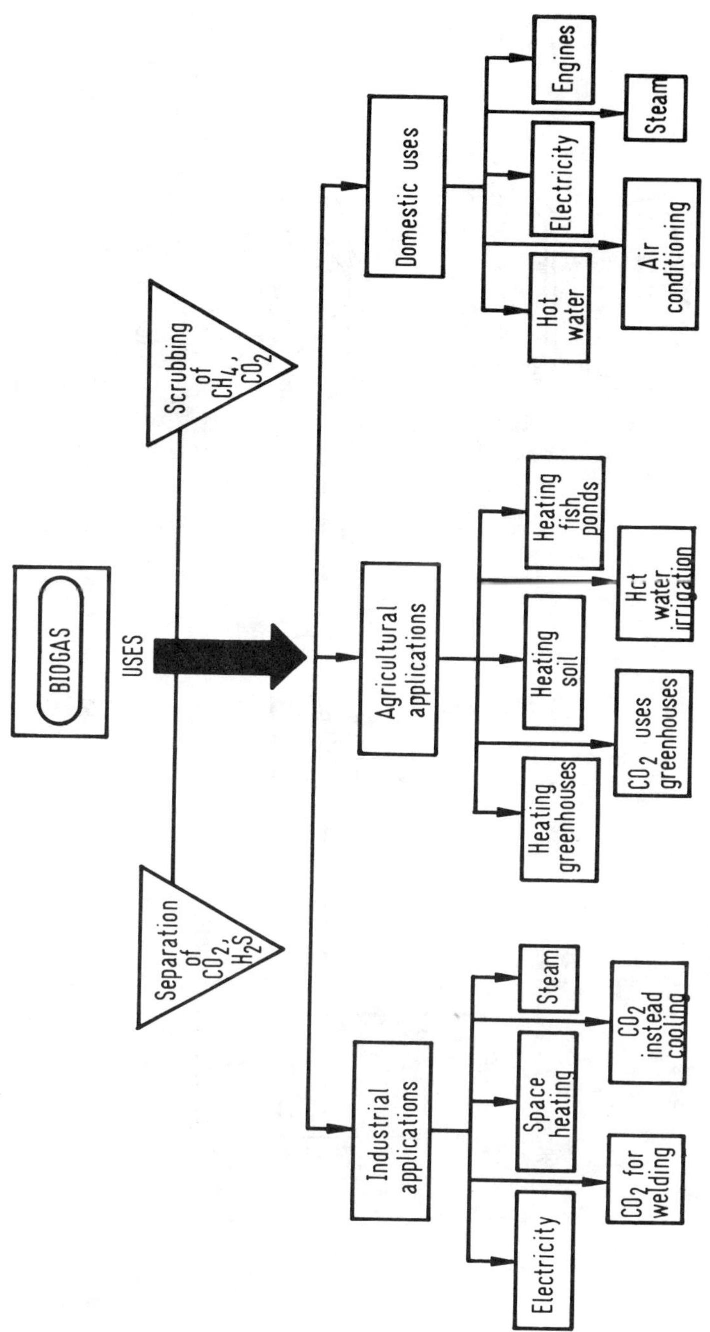

Fig. 5 - Biogas Utilization for Various Useful Purposes

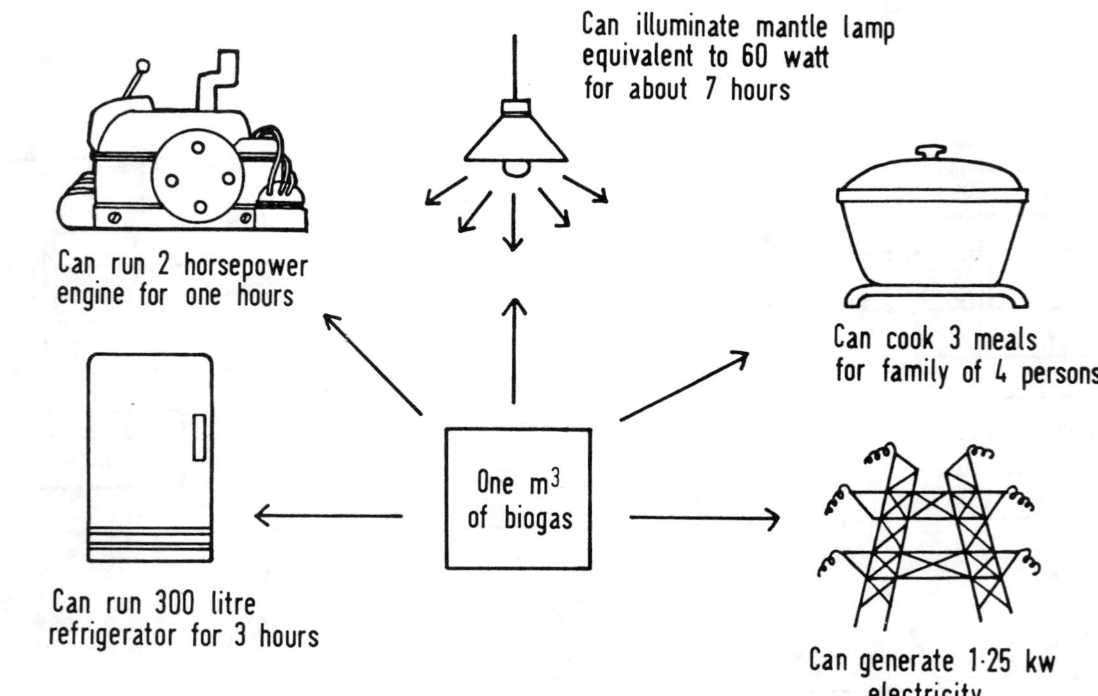

Fig. 6 - Possible applications of biogas

The same amount of biogas can generate about 1.25 kwh of electricity [7,10]. Biogas is also used extensively in agriculture amd horticulture especially for greenhouse heating and crop drying.

In addition, the slurry and the sludge from the digester are a good nutrient source and serve as an excellent soil conditioner. They can be applied directly to crops without further treatment. The use of biogas slurry as a feed for fish culturing has been used in many parts of Asia [5]. Ideally, an integrated farming system could be developed as represented in Fig. 4 involving the use of slurry (effluent) for growing algae and raising fish and ducks and eventually for growing of vegetables and other crops with the help of both slurry and sludge.

4. BIOGAS SYSTEMS IN ASIA

Biogas serves as an important energy throughout Asia both for traditional and innovative applications. Domestic use, especially for cooking, lighting and heating is of primary importance. A recent rise in the use of anaerobic digesters in Asian countries is helping to meet domestic, agriculture and industrial energy needs in addition to providing fertilizer for crops. In last twenty years, thousands of biogas plants of various designs have been constructed throughout Asian countries for the production of methane gas to be used for various beneficial purposes.

4.1 China

Mainland China is reported to be the world's largest biogas producer with about 9 million digesters. These plants are installed mainly in villages where agricultural crop wastes and animal and human excreta are converted into organic fertilizer. China has been dealing successfully with its energy problem through biogas production [5,6]. Biogas is China's answer to the energy crisis especially in rural areas. In a country where 70 percent of the rural people suffer from an acute shortage of fuel supply, biogas has become a boom to the rural household and has given a big push to modernizing agriculture and developing village communities. A majority of China's 957 million people live in rural areas. Firewood is still the main source of energy for 90 percent of the rural population. Rural areas consume about 500 million tonnes of firewood every year for fuel. To deal with this massive energy problem. Chinese authorities have been advocating the production and use of biogas for rural development. At present , China has over 9 million biogas plants all over the country benefitting about 35 million rural people. The most rapid expansion of biogas energy system has been registered in the province of Sichuan where there are 5 million biogas digesters and more than 100,000 trained personnel in biogas energy production. Raw materials used in the digesters include leaves, grass, agricultural, human and animal wastes along with woody biomass. The most common mixture used for fermentation is 40 to 50 percent raw materials with the remainder being water. The gas produced is used mainly for household cooking, heating, lighting and for generating electricity to process agricultural products.

4.2 India

India, where gobar (cow dung) gas programme has a target of 5,000 new digester units every year [11,12]. More than 70,000 biogas plants have already been established. Annual gas production from these units has been estimated at 152 million tonnes. Likewise, about 1.5 million tonnes of organic fertilizers are produced annually. Biogas is used for domestic cooking, lighting, heating and also for generating electricity for agricultural purposes. In addition, the organic fertilizer is used for crop production. To meet the ever increasing demands for the installation of biogas plants in the rural areas, the Khadi and the Village Industry Commission have plans to establish an additional 40,000 biogas plants.

4.3 Japan

Japan has pioneered some of the technologies for converting various biomass materials into energy [14]. It has been producing 750,000 to 2,500,000 m^3 of biogas per year with a methane content of 55 to 65 percent, mainly fermenting industrial wastes, particularly from the alcohol distilleries, in large digesters capacity ranging from 660 m^3 to 5,000 m^3 . The gas is used to heat boilers to produce steam and also partly it is used for generating electricity. In addition, Japan has started producing energy from other forms of biomass materials. For example, production of **Eucalyptus oil** and **Orange oil** for running small engines; production of methane gas from various agriculture, forestry and fisheries wastes, sewage sludge and other agro-based and agro-allied industrial wastes, and production of energy from waterhyacinth. At present, the dissemination of biogas technology is not very advanced, with 26 biogas digester systems (2620 m^3 capacity) in agriculture and 12 (10,500 m^3 capacity) in the field Of food industry. The reasons for this are: first, the price of oil is still sufficiently low to warramt biogas use; second, users feel insecure when using methane gas from digesters; and third, especially rural people do not appreciate the value of reclaiming wastes.

4.4 Philippines

Biogas technology in the Philippines has reached a fairly advanced level in research and development and has been demonstrated to be economically viable when applied on an industrial scale. Professor Romeo V Aliebusan of the pioneer researchers into biogas technology in the Philippines has recently stated **"The next step is for a nation-wide practical demonstration and acceptance of the biogas technology by the general public"**. The major research and development activities in biogas technology is centred at the National Institute of Science and Technology (NIST) at the University of Philippines at Los Banos and at Maya Farms in Manila [8,13]. More than 100 units of biogas digesters have been built under the guidance of NIST. Each of them costs around US$750 and need manure from 5 to 10 pigs diluted at 1:3 ratio. Each unit produces enough gas to supply the cooking needs of 5 to 7 people. The digester slurry and sludge are used as fertilizer. The **Maya Farm** is a private commercial enterprise with integrated livestock meat processing and canning operations. It covers 36 hectares of land and houses 70,000 pigs, 100 cattle and 20,000 ducks. To control pollution and to recycle organic wastes, there are 10 large biogas plants in continuous operation, producing 7075 m^3 (250,000 ft^3) of biogas/day. The complex is considered to be the largest single biogas establishment in Asia. The Farm attained its goal of 100 percent dependence on biogas energy in 1983. The gas is utilized to run alternately 6 units of 60 KVA generators, which supply electric power to a meat processing plant and a cannery, a food canteen, water pumps and air-conditioning units in various offices of the farm, workers dormitory, farm lights, refrigerators, cooking and laundry facilities. The sludge is added to the pig feed at the rate of 5 to 10 percent.

In addition, biogas digesters were constructed in 12 regions of the country for promotional and demonstration purposes. The continuous operation of these units has made it possible to appreciate the techno-economic value of biogas in Philippines. Biogas Farm uses cow manure as the raw material. In addition to the use of biogas for cooking, heating and lighting, the slurry is used for harvesting algae (chlorella), fish and vegetables and crops.

The Philippine government has introduced a simple but effective system for producing and promoting the acceptance of biogas as a source of cooking fuel for housewives and of bio-fertilizer for farmers. This has led to design and manufacture of a mini-digester (water-seal type) measuring 56 cm x 85 cm and with a full capacity of 210 L of raw materials. This unit costs US$95 complete with a fabricated single burner stove. It is light and portable and produces 0.23 m^3 (8.15 ft^3) of biogas per day. The main objective of this novel approach is to give rural people an opportunity to know and fully understand the basic principles of biogas production, its uses and advantages without incurring a big investment, thereby making biogas technology popular at grass-root level of the nation.

4.5 Korea

In Korea , it was recognised in the late sixties that tbere was a positive need to fine alternative energy sources especially for the farm households, since the main traditional energy sources in rural areas were crop residues and forestry products [7]. Therefore, between 1969 and 1975, around 29,000 household biogas units (6 m^3 capacity each) were installed to produce methane gas mostly from agricultural residues, livestock raising wastes, animal manure, urban refuse and human excreta to provide cooking fuel for farm households and to improve the modes of farmers' living standards. From 1976 onwards, a large number of village scale biogas plants instead of household units were constructed to overcome the shortcomings of the latter. The batch-load type biogas plant is very popular among the rural population. The amount of annual biogas production from agricultural wastes alone in Korea is projected at 2,283,857 x 10^3 m^3. Its thermal value is 12,990,591 million kcal which is equivalent to 1,082,548 tonnes of propane gas, 1,459,616 kcal of kerosene or 866,037 x 10^3 pieces (one piece weighs about 3.3 kg) of anthracitic briquettes.

4.6 Taiwan

There are nearly 7500 family sized biogas plants on the island. Small circular units (5 to 30 m^3 capacity) and large rectangular bag type units (50 to 100 m^3 capacity) of biogas digesters are commercially available. Among the 6697 hog farms in Taiwan, many use biogas digesters for biogas production and for pollution control. It has been reported that the anaerobic fermentation process used to treat hog wastes is practical and economical

[15]. In addition to anaerobic treatment, the Hsien-Tai Hog Farm at Ping-Thng also houses active sludge treatment as a secondary process to treat the waste from 20,000 hogs - the biogas produced from the digester being used to generate electricity for secondary treatment. The investment costs about US$10 perpig. The effluent from the digester is used to fertilize crops and for culturing spirulina while sludge is utilized as feed for fish. The biogas obtained is used to heat pig nursery, for cooking, to run water pumps and to generate electricity.

4.7 Malaysia

Malaysia has huge raw materials such as plants, vegetations, agricultural residues, palm oil shells, palm oil mill wastewaters, rubber processing wastes, rice husk, pineapple, coconut and other fruit wastes for the production of biogas. In Malaysia, out of a land area of about 14 million ha, more than 1 million ha are under oil palm cultivation. There are 210 palm oil mills and they generate more than 7.5 million tonnes of highly polluting effluent [16,18]. About 90 percent palm oil mills use digesters (4,000 to 15,000 m^3 capacity) for anaerobic decomposition of their effluents. Around 0.6 m^3 of biogas is generated from each kg of volatile solids added to the digester. The digestion is carried out between 44 to 52°C with a minimum retention time of 10 days using a high rate digestion system [16]. In case of palm oil mill processing, 60 tonnes of FFB (fresh fruit branch) per hour for 20 hours, a total of 20,000 m^3 of biogas can be produced per day. In Malaysia, the Sime Darby plantations have successfully utilized biogas generated from palm oil mill effluent (H?tE) for heating aed generating electricity [18]. Apart from biogas, the other valuable by-product from the anaerobic digestion of POME is the mother liquor which has a very high content of plant nutrient. When applied on to oil palm estates at controlled rate, it improves crop yields and soil properties. It is estimated that by the end of 1990, Malaysia will produce about 6 million tonnes of crude palm oil and about 15 million m^3 of POME. Around 420 million m^3 of biogas could be produced which could be capable of generating 756 million kwh of electricity. It is also estimated that around 23 million barrels of oil could be derived from the bio-wastes found in palm oil, rubber and coconut plantations. In view of the potentially large amount of biogas energy obtainable from these waste materials, the Malaysian government has set up a Biogas Steering Committee to promote the development and implementation of biogas technology all over Malaysia.

4.8 Thailand

Thailand is an agricultural country and has tremendous raw materials for biogas production. Biogas technology from India was first introduced in Thailand in 1966 [17]. Present biogas technology has been introduced from China, Taiwan and Philippines for use on large farms in the central Thailand, especially on the swine and dairy cattle farms. Currently, there are more than 500 biogas plants in the country. Family sized (3 to 5 m^3 capacity) units based on cattle and pig wastes and commonly used in rural areas are mode of concrete with a thin steel gas holder of 0.8 m^3 capacity floating on the digester slurry. The feeding rate is 20 to 40 kg/day at 1:2 to 1:5 dilution and the amount of gas produced is reported to meet the cooking needs of 6 to 8 persons. Currently, a new type of ferrocement biogas digester has been found to be most cost effective. In Thailand, there are four different ways in which biogas is used: (a) as a cooking fuel, as a substitute for charcoal and wood which are scarce and becoming unavailable in rural areas - it also helps to reduce the destruction forest trees; (b) as a fuel to be used in pumps. Supplying water to paddy fields and cleaning farm stables and also for electricity generation for lighting farm households; (c) as heating gas for rearing chicken aed piglets ; and (d) as a fuel gas for operating cold stores on farms.

4.9 Other Countries

Other Asian countries like Bangladesh, Indonesia, Hong Kong and Singapore are making use of biogas for experimental and training activities [8,10,19]. Iran, Pakistan, Nepal and Sri Lanka are harnessing biogas for energy, particularly as a cooking fuel for domestic uses and for feed and fertilizer production [8,10].

5. CONCLUSIONS

Biogas technology is well established in some Asian countries and is now generally accepted for the production of energy and fertilizer from various waste materials. There are many advantages associated with this process like (a) biogas production, (b) pollution control, (c) pathogen reduction, (d) odour reduction, (d) waste stabilization, (c) use of slurry as fertilizer, algae harvesting and fish farming, and (f) use of sludge as animal feed and/or soil conditioner. In many cases, it is difficult to place an economic value on the benefits of biogas technology. The value of biogas as an energy source is the most tangible financial return and has often been used as the only economic justification, but the value of the other benefits from this process can exceed the value of biogas and has frequently been ignored.

Biogas is an important alternative source of energy. The biogas technology has advanced significantly and some countries of this region like China, India, Philippines, Korea, Taiwan and Malaysia have been using this technology to solve their energy and fertilizer problems to a great extent. But in other countries of the region, like Bangladesh, Indonesia, Thailand and Sri Lanka, the application of this technology lies in the immediate time scale, limited only be the lead time required to develop the organizational infrastructure to plan, design, construct, operate and maintain such facilities. Biogas technology holds a good promise of supplying energy and useful materials to supplement the petroleum-based energy and fertilizer needs to millions of people of the Asian region. In many countries, biogas technology can contribute to the improvement of their balance of payments, gradual removal of dependence upon imported fuels and fertilizers, and saving of valuable foreign currency. Biogas technology is also a waste management technology. It is therefore important to develop safe, efficient, economically attractive and environmentally acceptable biogas plants for producing energy and useful materials.

Acknowledgement

The materials reported in this paper have been prepared from various publications and also from the author's own field aud laboratory investigations. The author gratefully acknowledges various authorities, organizations and publishers for reporting some of their materials in this paper.

REFERENCES

1. UN-ESCAP. Report of the Workshop on Biogas Technology and Utilization. Committee on Industry, Development and Environment. Bangkok, Thailand, 12-22 September 1975.

2. Barnett, A., Pyle, L. and Subramaniam, S.K., Biogas technology in the third world : A multidisciplinary review. TROC Report No. 120e, Ottawa, Canada, 1978 .

3. De Silva, E.G., Biogas : Fuel for future, AMBIO 9, 1:2-9, UNESCO, Place De Fontenoy, Paris, France, 1980,

4. Aziz, M. A. and Jalal, K.F., Biomass : Research and development for its effective utilization for energy and useful materials in the Asia-Pacific region. Proc. Workshop on Research and Technology Transfer for Asian Development. Bangkok, Thailand, 1-3, November 1983.

5. Development of alternative sources of energy: Technology for development in Southeast Asia and the Pacific. QuarterlyNewsletter, Singapore, Vol. 3, 1987 .

6. Ellias, T.G., Power technology : Potential of biogas in developing countries. National Development Journal , April 1988 .

7. Park, Y.D., Biogas research and utilization in Korea. Proc. Int. Symp. Alternative Sources of Energy for Agriculture, Taiwan, pp. 139-152, 4-7 September 1984,

8. Javier, S.V. Jr., Biogas promotional strategies : Philippines experience. Proc. Int. Symp . Alternative Sources of Energy for Agriculture. Taiwan, pp. 153-162, 4-7 September 1984.

9. Meynell, P.J., Methane: Planning a digester. Prism Press, London, UK, 1976.

10. Aziz, M.A., Biogas : An alternative source of energy. Proc. 6th Conf. ASEAN Fed. Eng. Orgns. (CASEO-6), Brunei, pp. 165-176, 13-15 January 1988.

11. Sathianathan, M.A., Biogas : Achievements and challenges. Association of Voluntary Agencies and Rural Development, New Delhi, India, 1975.

12. Gobar Gas, Directorate of Gobar Gas Scheme. Khadi and Village Industries Commission, Vol. 2, pp. 1-16, Bombay, India, 1979.

13. Alicbusan, R.V., Biogas technology - its prospects and problems. Mimeograph. Paper presented at the Seminar on Biogas Technology. Sponsored by the Bureau of Energy and Development. Govt. of Philippines, Port Bonifacio, 16 August 1979.

14. Maekawa, T., Dissemination of methane gas fermentation systems and associated problems in Japan. Proc. Int. Symp. Alternative Sources of Energy for Agriculture, Taiwan, pp. 163-174, 4-7 September 1984.

15. Hong, C.M. et al, Experience in the anaerobic treatment of hog wastes in Taiwan. Proc. Int. Symp. Alternative Sources of Energy for Agriculture, Taiwan, pp. 80-90, 4-7 September 1984.

16. Ahmed, B.I., High performance anaerobic digestion processes for the production of biogas from agro-industrial wastes in Malaysia. Proc. Int. Symp. Alternative Sources Energy for Agriculture, Taiwan, pp. 66-78, 4-6 September 1984 .

17. Kanarurak, P., Biogas production in Thailand. Proc. Int. Symp. Alternative Sources of Energy for Agriculture, Taiwan, pp . 175-177, 4-7 September 1984 .

18. Ma, A.N. and Ong, A.H.S., Production and utilization of biogas from palm oil mill effluent. Proc. Int. Symp. Alternative Sources of Energy for Agriculture, Taiwan, pp. 91-100, 4-6 September 1984.

19. Ibrahim, M.N., Research activities on alternative energy in Indonesia. Paper presented at the 6th Conf. ASEAN Fed. Eng. Orgns. (CAFEO-6), Brunei, 13-15 January 1988.

20. IFIC, Ferrocement biogas digester. Booklet No. 6, International Ferrocement Information Center, Asian Institute of Technology (AIT), Bangkok, Thailand, 1985.

BIOGASIFICATION OF AQUEOUS PHASES OBTAINED FROM LIQUEFACTION STUDIES OF SEKA SOLID WASTE AT 10 MPa INITIAL PRESSURE

F. Taner, U. Kimyonsen, G. Ersöz and A. Eratik
C. U. Arts and Sciences Faculty, Chem. Dept., 01330, Adana, Turkey

Abstract

In this study, the aqueous phases obtained from thermochemical treatment of SEKA solid waste with 15% acetic acid and NaOH at 523 K, 573 K, and 623 K, starting with 10 MPa initial pressure have been digested anaerobically. The anaerobic digestion has been carried out at 310 K, in lab-scale fementors for about 24 days. During digestion, sample were taken out at three-day intervals and analyzed for the determination of the COD, BOD5 and pH values. The total amount of biogas produced was also determined. It was found that the COD decreases and BOD5 increases as a whole with digestion time. It can be concluded that the aqueous phases can be treated anaerobically for both biogas production and organic pollutant removal.

1. INTRODUCTION

Worldwide interest in thermochemical treatment of organic waste material for combined purpose of energy production and pollution abatement continues to increase. Researches are aimed toward a better conversion of biomass into fuel; oil and biogas. Appell and others [1,2] have extensively studied the conversion of cellulosic wastes into liquid fuel in aqueous alkaline media. It has been reported that the anaerobic biodegradability of aqueous layers obtained from thermochemical conversion processes could be used for biogas production [3,4]. The potential of heat treatment for increasing anaerobic biodegradability was evaluated over a broad range of temperature and pH, but only using a 3.5 percent total solid concentration [5-7]. It has been shown that heat treatment can greatly enhence the degradability of waste activated sludge under mesophilic anaerobic conditions [8]. From the literature [9-13], it is apparent that high destruction of the substances can be achieved if the substrate is soluble in water.

In our laboratory, some liquefaction experiments have been applied to various lignocellulosic waste materials in aqueous solutions of acetic acid and sodium hydroxide [14-17]. It was determined that the aqueous phases, obtained from the thermochemical treatment, are rich in organic water-soluble compounds and has to be treated for the removal of the compounds which are considered as pollutants. For this purpose, the aqueous phases have been treated, anaerobically, and the water-solubles have been ozonized in our laboratories [18-19]. In this paper, the biodegradabilities of the aqueous phases obtained from the liquefaction of SEKA solid waste at 10 MPa initial pressure has been reported.

3. MATERIAL AND METHOD

SEKA is a pulp and paper industry where kraft pulp and paper are produced. This factory has a wastewater treatment plant. At suspended solid separation step, the fibrious material was separated from the wastewate by filtering. This material is named SEKA solid waste. About 10% of production is produced as waste. Daily production of solid waste (dry) is about 2 tons and there is no way to use this waste. In our laboratory, it was tried to find the liquefaction conditions of the SEKA solid waste to produce liquid fuels.

Thermochemical Treatment

The aqueous slurries of the SEKA solid waste (20 % by weight) have been treated with acetic acid and sodium hydroxide (15 % by weight of the solid) at 523 K, 573 K, and 623 K in a high pressure autoclave starting with 10 MPa initial nitrogen gas pressure for one hour [20]. After thermochemical treatment, the reaction mixtures were separated into solid-tar and aqueous phases by filtering. The solid-tar phases were washed with water several times to remove the water-soluble compounds adhered to the solid-tar. The washing filtrate and some of the aqueous phases were combined and saved for anaerobic digestion.

Anaerobic Biodegradation

For anaerobic biodegradation, about 10 mL and 25 mL bottles were used for each aqueous mixture obtained from each operating temperature. Ten of 10 mL bottles were used for the determination of the COD and BOD5 concentrations and pH of the digested solutions taken at three-day intervals during the digestion period. For the determination of the amount of biogas, only 25 mL bottle was used for each aqueous samples.

First of all, the pH's of the aqueous samples were adjusted to around 7 by using NaOH and HCl solutions, depending on the pH of the original aqueous samples, and then $NaHCO_3$ (5.39 g/L) was added into each sample solutions (150 mL) which will be digested. 8 mL and 20 mL of the sample solutions were poured into ten of 10 mL and one 25 mL bottles, respectively. All the bottles were sealed with metal covers having a hole at the center and rubber plug inside. Then all the bottles were flushed with CO_2 and nitrogen gas by inserting syringe nedles through the rubber plug to remove oxygen gas and to saturate the solution with CO_2 for buffering. Afterward, all the bottles were placed in the waterbaths which are adjusted to 310 K.

A stock culture solution was prepared by filtration of the fresh animal manure mixed with some water. The filtrate was used for the inoculation of the sample solutions. 0.8 mL, and 2.0 mL of the culture solution were added into each bottle by the aid of the syringe, anaerobically. After addition of the culture solution, biodegradability commenced. The 25 mL bottles were connected to the gas collection burettes that were filled with acidified water and set upside down in a reservoir, with plastic pipe to collect and to measure the amount of biogas which will form during the anaerobic digestion. Pikes for biogas measurements and analyses of the digested solution were also made in the same way by using the distilled water instead of addition of the sample solutions.

One of 10 mL bottles of each set was taken out of the waterbath at three-day intervals. The bottles were unsealed and pH's of the digested solutions were measured first, and then the digested solution were centrifuged (5 000 rpm), to remove microorganism produced during the digestion period. The COD and BOD5 concentrations of the supernatant were determined according to the Standard Methods (21). The amount of the biogas collected in the burettes have been measured every day, and the total amount of biogas produced was calculated at STP on a dry basis. The pH values with digestion time, and the total amount of biogas, calculated, were shown in TABLE I. The COD and BOD5 concentration changes with digestion time were plotted and shown in Fig.1-Fig.4.

3. RESULTS AND DISCUSSION

As seen in TABLE I, the pH's of the digested solutions do not change very much, and they all deviate around the pH, adjusted at the beginning of the digestion. It shows that the systems are buffered and there is biological activities in the digesters.

The amounts of the biogas produced from each of the aqueous sample obtained at 523 K and 573 K are differing, slightly. The amount of biogas obtained from the aqueous phase with acetic acid at 623 K is higher (47.73 mL) than the amounts, 2.15 mL and 9.03 mL at 523 K and 573 K, respectively. The amount of biogas increases, as the operating temperature of the thermochemical treatment increases. This shows that as the thermochemical treatment temperature increases, the water soluble compounds formed at low liquefaction temperature convert into biodegradable compounds that must be the low molecular weight compounds. Destruction of the big molecules is possible at high operating temperature. Therefore, the production yield of the biogas increases. It was also found that the oil yield of the solid-tar drops at 623 K when acetic acid was used [20].

When sodium hydroxide is used as chemical for liquefaction of SEKA solid waste, the amount of the biogas produced from each of the aqueous sample obtained at 523 K and 573 K are about the same. The amount of biogas obtained from the aqueous phase at 623 K is higher (121.3 mL) than the amounts, 1.8 mL and 3.4 mL at 523 K and 573 K respectively. The amount of biogas increases, as the operating temperature of the thermochemical treatment increases. This shows that as the thermochemical treatment temperature increases, the water soluble compounds formed at low liquefaction temperature convert into biodegradable compounds that must be the low molecular weight compounds. Destruction of the big molecules is possible at high operating temperature. Therefore, the production yield of the biogas increases. It was also found that the oil yield of the solid-tar drops at 623 K when sodium hydroxide was used [20].

When biogas yields of the aqueous samples of each chemicals at the same thermochemical treatment temperatures are compeared, it is seen that sodium hydroxide is more effective than acetic acid for conversion the water soluble substances into anaerobic biodegradable ones. This shows that the waste water obtained from the liquefaction processes can be treated anaerobically for the removals of the pollutants.

The COD and BOD5 changes in the aqueous layers obtained from thermochemical treatment of the solid with 15% acetic acid are shown in Fig.1 and Fig. 2, respectively. As seen in Fig.1 the COD concentrations of the digested samples obtained at the temperatures of 523 K, and 573 K drop very much on the days of 6-12, then they start to increase. As a whole, the COD decreases with digestion time relative to initial COD concentration of the aqueous layers. This shows that there is some degradation reactions occuring during the anaerobic digestion processes. For the aqueous phases, obtained at 523 K, the COD increases with digestion time as a whole. This shows that some polymer formed during the liquefaction processes converts into biodegradable compounds under anaerobic conditions, and the biodegradation products oxidize chemically. As seen in Fig.2, the BOD5 of each aqueous phases do not change very much, and the system conserves some values of BOD5, except the BOD5 concentration in the aqueous phases at 523 K with acetic acid, that increases, highly.

The COD and BOD5 concentrations of the digested samples obtained from the treatment of the solid with 15% NaOH at different temperatures with digestion time is shown in Fig.3 and Fig.4, respectively. The COD of the sample at 523 K increase generally, as digestion time increases. This shows that some organic compounds formed during thermochemical treatment become chemically oxidizable under anaerobic conditions that cause the molecules to hydrolyze. That may be the reason for increase of COD. The drop in COD shows that some compounds which are oxidizable chemically, biodegrades and biogas forms. The COD concentrations of the aqueous samples at 573 K and 623 K with NaOH, decrease as a whole (Fig.3).

As seen in Fig.4. the BOD5 concentrations of the digested

samples obtained from treatment of the solid with 15% NaOH increase with digestion time. This shows that the biochemical reactions occurs and biogas forms.

From the findings, it can be concluded that the wastewater obtained from liquefaction processes can be treated anaerobically for the partial removal of pollutants and for the production of methane gas.

REFERENCES

1. Appell, H.R.; Fu, Y.C.; Friedman, S.; Yavorsky, P.M.; Wander, I.; "Conversion of Cellulosic Wastes to Oil", U.S. Bureau of Mines Rep. of Investigation, No. 7560, p.20, 1971.

2. Appell, H.R.; Fu, Y.C.; Illig, E.G.; Stafgan, F.W.; Miller, R.D.; "Conversion of Cellulosic Wastes to Oil", U.S. Bur. of Mines Report of Investigation, No.8013, pp.27, 1975.

3. Taner, F., "Biodegradation of Aqueous Layers Obtained From Thermochemical Treatment of Cotton Stalk With Acetic Acid", Proc. of the 1986 Int. Cong. on Renew. Energy Sources. Edited by S.Terol, 18-23 May, p.439-446 Madrid, Spain.

4. Gül, A., "Effect of Thermochemical Treatment for Anaerobic Biodegradation of Cotton Stalk", Environment'89, 5th Env. Science and Technology Cong., June 5-9, 1989, Adana, Turkey.

5. Gossett, J.M.; McCarty, P.L., "Heat Treatment of Refuse for Increasing Anaerobic Biodegradability", Technical Rep. No. 192, Civil Eng. Dept., Stanford Univ., Stanford, California, Jan. 31, 1975.

6. Gossett, J.M., Healy, J.B., Young, L.Y., and McCarty, P., "Heat Treatment of Refuse for Increasing Anaerobic Biodegradability", Tech. Rep. No.198, Civil Eng. Dept., Stanford Univ., July 31, 1975a.

7. Gossett, J.M., and McCarty, P., "Heat Treatment of Refuse for Increasing Anaerobic Biodegradability", 68th An.Meeting, American Institute of Chem. Eng. Nov. 16-20, 1975.

8. Hauge, R.T., Stuckey, D. C., Gossett, J.M. and McCarty, P.L., "The Effect of Thermal Pretreatment on Digestibility of Organic Sludges", JWPCF, 50(1), 73, 1978.

9. McCarty, P. L.; Baba, D.J.; McDermott, G.N.; Weaver, P.J., "Treatability of Oily Wastewaters From Food Processing and Soup Manifacture", Proceeding of the 27th Conf. On Industrial Wastes, Purdue Univ., p.867, 1972.

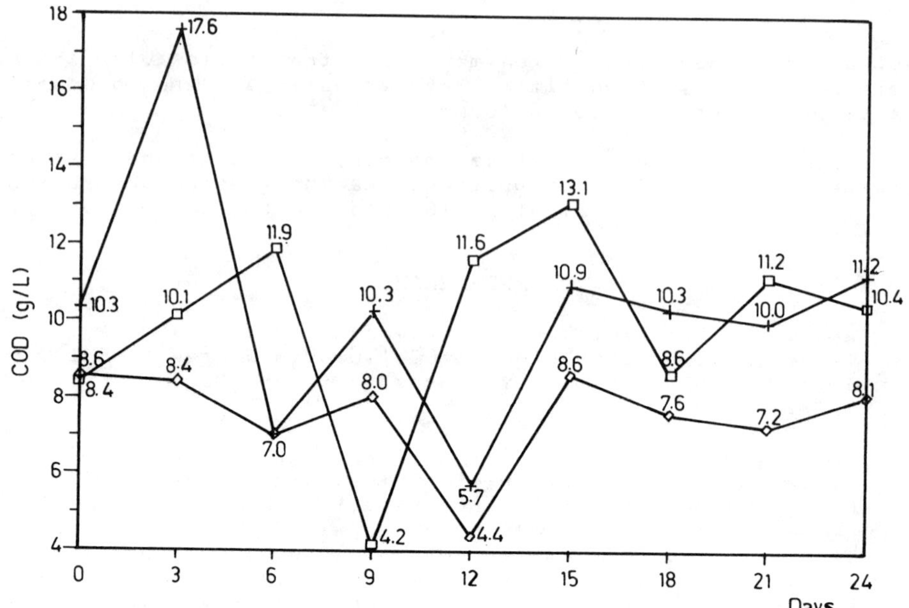

Fig. 1. Biodegradabilities of watersolubles in aqueous layer with digestion time (CH_3COOH) □ 523 K, + 573 K, ◇ 623 K

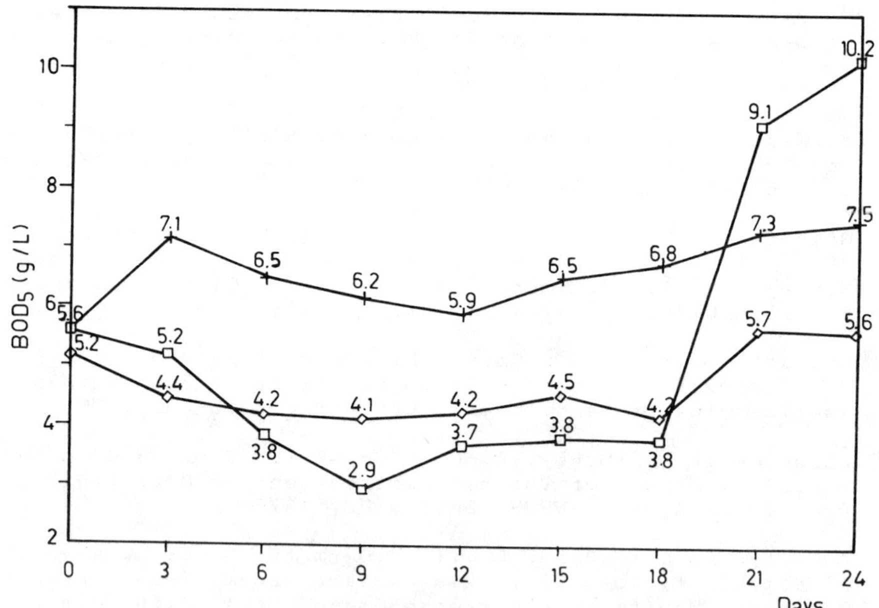

Fig. 2. Biodegradabilities of watersolubles in aqueous layer with digestion time (CH_3COOH) □ 523 K, + 573 K, ◇ 623 K

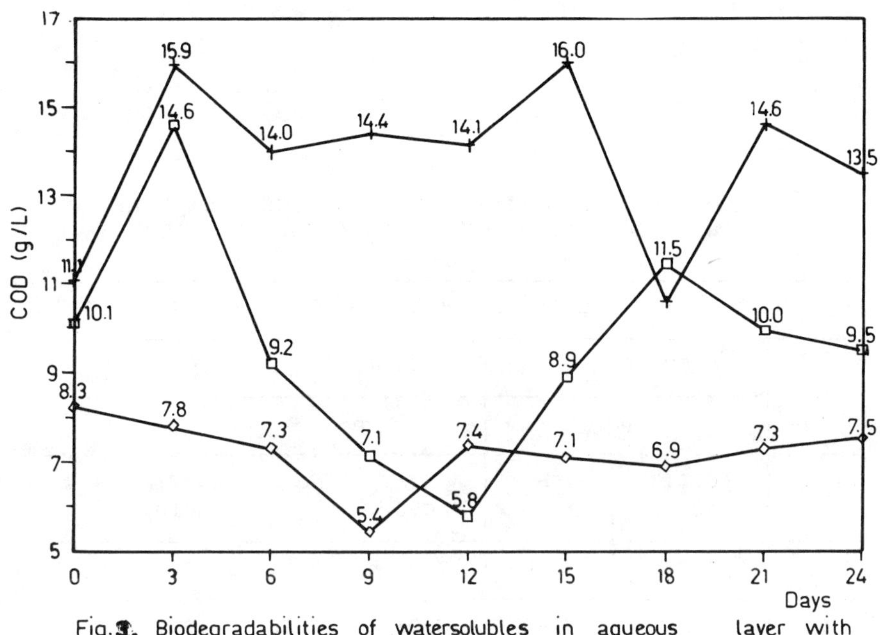

Fig. 3. Biodegradabilities of watersolubles in aqueous layer with digestion time (NaOH) □ 523 K, + 573 K, ◇ 623 K

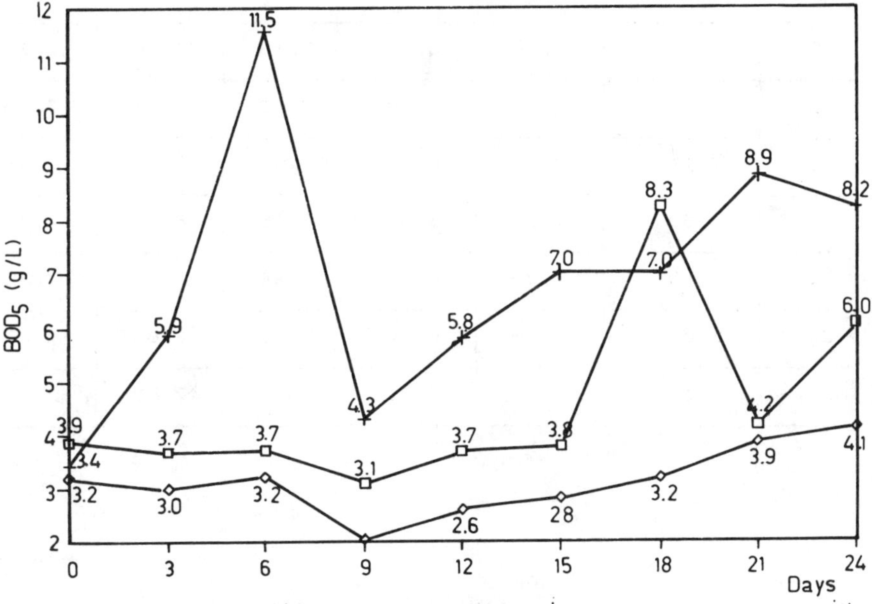

Fig. 4. Biodegradabilities of watersolubles in aqueous layer with digestion time (NaOH) □ 523K, + 573K, ◇ 623K

TABLE I: CHANGES OF pH WITH DIGESTION TIME AND THE AMOUNT OF BIOGAS (STP, DRY, mL)

LIQUEFACTION TEMPERATURES						
	523 K		573 K		623 K	
Time	pH		pH		pH	
days	ACETIC ACID	NaOH	ACETIC ACID	NaOH	ACETIC ACID	NaOH
0	7.05	7.36	7.87	7.21	7.06	7.40
3	6.87	7.40	7.01	7.22	6.28	7.37
6	6.99	7.39	7.00	7.22	6.86	7.35
9	6.85	7.34	6.96	7.16	6.75	7.38
12	6.95	7.65	6.90	7.23	6.87	7.40
15	7.27	7.79	6.52	7.23	6.91	7.02
18	7.30	7.26	6.79	7.18	6.77	7.31
21	6.69	7.30	6.82	7.94	6.83	7.32
24	6.68	7.29	6.90	7.15	6.84	7.27
mL gas	2.15	1.8	9.03	3.4	47.73	121.3

BIOGASIFICATION OF AQUEOUS PHASES OBTAINED FROM LIQUEFACTION STUDIES OF COTTON STALK AT 10 MPa IN THE PRESENCE OF SOME SALTS

F. Taner, U. Kimyonsen, G. Ersöz and A. Eratik
Ç. U. Arts and Sciences Faculty, Chem. Dept., 01330, Adana, Turkey

Abstract

The aqueous phases, obtained from liquefaction of cotton stalk with NaOH and Acetic acid (15% by weight) at 523 K, 573 K and 623 K starting with 10 MPa initial inert gas pressure for one hour, have been digested anaerobically at 310 K for 27 days in presence of some salts. During the digestion period, samples have been taken out at three-day intervals and analyzed to determine the pH, COD and BOD5 concentration changes with digestion time. On the other hand, the amount of biogas produced has also been determined. It was found that the addition of some salts enhances the biogas production. It can be concluded that the water soluble compounds produced by liquefaction processes can be used for biogas production, efficiently by the addition of some salts.

1. INTRODUCTION

Anaerobic conversion of solid organic wastes has not been very efficient in terms of volumetric loading capacities of reactors, solid retention times and extent of digestion [1,2]. For this reason application of anaerobic degradation to organic solid materials has been limited mainly to animal wastes and wastewater sludges. The low conversion rates probably due to a low rate of hydrolysis by physical, chemical or biological pretreatments of the waste materials [3,4]. Lignin-containing plant materials are believed to be the rate limiting components in methane production processes. Most studies to date have considered lignocellulose degradation under aerobic conditions only. Recently, Benner et al. [5] reported that the lignin component of intact softwoods and hardwoods is partially degraded to gaseous end products under anaerobic conditions. It has been shown that a consortium of anaerobic bacteria enriched from digester sludge is capable of cleaving the aromatic rings of eleven simple lignin derivatives [6-8]. Recent studies [9,10] have examined the biodegradation of several substituted aromatic compounds by enrichments and isolates from lake and marine muds. soluble lignin fragments can be released by natural and industrial processes. One of the industrial process on explorative stage is the conversion of biomass into fuel oil and biogas. Appell and others [11,12] have extensively studied the conversion of cellulosic wastes into liquid fuel in aqueous alkaline media. Same experimental studies like these [13,14] had shown that most of the solid material, about 65%, converted into water soluble compounds which are considered as a pollutants in wastewater. It has been found that the water soluble substances formed during liquefaction studies biodegrade anaerobically [15,16]. From the literatures [17,18], it is apparent that high destruction can be achieved if the substance is soluble in water. Eventhough, it was determined that the biodegradation of the water soluble substances is possible, it is important for technological application to get the higher yield of products. It was known [19] that some additional salts have to be used for the biological life alive.

In our previous work, it has been determined that the biodegradation of water solubles formed from liquefaction studies [20] is possible. In this study some additional salts [19] were used to determine the biogas yield of the water soluble compounds in aqueous phases obtained from the liquefaction studies.

2. MATERIAL AND METHOD

Raw Material

After harvesting the cotton fibrils (Delta pine 15/21) the cotton stalk left over the experimental field of Ç.U. Agricultural Faculty was collected and left dry in the open atmosphere. It was ground and sieved with 30 mesh screen. The particles passed through the 30 mesh screen were used for liquefaction studies.

Liquefaction Studies

100 g of cotton stalk dust (moisture and ash free) was slurried with 400 mL water containing 15 g NaOH and CH_3COOH (Both chemicals are analytical grade, Merck). The mixture was poured into a 2L high pressure autoclave (COOK vertical high pressure autoclave, Birmingham, UK). The autoclave was sealed and flushed with nitrogen gas by pressurizing the reactor upto 0.5 MPa and then releasing the gas three times. Then the reactor was pressurized upto 10 MPa with nitrogen gas and the reactor was started to heat at this initial pressure upto operating temperatures of 523 K, 573 K, and 623 K while stirring continuously. The liquefaction has been completed for one hour, after the reaching upto liquefaction temperature. The heating was stopped and the reactor was allowed to cool overnight while stirring continuously down to to about 373 K. On the next day, the inside pressure of reactor was recorded, and the gas formed was analyzed to determine CO, CO_2 and flammable contents. After venting the rest of the gas left in the reactor, the reactor was opened and the reaction mixture was taken out manually. The mixture was filtered through coarse filter paper to separate

solid-tar from the aqueous phases. Some analysis for the determination of some properties of the solid-tar have been made [21]. After separation of aqueous phases, the solid-tar was washed with water to remove all the water solubles in the solid-tar, and saved separately. For the anaerobic digestion the aqueous phases and wash waters, partially mixed, were used.

Anaerobic Biodegradation

For anaerobic biodegradation, about 10 mL and 25 mL bottles were used for each aqueous mixtures obtained from each operating temperature. Ten of 10 mL bottles were used for the determination of the COD and BOD5 concentrations and pH of the digested solutions taken at three-day intervals during the digestion period. For the determination of the amount of biogas only 25 mL bottle was used for each aqueous samples.

First of all, the pH's of the aqueous samples were adjusted to around 7 by using NaOH and HCl solutions, depending on the pH of the mixtures prepared for the anaerobic digestion. 250 mL of the sample solutions were prepared by adding the salt solutions to make the defined concentration for methanogenic media [19]; resazurine (0.001 g/L), ammonium phosphate (0.04 g/L), ammonium chloride (0.2 g/L), magnesium chloride (1.8 g/L), potassium chloride (1.3 g/L), manganese II chloride (0.02 g/L), Cobalt II chloride (0.03 g/L), copper II chloride (0.0027 g/L), boric acid (0.0057 g/L), sodium molibdate (0.0025 g/L), zinc chloride (0.0021 g/L), iron III chloride (0.3 g/L), sodium bicarbonate (2.64 g/L), and sodium sulfide (0.5 g/L). 8 mL and 20 mL of the sample solutions, hence prepared, were poured into ten of 10 mL and one 25 mL bottles, respectively. All the bottles were sealed with metal covers having a hole at the center and rubber plug inside. Then all the bottles were flushed with CO2 and nitrogen gas by inserting syringe needles through the rubber plug to remove oxygen gas and to saturate the solution with CO2 for buffering. Afterward, all the bottles were placed in the waterbaths which are adjusted to 310 K.

A stock culture solution was prepared by filtration of the fresh animal manure mixed with some water. The filtrate was used for the inoculation of the sample solutions. 0.8 mL, and 2.0 mL of the culture solution were added into each bottle by the aid of the syringe, anaerobically. After addition of the culture solution, biodegradability commenced. The 25 mL bottles were connected to the gas collection burettes that were filled with acidified water and set upside down in a reservoir, with plastic pipe to collect and to measure the amount of biogas which will form during the anaerobic digestion. Pikes for gas measurements and analyses of the digested solution were also prepared in the same way by using the distilled water instead of addition of the sample solutions.

One of 10 mL bottles of each set was taken out of the waterbath at three-day intervals. The bottles were unsealed and pH's of the digested solutions were first measured and then the digested solutions were centrifuged (5 000 rpm) to remove microorganism produced during the digestion period. The COD and BOD5 concentrations of the supernatant were determined according to the Standard Methods [22]. The amount of the biogas collected in the burettes have been measured every day, and the total amount of biogas produced was calculated at STP on a dry basis. The pH values with digestion time and the total amount of biogas calculated were shown in TABLE I.

3. RESULTS AND DISCUSSION

As seen in TABLE I, the pH's of the digested solutions do not change very much, and they all deviate around the pH, adjusted at the beginning of the digestion period. It shows that the systems are buffered and there is biological activities in the digesters.

The amount of the biogas (52 mL) produced from the aqueous sample obtained at 573 K when acetic acid was used as chemical is the higher than the amounts, 18 mL and 32 mL, from the aqueous phases obtained at 523 K and 623 K, respectively (TABLE I). This shows that liquefaction with acetic acid converts the cotton stalk in mosly biodegradable products at 573 K. The amount of biogas decreases, as the operating temperature of the thermochemical treatment increases. This shows that the biodegradable compounds formed at low temperature (573 K) polymerize and diminishe at high operating temperature (623 K). Therefore, the yield of the biogas drops at high operating temperature. It was also determined that the oil yield of the solid oil is higher at 623 K [21].

The amount of the biogas (354 mL) produced from the aqueous sample obtained at 523 K when sodium hydroxide was used as chemical is the higher than the amounts, 124 mL and 32 mL, from the aqueous phases obtained at 573 K and 623 K, respectively. This shows that liquefaction with sodium hydroxide converts the cotton stalk in mosly biodegradable products at 523 K. The amount of biogas decreases, as the operating temperature of the thermochemical treatment increases. This shows that the biodegradable compounds formed at low temperature (523 K and 573 K) polymerize and diminishe at high operating temperature (623 K). Therefore, the yield of the biogas drops at high operating temperature. Eventhough, the initial COD and BOD5 concentrations of the samples at 573 K and 623 K are higher than the values at 523 K, the biogas yield of the aqueous sample obtained at 523 K is higher than the amounts obtained from the samples at 573 K and 623 K. It was also determined that the oil yield of the solid oil is higher at 623 K [21].

It was reported that when the same aqueous samples were subjected to biodegradation without enrichments [23], the yield of biogas is much more lower the yields obtained from the biogasification of the aqueous sample with addition of some salts.

From the findings, it can be concluded that the wastewater obtained from liquefaction processes can be treated anaerobically for the partial removal of pollutants and for the production of methane, by addition of some salts necessary fo r methanogenic media.

4. NOMENCLATURE

COD : Chemical Oxygen Demand.
BOD5 : Biochemical Oxygen Demand in 5 day.

REFERENCES

1. Hobson, P.N.; Bouspield, S.; Summers, R., 1974, "CRC Crit. Rev. Environ. Control" 4, p.131-191 .

2. Weland, A.F.; Cheremisinoff, P.N.,1975, "Water Sewage Works" 122, p.45-47.

3. Miller, M.A.; Baker, A.J.; Satler, L.D., 1975, "Biotechnol. Bioeng Symp." 5, p.193-219.

4. Han, Y.W.; Pence, J.W.; Anderson, P.W., 1975, "Appl. Microbiol" 29, p.708-709.

5. Benner, R.; Maccubbin, A.E.; Hobson R.E., 1984, "Appl. Environ. Microbiol." 47, p.998-1004 .

6. Healy, J.B.; Young, L.Y., 1978, "Appl. Environ. Microbiol.", 35, p.216-218.

7. Healy, J.B.; Young, L.Y., 1979, "Appl. Environ. Microbiol." 38, p.84-89.

8. Healy, J.B.; Young, L.Y.; Reinhard, M., 1980, "Appl.Environ. Microbiol." 39, p.436-444.

9. Kaiser, J.P.; Hansclmann, K.W., 1982, " Arch. Microbiol." 185, p.185-194.

10. Schink, B.; Pfenning, N., 1982, "Arc. Microbiolog." 133, p.195-201.

11. Appell, H.R.; Fu, Y.C.; Friendman, S.; Yavorsky, P.M.; Wender, I., 1971, " U.S. Bur. of mines Rep." No: 7560.

12. Appell, H.R.; Fu, Y.C.; Miller, R.D., 1975, "U.S. Bur. of mines Rep." No: 7560.

13. Taner,F., 1986, Proc. of the 1986 Int. Congrees on Renewable Energy sources", Edited by Terol, S., Vol.I, P.426, Madrid, 18-21 May.

14. Taner, F., 1986,"Proc. of the 3th Int. Symposium", Edited by Buekens, E., Anwerp, Belgium, p.7-67, 18- 20 March.

15. Taner,F., 1986, Proc. of the 1986 Int. Congrees on Renewable Energy sources", Edited by Terol, S., Vol.I, P.439, Madrid, 18-21 May.

16. Gül, A.; Taner, F., 1989, "The Effect of Thermochemical Treatment For Anaerobic Biodegradaition of Cotton Stalk", Environment'89 5th Scientific and Technical Environmental Congress, 5-9 June, 1989, Adana, Turkey.

17. Mc Carty, P.L.; Hahn, D.J.; Mc Dermott, G.N.; Weaver, P.J., 1972, "Proc. of the 27th Conf. on Industrial Wastes", Purdue Univ., p.-867.

18. Van den Berg, L.; Lentz, C.P., 1972, " Proc. of the 27th Conf. on Ind. Wastes", Purdue Univ. p.313.

19. Gossett, J.M.; Healy, J.B.; Stuckey, D.C.; Young, L.Y.; Mc Carty, P.L.; 1976, "Heat Treatment of Refuse for Increasing Anaerobic Biodegradability", Stanford Univ. Civil Eng. Tech. Report, No. 205 January 31, p.59.

20. Taner, F.; Kimyonşen, U., Boztepe, H., 1989, "Thermochemical Treatment Solid Waste for Biodegradation", VTT Symp.103, Non-Waste Technology, Vol.II, p.199-206

21. Taner, F.; Kimyonşen, U., 1989, "Liquefaction of Cotton Stalk With 15% Acetic Acid and NaOH at 10 MPa" Environment'89 5th Scientific and Technical Environmental Congress, 5-9 June, 1989, Adana, Turkey.

22. American Public Health Association, 1980, Standart Methods for the Examination of Water and Wastewater, 17th edition, Washington, D.C.

23. Taner, F., 1988, "Ligno-selülozik Katı Atıkların Enerji Kaynağı Olarak Kullanımı; yag ve Biyogaza Dönüştürme Koşullarının Saptanması", Turkish Scientific and Technical Research Council (TUBİTAK), Proje No: ÇAG79, p.46, Adana, Turkey.

TABLE I: COD, BOD5 AND PH CHANGES OF THE AQUEOUS PHASES (ACETIC ACID) WITH DIGESTION TIME

LIQUEFACTION TEMPERATURES									
Time	523 K			573 K			623 K		
Days	COD g/L	BOD5 g/L	pH	COD g/L	BOD5 g/L	pH	COD g/L	BOD5 g/L	pH
0	4.6	3.0	6.4	5.5	2.6	6.5	8.1	4.1	6.8
3	6.2	2.9	6.6	5.0	2.0	6.0	7.1	4.0	6.9
6	5.9	3.1	6.6	3.8	2.1	6.5	8.6	3.3	6.8
9	5.1	3.0	6.6	4.1	2.7	6.6	7.7	4.0	6.8
12	5.9	3.3	6.5	5.3	2.5	6.3	8.6	3.9	6.8
15	7.1	1.4	6.5	4.3	2.1	6.2	4.9	2.3	6.5
18	5.8	3.1	6.1	5.0	2.5	6.3	6.7	3.3	6.5
21	5.7	3.1	6.2	4.3	1.5	6.3	7.8	3.8	6.5
25	6.1	2.5	6.2	8.1	3.4	6.3	5.2	4.2	6.5
27	5.1	2.0	5.6	4.3	1.6	6.8	7.2	3.5	6.6
Total gas STP, dry, mL	18			52			32		

TABLE II: COD, BOD5 AND PH CHANGES OF THE AQUEOUS PHASES (SODIUM HYDROXIDE) WITH DIGESTION TIME

LIQUEFACTION TEMPERATURES									
Time	523 K			573 K			623 K		
Days	COD g/L	BOD5 g/L	pH	COD g/L	BOD5 g/L	pH	COD g/L	BOD5 g/L	pH
0	4.9	2.2	7.0	7.2	3.7.	7.	7.8	3.2	6.8
3	4.4	1.3	7.0	8.7	3.3	7.0	7.2	2.5	6.9
6	5.0	1.6	7.0	9.1	3.1	6.9	7.4	2.7	6.8
9	3.7	1.7	7.0	8.0	3.0	7.0	6.5	2.9	6.8
12	4.6	1.7	7.0	8.6	3.5	7.0	7.6	3.0	6.8
15	4.8	1.6	7.0	7.7	2.6	6.6	7.1	2.4	6.5
18	4.6	1.5	6.7	8.9	3.1	6.6	7.2	2.7	6.5
21	5.0	2.2	6.7	8.3	3.0	6.6	6.6	2.7	6.5
24	4.0	1.1	6.6	9.0	2.6	6.7	7.3	2.5	6.5
27	3.1	1.0	6.8	7.1	1.8	6.8	5.6	1.4	6.6
Total gas STP, dry, mL	354			124			32		

ENHANCED LIQUID FUEL YIELD FROM A SINGLE BIOMASS SPECIES NAMELY CALOTROPIS PROCERA

Ratna Choudhury and Ritu Singh
Centre for Energy Studies, Indian Institute of Technology,
New Delhi-110016, India

ABSTRACT

Solvent extraction of Calotropis Procera was conducted with a variety of solvents. Toluene which was immiscible with water was found to be the most effective in soxhlet apparaqtus giving 7.2% extraction. Modification of design to continuous solvent extraction with toluene and recycle of toluene after elimination of water was applied. This raised the extraction yield to 13%. To further add to liquid content, the chlorophyll in the extract was broken down to phytol and nitrogen containing base under alkaline conditions.

The upgradation of these liquids to gasoline and diesel range hydrocarbons by disproportionation with hydrogen donors or by dehydration (as in Mobil Gasoline Synthesis) has been presented and discussed.

Upgradation of a natural extract to a liquid fuel under atmospheric conditions leads to great possibilities in the age of alternate sources of energy.

Key words: Hydrocarbon from Calotropis Procera, Chlorophyll Distintegration, Disproportionation, Continuous solvent extraction with water elimination.

The existing supply of petroleum would eventually be exhausted in spite of comments that if more money was spent, more oil would be obtained. Oil and natural gas supplies are the fossilized remains of biomaterial photosynthesised over the ages. Since petroleum is used up faster than it is generated, it is easy to conceive that there would be a limit to its availability.

In this piece of work, liquid fuel is looked upon in the form of a renewable resource that can be grown each year. Most of the plants store the solar energy as half reduced carbon in the form of carbohydrates (wood, cellulose, starch or sugar). But some species of plants reduce the caron dioxide all the way down to hydrocarbons, containing no oxygen at all. These hydrocarbons are present as a white milky emulsion in the plants and popularly called latex. The species chosen for this project, was Calotropis Procera, (a member of Asclepiadaceae family) which is found growing widely in Indian climate even in the arid zones. So it does not encroach upon the land used for food crops.

Methods and Materials

Since calotropis procera was not amenable to continuous tapping, therefore whole plant was cut and dried at room temperature to a constant weight. The dried material was ground and thoroughly mixed.

The extraction of plant was done in soxhlet with a wide range of solvents like cyclohexane, toluene carbon tetrachloride, hexane, etc. It was found that toluene gave maximum percentage extraction on dry weight basis. The results of solvent extraction are presented in Table-1.

The efficiency in extraction versus modes of extraction can be seen in Table-II.

A study was made to find out a better solvent extraction technique of the whole of calotropis Procera with either single or multiple solvents. The results are presented in Table-III. The Biomass taken initially was either as received or room dried to constant weight (about 30 days).

Results and discussion

Percentage extraction of dried Calotropis Procera in Soxhlet Extractor for a period of 12 hrs was found to be highest (7.12) with toluene. This was followed by Heptane,

Benzene, Hexane, Cyclohexane and carbon tetrachloride respectively (Table-1).

Soxhlet extractor has the advantage of recycling fresh solvent every time, whereas in-phase extraction leads to overheating of material and loss of solvent while recovering the material. A comparative study has been presented in Table-II.

Since drying of the plant material was an additional operation and since toluene had the property of entraining water vapour in the vapour phase which could be separated out as an immiscible layer in the liquid phase, it was of interest to conduct extraction with toluene in the as received state of the plant material. This would eliminate the drying procedure. It was expected that percentage moisture would tally with the moisture determinations conducted earlier ~~with~~ the plant on wet weight basis.

When the toluene and water layer were separated in the continuous extractor (c, Fig.1) the moisture determined was found to tally. Contrary to this, Soxhlet extraction of plant material on fresh weight basis was found to be higher. This could be explained on the theory that the water extracted by toluene was being recycled through the plant bed thereby also resulting in the extraction of water soluble components in the plant material. This was checked by conducting a simple experiment. The water layer from C of the continuous extractor was evaporated to dryness and the residue weighed. As expected it was found to tally with the difference in extraction value of 1 & 3 (in Table-III).

In conclusion a stage has been reached where efficient extraction of hydrocarbon portions of Biomass could be repeatedly obtained. Further studies on ascertaining the hydrocarbon chain size by GCMS and disproportionation reaction in the presence of hydrogen donor solvents are being conducted.

From the GCMS of the extract, it was found that the breaking pattern tallied with that of straight chain hydrocarbons giving differences of -14- mainly. Further experimentation is in progress to prove the assumptions.

REFERENCES

1. Nemethy E.K., Biomass As a source of liquid fuel. Submitted to CRC Critical Review in Plant Biology University of California. Oct, 1983.

2. Calvin, M. Fuels Oils from Euphorbs and other Higher Plants. Presented at the international Symposium on Chemistry, Taxonomy, and Economic Botony of Euphorbiaceae, Royal Botonic Gardens, Kew, England, April 2-4, 1986.

3. Calvin, M. Oil From Plants presented at BARC science seminar, Beltsville Agricultural Research Centre, U.S. Department of Agriculture, Beltsville, MD, September 8, 1982.

4. Nemethy, E.K., J.W. Otvos and M. Calvin. Analysis of Extractable from one Enphorbia. J. Amer, Oil chem. Soc. (1979) 56, 957.

5. Janardanarao, M. Liquid Fuels from Biomass Thermal Coversion Process. Urja (1981) 9(3) 185-87.

6. Calvin, M. New Sources for fuel and Materials Science, 1983: 219, 24.

7. Nemethy E.K. & M. Calvin Terpenes from Pittosporaceac, Phyto Chem (1982) 21, 2987.

TABLE I

Solvent extraction with Dried plant in Soxhlet Extractor

S.No.	Name of Solvent	Time of Extraction	Percentage of Extraction %
1.	Heptane	12 hrs	4.6 %
2.	Hexane	12 hrs	3.68%
3.	Carbon tetrachloride	12 hrs	3.12%
4.	Cyclohexane	12 hrs	3.49%
5.	Benzene	12 hrs	3.89%
6.	Toluene	12 hrs	7.12%

TABLE II

Modes of Solvent extraction of dry plant with Toluene

S.No.	Mode of Extraction	Time of Extraction	% of Extraction
1.	Inphase Extraction	12 hrs	2.85 %
2.	Soxhlet Extraction	12 hrs	7.12 %

Table-3

S.No.	Biomass Basis	Modes of Extraction	Solvent	Time 1 hrs)	Temp.	% Extraction
1.	Fresh plant	Soxhlet	1) Toluene	12	110°C	88.75
2.1	Fresh plant (m)	Soxhlet	1) Toluene	12	110°C	87.2
2.2	Room Dry (m)	Soxhlet	Toluene	12	110°C	15.18
2.3	Residue of 2.2		2) Acetone	12	68°C	6.72
3.1	Fresh plant	Continuous Biomass Extractor (Solvent Manually recycled after separating the moisture)	1) Toluene	12	110°C	87.3%
3.2	Residue of 3.1		2) Acetone	12	68°C	6.6

EVALUATION OF RESEARCH IN PLANT BIOMASS PRODUCTION FOR LIQUID FUEL CONVERSION: THE CASE OF INDIA, BRAZIL AND JAPAN

Sandra M. Thomas
Science Policy Research Unit, University of Sussex, Falmer
Brighton BN1 9RF, East Sussex, UK

Abstract

The aims of this study were to identify research activities in the field of plant biomass production for liquid fuel conversion and to evaluate research performance in selected groups.

Results are presented for three countries: Japan, India and Brazil. Research groups were identified from a range of information sources. Data were collected by interview and related to funding, information access, staffing, publication policy and degree of awareness of other research groups in the field. Bibliometric analysis and peer review were used as indicators in an attempt to assess research output. The findings are discussed in relation to agro-industrial policy in Japan, the use of marginal land in India and the Proalcohol programme in Brazil.

1. INTRODUCTION

Most countries were adversely affected by the oil crises of 1973 and 1979. In response to the resulting rises in oil prices, many governments launched R&D programmes on alternative, renewable energy. The use of plant biomass for conversion to liquid fuels which could substitute for transport fuels aroused wide interest in several countries in the mid and late 1970's. Biomass can be defined as all matter that is derived directly or indirectly from photosynthesis. The essential resource is carbon which is fixed on a renewable basis via the photosynthetic process. The resource is vast - some 170 billion tons of dry biomass are produced worldwide each year, while the current global consumtion by mankind is only 1.3 billion tons.[1] The production of fuel ethanol from biomass represented an attractive option to those countries which had a surplus of sugar and substantial oil imports. Ethanol is easily produced by well-known fermentation processes and up to 20% can be added to petrol without any significant adverse effects on existing engines.

Investigations of the potential of plant biomass to produce liquid fuels have also focused on the production of diesel substitutes. Ethanol does not perform well in diesel engines and other substitutes have therefore been sought. Many studies have demonstrated the efficiency of plant oils as fuels in diesel engines, particularly when trans-esterified.

Brazil is the only country with a well developed national fuel ethanol programme although several other countries, namely the United States, Argentina, Zimbabwe, Mali and Paraguay, produce fuel ethanol for petrol blends which meet at least some of the national transport fuel requirements. With the drop in the real price of oil which began on the international markets in 1982, many plant biomass research projects were terminated or greatly scaled down.[2] Most R&D in the field of plant biomass production has now been completed or discontinued in countries which do not currently produce fuel ethanol. A notable exception is the European Community (EC) which still funds significant numbers of plant biomass projects.[3] This is largely a consequence of the over-production of food and the need to develop alternative crops.[4]

This research project, commissioned by Shell Research Ltd. UK, had two specific aims: to identify research activities in the field of plant biomass production for liquid fuel conversion in areas outside of the United States (USA) and the EC and secondly, to evaluate research performance in selected groups. The rationale for excluding the United States and the EC concerns the way in which scientific information is disseminated in different parts of the world. In general scientific publications from the West, namely North America and Western Europe, are well-represented in published abstracts and on-line databases.[5] This is in contrast to the situation for newly-industrialized countries (NICs) and developing countries (DCs). The project aimed, therefore, to explore methods of gaining access to the output of research groups. The fast-moving nature of scientific research makes it essential for researchers to keep abreast of developments if duplication of research is to be avoided. Once research activities had been identified, the information was synthesized to provide a comprehensive overview of research activity in the field.

Regarding the second aim, several methods of assessing research have been developed in recent years, including the adoption of bibliometric indicators - i.e. indicators derived from the analysis of publications and citations - which have been successfully used in conjunction with other approaches such as peer review. These methods of evaluation have been applied to matched research groups working in industrialized countries, particularly those engaged in 'big science'.[6][7][8] The project has attempted to explore the application of such methods to a heterogeneous research field which is spread across developed, developing and newly-industrialized countries.

The focus of this research was on plant biomass production rather than conversion and all types of plant material were considered although the coverage of woody biomass was restricted in view of the extensive research on fuelwoods. With the exception of methanol, research on most types of biomass-derived liquid fuels was included. Nine countries were selected for detailed study and results for three - Japan, India and Brazil - are presented here.

2. METHODOLOGY

Information Sources

Five methods were employed to locate relevant information concerning research on plant biomass production: searching of computerized databases; consultation of energy and biomass directories; use of national and international networks (i.e. national and international societies and associations); manual searching of the literature; and personal contacts. In general, databases were found to be heavily dominated by American and European research literature. Papers from DCs and NICs, in contrast, were markedly under-represented and led to very few contacts being established. The other four methods, particularly the use of directories, proved to be more successful and provided a relatively efficient means of making contact with research institutes and government departments concerned with plant biomass R&D.

After initial contact had been made, research groups were interviewed by means of a structured questionaire which sought information on the following areas: academic and employment experience; project objectives; project funding; information access; publication policy; and awareness of international research groups. The aim of the interviews was to gather information to evaluate the project and to assess research group performance. Interviewees were also asked to undertake a peer review exercise, ranking the performance of various biomass groups around the world. Ten projects were assessed in Japan, 19 in India and 10 in Brazil (Tables I-IV). The questionaire data was transformed into 104 variables and computerized using SPSS PC (Statistical Package for the Social Sciences, SPSS inc., Chicago, USA) software and currency data was converted into 1985 constant US dollars using GDP deflators supplied by the International Monetary Fund (IMF).[10] Data for 'total funding' was divided into three categories for ease of presentation (Table V).

Choice of Indicators

Two classes of output indicators were used in this study: bibliometric indicators and peer review. In the broad area of research being studied, the scientists do not form a homogeneous research community where, for example, members tend to publish in the same journals, attend the same conferences and exhibit broad familiarity with each other's work. Consequently publication patterns at the outset of the project were largely unknown. Bibliometric indicators frequently used in developed countries cover only the formal research literature, omitting publications in the form of reports and popular articles which could be significant in DCs and NICs. Detailed publication data were collected during interview to maximize the options available for data analysis. Researchers were asked to classify all publications arising from the project under consideration into journal papers, conference papers, technical reports, book chapters, and popular articles.

The peer review component of the study involved asking researchers to rank eight centres for plant biomass production drawn from seven countries. The list comprised internationally known centres such as the Forest Research Institute in New Zealand as well as lesser known agricultural centres such as Kasetsart University in Thailand. Researchers were asked to assess the quality of the research on a scale of one to four and to comment on its relevance to industry. Peer review proved to be unsuccessful as researchers generally felt that they were insufficiently familiar with the centres to comment. A second approach was then explored in two countries,

India and Brazil. Researchers were asked to make the same assessment on a list of research institutes in their own countries.

3. RESULTS

Research relating to plant biomass production for energy in each of the three countries during the period 1980-89 is summarized in Tables II-IV amd discussed in Section 4. Results from the bibliometric analysis and peer review are presented below.

Bibliometric indicators

Of the 194 publications produced from a total of 43 projects drawn from the three countries, only 30 (i.e. 16%) were scientific papers published in ISI-recognized[1] journals (Table V). This low total precluded the use of citation analysis to provide citation counts as an output indicator. Over twice this number, 67 (i.e. 35%), were published in non-ISI-recognized journals. The distribution of papers published in ISI-recognized-journals and non-ISI-recognized journals revealed considerable variation between countries. In Japan, papers were published only in non-ISI journals and the output was relatively low at 1.1 papers/project. By contrast, in India 1.2 papers/project were published in ISI and 2.2 non-ISI journals respectively. The publication pattern in Brazil was intermediate between these countries with 0.4 papers and 1.6 papers/project in ISI and non-ISI journals respectively.

A total of 86 (i.e. 44%) papers were published in the form of conference proceedings, slightly less than the total number of scientific papers (i.e. 97 or 50%). In Japan and Brazil, the number of papers published in conference proceedings as compared to the number of papers published in non-ISI journals was close while in India, 14% more conference papers were published. The percentage of publications from other categories was small.

These results clearly demonstrate the tendency of Japanese and Brazilian researchers in this field not to publish in mainstream journals but rather to publish in journals of more restricted circulation. Publication output was generally low. In India, on the other hand, researchers publish more frequently in journals of a wider circulation and are more prolific in their publication output. These results question whether "success" in terms of yield of publication is associated with particular factors such as those relating to resources and the environment in which the project was carried out.

The next task therefore was to investigate the relationship between the number of publications produced and possible factors affecting research performance. As the total number of journal papers

---

[1] ISI is the Institute for Scientific Information, Philadelphia, USA. The ISI is responsible for scanning 3500 of the world's leading scientific journals (ISI-recognized-journals) and publishes the **Science Citation Index** which contains citation data for papers in ISI-recognized-journals.

produced in these three countries was so low, publication data for all nine countries was merged for analysis. Projects were sorted into countries and divided into those with or without publications (projects which had been in progress for one year or less were excluded). Association between propensity to publish and 19 selected variables was tested using Chi-square analysis.

There was a significant association (P <0.05) between propensity to publish and the overseas experience of the researchers involved during the current post. Of cases having overseas experience, 84% produced a publication compared to 52% in cases which had no overseas experience. These results suggest that overseas experience is likely to stimulate the publication of research results.

A significant association was also obtained (P <0.05) between propensity to publish and total funding. For example, in the highest of the three categories of funding, 88% of cases published a paper compared to 40% in the lowest category. This may indicate that the most productive researchers are more able to attract substantial research funds. The third and final significant association involved the variable 'attendance at conferences'. Of those cases where researchers regularly attended conferences, 86% had produced a publication in comparison to 34% of those who did not regularly attend. Researchers taking part in conferences are more likely to be informed of developments in their field and this may stimulate the publication of research work.

There was no association to publish and the following variables:

type of institution
type of post
duration of project
source of funding
number of researchers involved in the project
number of technicians involved in the project
collaboration with industry
collaboration with other institutes
collaboration with overseas organizations
use of networks
perceived lack of equipment/spare parts/servicing, funding, expertise, information

It must be emphasized that these results apply to the whole range of projects drawn from nine countries.

Peer Review

As previously mentioned, peer review was unsuccessful with less than 1% of cases able to provide any useful comment. Scientists generally felt insufficiently familiar with the named institute to comment on the quality of the research. The numbers of cases providing information on peer review of Indian and Brazilian institutes was too low to permit statistical analysis.

Factors Affecting Research Performance

Selected information relating to assessment of research performance is presented in Table VI for the following factors: collaboration with industry; collaboration with overseas organizations; use of networks (i.e. national and international scientific organizations); and total funding (i.e. funds provided for the project, but excluding salaries provided by the institution).

In Japan, the data suggest that the researchers work in relative isolation with limited funds. Only one project involved collaboration with industry and none with overseas organizations. Use of networks was very infrequent. The majority of projects were funded by $US 30,000 or less, only one being supported by funds in the middle category ($US 30,000 - 200,000).

In contrast, researchers from India appear to be better integrated within the national and international research communities. Nearly a third of projects involved collaboration with overseas organizations and over half of those interviewed made use of networks. With regard to funding, projects mainly occurred in the two smaller categories, although three out of 22 received very substantial amounts (i.e. over $US 200,000). Sponsorship in these cases was mainly by overseas agencies such as the EC and the United States Agency for International Development (USAID).

At the national level, Brazilian scientists working on plant biomass production appear to belong to a more integrated scientific community. For example, collaboration with industry was a feature of over half of the projects surveyed and use of national networks was made in nearly all cases. However, collaboration with overseas organizations and use of international networks was absent in all the projects surveyed. Data relating to funding was difficult to obtain but at least two projects were each supported by funds in excess of $US 200,000 (Table VI).

4. DISCUSSION

Of all the industrially advanced countries, Japan is the most dependent on crude oil and since most of the nation's oil supplies are largely imported, its energy supply is potentially precarious. In 1980, Japan launched a number of major research initiatives aimed at increasing her self-sufficiency in energy. The New Energy Development Organization (NEDO) was established to act as a coordinating organization for the promotion of R&D on oil-alternative technologies. Plant biomass conversion, which occupies a relatively small part of the New Energy Programme, is regarded by NEDO as a promising source of ethanol which could be substituted directly for petrol.

The Special Project on Energy, carried out in Japanese universities and funded by the Ministry of Education, Science and Culture, was completed in 1987. Here, research on plant biomass has focused on the use of high energy crops and plants and the conversion of forest waste. The list of plant species given in Table II is dominated by a broad selection of non-woody crop species which reflects the exceptionally wide range of agroclimatic conditions which exist in Japan.

Regarding agronomic research into plant biomass, the view is that much of the basic research has now been done and its development awaits an appropriate economic climate. Sweet sorghum and Napier grass in particular showed promise as potentially high-yielding sources of plant biomass which could be converted to liquid fuels.[10] However, interpretation of yield data should be approached with caution as most of the agronomic trials described here were on a relatively small scale. In contrast, R&D on the conversion of plant biomass is continuing for a number of reasons. Not only does fermentation technology play an essential role in biotechnological processes, but advances in genetic manipulation of micro-organisms are also continually improving the efficiency and hence the economic viability of plant biomass conversion into a variety of chemicals and materials. Moreover, as with all the alternative energy projects under development at NEDO, Japan intends to export those which prove economically viable and her highly regarded fermentation technology has already established several overseas markets.

Agricultural land is very limited in Japan and much is already devoted to rice production. This scarcity of land raises questions about the likely implementation of plant biomass production in the future. However, it seems probable that in a climate of reduced oil usage, Japan would not only seek to provide some ethanol feedstocks herself, but would also import significant quantities from neighbouring countries.

In the plant biomass production field, researchers from Japan appeared to work in relative isolation. Very little use was made of networks and a lack of collaboration was evident with industry and overseas organizations. Publication output was low with papers being published mainly in Japanese non-ISI-recognized journals. Researchers, working mainly in the universities, appeared relatively underfunded.

In India, the oil crises of the 1970's did not produce a major R&D initiative in alternative liquid fuels such as those that were launched in Japan, Australia and New Zealand. Rather, the field of plant biomass research has involved diverse organizations with varying aims. The principal reason for this difference lies with the complex energy requirements of a large and predominantly rural developing country. One of the most pressing needs in terms of energy is adequate provision of firewood or alternative fuels for India's largely rural population of 685 million.

In an attempt to increase production of food and fuel, India is putting growing emphasis on the utililization of marginal land. Approximately 93 million hectares of land are unsuitable for agriculture and development depends to a large extent on the selection of tolerant, multi-purpose plant species which can not only survive the often harsh environmental conditions but also facilitate soil improvement.[11]

Much of the research on oil-alternative energy is funded by the Department of Non-Conventional Energy Sources (DNES) at the Department of Energy. The DNES was created in 1983 to develop non-conventional and renewable energy sources to meet the growing energy needs of the country. This is achieved mainly through the agency of public and private research institutes and universities which receive research sponsorship from the DNES for selected projects.[12] Although the majority of the projects are concerned with the development of fuelwoods for domestic consumption, there are a number investigating latex-containing species or 'petrocrops', algal hydrocarbons and plant oils as potential diesel substitutes (Table III). Pilot plants for the production of alcohol from lignocellulosic residues, cassava flour and sweet sorghum have also been established with DNES funding.

It seems unlikely however that current R&D in the field of starchy feedstocks for alcohol conversion will expand in the climate of present oil prices. In India, alcohol is produced almost entirely from molasses of which 2,4000,000 tonnes is produced anually. Many of the fuelwood provenance trials should be completed in the next five to ten years, culminating in the identification of the best performing accessions. The intention is that 'bulked up' seed from elite trees will be distributed by agencies such as the Forest Service and planted on marginal land. Research is likely to continue in the fields of genetic improvement, micropropagation, mycorrhiza and nitrogen fixation. The application of micropropagation for some woody species may become established practice but the field of plant tissue culture is likely to expand regardless since the techniques are an essential prelude to plant transformation, a field in which India has, as yet, little experience.

Although the number of scientific papers published in ISI-journals is low compared to many countries in the West, the Indian research community appears to be better integrated with the international community than Japanese researchers in the same field. Collaboration with overseas organizations and use of networks were evident in several research groups. Research funding was also distributed more widely spread with emphasis being placed on a greater application of resources in the middle range.

The Brazilian National Alcohol Programme is a unique example of a nationally coordinated scheme for the production of liquid fuels from plant biomass. The alcohol programme, like several others, originated in response to the leap in oil prices of 1973. By 1975, Brazil had adopted a new energy strategy, one element of which was the replacement of petrol with ethanol produced from sugar cane. The implementation of this energy policy has resulted in the production of over 15 billion litres of ethanol from sugar cane in 1988, supplying 3.5 million cars with pure alcohol and eight million with a 78/22 petrol/alcohol blend.[13]

Current conditions have nevertheless made the functioning of the programme more difficult. The Instituto do Açucar e do Alcool (1AA) is charged with regulating the sale of sugar either for alcohol or on the sugar markets. In 1989, sugar producers in the northern regions took matters into their own hands and sold their sugar, earmarked for alcohol production, as sugar on the international markets, where it fetched higher prices. This turn of events resulted in a shortfall of alcohol supplies by late spring, precipitating an early sugar cane harvest in June.

Since its inception, the alcohol programme has used sugar cane as an ethanol feedstock to the exclusion of virtually all other sources of plant biomass. However, several other plant materials have been under consideration as potential ethanol or diesel substitutes over the past 15 years. The failure of cassava to provide an alternative ethanol feedstock was caused largely by poor agronomic performance and disease problems. Sweet sorghum has been investigated as a complementary crop to sugar cane and a number of projects have considered the use of vegetable oils as diesel substitutes. It now seems highly improbable that any alternative or complementary alcohol feedstocks will be used in Brazil for some years to come. Although the need for a diesel substitute is more acute, potential substitutes such as palm oil or other vegetable oils are too expensive to produce given current crude mineral oil prices.

Publication in ISI-journals was very limited in this field of research; scientists tended to publish in Brazil in Portuguese-language journals, leading to a restricted circulation of their work. Overseas collaboration was also lacking although use of Brazilian networks was very frequent. Because of the Proalcohol programme, plant biomass research in Brazil generally exhibited closer collaboration with industry when compared to Japan and India. Industrial sponsorship of sugar cane biomass production was also very substantial in a small number of projects.

In conclusion, it should be emphasized that plant biomass research for energy is an applied field of scientific research. Consequently, there is less emphasis on the publication of scientific papers than in similar fields of more basic research, leading in turn to reduced coverage in most databases. Furthermore the field is not characterized by an internationally homogeneous research community where scientists publish in the same journals, attend the same conferences and exhibit a reasonable familiarity with most other work in the area. Instead, it consists of a regionally fragmented research community made up of largely non-interacting research groups. Even within a single country like India, research groups do not appear to form a very cohesive community. One of the principal causes of this heterogeneity lies with the multidisciplinary nature of the field. This is particularly evident in the division of scientists into 'forestry' and non-forestry' groupings.

The use of a peripatetic expert to evaluate research groups had several advantages over the two other methods (ie. bibliometric analysis and peer review) and proved to be a reasonably successful means of assessment. It was possible to distinguish between groups undertaking relatively mundane work and those engaged in more innovative research, and to obtain considerable insight into the quality of the research. However, it must be emphasized that such assessments are hard to justify objectively.

The implication of these findings is clear: namely, that the use of the scientific literature and databases as a means of keeping abreast of developments in certain research fields is inadequate. This point is of particular relevance to research in developing, newly-industrialized and non-English language-speaking countries. Although database coverage of publications from these areas is likely to improve, other methods of locating information such as the monitoring of key conference proceedings, the maintenance of ongoing personal contacts with leading institutions in selected countries and the use of a peripatetic expert, need to be employed. The reappraisal of biomass-derived liquid fuels as substitutes for fossil fuels in the context of current environmental issues such as global warming is already under way. The need for effective monitoring and evaluation of key research developments in the field is therefore likely to assume greater importance.

Acknowledgements

Sponsorship for this research project by Shell Research Ltd. is gratefully acknowledged. I am also grateful to the Economic and Social Research Council, UK (the Science Policy and Research Evaluation Programme) for funds to assist with preparation of this manuscript and to Ben Martin for his valuable comments.

REFERENCES

1. International Energy Authority (IEA). 1987. Renewable Sources of Energy. IEA, Organization for Economic Co-operation and Development (OECD), Paris, France.

2. Thomas, S.M. 1989. Evaluation of Research in Plant Biomass Research for Liquid Fuels. Vol.1. Project Report, Science Policy Research Unit, University of Sussex, UK.

3. Grassi, G. Opening Speech. In the Proc. of the 5th European Biomass Conference: Biomass for Energy and Industry, 9-13 October, Lisbon, Portugal. Elsevier Applied Science Publishers, London, UK, (in press).

4. Rexen, F.P. Agricultural Raw Materials for the Chemical Industry. In Proc. 5th European Biomass Conference: Biomass for Energy and Industry, 9-13 October, Lisbon, Portugal. Elsevier Applied Science Publishers, London, UK, (in press).

5. Morita-Lou, H. (ed.) 1985. Science and Technology Indicators for Development. United Nations Science and Technology for Development Series, Westview Press, Boulder and London, p.13.

6. Irvine, J & Martin B.R. 1983. Social Studies of Science, 13, p.49.

7. Irvine, J & Martin B.R. 1983. Research Policy, 12, p.61.

8. Irvine, J & Martin B.R. 1984. Research Policy, 13, p.247

9. IMF. 1988. International Financial Statistics, XLII, (4). IMF, Washington, USA.

10. Ministry of Education, Science and Culture. 1987. Research on Energy from Biomass. Reports of Special Project Research on Energy under Grant-in-Aid of Scientific Research of the Ministry

of Education, Science and Culture, Japan. Ministry of Education, Science and Culture, Tokyo, Japan.

11. Vimal, O.P. & Tyagi, P.D. 1985. In Proc. Nat. Workshop on Utilization of Wastelands for Bio-Energy, held in Pune, April 27-29, 1985, p.94.

12. DNES. 1988. Ann Rep., 1987-88. DNES, Ministry of Energy, New Delhi, India.

13. Rosillo-Calle, F. & Hall, D. 1988. New Scientist, 19 May 1988.

TABLE 1: DETAILS OF ASSESSED PROJECTS

	COUNTRY			
	BRAZIL	INDIA	JAPAN	TOTAL
Number of Projects Surveyed				
Total	10	22	11	43
Number of Researchers Interviewed				
Total	14	30	13	57
Type of Project				
Production	8	22	10	40
Conversion	1	0	0	1
Both	1	0	1	2
Plant Material Studied				
Woody	2	13	2	17
Non-Woody	8	9	9	26
Type of Plant Studies				
Crop	8	3	9	20
Wild	1	3	1	5
Semi-Wild	1	6	1	8

TABLE II: SUMMARY OF RESEARCH PROJECTS RELATED TO PLANT BIOMASS PRODUCTION FOR ENERGY IN JAPAN DURING THE PERIOD 1980-89.

Plant material	Researchers/ Coordinators	Institute
Sorghum bicolor var.saccharum[1]	K. Hoshikawa	Tohoku University
Ipomea batatus[1,3]	K. Miyazato et al.	University of the Ryukyus
Manihot esculenta[1,3]	S. Murayama et al.	University of the Ryukyus
Helianthus anuus[1,3]	A. Kimura et al.	University of Tokyo
Saccharum officinale[1,3]	A. Nose et al.	University of the Ryukyus
Solanum tuberosum[1] Beta vulgaris	A. Tanaka et al.	Hokkaido University
Canna edulis[3]	K. Imai et al.	University of Tsukuba
Pennisetum purpureum[1,3]	T. Takeda et al.	Kyushu University
Eucalyptus citriodora[1] E.radiata, Mentha Copaifera	H. Nishimura et al.	Hokkaido University
Botyococcus braunii	O. Nakayama	Yamanshi University
Metaxylon, Saccharum officinale[3]	Y. Takamura	University of Kyoto
Nypa fruticans[2]	F. Yanagida et al.	Tokyo University of Agriculture
Populus italica[3]	S. Sasaki	University of Tokyo
Acacia mearnsii[3]	S. Asanuma	IFFP

[1] denotes funding from Ministry of Education, Science and Culture. Sub-project: Studies on High Energy Crops and Plants in Regard to Their Screening, Cultivation and Control as Energy Resources.

[2] denotes funding from Ministry of Science, Education and Culture. Sub project: Studies on Conversion of Forest Products to Energetically Effective Substances.

[3] denotes project assessed.

TABLE III: SUMMARY OF RESEARCH PROJECTS RELATED TO PLANT BIOMASS PRODUCTION FOR ENERGY IN INDIA DURING THE PERIOD 1980-89.

Plant material	Reseachers/ Coordinators	Institute
Beta vulgaris[1 3]	V.K. Garg et al.	NBRI[2], Lucknow
Bambusa[3] Dendrocalamus strictus Melocana bambusoides	U. Rao	University of Delhi
Saccharum[1 3] munja	P. Vasudevan G.S. Gujral	IIT, New Delhi
Jatropha curcas[1 3]	V. Deshpande	BAIF, Pune
Jatropha curcas[1 3]	A. Kumar, S. Roy	University of Rajasthan
Euphorbiaceae[1 3]	S.Roy et al.	University of Rajasthan
Fuelwood species[1 3]	A. Mathur, I. Hussein	Agricultural University of Rajasthan[2]
Euphorbiaceae[1 3]	G.S. Srivastra S.S. Baghel K.M. Ratti	NBRI[2], Lucknow
Parthenium argentatum[1]	G.S. Srivastra & P. Narain	NBRI[3], Lucknow
Parthenium argentatum E.antisyphyllitica		CAZRI, Jodhpur
Euphorbiaceae[1 3]	A. Mathur, I. Hussein	Agricultural University of Rajasthan[2]
Eucalyptus[2]	M. Pal	FRIC, Dehra Dun
Euphorbiaceae[1 3]	A.K.Bhatnagar	University of Delhi
Sesbania speciosa[1 3]	N. Seth	University of Delhi
Leucaena[1 3] leucocephala	S. Jain	University of Delhi

Botryococcus braunii[1] B. protruberance		Banaras Hindu University, Varansi
Madhuca indica M.latifolia Bassia latifolia B. longifolia	P.K. Agrawal	National Sugar Institute, Kampur
Eucalyptus[1 3] Salvadora persica	A. Mascarenhas	NCL, Pune
Fuelwood species[1 3]	P.V. Sane et al.	NBRI,[3] Lucknow
Eucalyptus species[3] and hybrids	R. Swarup M.L. Kapoor	FRIC, Dehra Dun
N-fixing fuelwood spp.[1 3]	V. Dhawan et al.	TERI, New Delhi
N-fixing fuelwood species [1 3]		BAIF, Pune

[1] denotes funding from DNES.
[2] denotes Biomass Research Centre.
[3] denotes project assessed.

Abbreviations:

NBRI	National Botanical Research Institute
IIT	Indian Institute of Technology
BAIF	Bharatiya Agro-Industries Foundation
CAZRI	Central Arid Zone Research Institute
FRIC	Forestry Research Institute and Colleges
NCL	National Chemical Laboratory
TERI	Tata Energy Research Institute

TABLE IV: SUMMARY OF RESEARCH PROJECTS RELATED TO PLANT BIOMASS PRODUCTION FOR ENERGY IN BRAZIL DURING THE PERIOD 1981-89.

Plant material	Reseachers/ Coordinators	Institute
Saccharum officinale[1]	G. Cardoso et al.	CTC
Saccharum officinale[1]	S. Matsuoka et al.	Planalsucar
Saccharum officinale[1]	R. Alvarez	ICA
Saccharum officinale[1]	O. Crocomo et al.	CEBTEC
Manihot esculenta[1]	J.O. Lorenzi et al.	ICA
Sorghum bicolor[1] var. saccharum	F. Giacomini et al.	CNPMS
Obignya		CENARGEN INEB
Eucalyptus[1]	A.N. Goncalves	ESALQ
Bamboo[1]	A. Azzini	ICA

[1] denotes project assessed.

Abbreviations:

CTC	Copersucar Technology Centre
ICA	Instituto Agronômica de Campinas
CNPMS	Centro Nacional de Pesquisa de Milho e Sorgo
CENARGEN	Center for Genetic Resources
INEB	State Institute for Babassu Research
CEBTEC	Centro de Biotecnologia Agrícola
ESALQ	Escola Superior de Agricultura "Luis de Queiroz", University of Sâo Paulo, Piracicaba.

Table V: BIBLIOMETRIC DATA: NUMBERS OF PUBLICATIONS

	COUNTRY			
	BRAZIL	INDIA	JAPAN	TOTAL
Number of projects surveyed Total:	10	22	11	43
Number of articles in ISI-recognized-journals Total:	4	26	0	30
Number of articles in non-ISI-recognized-Journals Total:	16	41	10	67
Number of conference papers published Total:	18	55	13	86
Number of technical reports/ book chapters published Total:	2	9	0	11

TABLE VI: FACTORS AFFECTING RESEARCH PERFORMANCE: INTERVIEW DATA

		Brazil	India	Japan
Collaboration with industry	NO	5	19	10
	YES	5	3	1
Collaboration with overseas organizations	NO	10	17	11
	YES	0	5	0
Use of networks	NO	1	10	10
	YES	8	11	1
Total funding ($US)				
up to 30,000		0	5	9
30,000 - 200,000		0	7	1
over 200,000		2	3	0
Total number of projects		10	22	11

Note: figures refer to number of projects

STUDIES ON SIMULTANEOUS SACCHARIFICATION AND FERMENTATION OF STARCH IN A CONTINUOUS FLOW REACTOR

N. Chithra and A. Baradarajan
Department of Chemical Engineering, Indian Institute of Technology, Madras - 600036, India

Abstract

The direct conversion of starch hydrolysate(15 Dextrose Equivalent) to ethanol in a continuous flow reactor using a coimmobilizate of amyloglucosidase and *Saccharomyces cerevisiae* is reported. The highest ethanol productivity of 27.5 $gL^{-1}h^{-1}$ was obtained at 1500 ml h^{-1} at an initial substrate concentration of 100 g L^{-1}. A model developed for the performance of the continuous flow reactor includes inter and intraparticle mass transfer effects and accounts for the nonideality of the reactor by axial dispersion. It assumes that each step in the reaction follows inhibition limited Michaelis Menten type kinetics. The model matches the experimental data quite satisfactorily.

1. INTRODUCTION

Ethanol from renewable energy resources like starch, cellulose etc. is gaining industrial and technological importance. Extensive research work has been carried out on alcohol fermentation from glucose using immobilized whole cells. There has been an increasing trend on the use of coimmobilized enzyme-cell systems for multistep reactions and the work is mainly focussed on alcohol fermentation from xylose[1], cellobise[2], lactose[3] and whey permeate[4]. Coimmobilization has the additional advantages of higher efficiency and the system attains steady state faster [5] compared to separately immobilized biocatalysts. Since work on direct conversion of starch to ethanol is limited to soluble systems[6,7] only, the present study is focussed on the simultaneous saccharification and fermentation of starch hydrolysate to ethanol using a coimmobilizate of amyloglucosidase and *Saccharomyces cerevisiae* in a continuous flow reactor.

A model which describes the performance of a packed bed reactor for this particular system has been discussed. The model includes all the resistances controlling the rate of the reaction and the additional effect of axial dispersion in the bulk liquid. Each step in the reaction follows Michaelis Menten type kinetics with complexities due to substrate and product inhibition. The method of solution to the problem is based on the explicit finite difference method to both the catalytic bead and the bulk liquid.

2. MATERIALS AND METHODS:

Materials:

The enzyme amyloglucosidase from A.niger (EC 3.1.2.3) with a protein content of 60 mg/g of enzyme. was purchased from M/s Triton Chemicals, Mysore, India. The yeast Saccharomyces cerevisiae NRRLY - 132 was obtained from the culture collections of National Chemical Laboratory (3050), Pune, India. The culture was maintained on agar slants of glucose medium consisting of 1 % glucose, 0.5 % peptone, 0.3 % yeast extract and 0.3 % malt extract. When required in large quantities,it was grown in a tapered conical reactor of 3 litre capacity in the following medium:
Glucose - 2.0 % (w/v), KH_2PO_4 - 0.1 % (w/v), NaCl - 0.1 % (w/v) $MgSO_4$ - 0.07 % (w/v), $(NH_4)_2SO_4$ - 0.4 % (w/v), Yeast extract - 0.2 % (w/v).

Sodium alginate was purchased from Riedel-De Haenag Seelze - Hannover. Starch hydrolysate (15 Dextrose Equivalent) was purchased from Loba. All other materials used were Analar grade.

Analytical Methods:

Starch was estimated by the anthrone method[8]
Reducing sugars were estimated by the Dinitrosalicylic acid method [9]
Glucose was estimated by Glucose oxidase method[10]
Ethanol was estimated by the gas chromatographic method using the activated chromosorb 101 (80/100 mesh) column with Flame Ionization detector[11].

Coimmobilization procedure

The commercial enzyme was dissolved in 0.05 M sodium acetate buffer. The inerts were removed by centrifugation. The clear enzyme solution was used for immobilization. The stationery phase cells grown in the tapered reactor were harvested by centrifugation. First the enzyme was covalently attached to calcium alginate matrix. Then the cells were added to the enzyme immobilized alginate and extruded in calcium chloride solution as described elsewhere[12]

Continuous flow reactor :

The flow reactor made of glass had an inner diameter of 6 cm and a height of 80 cm. The bottom of the reactor was packed with glass beads of same diameter as that of the coimmobilized beads upto a height of 5 cm to eliminate the entrance effects. The column was then packed with coimmobilized beads of 0.25 cm diameter to a height of 65 cm, with extreme care to prevent entrapment of air bubbles. Stainless steel sieve plates were used to separate the glass and coimmobilized beads at the bottom of the reactor and to stop any possible entrainment of coimmobilized beads at the top of the reactor at flow rates under study. Sample outlets were provided at different heights and an outlet for carbon dioxide was provided at the top of the reactor. The porosity of the bed was estimated as 0.49. The substrate solution at a selected concentration was pumped using a Watson - Marlow peristaltic pump at a selected flow rate. The pH of the medium was maintained at 5.0 and temperature at 30°C. Samples were taken and analysed at frequent intervals until steady state was reached. The experiments were repeated for different flow rates, bed heights and initial substratre concentration. The range of variables studied are given in table I.

3. RESULTS AND DISCUSSIONS:

The operating temperature was chosen as 30°C since ethanol tolerance of the yeast is higher at lower temperatures[13]. The flow rate and the height of catalyst bed influence the extent of conversion in a flow reactor.

Effect of bed height:

The effect of bed height on the conversion of starch hydrolysate was studied by conducting the experiments at different bed heights of 25 cm, 35 cm, 45 cm, 55 cm and 65 cm. The conversion of starch hydrolysate to ethanol increased with increase in bed height. At the same residence time, the linear velocity will be greater in the deeper bed. In the presence of external mass transfer resistances, the extent of conversion will be higher for deeper bed as shown in figure 1.

Effect of flow rate:

Figure 2 shows the effect of flow rate on ethanol yield The yield is maximum at lower flow rates and maximum for all the flow rates at a substrate concentration of 20 g L^{-1}. At flow rates of 100 ml h^{-1} and 500 ml h^{-1}, upto an initial substrate concentration of 100 g L^{-1} the yield is maximum and decreases with an increase in initial substrate concentration. This may be due to the inhibitory effect. At higher flow rates the yield decreases and as initial substrate concentration increases the yield is drastically reduced. This may be due to less residence time and the inhibitory effect.

Effect of initial substrate concentration:

Experiments were conducted at different initial substrate concentrations in the range of 20 g L^{-1} to 200 g L^{-1}. Figure 3 shows the results of studies on the effect of initial substrate concentration. The data indicates that the productivity increases with increase in initial substrate concentration upto a flow rate of 1200 ml h^{-1}. Beyond this flow rate, the productivity increases with increase in initial substrate concentration upto 100 g/L^{-1} and it drops with further increase in initial substrate concentration. The highest ethanol productivity of 27.5 g L^{-1} h^{-1} was obtained at 1500 ml h^{-1} giving 34 g L^{-1} of ethanol at an initial substrate concentration of 100 gL^{-1}. The decrease in productivity above an initial substrate concentration of 100 g L^{-1} may be attributed to the inhibitory effects.

4. MATHEMATICAL MODELLING:

The performance characteristics of the packed-bed reactor have been analysed taking into consideration both the inter and intra particle resistances and the additional effect of axial dispersion in the bulk liquid. Each step in the consecutive reaction is assumed to follow Michaelis Menten type kinetics with added complexities due to substrate inhibition for step1 and substrate and product inhibition for step2. The enzyme complex of the cell is assumed to be a single pseudoenzyme and that both enzyme and microorganisms are distributed homogeneously inside the pellet. A general model for a coimmobilized system and the effect of various parameters have already been discussed elsewhere [14] Here the equations pertaining to this system are written as follows.

The analysis is based on the consecutive reaction scheme :

$$\text{Starch hydrolysate} \xrightarrow{\text{step 1}} \text{Glucose} \xrightarrow{\text{step 2}} \text{Ethanol} + CO_2$$

$$S \xrightarrow{\text{step 1}} P_1 \xrightarrow{\text{step 2}} P_2 + P_3$$

taking place inside a spherical pellet of radius R which is permeable to all the species.

Formulation of the equations :

The fluid concentration change along the length of the reactor is characterised by the following differential equation:

$$D_L\ (\delta^2 S/\delta z^2) - u(\delta S/\delta z) = \epsilon(\delta S/\delta t) + k_s a(S-S_s) \qquad \text{---- (1)}$$

$$D_L\ (\delta^2 P_1/\delta z^2) - u(\delta P_1/\delta z) = \epsilon(\delta P_1/\delta t) + k_p a(P_1-P_{1s}) \qquad \text{---- (2)}$$

The differential equation governing the transport by diffusion and disappearance by reaction inside the pellet is given by

$$(D_s/r^2)[\delta/\delta r\{r^2(\delta S_p/\delta r)\}] = [V_{max1}\ S_p/\{K_{m1}+ S_p(1+S_p/K_{I1})\}] + \delta S_p/\delta t \quad ---- (3)$$

$$(D_p/r^2)[\delta/\delta r\{r^2(\delta P_{1p}/\delta r)\}] = [V_{max2}\ P_{1p}/\{K_{m2}+P_{1p}(1+P_{1p}/K_{I2})\}]\{1-P_{2p}/P_m\} - [V_{max1}\ S_p/\{K_{m1}+S_p(1+S_p/K_{I1})\}] + \delta P_{1p}/\delta t \quad ---- (4)$$

The boundary conditions are:

$$\text{At } t \geq 0,\ z = 0,\ D_L(\delta S/\delta z) = u(S_{(z=0+)} - S_0);\quad P_1 = 0 \quad --- (5)$$

$$\text{At } t \geq 0,\ z = L,\ \delta S/\delta z = 0;\quad \delta P_1/\delta z = 0 \quad --- (6)$$

$$\text{At } r = 0,\ t \geq 0,\ \delta S_p/\delta r = 0;\quad \delta P_{1p}/\delta r = 0 \quad --- (7)$$

$$\text{At } r = R,\ t \geq 0,\ D_s(\delta S_p/\delta r) = k_s(S-S_s);\quad D_p(\delta P_{1p}/\delta r) = k_p(P_1-P_{1s}) \quad --- (8)$$

Introducing the following dimensionless parameters,

$c = [S/S_0]$; $p_1 = [P_1/S_0]$; $p_2 = [P_2/S_0]$

$c_p = [S_p/S_0]$; $p_{1p} = [P_{1p}/S_0]$; $p_{2p} = [P_{2p}/S_0]$

$x = [z/L]$; $y = [r/R]$; $q = [D_p/D_s]$

$$\phi_1^2 = [V_{max1} R^2/K_{m1} D_s] \; ; \qquad \phi_2^2 = [V_{max2} R^2/K_{m2} D_s]$$

$$\Gamma_1 = [S_0/K_{m1}] \; ; \quad \Gamma_2 = [S_0/K_{m2}]$$

$$\Gamma_{I1} = [S_0/K_{I1}] \; ; \quad \Gamma_{I2} = [S_0/K_{I2}] \; ; \; \Gamma_{I3} = [S_0/P_m]$$

$$\tau = [ut/L\epsilon] \; ; \quad N_1 = [R^2 u / D_s\epsilon L]$$

$$\alpha_1 = [k_s a L/u] \; ; \quad \alpha_2 = [k_p a L/u] \; ; \; Pe = [uL/D_L]$$

$$Sh_s = [k_s R/D_s] \; ; \quad Sh_p = [k_p R/D_p]$$

The equations for the bulk liquid become :

$$(1/Pe) \{\delta^2 c/\delta x^2\} - [\delta c/\delta x] - \alpha_1(c-c_s) = \delta c/\delta\tau \quad \text{--- (9)}$$

$$(1/Pe)\{\delta^2 p_1/\delta x^2\} - [\delta p_1/\delta x] - \alpha_2(p_1-p_{1s}) = \delta p_1/\delta\tau \quad \text{--- (10)}$$

with the boundary conditions ,

$$\left.\begin{array}{l} \text{At } x = 0 \text{ , } \tau \geq 0 \text{ , } 1/Pe(\delta c/\delta x) = (c_{0+} - 1) \\ \qquad p_1 = 0 \end{array}\right] \quad \text{--- (11)}$$

$$\left.\begin{array}{l} \text{At } x = 1 \text{ , } \tau \geq 0 \text{ , } [\delta c/\delta x] = 0 \\ \qquad [\delta p_1/\delta x] = 0 \end{array}\right] \quad \text{--- (12)}$$

The particle equations become :

$$1/y^2[\delta/\delta y\{y^2(\delta c_p/\delta y)\}] = [\phi_1^2 c_p/\{1+\Gamma_1 c_p(1+\Gamma_{I1} c_p)\}] + N_1\delta c_p/\delta\tau \quad \text{---- (13)}$$

$$q/y^2[\delta/\delta y\{y^2(\delta p_{1p}/\delta y)\}] = [\phi_2^2 p_{1p}/\{1+\Gamma_2 p_{1p}(1+\Gamma_{I2} p_{1p})\}] [1 - \Gamma_{I3} p_{2p}] - [\phi_1^2 c_p/\{1+\Gamma_1 c_p(1 + \Gamma_{I1} c_p)\}] + N_1\delta p_{1p}/\delta\tau \quad \text{---- (14)}$$

The boundary conditions are :

$$\text{At } y = 0,\ \tau \geq 0,\quad [\delta c_p/\delta y] = 0,\quad [\delta p_{1p}/\delta y] = 0 \qquad \text{--- (15)}$$

$$\text{At } y = 1,\ \tau \geq 0,\quad [\delta c_p/\delta y] = Sh_s(c-c_s),\quad [\delta p_{1p}/\delta y] = Sh_p(p_1-p_{1s}) \qquad \text{--- (16)}$$

Method of Solution :

The differential equations (9) to (16) can be recognized to be belonging to the case of second order parabolic differential equations of the boundary value type coupled with one another. These equations were solved by the explicit finite difference method as described elsewhere[14].

Estimation of physical constants:

Typically the axial dispersion number is found by an appropriate tracer study. However, for the packed bed fermenter sytem almost any substance used as tracer would adsorb into the highly porous calcium alginate beads. This introduces great uncertainty into the tracer study analysis. Hence the dispersion number was found using the charts[15].

External mass transfer was determined by the correlations proposed by Wilson and Geankoplis[16] which covers the range of conditions normally encountered with immobilized enzyme systems.

Effective diffusivity was determined experimentally.

Comparison of model to experiment :

Figures 4 to 7 show the comparison of the predicted and experimental dimensionless concentration of starch hydrolysate, glucose and ethanol. The proposed model predicts the concentrations quite closely. The set of parameter values based on unit reactor volume, which match the experimentally observed concentrations are :

V_{max1} = 471.86 μg sec^{-1} cm^{-3}

V_{max2} = 263.87 μg sec^{-1} cm^{-3}

K_{m1} = 5.0 mg ml^{-1}

K_{m2} = 11.0 mg ml^{-1}

K_{I1} = 260 mg ml^{-1}

$K_{I2} = 310 \text{ mg ml}^{-1}$

$P_m = 145 \text{ mg ml}^{-1}$

5. CONCLUSIONS:

In conclusion the productivity increased with increase in flow rate upto 1200ml h^{-1} above which it decreased, for initial substrate concentrations greater than 100 g L^{-1}. The highest ethanol productivity of 27.5 g $L^{-1}h^{-1}$ was obtained at 1500 ml h^{-1}, giving 34 g L^{-1} of ethanol at an initial substrate concentration of 100 g L^{-1}. At lower flow rates the yield was 44 %.

The model developed for the coimmobilized bio reactor matches the experimental data satisfactorily. However, slight deviations between the model and the experiment towards the end of fermentation is noticed. This may be attributed to the fact that cabohydrates are also utilised for the maintenance of the cells[17,18]. The model assumes that the entire carbohydrates are available for the reaction and hence predicts a higher value than what is realised by the experiment.

6. NOMENCLATURE:

a - External surface area of the catalyst per unit volume of the reactor, ($cm^2\ cm^{-3}$)

c, c_p - Dimensionless substrate concentration in the bulk liquid, (S/S_0) and inside the particle, (S_p/S_0) respectively

c_s - Dimensionless substrate concentration at the surface of the particle, (S_s/S_0)

D_L - Dispersion coefficient, $cm^2\ sec^{-1}$

D_s, D_p - Diffusion coefficient of the substrate S and intermediate P_1 inside the particle respectively, $cm^2\ sec^{-1}$

K_{m1}, K_{m2} - Michaelis constants for the first and second reaction respectively, mg ml^{-1}

K_{I1}, K_{I2} - Substrate inhibition parameters for first and second reaction respectively, mg ml^{-1}

k_s, k_p - External mass transfer coefficients for the transport of substrate S and intermediate P_1 respectively, cm sec^{-1}

L - Length of the reactor, cm

N_1 - Dimensionless parameter, $R^2u/D_sL\epsilon$

Pe	- Peclet number, uL/D_L
P_m	- Product inhibition parameter for the second reaction, mg ml^{-1}
P_1, P_{1p}, P_{1s}	- Concentration of the intermediate in the bulk, inside the particle and at the surface of the particle respectively, g L^{-1}
P_2, P_{2p}, P_{2s}	- Concentration of the product P_2 in the bulk, inside the particle and at the surface of the particle respectively, g L^{-1}
p_1,	- Dimensionless intermediate concentration in the bulk liquid, $(P_1/S_\emptyset)$
p_{1p}	- Dimensionless intermediate concentration inside the particle, $(P_{1p}/S_\emptyset)$
p_{1s}	- Dimensionless intermediate concentration at the surface of the particle, $(P_{1s}/S_\emptyset)$
p_2	- Dimensionless product concentration in the bulk liquid, $(P_2/S_\emptyset)$
p_{2p}	- Dimensionless product concentration inside the particle, $(P_{2p}/S_\emptyset)$
p_{2s}	- Dimensionless product concentration at the surface of the particle, $(P_{2s}/S_\emptyset)$
q	- Ratio of diffusion coefficients, D_p/D_s
r	- Radial position inside the particle, cm
R	- Radius of the pellet, cm
S	- Substrate concentration in bulk liquid, g L^{-1} or mg ml^{-1}
$S_\emptyset$	- Initial substrate concentration in the bulk liquid, g L^{-1} or mg ml^{-1}
S_p, S_s	- Substrate concentration inside the particle and at the surface of the particle respectively
Sh_s, Sh_p	- Sherwood numbers for the substrate and intermediate, (k_sR/D_s) and $(k_p R/D_p)$ respectively
t	- Time, sec
u	- Superficial velocity of the liquid

V_{max1}, V_{max2} - Maximum reaction velocity for the first and second reactions respectively

x - Dimensionless distance, (z/L)

y - Dimensionless radial distance, (r/R)

z - Axial distance from the entrance of the reactor

Greek letters:

α_1 - Parameter defined as k_s a L/u

α_2 - Parameter defined as k_p a L/u

Γ_1 - Parameter, S_0/K_{m1}

Γ_2 - Parameter, S_0/K_{m2}

μ - Viscosity, g cm^{-1} sec^{-1}

τ - Dimensionless time defined as $ut/L\epsilon$

ϕ_1^2 - V_{max1} R^2/K_{m1} D_s

ϕ_2^2 - V_{max2} R^2/K_{m2} D_s

ϵ - Voidage of the bed

Acknowledgement

The authors thank Council of Scientific and Industrial Research, India for the financial support given.

REFERENCES

1. Chiang, L.C., Hsiao, H.Y., Veng,P.P., Chen, L.F. and Tsao, G.T.,Third Symposium on Biotechnology in Energy Production and Conservation, (Scott, E.D., Ed.) Wiley, New York, 263

2. Hägerdal, B. (1984) Biotechnol. Bioeng., **26**, 771.

3. Hartmeier W., Jankovic E.D., Forster, U., and Tramm-Werner Biotech. Europe 84 1984 , Vol.1 ,p415 ,(Pinner (GB): On line Publications).

4. Hahn Hägerdal,B. (1985) Biotechnol. Bioeng. , **27**,914.

5. Mosbach, K., In: Biotechnological Applications of Proteins and Enzymes, (Bohak, Z., and Sharon, N., Eds.) Academic Press, 1977, pp141-152.

6. Lee, J.H., Pagan, R.J., and Rogers P.L., Biotechnol. Bioeng., **25**, 659 (1983).

7. Ueda,S. , C.T.Zennin, D.A.Monteiro and Y.K.Park Biotechnol.Bioeng., **23** 291-299 (1981).

8. Hall R.D., J. Inst. Brewing, 1956, **62**, 222.

9. Miller G.L., Anal. Chem.,1959, **31** , 426

10. Bergmeyer, Hans Ulrich,Methods of Enzymatic Analysis, 1971, Vol.3, p 1205.

11. Ghose,T.K., and Bandhyopadhyay, Biotechnol. Bioengng., 1980,**22**, 1489

12. Chithra, N.and A.Baradarajan, Process Biochemistry, to be published in Dec., 1989.

13. Hayashida, S. and K.Ohta, J.Inst. Brew., **87**, 42, 1981

14. Chithra, N. and A.Baradarajan , J. Chem. tech. and Biotech. , in Print.

15. Bischoff, K.B., Ph.D. Thesis, Illinois Institute of Technology, (1970).

16. Wilson, E.O., and Geankoplis, C.J., Ind. Eng. Chem. Fundamentals, **5**, 9 (1966).

17. Maiorella, B.L., Comprehensive Biotechnology , Murray - Moo Young Ed., **3**, Pergamon press, 861, (1985)

18. Hodge,H.M., and Hildebrandt, F.M., in Industrial Fermentations, L.A.Underkofler and R.J.Hicky, Eds., 1, 73 (1954)

TABLE I: RANGE OF VARIABLES STUDIED IN CONTINUOUS FLOW REACTOR

Parameter	Range of study
Bed height	25 cm to 65 cm
Flow rate	100 ml h^{-1} to 2000 ml h^{-1}
Initial substrate concentration	20 g L^{-1} to 200 g L^{-1}

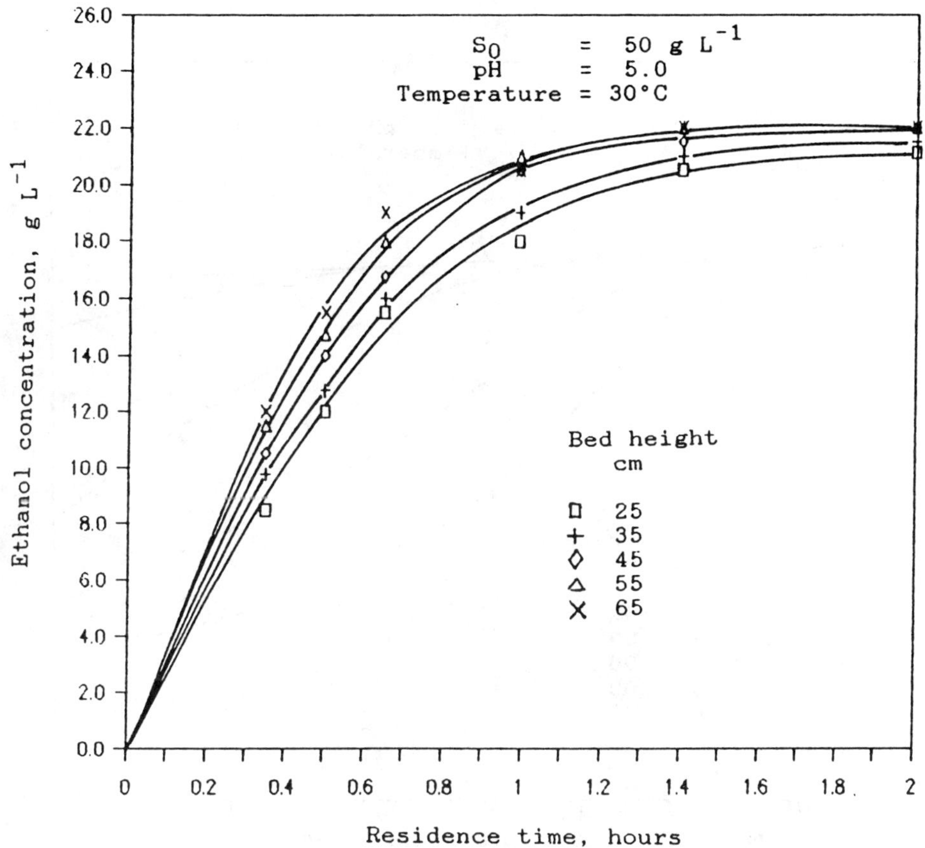

FIG. 1. EFFECT OF BED HEIGHT ON ETHANOL CONCENTRATION

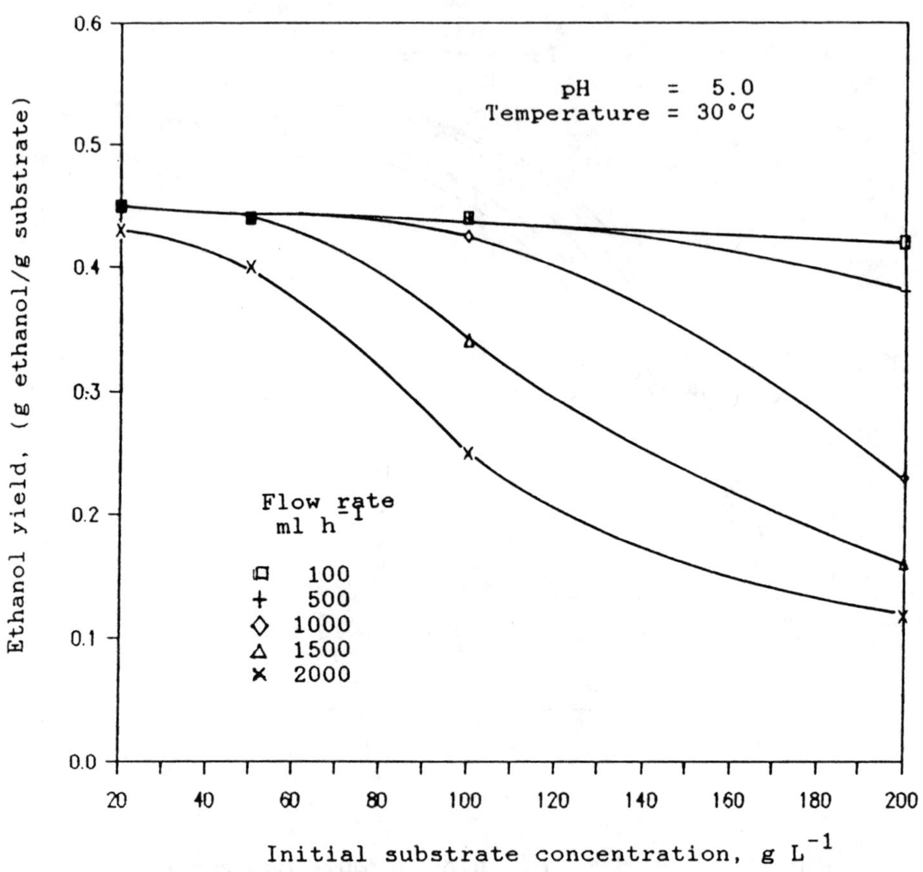

FIG. 2. EFFECT OF FLOW RATE ON ETHANOL YIELD

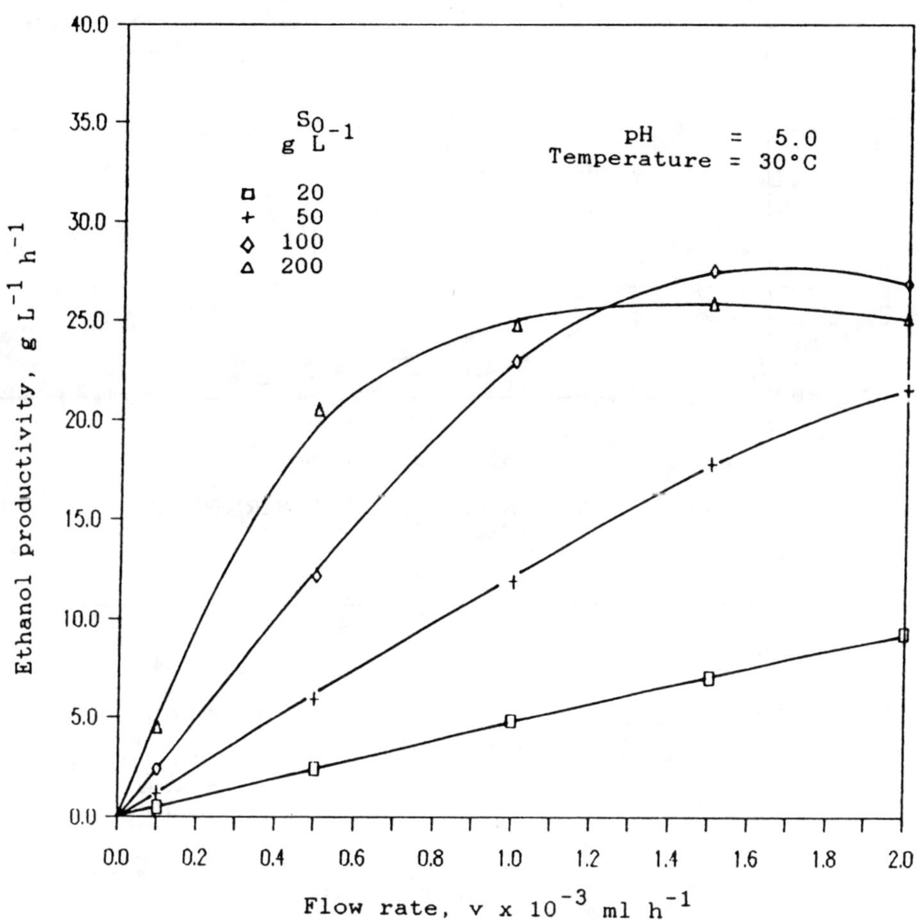

FIG. 3. EFFECT OF INITIAL SUBSTRATE CONCENTRATION ON ETHANOL PRODUCTIVITY

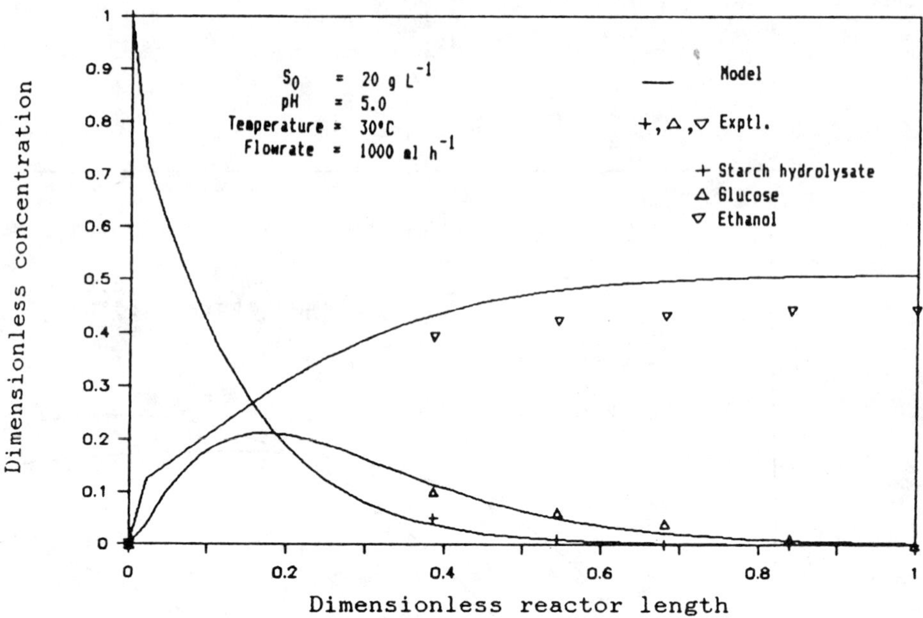

FIG. 4 COMPARISON OF THE EXPERIMENTAL DATA WITH THE THEORETICAL MODEL

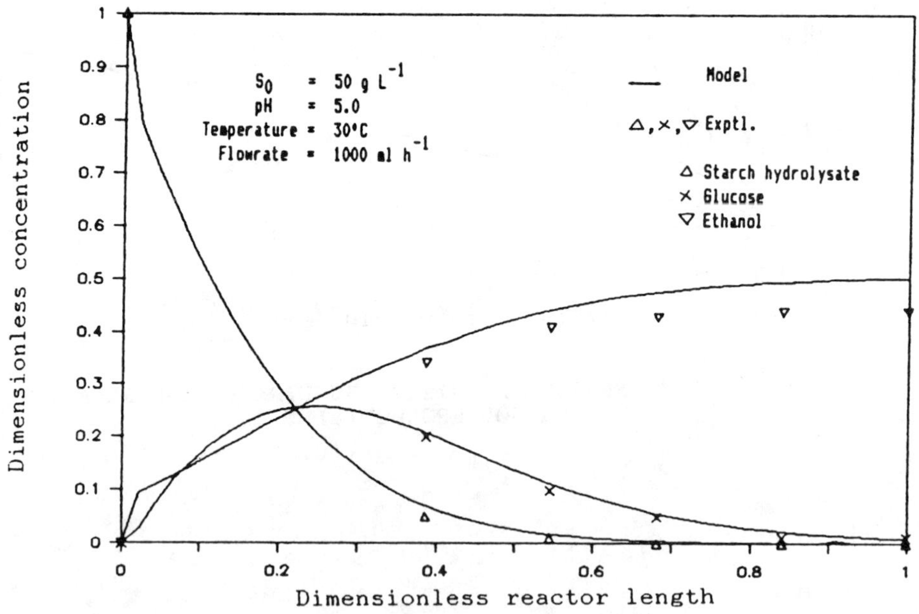

FIG. 5 COMPARISON OF THE EXPERIMENTAL DATA WITH THE THEORETICAL MODEL

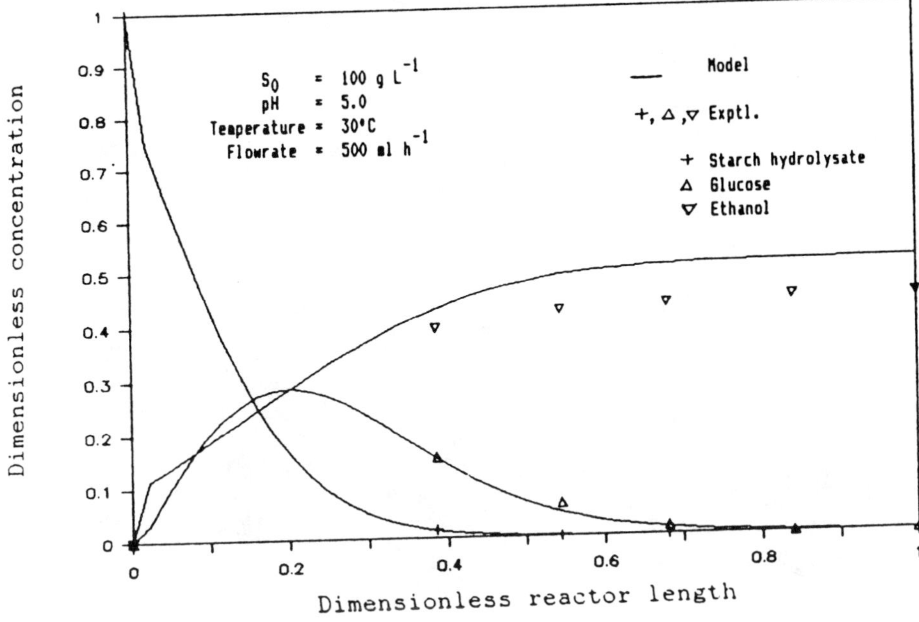

FIG.6 COMPARISON OF THE EXPERIMENTAL DATA WITH THE THEORETICAL MODEL

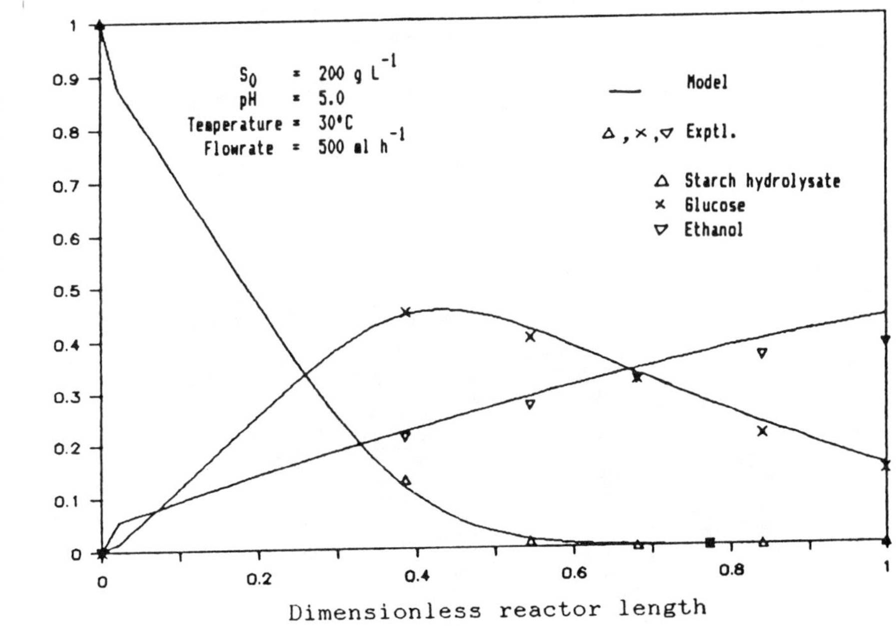

FIG.7 COMPARISON OF THE EXPERIMENTAL DATA WITH THE THEORETICAL MODEL

EFFECT OF TEMPERATURE ON BIOGAS PRODUCTION FROM AQUATIC FERN SALVINIA

S.A. Abbasi* and P.C. Nipaney
Salim Ali School of Ecology
Pondicherry (Central) University, Pondicherry 605 001, India

ABSTRACT

Anaerobic digestion of aquatic fern salvinia (*Salvinia molesta* Mitchell) was carried out at different temperatures(29°-40 °C) and hydraulic retention times (0 - 35 days). The gas production increased sharply as the temperatures were raised above 29°C, attained maxima at 37°C, and fell thereafter. The pattern of rise in gas production with temperature was different from the one reported by others for animal dung but was similar to the one reported for aquatic weed *Pistia stratiotes*. The temperature for maximum gas production was 37°C rather than 35°C reported commonly as the optimum temperature for the anaerobic digestion of zoomass.

---

* Concurrently : Adjunct Professor, California State University, Sonoma 94928, USA.

INTRODUCTION

The aquatic fern *Salvinia molesta* (Mitchell) has earned the sobriquet of the 'Worlds most intransingent weed' (Russell, 1987) due to its explosive growth rate and its propensity to outcompete other aquatic weeds. It is very difficult to destroy the weed without damaging other, non-target, organisms (Abbasi & Nipaney, 1986; Abbasi *et al.*, 1988) but the very high productivity of the weed (Abbasi & Nipaney, 1989) may be turned to advantage by utilising it in producing energy (methane) through anaerobic digestion (Abbasi, 1987).

The present paper describes results of the studies on the effect of temperature on biogas yield from salvinia. Temperature is the most important of the physical factors which profoundly influence anaerobic digestion. A knowledge of biogas production as a function of temperature is therefore basic to energy efficient anaerobic digester design.

A survey of literature reveals a few systematic studies pertaining to the effect of temperature on the anaerobic digestion of animal wastes (Chen *et al.*,1980; Hashimoto, 1984; Hawkes *et al.*, 1984; Erdman, 1985; Peck *et al.*, 1986), but there is only one previous report on phytomass (water lettuce: Nipaney & Panholzer, 1987) and none on salvinia.

EXPERIMENTAL

Sampling of Salvinia

The Salvinia (whole plants) was collected from natural ponds on the outskirts of Calicut, India. It was washed liberally with water to remove attached coarse sediments. It was then washed with 50 g l^{-1} of EDTA solution, followed by deionised water, to remove dust particles absorbed on the plant surface (Abbasi,1988). After draining off the water, the plants were carefully wiped with filter paper and a set of six randomly picked samples were taken for analysis; they were dried at 105°C overnight in hot-air oven to constant weights ('dry weight') and stored in sealed polythene bags. The rest of the plants were weighed and used for charging the digesters.

Chemical Analysis

Volatile solids (VS) were estimated by the combustion of known quantities of the 'dry' weeds at 620°C in muffle furnace to constant weights. Total kjeldahl nitrogen, and phosphorous were determined spectrophotometrically (Rand *et al.*, 1985) using Hitachi 220 spectrophotometer. Total organic carbon (TOC) was determined volumetrically by the method of Piper (1966). Copper and nickel were analysed by atomic absorption spectrometry using IL 951 double beam atomic absorption/emission spectrometer (Rand *et al.*, 1985).

Anaerobic Digestion of Salvinia

Thermostated all-glass air tight digesters of 1000 ml capacity attached with gas sampling and measuring units (Figure 1) were employed. The digesters were charged with 100 g biomass (whole plants) and 250 ml water; duplicates were run at each temperature except at 34°C at which triplicates were run. The methane content of the biogas was determined using Tracor 590 gas-liquid chromatograph.

RESULTS AND DISCUSSION

The chemical compostion of salvinia - average of six random samples with relative error less than 5 % - was : moisture 94.2%, total solids 5.8%, volatile solids 4.9%, total kjeldahl nitrogen 0.058%, phosphorus 0.006%, organic carbon 1.624%, C/N ratio 28,and C/P ratio 280.

A typical set of observations reflecting the reproducibility of the digester preformance is presented in Table 1. Considering the heterogenous nature of the digester feed, the agreement between the results obtained from the replicates is good. During the first five days of digestion the methane content of the biogas was below 40 % and the gas was not combustible. The methane content build up to 58 % by the sixth day and remained in the range of 55 - 65% at all temperatures. The overall results are summarised in Tables 2 and 3, and Figure 2.

The average biogas yields in 35 days were only 2.328 and

2.690 l kg^{-1} fresh weightt at 29°C and 31°C respectively (Table 2). The yield increased sharply at 34°C (9.460 l kg^{-1}) and at 37°C the yield (20.096 l kg^{-1}) was nearly nine times the value obtained at 29°C. The yield did not increase further when the digester temperatures were raised to 37.5°C but thereafter the yield began to decrease. The optimal digestion temperature for salvinia is therefore 37°C.

Some authors, working on dairy wastes, have reported that maximum gas production in the mesophilic temperature range (10°C - 40°C) occurs at 35°C (Persson *et al.*, 1979 ; Roediger, 1967; Cowley & Wase, 1981). Ghosh *et al.* (1980) have placed the optimum digestion temperature 'usually near 35°C to 37°C'. Wolverton & McDonald (1981), Dhavises *et al.* (1985), and Gunnarson *et al.* (1985) have carried out anaerobic fermentation of phytomass at 37°C without explaining why this temperature was chosen by them. Joubert *et al.* (1985) working with synthetic biowaste at 35°C and 37°C found that the latter temperature gave better biogas as well as methane yield. The findings of the present investigation support the contention of Joubert *et al.* that 37°C may be the operating temperature for maximum biogas yield from anaerobic digesters rather than 35 ° C, in the mesophilic temperature range.

An examination of biogas production as a function of hydraulic retention time (HRT) indicated that at 29°C the gas production was insignificant till the 21st day of digestion (Figure 1) and picked up only from 22nd day onwards. At 31°C the gas production was very low till the 10th day and increased significantly thereafter. At higher temperatures this initial lull was not seen. Evidently the mesophilic microbial population built up very slowly at 29°C, needing 21 days for significant activity; at 31°C it took 10 days for the build-up to reach significant levels. At higher temperatures the gas production increased linearly with the retention time upto HRT 20 d, thereafter the curves tilted towards the abscissa upto HRT 35 d, indicating a fall in the rate of gas production per day after HRT 20 d; the tilt increased as the digestion temperature increased from 34°C to 37°C. It may be seen that operating the salvinia-fed digesters at 37°C and HRT 20 d produces as much gas per unit feed per day as the digester operating at 35°C and HRT 35 d, and more gas per unit feed per day than the digester operating at 34°C and HRT 35 d. These observations highlight the magnitude of the importance of maintaining the optimum temperature - 37°C - in the digestion of salvinia. The impact of temperature on the biogas yield from salvinia is summarised in Table 3.

The patterns of change in gas production with temperature are presented in Figure 3, for animal waste (Curve A; from Persson *et al.*,1979), aquatic plant *Pistia stratiotes* (Curve B; from Nipaney & Panholzer, 1987) and *Salvinia* (Curve C; present work). The curves indicate that lowering of digestion temperature below 37°C causes a much sharper fall in gas production from phytomass feeds compared to the zoomass feed. Perhaps the community structure of the microorganisms and the dominant species operating in the two cases is different. There may also be subtle differences in the biochemical reactions governing the cellulolytic, acidogenic and methanogenic phases leading to biogas production in the two cases.

ACKNOWLEDGEMENT

The authors are grateful to Dr. K. Venkatasubramanian, Vice-Chancellor, Pondicherry University, for inspiring guidance. Thanks are also due to Ms. V. Gomathi for secretarial assistance. Financial support from Department of Non-Conventional Energy Sources, Government of India, New Delhi (in part) and CSIR, New Delhi(in part) is gratefully acknowledged.

REFERENCES

Abbasi, S.A. (1987). Renewable energy from biomass. In : S.Terol (Ed),*Proceedings of the 1986 International Congress on Renewable Energy Sources*, Consejo Superior De Investigaciones Cientificas, Madrid, 60-69.

Abbasi,S.A. (1988). Environmental analysis of cerium.*Inter. J. Environ. Anal. Chem.*34, 181-190.

Abbasi, S.A. & Nipaney, P.C. (1986). Infestation of the fern genus *Salvinia* : its status and control. *Environmental Conservation*, 13, 235-241.

Abbasi,S.A. & Nipaney, P.C. (1989). *Salvinia molesta* (Mitchell) as a solar crop.*Journal of Solar Energy Society of India*. In press.

Abbasi, S.A., Nipaney, P.C. & Soni, R. Aquatic Weeds : Distribution, impact and control. *Journal of Scientific and Industrial Research*. 47, 650-61.

Cowley, I.D., & Wase, D.A.J. (1981). Anaerobic digestion of animal waste.*Process Biochemistry*, 16, 28.

Chen, Y.R., Varel, V.H. & Hashimoto, A.G. (1980). Effects of temperature on methane fermentation kinetics of beef cattle manure. *Biotechnology and Bioengineering Symposium* No.10, 325-339 (1980).

Dhavises, G., Sriprasertsak, P., Tanaka, T., Taniguchi, M., & O.S. (1985). Mesophilic and thermophilic methane fermentation of agro-wastes and grasses. *J. Ferment. Technol.*, 63, 45-49.

Erdman, M.D. (1985). Mesophilic fermentation of low-solids dairy-cattle waste. *Agricultural Wastes*, 13, 115-129.

Ghosh, S., Klass, D.L., Edwards, V.H., & Christophers, R.W., (1980). Thermophilic biogasification of biomass.In: *Seventh Energy Technology Conference and Exposition*, Washington D C, March 24-26,1980.

Gunnarson, S., Malmberg, A., Mathisen, B., Theander, O., Thyselius, L., & Wiresche, U., (1985). Jerusalam artichoke (*Helianthus tuberosus* L.) for biogas production. *Biomass*, 7, 85-97.

Hashimoto, A.G. Methane from swine manure : Effect of temperature and influent concentration on kinetic parameter (K). *Agricultural Wastes*, 9, 299-308.

Hawkes, F.R., Rosser, B.L., Hawkes, D.L. & Statham, M. (1984). Mesophilic anaerobic digestion of cattle slurry after passage through a mechanical separator : Factors affecting gas yields. *Agricultural Wastes*, 10, 241-256.

Jourbert, W.A., Britz, T.J. & J.L.F. Kock (1985). Improved anaerobic digester performance using a simple statistical technique.*J. Ferment Technol.*, 63 575-578.

Mah, R.A. (1981). The methanogenic bacteria, their ecology and physiology. In:*Trends in the Biology of Fermentations for Fuels and Chemicals*. Basic Life Sciences, Plenum Press, new York.

Nipaney, P.C. & Panholzer, M.B. (1987). Influence of temperature on biogas production from *Pistia stratiotes*. *Biological Wastes*, 19, 267-274.

Peck, M.W., Skilton, J.M., Hawkes, F.R. & Hawkes, D.L. (1986). Effects of temperature shock treatments on the stability of anaerobic digesters operated on separated cattle slurry. *Water Research*, 20, 453-462.

Persson, S.P.E., Bartlet, H.D., Branding, A.E. & Regan, R.W. (1979).*Agricultural Anaerobic Digesters: Design and Operation*. The Pennsylvania State University, pennsylvania, Bulletin 827, 50.

Piper, C.S. (1966). *Soil and Plant Analysis*. Interscience Publication Inc., New York.

Rand, M.C., Greenberg, A.R. & Taras, M.J. (1985). *Standard Methods for the Examination of Water and Waste Water*. 16th edn., APHA, Washington DC.

Roediger, H. (1967). *The anaerobic basic sludge digestion*. Wasser-Abwasser, H.1, Verlag R. Oldenbourg, Munchen U. Wien. 3 Auflage.

Russell, A. (1987). *Guinees Book of World Records*. Bantom Books, New York, 105.

Wolverton, B.C. & McDonald, R.C. (1981). Energy from vascular plant wastewater treatment systems. *Economic Botany*, 35, 224-232.

TABLE 1

Biogas production from salvinia at 34°C

Salvinia : 100 gm   Temperature : 34°C
Water : 250 ml

Number of Days	Biogas Production l kg^{-1} (Fresh Weight)			
	Digester I	Digester II	Digester III	Average
7*	0.987	1.131	0.919	1.012 ± 0.119
9	2.137	2.221	2.123	2.160 ± 0.061
11	3.055	3.141	3.219	3.138 ± 0.081
13	4.042	4.012	4.048	4.034 ± 0.014
15	4.702	4.572	4.648	4.641 ± 0.062
17	4.902	4.792	4.848	4.847 ± 0.055
19	5.662	5.462	5.388	5.504 ± 0.158
21	6.322	6.017	5.938	6.092 ± 0.237
23	6.663	6.316	6.283	6.421 ± 0.243
25	6.913	6.644	6.543	6.770 ± 0.213
27	7.503	7.104	6.703	7.103 ± 0.400
29	7.843	7.544	7.433	7.607 ± 0.173
31	8.460	8.070	8.160	8.230 ± 0.493
33	9.080	8.680	8.770	8.843 ± 0.224
35	9.670	9.300	9.410	9.460 ± 0.210

* The gas produced in the first 5 days was not combustible and has been omitted in computing the total gas production.

TABLE 2

Biogas production from salvinia at various temperatures and hydraulic retention times

Retention time, days	Biogas production									
	29°C		31°C		34°C		35°C		37°C	
	litres Kg^{-1}, Dry weight	litres Kg^{-1}, VS	litres Kg^{-1}, Dry weight	litres Kg^{-1}, VS	litres Kg^{-1}, Dry weight	litres Kg^{-1}, VS	litres Kg^{-1}, Dry weight	litres Kg^{-1}, VS	litres Kg^{-1}, Dry weight	litres Kg^{-1}, VS
5	3	4	6	7	9	11	41	49	60	71
10	3	4	8	9	46	55	84	100	114	135
15	4	5	13	15	80	95	123	145	175	207
20	5	6	30	35	101	119	162	191	240	284
25	9	11	39	46	116	137	185	219	258	306
30	27	32	43	51	137	162	209	248	305	361
35	40	48	46	55	163	193	244	288	347	410

TABLE 3

Impact of increase in temperature on the biogas yield from salvinia

Increase in Temperature			Effect on gas yield
From		To	
29	-	31°C	1.2 times higher
29	-	34°C	4.1 times higher
29	-	37°C	8.6 times higher

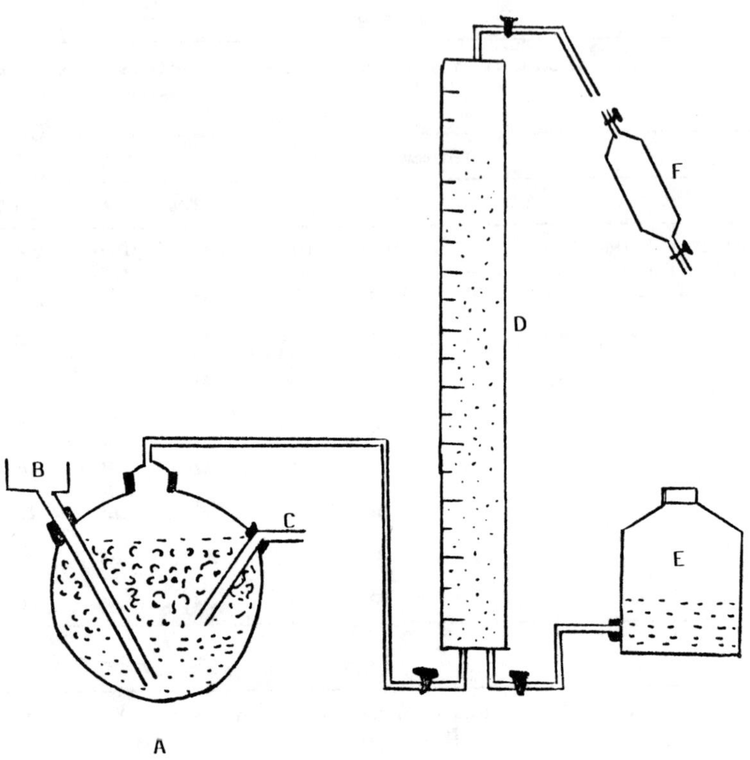

Figure 1 : **Digester assembly** : A : digester; B : inlet; C : outlet; D : gas collecting tube; E : displaced water collecting jar; F : gas sampler for chromatography

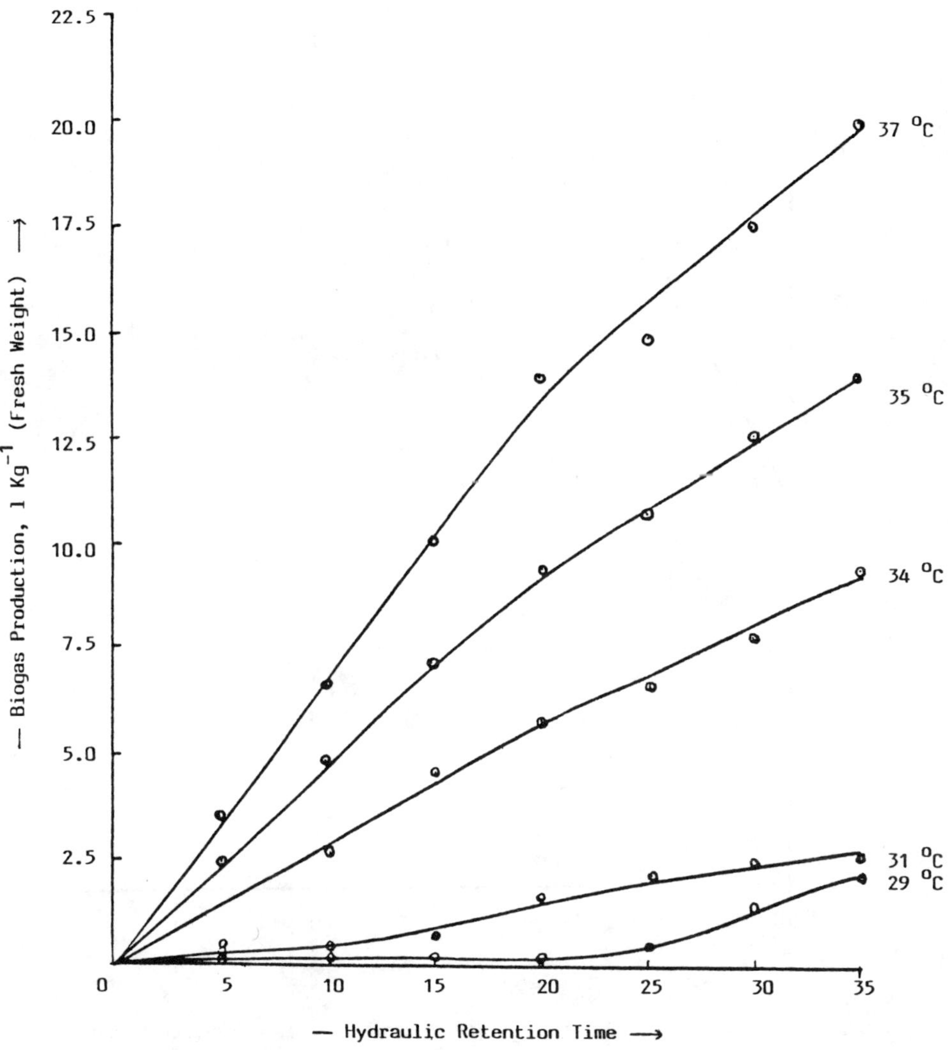

Figure 2 : Biogas production from *Salvinia molesta* (whole plants) as a function of temperature and hydraulic retention time

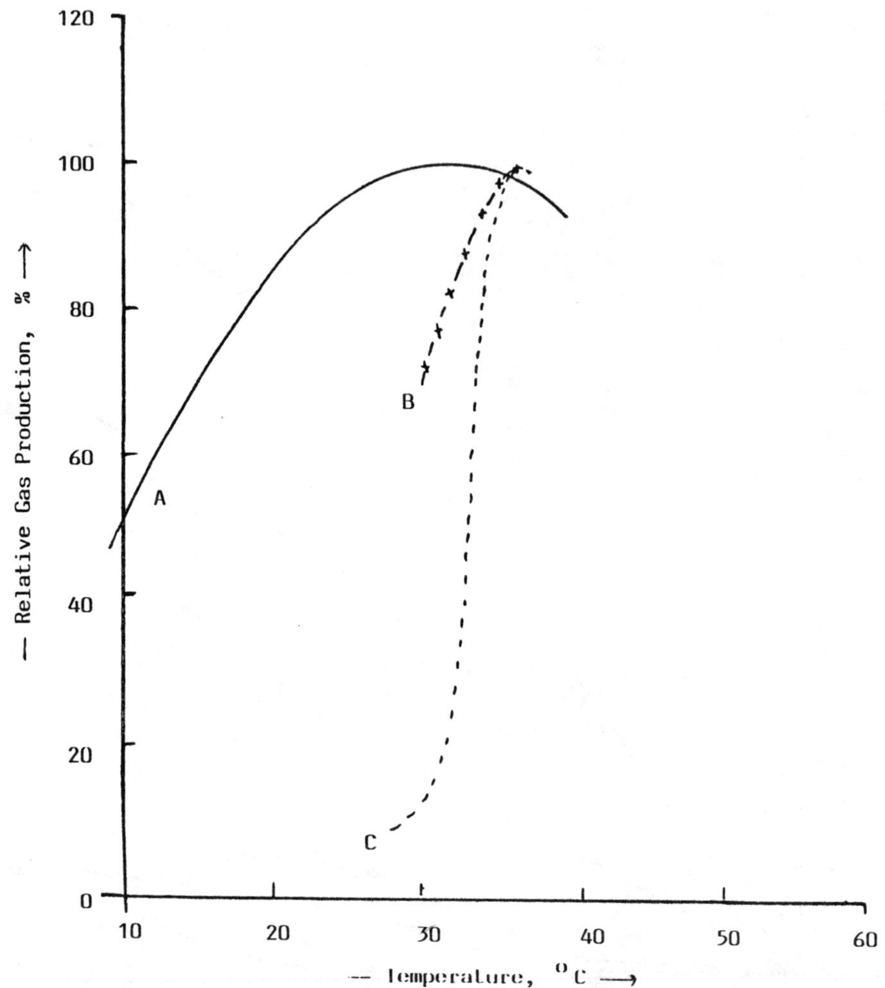

Figure 3 : **Patterns of change in relative gas production with temperature.** Curve A : animal dung (Persson *et al.*, 1979); Curve B : aquatic plant *Pistia stratiotes* (Nipaney & Panholzer, 1987); Curve C : aquatic plant *Salvinia molesta* (this worrk)

BIOENERGY POTENTIAL OF EIGHT COMMON AQUATIC WEEDS

S.A. Abbasi, P.C. Nipaney, and G.D. Schaumberg*
Salim Ali School of Ecology
Pondicherry (Central) University, Pondicherry 605 001, India

ABSTRACT

Eight common aquatic weeds *Salvinia molesta, Hydrilla verticillata, Nymphaea stellata, Azolla pinnata, Ceratopteris sp, Scirpus sp, Cyperus sp* and *Utricularia reticulata* were digested anaerobically to produce methane. At the hydraulic retention time (HRT) of 15 days the energy yield followed the trend *Salvinia* > *Ceratopteris* > *Hydrilla* > *Azolla* >*Utricularia* > *Nymphaea* > *Scirpus* > *Cyperus*. At higher retention times salvinia, ceratopteris, and cyperus maintain their first, second, and last positions vis a vis energy yield while the positions of other weeds keep changing. The carbon to nitrogen (C/N) ratio, carbon to phosphorous (C/P) ratio, and the volatile solids (VS) content of the weeds varied widely. No trend between these factors and the methane yield was discernable; the possible reasons are discussed.

The energy potential of the weeds per unit area of the weed crop was worked out. Natural stands of salvinia, such as the one employed in the present investigation, would yield energy (methane) of the order of 10^8 K cal ha^{-1} yr^{-1}.

---

*Division of Natural Sciences, Sonoma State University, California 94928, USA.

INTRODUCTION

Salvinia (*Salvinia molesta*, Mitchell), hydrilla (*Hydrilla verticillata*, casps.), water lily (*Nymphaea stellata*, Wild.) azolla (*Azolla pinnata*), water sprite (*Ceratopteris sp*), bulrush (*Scirpus sp*), papyrus (*Cyperus sp*), and bladder wort (*Utricularia reticulata*) are amongst the aquatic weeds commonly occurring in tropical and sub- tropical regions of the World (Abbasi et al., 1988). *Salvinia*, a free-floating water fern indigenous to South America, is currently recognised as the world's most intransigent weed (Russell, 1987) mainly because of its propensity to grow very rapidly at the expense of otherr weeds, even water hyacinth (*Eichhornia crassipes*, Mart). *Hydrilla* is the most problematic amongst the submerged aquatic weeds; during summer months it grows several inches a day and can attain biomass densities of 15 tons per hectare (Martyn & Snell, 1979). Water lily (*Nymphaea stellata*) is a rooted floating plant. It has been reportedly inundating large areas of stagnant waters used otherwise for drinking water supply, irrigation, pisciculture, or recreation in tropical countries like India (Gupta, 1979) and Philippines (Gangstad, 1978). *Azolla pinnata* has become a menacing aquatic weed in Africa, Asia, Australia, and South America. The small floating plants of *A.pinnata* can pass through intake screens and clog the pumps and control equipment of irrigation and drainage systems (Varshney & Rzoska, 1976; NAS; 1976).

Water sprite (*Ceratopteris thalictroides*) is commonly found wild in swamps,rice paddies, and ditches throughout much of tropical and subtropical Asia, the East Indies, and Oceana (NAS, 1976).

The infestation of *Scirpus sp* has been reported in the Lakes of Indonesia (*Scirpus grosses*) and East Africa. It invades paddy fields to the detriment of the target crop in the Eastern and Southern India (Gupta, 1979).

Cyperus (*Cyperus papyrus, C. antiquorum*) form, vast stands in swamps, in shallow lakes, and along stream banks throughout Africa (NAS, 1976). In some African waters (for example Lake Victoria, the White Nile, the Congo, and the Okavango swamp) large clumps detach from the land and form floating islands that obstruct navigation and water flow (NAS, 1976). Kisumu Harbour on Lake Victoria, Kenya, is constantly plagued with floating islands (NAS, 1976). Uganda has about 6500 km^2 of permanent swamp, much of it covered with *C. papyrus*. *Cyperus* has also invaded Lakes and streams in the tropical Asia, including Kerala, India (NAS, 1976).

Utricularia sp has been reported in ponds, lakes and irrigation canals in the Indian sub-continent. It has also been reported elsewhere in the tropics (Gupta, 1979; NAS, 1976).

The aquatic weeds, especially salvinia, azolla, hydrilla and *Cyperus* are highly productive, and can easily attain biomass yields of 10 dry tons ha^{-1} yr^{-1} or more (Reddy and DeBusk 1985). It is difficult to control or destroy the weeds through chemical or biological agents owing to the high costs and environmental backlashes (Abbasi & Nipaney, 1986; Abbasi, *et al.*,1988). Periodical harvesting and utilisation is apparently the best strategy for keeping the weeds under control and amongst the various utilisation options, the one involving anaerobic digestion to prodduce energy (methane) appears to be the most promising (Abbasi, 1987; Abbasi *et al.*, 1988).

In this paper studies are presented on the anaerobic digestion of the eight weeds mentioned above. The energy potential of the weeds per unit area of the weed crop has been worked out.

EXPERIMENTAL

Sampling of the Weeds

The weeds (whole plants) were collected from ponds, irrigation canals, and paddy fields in and around Calicut, India. They were washed liberally with water to remove attached coarse sediments. They were then washed with 50 g l^{-1} of EDTA solution, followed by deionised water, to remove dust particles absorbed on the plant surface (Abbasi, 1988). After draining off the water, the weeds were carefully wiped with filter paper. From the weeds thus washed and wiped, whole plants were randomly picked for analysis. They were weighed ('wet weight'), dried at 105°C in hot-air oven to a constant weiight ('dry weight'), and stored in sealed polythene bags. The rest of the plants were weighed and chopped for feeding the digesters.

Chemical Analysis of the Weeds

Volatile solids (VS) conyeny was estimated by the combustion of known quantities of the dried weeds at 620°C in muffle furnace to constant weights. Total Kjeldahl Nitrogen, and phosphorous were determined spectrophotometrically (Rand *et al.*, 1985) using Hitachi 220 spectrophotometer. Total organic carbon (TOC) was determined volumetrically by the method of Piper (1966). Copper and nickel were analysed by atomic absorption spectrometry using IL 951 double beam atomic absorption/emission spectrometer (Rand *et al.*, 1985).

Anaerobic Digestion of the Weeds

All-glass air tight digesters of 1000 ml capacity attached with gas sampling and measuring units (Figure 1) were employed. The digesters were charged with 100 g chopped biomass and 250 ml water; duplicates were run for each weed. The digesters were housed in a chamber of which the temperature was maintained at 37 ± 1.0 °C. The methane content of the biogas was determined using Tracor 590 gas-liquid chromatograph.

RESULTS AND DISCUSSION

The chemical analysis of the aquatic weeds is presented in Table 1. The water content of the weeds was high, in the range 90.2 ± 2.9% and is typical of aquatic plants (Gupta, 1979). The VS, nitrogen, phosphorous, copper and nickel content varied widely amongst the weeds but no trend was discernable. Likewise the C/N and C/P ratios also varied widely but showed no relationship with each other; for example azolla and *utricularia* had similar C/N ratios but thier C/P ratios differed markedly.

A typical set of results on biogas production is given as Table 2. Considering the heterogeneous nature of the digester feed, the agreement between the results obtained from the pair of digesters is good. The biogas production in litres kg^{-1} (fresh weight) as a function of hydraulic retention time (HRT) is presented in Figure 1 while the biogas production in litres kg^{-1} (VS) as a function of HRT is presented in Figure 3. In case of azolla (Figures 2 & 3, A) there is initially a sag in the biogas production curve (HRTs of 0-10 days) but at higher HRTs the biogas production increases almost linearly. *Scirpus* (Figure 2 & 3, C), *Ceratopteris* (Figure 2 & 3, F) and, to a lesser degree, *Utricularia* (Figure 2 & 3, B) and *Cyperus* (Figure 1 & 2, H) show similar trends; the sag occurring upto HRT 20 d in *Scirpus, Ceratopteris*, and *Cyperus*, and upto HRT 25 d in *Utricularia*. A reverse trend is manifested in case of *Nymphaea* (Figure 2 & 3, D), hydrilla (Figure 2 & 3, E) and salvinia (Figure 2 & 3, G). These weeds produce biogas at a higher rate during low HRTs; the gas production slows down when the HRT is increased above 15 d for hydrilla and 20 d for *Nymphaea* and salvinia. It is possible that in azolla, *Scirpus* ,*Caratopteris*, and *Utricularia* the microbial population slowly builds up and reaches a maximum in 20-25 days, causing the initial sag and the subsequent linear gas production. In *Nymphaea*, salvinia and hydrilla the biogas production is not apparently inhibited by low microbial density - perhaps due to more easily degradable organic matter in the biomass of these weeds - and the tapering in gas production after 15-20 days of digestion is perhaps due to fall in VS concentration after the earlier brisk utilization.

The results on biogas, methane, and energy equivalent of methane produced by different weeds as a funcion of HRT are summarised in Table 3. The energy equivalent was calculated on the basis of reported (Meyers, 1983) calorific value of methane (213 K Cal l^{-1} at NTP). At HRT 15 d the energy yield follows the trend salvinia > *Ceratopteris* > hydrilla > azolla > *Utricularia* > *Nymphaea* > *Scirpus* > *Cyperus*. At higher retention times salvinia, *Ceratopteris*, and *Cyperus* maintain their first, second, and last positions vis a vis energy yield while the positions of other weeds keep changing; the final trend at HRT 35 d is salvinia > *Ceratopteris* > azolla = *Utricularia* > hydrilla > *Scirpus* > *Nymphaea* > *Cyperus*.

Several factors are known to influence anaerobic digestion and methane yield from biomass, including C/N ratio, (Meynell, 1976, Chynoweth & Srivastava, 1980) Ghosh *et al*., 1985), C/P ratio (Ghosh, 1981), nature of organic matter constituting VS (Ghosh, *et al*.,

1981), iron (Speece & Perkin, 1985), vitamins (Scherer & Sahm, 1981), trace nutrients (Sherer & Sahm, 1981), and natural inhibitants (Ghosh, *et al.*, 1981) present in the biomass. An examination of Tables 1 and 3 indicates that no single factor dominates the gas yield. *Cyperus* has higher VS content than other weeds (Table 1) and should have produced maximum biogas but it in fact produces the least (Table 3). Perhaps the advantage of having the highest VS content by *Cyperus* is nullified by having the highest C/N ratio and lowest C/P ratio. Moreover There is no agreement in literature on the 'ideal' C/N or C/P ratio. It has been suggested (Meynell, 1976) that an optimum C:N ratio is between 20:1 and 30:1, although it was reported by Sanders and Bloodgood (1965) that for the digestion of an experimental feed containing pig food there was a minimum C:N ratio of 16:1 and increasing it by nitrogen addition did not appear to improve digestion. Kaplovsky (1952) found that to achieve satisfactory decomposition of white water it was necessary to add ammonium carbonate in sufficient quantities to give a C:N ratio of 40:1. Experiments with paper pulp (De Renzo, 1952) with raw sewage as the nitrogen source showed that digestion of different mixtures was feasible upto C:N ratio 45:1; digester failure occurred when the ratio reached 52:1. However, digestion of paper pulp to which chicken manure was added as the nitrogen source proceeded without difficulty upto a ratio as high as

70:1. Chynoweth & Srivastava (1980) reported that anaerobic digestion of Kelp (*Macrocystis pyrifera*) suffered when C/N ratios exceeded 14 but Ghosh, *et al* (1981) found that changing C/N ratio from 17 to 12 and C/P ratio from 96 to 27 did not enhance biomethanation of Kelp. According to Habig *et al* (1984), digesters containing low - nitrogen seaweed as a substrate did not exhibit symptoms of nitrogen limitation, despite C/N ratios reaching 30. A report from Parnas (1975) supports this view. On the other hand Fannin *et al* (1982) has suggested that a C/N ratio exceeding 15 represented nitrogen limitation in *M. pyrifera* digesters; a view shared by Wilkie *et al* (1986).

It is obvious that several factors besides C/N and C/P ratios contribute to the biomethanation. A major factor is the presence or absence of adequate levels of micronutrients such as nickel, cobalt, molybdenum, copper, zinc and selenium. Abbasi and Nipaney (1984) reported enhanced biogas production from salvinia-fed anaerobic digesters when the latter were spiked by zinc and copper. Wilkie *et al* (1986) obtained 'immediate enhancement' in biogas production when a semi-continuous digester running for 142 days with napiergrass (*Pennisetum purpureum* Sehum) as feed was supplemented with a nutrient solution containing nickel, cobalt, molybdenum, and selenium. The findings of the present investigation also indicate that no direct correlation exists between methane yield from biomass and the latter's VS content, C/N ratio or C/P ratio. A correlation may be possible only if no other limiting factor is operative.

Table 4 presents figures for energy yield from weeds as a function of crop area and annual biomass production in typical natural stands. The calculations have been limited to the weeds of which annual productivity figures are available in literature.Salvinia would yield energy of the order of 10^8 K cal ha^{-1} yr^{-1} while azolla, *Scirpus*, hydrilla, and *Nymphaea* have energy potential of order of 10^7 K cal ha^{-1} yr^{-1}.

ACKNOWLEDGEMENT

The authors are grateful to Dr. K. Venkatasubramanian, Vice-Chancellor, Pondicherry University, and Dr D.D.Ferish,Dean Division of Natural Sciences, Sonoma State University, California, for inspiring guidance. Thanks are also due to Mr Scott Heidike,BS Student, for valuable assistance. Financial support from Department of Non-Conventional Energy Sources, Government of India, New Delhi (in part) and CSIR, New Delhi (in part), is gratefully acknowledged.

REFERENCES

Abbasi, S.A. (1987). Renewable energy from aquatic biomass, In : *Proceedings of the 1986 International Congress on Renewable Energy Sources*, CSIC, Madrid. 60-69.

Abbasi, S.A. (1988). Environmental analysis of cerium. *Intern. J. Environ. Anal. Chem.* 34, 181-190.

Abbasi, S.A. & Nipaney, P.C. (1984). The catalytic effect of copper (II), zinc (II) and nickel (II) on the anaerobic digestion of *Salvinia molesta* (Mitchell). In: *Energy Development New Forms, Renewables, Conservation*, F.A. Curtis (Ed), Pergamon Press, Oxford, 237-242.

Abbasi, S.A. & Nipaney, P.C. (1986). Infestation of the fern genus *Salvinia* : its status and control *Environmental Conservation*, -13, 235- 241.

Abbasi, S.A. & Nipaney, P.C. (1989). Productivity of *Salvinia molesta* (Mitchell). Unpublished work, communicated to Aquatic Botany.

Abbasi, S.A., Nipaney, P.C. & Soni, R. (1988). Aquatic weeds: distribution, impact and control. *Journal of Scientific and Industrial Research*, 47, 650-61.

Boyd, C.E. (1968). Freshwater Plants : a potential source of protein. *Economic Botany*. 22, 359.

Chynoweth, D.P. & Srivastava, V.J. (1980). Methane production from marine biomass. In: *International Symposium on Biogas, Microalgae, and Livestock Wastes - 1980*, Taipei, Taiwan, R.O.C. September 15-17, 26.

De Renzo, D.J. (1977). *Bioconversion of Solid Waste and Sewage Sludge in Energy from Bioconversion of Waste Material.* Noyes Data, Dark Ride, N.J.

Fannin, K.F., Srivastava, V.J. & Chynoweth, D.P. (1982). Unconventional digester, designs for improved methane yields from sea kelp. In: *Proc. I.G.T. Symp. Energy from Biomass and Wastes*, Lake Buena Vista, Florida, January 1982.

Gangstad, E.O. (1978). *Weed Control Methods for River Basin Management*. CRC Press Inc., Florida.

Ghosh, S. (1981). Biomethanation of peat, In: *International Peat* Symposium, *Benidji State University, Benidji, Minnesota.*

Ghosh, S., Henry, M.P. & Christopher, R.W. (1985). Hemicellulose conversion by anaerobic digestion, *Biomass*, 6, 257.

Ghosh, S.,Klass, D.L. & Chynoweth, D.P. (1981). Bioconversion of *Macrocystis pyrifera* to methane. *J.Chem. Tech. Biotechnol.*, 31, 791-807.

Gupta, O.P. (1979). *Aquatic Weeds : their Menace and Control- A Textbook and Manual.* Today and Tommorrows Printers and Publishers, New Delhi 279.

Habig, C., DeBusk, T.A. & Ryther, J.H. (1984). The effect of nitrogen content on methane production by the marine algae *Gracilaria tikvahiae* and *Ulva sp. Biomass*, 4,. 239-251.

Kaplovsky, A.J. (1952). Volatile acid production during digestion of several industrial wastes. *Sewage Ind. Wastes*, 24, 194.

Lakshman, G. (1987). Ecotechnological opportunities for aquatic plants - a survey of utilization options. *In* : *Proceedings of Aquatic Macrophytes for Water Treatment and Resrouce Recovery*, K.R. Reddy & W.H. Smith (Ed), Magnolia Publishing, Orlando, Florida, 49.

Martyn, R.D. & Snell, W.W. (1979). *Lake Conroe Aquatic Vegetation Survey I Aerial Color Infrared Photography - Base - Line Map - 1979.* MP - 1502, 2M - 2 - 82.

Meyers, R.A. (1983). *Handbook of Energy Technology and Economics.* John Wiley & Sons, New York, 257.

Meynell, P.J. (1976). *Methane : Planning a Digester.* Prism, Detroit.

National Academy of Sciences (1976). *Making Aquatic Weeds Useful : Some Perspective for Developing Countries.* Report, Board on Science and Technology for International Development, Commission of International Relations, Washington, DC, 175.

Parnas, H. (1975). Model for decomposition of organic material by micro-organisms. *Soil Biol. Biochem.*, 7, 161-9.

Piper, C.S. (1966). *Soil and Plant Analysis.* Inter Science Publication Inc., New York.

Rand, M.C., Greenberg, A.R. & Taras, M.J. (1985). *Standard Methods for the Examination of Water and Wastewater.* 16th edn., APHA, Washington D.C.

Reddy, K.R. & Bagnall, L.O. (1981). Biomass production of aquatic plants used in agricultural drainage water treatment. *In* : *International Gas Research Conference Proceedings*, Govt. Institute, Inc., Rockville, MD, 660.

Reddy, K.R. & DeBusk, W.F. (1985) Growth characteristics of aquatic macrophytes cultured in nutrient-enriched water: II. azolla, duckweed and salvinia *Economic Botany*, 39, 200.

Russell, A. (1987). *Guinness Book of World Records*. Bantom Books, New York, 105.

Sanders, F.A. & Bloodgood, D.E. (1965). The effect of nitrogen to carbon ratio on anaerobic decomposition. *J. Water PPollut. Control Fed.*, 37, 1741.

Scherer, P. & Sahm, H. (1981). Effect of ttrace elements and vitamins on the growth of *Methanosarcina barkeri*. *Acta Biotechnologica*, 1, 57.

Speece, R.E. & Perkin, G.F. (1985). Nutrient requirements for anaerobic digestion. In : *Biotechnological Advances in Processing Municipal Wastes for Fuels and Chemicals*. In: A.A. Antonopoulos (ed), ANL/CNSV-TM-167, Argonne National Laboratory, Argonne, Illinois, 195.

Varshney, C.K. & Rzoska, J. (1976). *Aquatic Weeds in South East Asia*. Dr. W. Junk BV Publishers, The Hague.

Wilkie, A., Goto, M., Bordeaux, F.M. & Smith, P..H. (1986). Enhancement

of anaerobic methanogenesis from napiergrass by adddition of micronutrieeeonts. *Biomass*, 11, 135-146.

TABLE 1
Chemical analysis of the aquatic weeds

Weed	Total solids (TS), %	Volatile solids (VS), %	Moisture, %	Dry weight basis (g/g dry weight)					C/N Ratio	C/P Ratio
				Total kjeldahl nitrogen	Phosphorus	Copper*	Nickel*	Organic carbon		
Salvinia	5.8	4.9	94.2	0.01	0.001	0.031	0.002	0.28	28.0	280
Azolla	5.9	4.8	94.1	0.031	0.002	0.030	ND	0.43	13.7	215
Utricularia	3.9	3.2	96.1	0.031	0.001	0.010	0.0007	0.43	13.9	430
Cyperus	15.7	12.6	84.3	0.012	0.003	NT	NT	0.42	35.0	140
Nymphaea	10.5	8.5	89.5	0.017	0.002	NT	NT	0.42	24.9	210
Hydrilla	8.0	6.5	92.0	0.03	0.003	0.015	0.003	0.43	14.2	143
Ceratopteris	4.1	3.3	95.9	NT	NT	NT	NT	NT	NT	NT
Scirpus	14.0	11.2	85.8	NT	NT	NT	NT	NT	NT	NT

ND = not detectable (below 10^{-5} ppm)
NT = Not tested
*µg/g dry wt.

TABLE 2

Biogas production from *Nymphaea stellata*

Biomass : 100 gm Temperature: 37±1°C
Water : 250 ml

Number of days	Biogas production, l kg^{-1} (fresh weight)		
	Digester I	Digester II	Average
1	0.6	1.0	0.80±0.20
3	0.8	1.1	0.95±0.15
5	1.2	1.6	1.40±0.20
7	1.5	1.8	1.65±0.15
9	2.0	2.2	2.10±0.10
11	2.9	3.1	3.00±0.10
13	3.4	3.9	3.65±0.25
15	3.8	4.4	4.10±0.30
17	4.3	4.9	4.60±0.30
19	4.7	5.2	4.95±0.25
21	5.5	5.9	5.70±0.30
23	5.5	6.1	5.80±0.30
25	6.0	6.3	6.15±0.15
27	6.2	6.5	6.35±0.15
29	6.4	6.7	6.55±0.15
31	6.5	6.9	6.70±0.20
33	6.7	7.0	6.85±0.15
35	6.8	7.2	7.00±0.20

TABLE 3

Biogas Production and energy equivalent of different aquatic weeds

Plant	Hydraulic retention time, days	Biogas production		Methane production		Energy equivalent
		$l\ Kg^{-1}$ Dry weight	$l\ Kg^{-1}$ VS	$l\ Kg^{-1}$ Dry weight	$l\ Kg^{-1}$ VS	K cal Kg^{-1} Dry weight
Salvinia	15	175	207	103	122	928
	20	240	284	142	168	1279
	25	258	306	152	180	1370
	35	347	410	205	242	1847
Azolla	15	64	79	41	51	369
	20	83	102	54	66	487
	25	111	137	72	88	649
	35	167	205	108	132	973
Ceratopteris	15	75	93	55	68	496
	20	101	125	74	91	667
	25	131	163	96	119	865
	35	225	280	164	204	1478
Scirpus	15	18	23	15	20	135
	20	26	32	22	27	198
	25	38	48	32	41	288
	35	63	78	54	66	487
Cyperus	15	19	24	11	13	99
	20	25	31	14	17	126
	25	33	42	19	24	171
	35	54	67	30	38	270
Utricularia	15	64	78	38	46	342
	20	85	104	51	62	460
	25	108	132	64	79	577
	35	182	222	108	132	973
Hydrilla	15	76	94	48	59	433
	20	92	113	58	71	523
	25	106	130	66	81	595
	35	138	170	86	106	775
Nymphaea	15	39	48	22	26	198
	20	51	63	28	35	252
	25	59	72	33	40	297
	35	67	82	37	45	333

TABLE 4

Energy equivalent of weeds per unit crop area per year*

Plant	Medium	Productivity t ha^{-1} yr^{-1} (dry weight)	Reference	Energy equivalent, K cal ha^{-1} yr^{-1}
Salvinia molesta	Natural ponds	57	Abbasi & Nipaney, 1989	10.53×10^7
Azolla pinnata	Nutrient medium	11	Reddy & DeBusk, 1985	1.07×10^7
Scirpus	--	2.5-25.0	Lakshman, 1987	$0.12\text{-}1.22 \times 10^7$
Hydrilla verticillata	Nutrient medium	15	Reddy & Bagnall, 1981	1.16×10^7
Nymphaea sp	--	12**	Boyd, 1968	0.40×10^7

* Calculated for weeds of which annual productivity has been reported in literature

** The annual productivity calculated on the basis of 1.8 t/ha (dry weight) standing crop

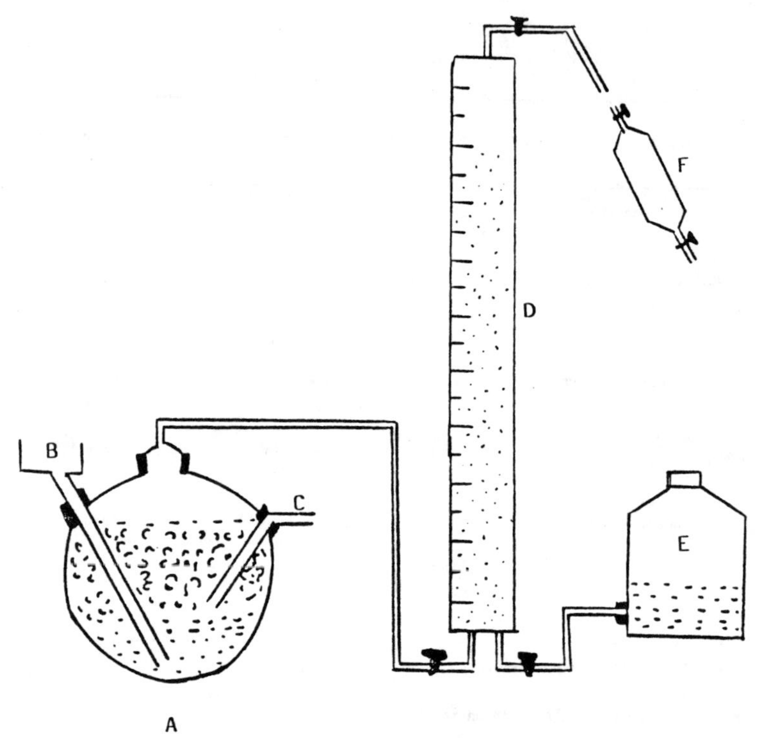

Figure 1 : Digester assembly. A : digester; B : inlet; C : outlet; D : gas collecting tube; E : displaced water collecting jar; F : gas sampler for chromatography

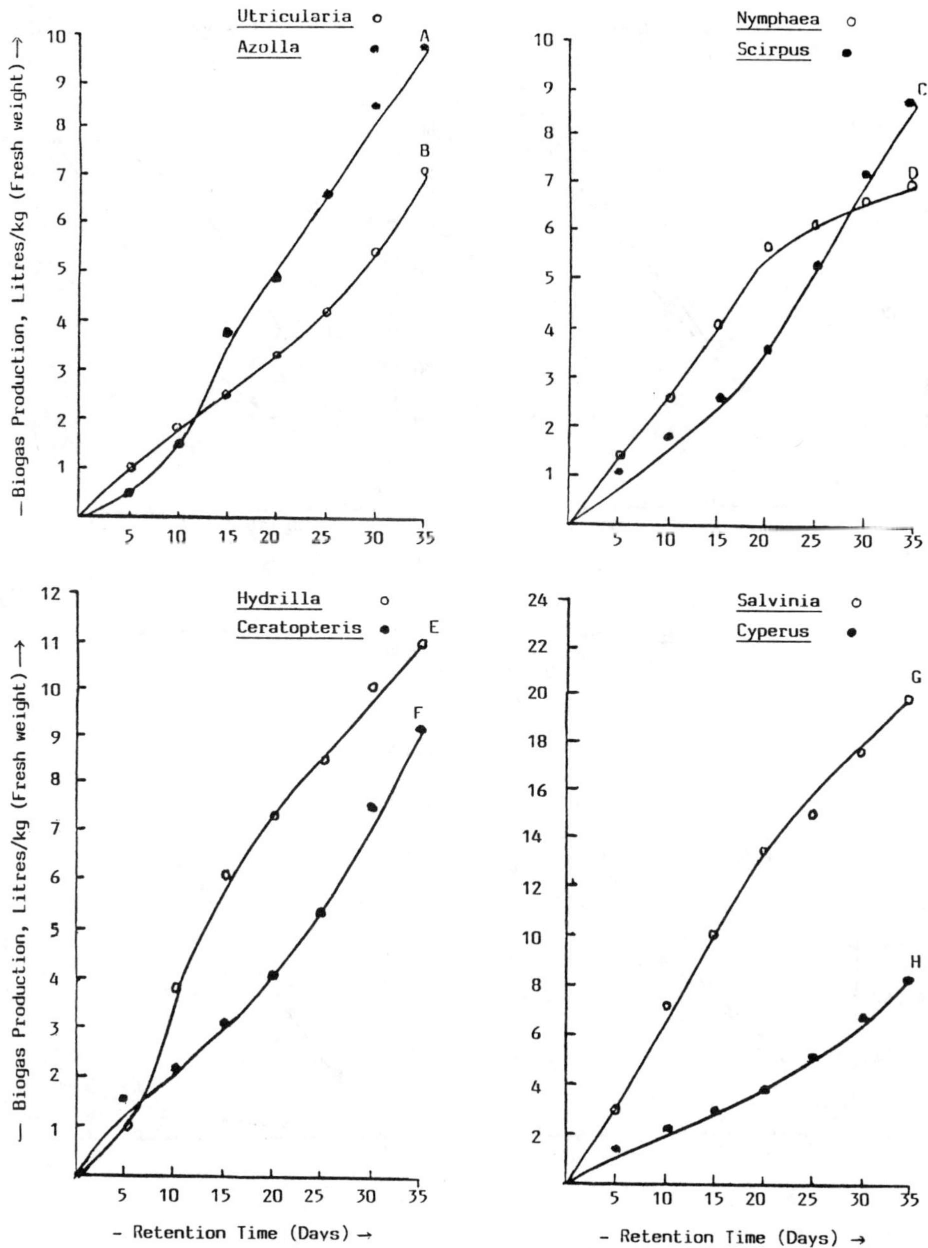

Figure 2 : Biogas production from fresh weight off the biomass as a function of hydraulic retention time

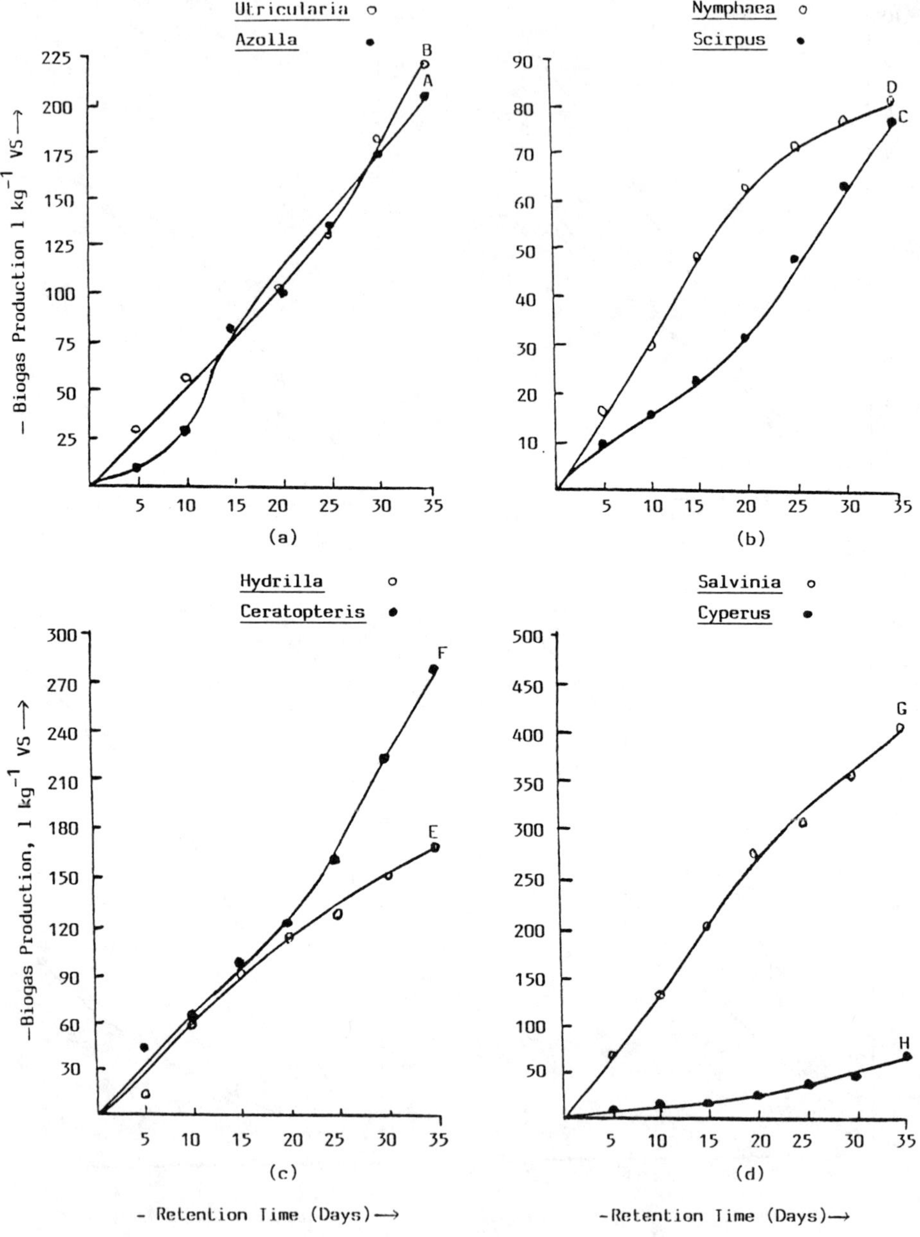

Figure 3 : Biogas production as a function of volatile solids content of the biomass and hydraulic retention time

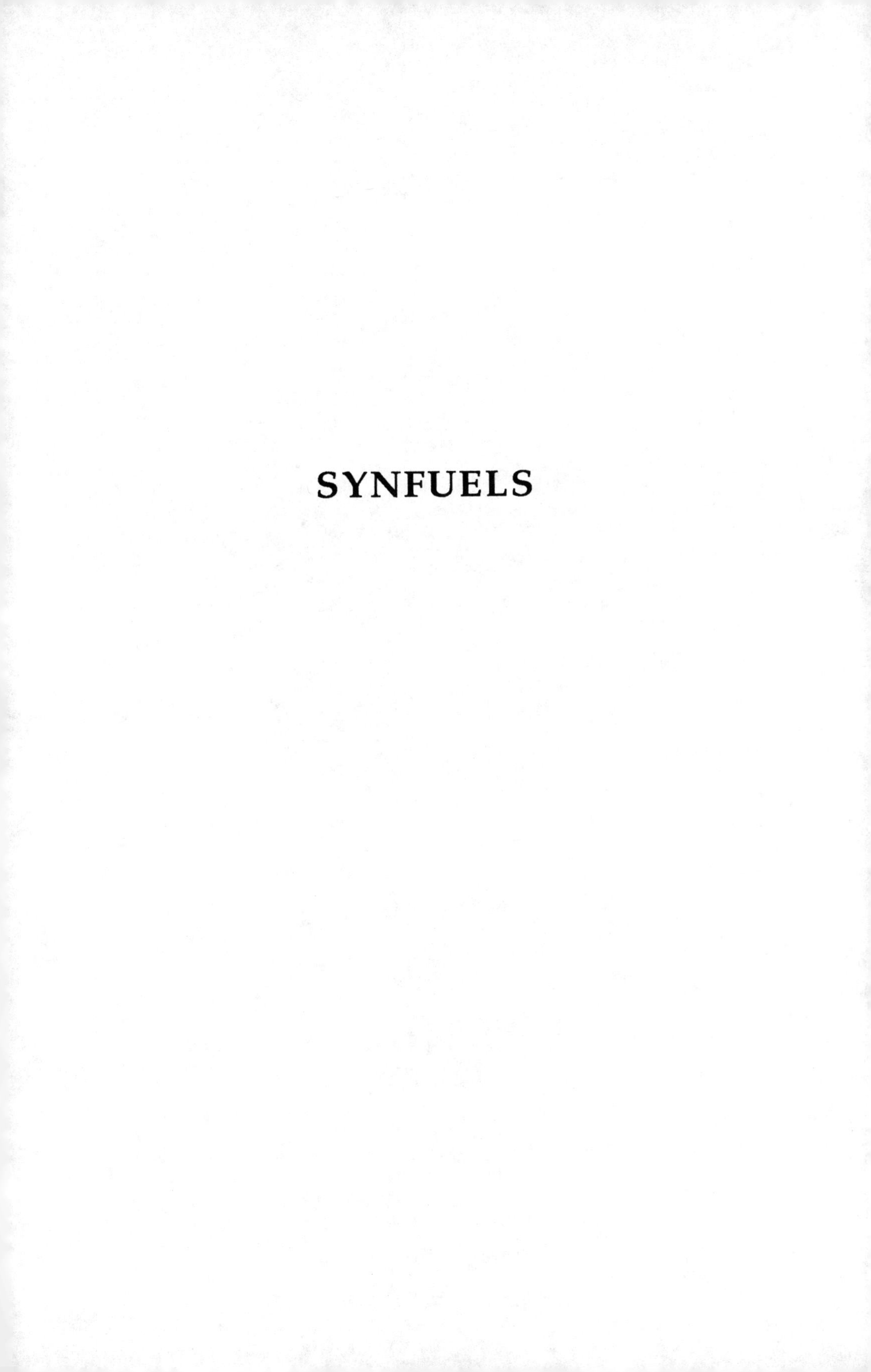

SYNFUELS

KINETICS OF METHANE REFORMING REACTION AT LOW TEMPERATURES AND PRESSURES

Süha Atamer
Department of Chemical Engineering, Middle East Technical University
06531 Ankara, Turkey

Abstract

A commercial hydrocarbon reforming catalyst has been used for the kinetic study of methane-steam reaction conducted in an integral reactor.

Experimental determinations of catalyst surface area and pore size distribution and of reaction rates for a range of temperatures, pressures and steam to methane ratios have been made.

The contributions made by various rate processes have been examined on cylindrical pellets presenting a definite external area of contact to an up-flowing stream of reacting gas. The rate of flow of gas and the annular gap between catalyst and reactor wall have been varied so that the mass and external area of catalyst exposed to the gas had a range of values.

Although no exact evaluation of catalyst's effectiveness factor was performed it was inferred that its value could not be greater than 0.1 and hence, as would be expected from the determined pore sizes, there is appreciable pore diffusion resistance.

Experiments over a range of pressure and initial reactant concentrations provided evidence that adsorption of none of the reactants was rate controlling.

It is concluded that most rate expressions are adequate only in a narrow temperature and composition range and that probably no single rate expression can be applied in the whole range of industrial conditions.

Two semi-empirical expressions were developed for the initial rate of the reaction, and a further correlation as a best-fit of the integral rate data.

1. INTRODUCTION

Catalytic steam reforming of methane is a basic process for large scale production of hydrogen and synthesis gas. Its applications are wide-spread in ammonia and methanol syntheses as well as in ferrous metallurgy and clean energy related industry (fuel cells). An integrated treatment of different aspects of the process is presented by Rostrup-Nielsen [1] and El-Nashaie et al. [2].

In general, the stoichiometry is represented by the following reversible reactions:

$$CH_4 + H_2O \rightleftharpoons CO + 3\,H_2$$
$$CO + H_2O \rightleftharpoons CO_2 + H_2$$

For the design and efficient operation of compact reformer tubes the kinetics of apparent and intrinsic rates of the reactions, including the deactivation of catalyst, must be known. In spite of innumerable investigations in this field there still exists a significant diversity of opinion concerning the mechanism and sequence of reactions involved [3,4,5,6,7,8,9]. Similar disagreement extends also into the conclusions that have been drawn on the primary species and stable products being formed in the complex system of reactions. Whilst most of the suggested rate expressions are based on Langmuir-Hinshelwood-Hougen-Watson (L.H.H.W.) - type postulations, few hybrid rate equations as combination of L.H.H.W. and power law have also been recently published [8]. The types of reactors used in the investigations, the ranges of operating temperature and pressure, reactant concentrations in the feed, the accuracy of analytical techniques, and the chemical, physical and crystal lattice properties of catalysts depending on the method of preparation and their treatment are the principal reasons responsible for the conflicting conclusions.

The primary objective of this study is to investigate on a commercial reforming catalyst which of the oxides of carbon is being formed in conjunction with mass transfer limitations.

2. REFORMING REACTOR and PROCEDURE

The construction of the equipment could enable a large range of experimental conditions related to the type of low molecular weight hydrocarbons, steam : hydrocarbon ratio, feed rate, temperature, and pressure. Two types of reactors were used for the experimental runs according to type of catalyst. One reactor was an integral type fixed bed flow reactor specifically to test catalyst pellets in form of hollow cylinders with definite external diameter (6.4 mm I.D., max. 16.0 mm O.D.). It consisted of a stainless steel tubing of 18.3 mm I.D. . A 6.3 mm O.D. tube, placed concentrically in the reactor main body served as the catalyst holder. The location of the catalyst pellets within the reactor could be adjusted by this catalyst support (Figs. 1 and 2). A second tubular reactor was developed for experiments carried out with catalyst in granular or powdered form to investigate the effects of particle size (Fig.3). It consisted of a 6.3 mm O.D. stainless steel tube, both ends fitted high pressure joints. A layer of stainless steel wire mesh was used as the catalyst support. Three moveable thermocouples of 0.51 mm diameter enabled the internal axial temperatures to be measured and controlled throughout the whole reactor length. A schematic diagram of the set-up is shown in Fig.4. CH_4, H_2, and N_2, the latter to provide an inert atmosphere for purging purposes and for pressure testing of the equipment, were all of high purity (99.99 %). The gases were metered with rotameters within the range $(0.1)(10^{-3})$-$(15.0)(10^{-3})$ moles/min. They then passed through a preheating furnace, where the temperature was increased to 300-350°C, and into the evaporator furnace which was fed with liquid water by a microfeed pump. The homogeneous mixture of methane-steam then passed into two successive superheating furnaces where its temperature gradually increased to the reactor temperature. On leaving the reactor the reformed gas passed into a cocurrent water cooled condenser for separation and collection of unreacted steam, underwent a pressure reduction in cases of high pressure runs and branched into two streams, one of which led to the gas chromatograph to provide a continuous supply of sample, while the other to the flare. The construction material was stainless steel (type 304) throughout with the exception of glass U-tubes containing $CaCl_2$ for drying the gas samples. It was confirmed by blank runs prior to the systematic study, that the catalytic effect of nickel in stainless steel on the pyrolysis and reforming of methane at different temperature levels was negligible; at relatively long contact time was the conversion 0.2 - 0.3% coupled with slight carbon formation. The thermocouples were situated in thermowells and extended along the whole height of the feedstock preheater, the evaporator, the superheaters, and the reactor, permitting internal temperatures to be scanned. The heat loads on all of the furnaces were adjusted by Variac type regulators controlling separately the bottom, middle, and top resistances. The reactor furnace had two windings, the one carried the base load while the other was a supplementary one wound along the middle third of the furnace. Both were equipped with temperature controllers in addition to the voltage regulators. The temperature fluctuations could be evened out to within $\pm$ 1°C around the set point of the controllers and recorded on a six-point recorder. Reactor temperatures were measured at 3 internal positions and 2 wall temperatures, all five thermocouples being capable of movement in vertical direction. In the course of experiments the temperatures measured by thermowells located on the reactor outside wall and inside the catalyst pellet along the bed showed differences of 2-6 ^{o}C. The flow rates of gases as obtained for given scale readings on feed flowmeters were checked with reactor exit gas and chromatograph sample gas flowmeters using the whole apparatus operated at ambient temperature and different pressures as blank runs. In all cases the maximum deviation was found to be not greater than $\pm$ 0.8 %. The variation of the water feed rate was obtained by the micrometer adjustment of the plunger stroke. The pumping rates were not affected by the line pressure.

In order to minimize the conditions which may invalidate the plug flow assumption, the catalyst pellets to be tested were sandwiched in the reactor between stainless steel pellets of the same diameter as the catalyst. The heights of columns formed by stainless steel pellets were each twice the height of the catalyst pellets. Moreover, a stainless steel cone located underneath the bottom pellet gave a gradual change of flow cross-section so that plug flow conditions should prevail and any effect of back-mixing or axial dispersion may be minimized.

The analyses of the reaction products were performed using a gas chromatograph in conjunction with a potentiometric recorder having one second response and 1 mV span. The column packing was Porapak Q, 6.4 m long.

3. THE CATALYST

The catalyst used was standard commercial catalyst which was formulated particularly for naphtha reforming. It was used in the present work for methane reforming. The chemical analysis of the catalyst on moisture free basis is:

Ignition loss	8.70 %
SiO_2	13.95 %
Al_2O_3	26.05 %
NiO	21.34 %
CaO	11.87 %
MgO	10.29 %
K_2O	7.80 %

The apparent density of the fresh, not activated catalyst was found to be 2.371 g/cm^3. The true density was determined using helium displacement technique as 3.246 g/cm^3. The porosity of the pellet was calculated from the apparent and true densities as 26.9 %. The nitrogen isotherms obtained from a volumetric Brunauer, Emmett, Teller (B.E.T.) apparatus yielded specific surfaces of 32.05 - 32.84 m^2/g for fresh, not activated catalyst, 28.88 - 29.80 m^2/g for catalyst after reduction at 650°C, and 25.29 - 25.86 m^2/g for catalyst used for reforming at 540°C, 1 atm abs. with a H_2O : CH_4 ratio of 3 for a period of 14 hours. Pore volume and pore radii distribution of fresh, unreduced catalyst determined from desorption isotherms based on the procedure developed by Barrett, Joyner, and Halenda showed that the micropore area (18-30 Å) of the not activated, fresh catalyst was 20.8 m^2/g. The remaining portion seemed to be distributed among the pores with larger radii up to 330 Å. The second peak traced by the distribution curve within the macropore range (170 - 330 Å) was found insufficiently significant to consider the catalyst to be a bi-dispersed pore system. (Figs. 5 and 6).

Each series of experiments was run with a new catalyst charge. The activation schedule to every new catalyst consisted of

i. increase of reactor temperature to 300°C under a steady flow of dry hydrogen at a rate of $(40)(10^{-3})$ moles/min,

ii. at 300°C addition of steam to H_2 stream at a rate of $(6)(10^{-3})$ moles/min,

iii. increase of reactor temperature to 600°C at a rate of 80 - 100 °C/min keeping the flowrate of H_2 and steam at the same level,

iv. adjustment of H_2 : steam molar ratio to 9,

v. increase of the catalyst bed temperature to 650 ± 2^oC.

The catalyst was exposed to the steady flow of reducing atmosphere at this temperature for 16 hours. The existence of constant activity was checked to set a time limit for the duration of the experiments. In order to assess the effect of time on catalyst activity under reforming conditions, experiments at the same temperature, pressure, steam : methane ratio and feed rate were extended to 14 hours. The samples of product gas were analyzed every two hours. The conversions calculated from analytical results indicated no progressive decline of activity within this period.

4. RESULTS and DISCUSSION

4.1. Interpellet Mass Transfer Limitations

In establishing the operational limits for the elimination of film diffusion impedences it was observed that the external surface of the pellet (S_c) exposed to the reacting gas had a dominating effect on the extent of conversion (x_{CH_4}) rather that the mass of pellets (W). Consequently the parameter S_c/F_T ratio was used in graphical presentations instead of W/F_T). The inconsistencies of the parametric pellets of conversion against linear gas velocities for constant values of S_c/F_T ratio indicated that the linear velocity or gas flow rate was not the appropriate coordinate for the estimation of onset of film diffusion resistances. The irregularities were eliminated by substituting for velocity the Reynolds number (N_{Re}). A further advange of N_{Re} as the basic coordinate appeared in evaluating the effect of pressure on conversion. It was possible to infer the limiting value of N_{Re} at which the external mass transfer resistance at any S_c/F_T value influences the rate.

4.2. Intrapellet Mass Transfer Limitations

The clear dependence of the conversion on the external surface of the pellet exposed to gas flow was a significant indication that the observed rate was markedly influenced by pore diffusion resistances. To eliminate the extent of intrapellet mass transfer limitations a series of runs has been performed with ground and seized catalyst at a relatively low temperature level (540 ^oC) to diminish the intrinsic rate constant so that the potential required to provide the diffusion flux in the particles becomes insignificant and intrinsic kinetics could be observed. The particle sizes of the catalyst were changed for the ranges 0.25 - 0.30 mm, 0.425 - 0.70 mm, 0.353 - 0.60 mm, and 0.5 - 0.6 mm. The procedure for catalyst activation and adjustment of the flow rates were identical to those applied in the runs carried out with cylindrical pellets.

All the runs with ground catalyst resulted in massive carbon deposition independent of the feed rate of reactants even the H_2O : CH_4 ratio was far above the thermodynamic limit to prevent the "equilibrium carbon" formation. The formation of this "kinetic carbon" started apparently as soon as steam-H_2 mixture of the activation stage was replaced by the steam-CH_4 mixture, since even the first sample gave a very low conversion compared to that obtained with cylindrical pellets at identical catalyst mass and gas flow rates. Furthermore mass balances resulted in large discrepancies. The blocking of the catalyst bed with deposited carbon and the decrease of conversion at identical operating conditions were enhanced with the decrease of particle size. The carbon build-up was extremely accelerated when the particle diameter to reactor diameter ratio (d_p : D_R) was 1 : 34 or less. Coarse particles gave higher conversion, and blocking of the bed took relatively longer time. Prolonged operation resulted in the build-up of a back pressure in the reactor up to 0.8 atm which led to the conclusion that the catalyst bed was chocked. It is worthwhile noting that an increase of H_2O : CH_4 ratio in the feed from 3.0 to 7.0 without interrupting the operation did not restore the activity of the catalyst nor did it prevent blocking of the bed. The microscopic examination of the particles indicated that some size enlargement had also occurred. This observation solely and the low operating temperature were not sufficient to make a definite distinction whether carbon deposited was whisker carbon or encapsulating type carbon [1, 4]. On the other hand, numerous investigations have been carried out using catalyst with different Ni contents and support material at particle sizes changing between 0.25 - 0.63 mm and at various operating temperatures and pressures [3,6,9,10,12,13]. The results have even led to certain rate expressions deduced by L.H.H.W. - approach.

The experimental data obtained with cylindrical catalyst pellets with different diameters and heights at 540 ^oC, 1 atm abs. pressure provided sufficient evidence to conclude that the effectiveness factor was approximately 0.1 which could be verified from the asymptotic relationship $\eta = 1/\phi$, since the observed reaction rates based on volume of the pellet ($-r_{CH_4, obser., v}$) became almost identical with the inverse ratio of the length parameters (L) of the pellet. Values as low as 0.01, but increasing to 0.1 - 0.15 for effectiveness factor of commercial catalysts are reported elsewhere [1].

4.3. Effect of Initial Reactant Concentrations

The effect of initial concentrations of the reactants on the extent of conversion was studied by conducting series of runs each at six different values of H_2O : CH_4 ratio at 4 temperature levels (540, 600, 650, and 700^oC) all at atmospheric pressure. The runs were carried out with cylindrical pellets and the experimental conditions were:

Temperature, ^oC	: 540, 600, 650, 700
Pressure, atm abs.	: 0.987 - 0.992
Catalyst mass, g	: 5.6724
	5.7294
	5.8183
	5.7142
	5.7313
Total feed rate, 10^{-3} moles/min	: 56.74 - 6165
	28.34 - 31.21
	19.38 - 21.17
	14.29 - 15.46

W/F_T ratio, 10^3 (g)(min)/moles	: 0.100 - 0.092
	0.202 - 0.184
	0.300 - 0.275
	0.400 - 0.369
S_c/F_T ratio, 10^3 (cm^2)(min)/moles	: 0.135 - 0.124
	0.270 - 0.245
	0.397 - 0.363
	0.538 - 0.497
H_2O : CH_4 molar ratio	: 2.0
	3.0
	4.0
	5.0
	6.0
	7.0

Even the lowest H_2O : CH_4 ratio in each series of runs was kept above the thermodynamic minimum of 1,25 for prevailing temperature and pressure conditions to prevent the possible formation of kinetic carbon on active sites during the sequence of reactions. The results obtained are presented as plots of conversion (x_{CH4}) versus S_c/F_T values with H_2O : CH_4 ratio as parameter for 540 ^oC (Fig.7). It is evident from this figure that an increase of partial pressure of CH_4 in the feed caused an increase of conversion. The derivatives of the curves in Fig.7 at different S_c/F_T ratios represent the values of observed reaction rates based on external surface of the pellet. These values are plotted versus S_c/F_T (Fig.8) and are extrapolated to S_c/F_T = 0 to obtain the initial rates. The variation of the initial rates with the initial partial pressure (mole fraction) of CH_4 is shown in Fig.9. Any run corresponding to $y_{CH4,i} > 0.33$ has not been attempted, since in the region of very low water feed rates operational restrictions of the pumping system affected the uniformity of the feed composition. Furthermore, at the operating conditions relevant to the runs, the upper limit for $p_{CH4,i}$ from the thermodynamic point of view was calculated to be 0.44 (at 540 ^oC and 1 atm) beyond which a continuous deposition of equilibrium carbon would take place. Although the catalyst mass in the reactor and the flow conditions were not adequate to enable a close approach to equilibrium, an increase of $p_{CH4,i}$ to the limiting value would definitely have promoted the formation and probable build-up of kinetic carbon. In spite of the termination of the curve at $p_{CH4,i}$ = 0.333, the plot tends to give a maximum.

4.4. Oxides of Carbon Formed

The experimental data provided indications as to which of the oxides of carbon is being formed first. Plotting the ratio $CO_2/(CO + CO_2)$ against conversion x_{CH4}(Fig. 10) and versus S_c/F_T (Fig. 11) for operating conditions 540^oC, 1 atm abs., steam ratio 3.0, and for different pellet dimensions, resulted in curves which, if extrapolated to zero conversion gave $CO_2/(CO + CO_2)$ values very close to unity. This fact furnished evidence to conclude that the primary reaction between steam and methane has the stoichiometric form

$$CH_4 + 2\ H_2O \rightarrow CO_2 + 4\ H_2$$

and a stoichiometry of the type

$$CH_4 + H_2O \rightarrow CO + 3\ H_2$$

is not representative of the reforming reaction. CO is being formed by the water gas shift reaction taking place in the reverse direction

$$CO_2 + H_2 \rightarrow CO + H_2O$$

An additional graph (Fig. 12) illustrates the variation of the ratio $CO_2/(CO+CO_2)$ with conversion x_{CH_4} at four different temperatures and pressures. The values of $CO_2/(CO + CO_2)$ at equilibrium conversions at different temperatures and 1 atm abs. pressure with $\alpha = 3.0$ are shown in Fig. 13. It can be seen from Fig. 12 that the pressure had no effect on the relative concentrations of carbon oxides, since the values of $CO_2/(CO + CO_2)$ ratio for given x_{CH_4} were almost the same at 600°, 650°, and 700^oC independent of pressure. The relative location of the curves in Fig. 12 indicates the sensitivity of the rate of the reverse shift reaction to temperature. In fact, for the same extent of conversion of CH_4, the value of $CO_2/(CO + CO_2)$ decreases rapidly if the temperature is increased from 540° to 700^oC.

In view of these facts it is possible to explain the contradiction between established opinion and the conclusions reached in this investigation with respect to the primary reforming reaction. Bodrov et al. [12, 13] studied the kinetic of methane reforming on metallic nickel foil and with nickel on alumina catalysts at 800 - 900^oC, barometric pressure and different $H_2O : CH_4$ mole ratios, and found CO very much in excess over CO_2. On this basis, a kinetic mechanism was postulated for the reforming reaction with CO as the primary reaction product which was assumed to be oxidized subsequently to CO_2. Akers and Camp [11] also conducted experiments for reforming of CH_4 at different temperatures and barometric pressure. From the concentration of CO and CO_2 in the reformed gas obtained at 638 ^oC catalyst bed temperature, they concluded that both oxides of carbon were primary reaction products, but that CO_2 was being formed at an initial rate of between 2 to 9 times the rate of CO formation, depending on $H_2O : CH_4$ molar ratio in the feed. These interpretations and conclusions are not surprising since at temperatures 800 - 900 ^oC the rate of the reverse shift reaction was increased to such an extent that CO_2, having been formed as the initial reaction product on active sites, had been instantaneously converted to CO, so that CO appeared virtually as the primary product. The temperature 638 ^oC of the experiments conducted by Akers and Camp was such that both CO_2 and CO existed in significant concentrations in the reformed gas and led the investigators to inconclusive interpretation. If the initial oxide of carbon were CO corresponding to the stoichiometry

$$CH_4 + H_2O \rightarrow CO + 3\ H_2$$

then the curve of $CO_2/(CO + CO_2)$ against S_c/T_F such as Fig. 10 would have approached the value of $CO_2/(CO + CO_2)$ corresponding to the equilibrium conversion from the opposite side to that obtained in the experimental work as shown in Fig. 14.

4.5. Rate Equation

Although numerous kinetic mechanisms and rate equations for CH_4 reforming have been proposed, these tend to be rather conflicting due to the differences in operating conditions, catalyst formulation and activation procedure, and reactors geometry [1 - 14]. There is a general agreement, however, that the reaction is first order with respect to CH_4. Values reported for the activation energy cover a large span from 4.800 to 43,800 cal/mole, which may be explained by different intrapellet mass transfer impedences. Reaction orders with respect to other components are changing in a very dissimilar nature showing in some cases a negative order for steam which indicates a non-monotonic dependence [2] of the rate on H_2O concentration. In some rate expressions developed, the absence of CH_4 partial pressure in the denominator suggest that it is not being adsorbed, which is very unlikely since it is well known that CH_4 is being dissociatively adsorbed on Ni to give CH_2 - * radicals.

In view of the uncertainties about the real nature of the complex surface kinetics, the rate equation developed in this study is restricted to an expression in macrokinetic terms which would satisfactorily fit the data. The experiments at 540, 600, 650, and 700 ^oC with freshly activated catalyst and at varying H_2O : CH_4 ratios were correlated by regression and the following rate equation was obtained for the initial reaction based on mass of catalyst

$$- r_{CH_4,\ init} = k\ p_{CH_{4,i}}^{0.973}\ p_{H_2O,i}^{1.094}$$

The activation energy is calculated to give 17,640 cal/mole with the frequency factor $k_o = (0.372)(10^2)$. Confidence limit of correlation was 99 %. The sum of the square of approximation errors and the residual error were $(0.123)(10^{-3})$ and $(0.642)(10^{-2})$, respectively.

An attempt to further correlate the data at 540 ^oC by a LHHW - type expression which involves the simultaneous adsorption of CH_4 and H_2O resulted in a semi-empirical equation for the initial reaction rate

$$- r_{CH_4,init.} = \frac{(0.02412)\ p_{CH_4,i}\ p_{H_2O,i}^{2}}{\left[1 + (1.422)p_{CH_4,i}^{0.5} + (1.96)\ p_{H_2O,i}^{1.5}\right]^3}$$

The sum of the squares of approximation errors was 0.0077.

5. CONCLUSION

The steam reforming reaction of CH_4 has been examined on commercial catalyst pellets exposing a definite external area of contact to an up-flowing stream of gas. All experimental rate determinations were made on freshly reduced catalyst within 14 hours after activation, during which the existence of constant activity was confirmed. It was found that the extent of conversion was clearly dependent on the external surface and pore diffusion resistances limit the rate of reaction. It was established taht the gas film resistance used to have any effect for $N_{Re} > 4$ at 540 ^oC. The effectiveness factor of the catalyst was found to be no more than 0.1 at 540 ^oC and even less at higher temperatures as the intrinsic kinetic rate increased. Experiments performed with sized fractions of ground catalyst resulted in massive carbon deposition and blocking of the bed. Evidence has been obtained for the formation of CO_2 as the initial stable oxide of carbon instead of CO. Empirical expressions for the initial reaction rate have been derived by the least mean square approximation for the operation at 540 ^oC and 1 atm pressure:

$$-r_{CH_4,init.} = (0.000673)\, p_{CH_4,i}^{0.973}\, p_{H_2O,i}^{1.094}$$

$$-r_{CH_4,init.} = \frac{(0.02412)\, p_{CH_4,i}\, p_{H_2O,i}^{2}}{\left[1 + (1.422)\, p_{CH_4,i}^{0.5} + (1.96)\, p_{H_2O,i}^{1.5}\right]^3}$$

Conversions for given S_c/F_T or W/F_T did not change with pressure within the range 1 - 13 atm abs. at 540° - 700 ^oC which implied that the adsorption of none of the components is likely to be a rate controlling factor for the reaction.

NOMENCLATURE

D_R	Diameter of the reactor, cm
F_T	Total feed rate of the reactant mixture, moles/min
L	Length parameter of the pellet, (volume of pellet/external surface), cm^3/cm^2
N_{Re}	Reynolds number
S_c	External surface of the catalyst pellet, cm^2
W	Mass of the catalyst charge, g
d_p	Diameter of the ground catalyst particle, cm
p_i	Partial pressure of component i, atm.
$-r_{CH_4,m}$	Reaction rate of methane based on catalyst mass, moles of CH_4/(g of cat)(min).
$-r_{CH_4,s}$	Reaction rate of methane based on external surface of catalyst pellet, moles of $CH_4/(cm^2)$(min).

$-r_{CH_4,v}$	Reaction rate of methane based on volume of catalyst pellet, moles of $CH_4/(cm^3)(min)$.
$-r_{CH_4,obs}$	Observed reaction rate of methane, moles of $CH_4/(cm^3)(min)$.
x_{CH_4}	Moles of CH_4 reacted per mole of total feed.
y_i	Mole fraction of component i.
α	Steam : methane molar ratio in the feed to the reactor.
η	Effectiveness factor, dimensionless.
ϕ	Thiele modulus, dimensionless.

REFERENCES

1. Rostrup-Nielsen, J.R. ; **Catalysis Sci. and Tech.;** 5, 1, Springer Verlag, Berlin (1984).
2. El-Nashaie, S.S.E.H. ; A.S. AR-Ubaid ; M.A. Soliman ; A.M. Adris ; **Jour. Eng. and Sci., King Saud Univ.** ; 14, 2, 247 (1988).
3. Ross, J.R.H. ; M.C.F. Steel ; **Jour. Chem. Soc. Faraday Trans. I** ; 69, 10 (1973).
4. Trimm, D.L; **Catal. Rev. Sci. and Eng.** ; 16, 155 (1977).
5. Van Hook, J.P. ; **Catal. Rev. Sci. and Eng.**; 21, 1 (1981).
6. Agnelli, M.E.; E.N. Ponzi; A.A. Yeramian ; **Ind. Eng. Chem. Res.** ; 26, 1707 (1987).
7. Soliman, M.A.; S.S.E.H. El-Nashaie ; A.S. Al-Ubaid ; A.M. Adris ; **Chem. Eng. Sci.**; 43, 8 (1988).
8. Numaguchi, T.; K. Kikuchi ; **Chem. Eng. Sci.**; 43, 8 (1988).
9. De Deken, J.C. ; E.F. Devos ; G.F. Froment ; **Steam Reforming of Natural Gas ; Amer. Chem. Soc. Symposium Series** ; 181-197, Boston (1982).
10. Kopsel, R.; A. Richter ; B. Meyer ; **Chem. Techn.; 32** 9, 460 (1980).
11. Akers, W.W.; D.P. Camp ; **A.I.Ch.E.Jour.**; 1, 471 (1955).
12. Bodrov, I.M.; L.O. Apel'baum ; M.I. Temkin ; **Kinet. i Katal.** ; 5, 696 (1964).
13. Bodrov, I.M.; L.O. Apel'baum ; M.I. Temkin ; **Kinet. i Katal.** ; 8, 832 (1967).
14. Bodrov, I.M.; L.O. Apel'baum ; M.I. Temkin ; **Kinet. i Katal.** ; 8, 877 (1967).

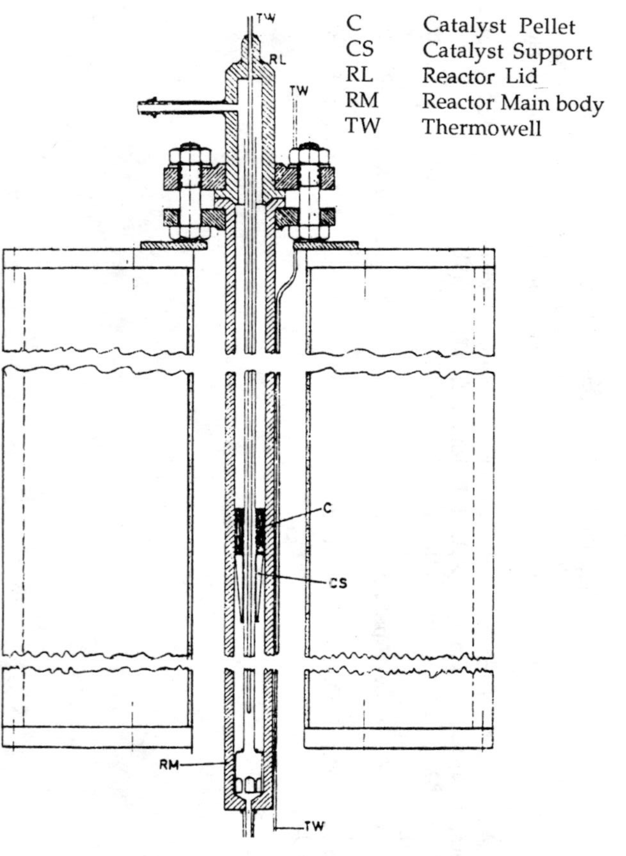

Fig. 1. Reactor No. 1

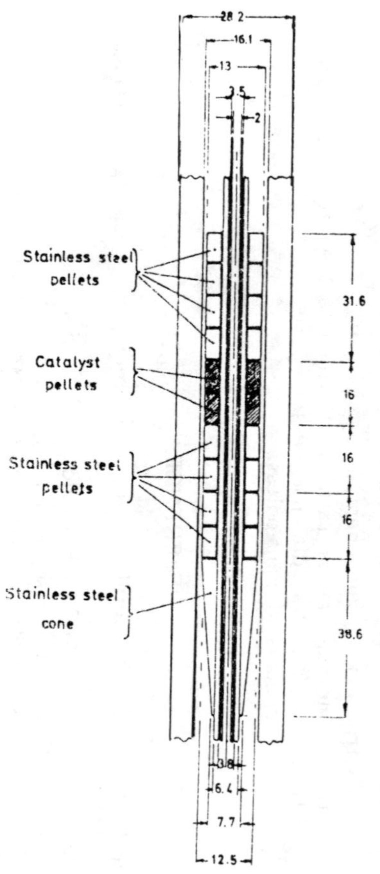

Fig. 2. Catalyst Bed of Reactor

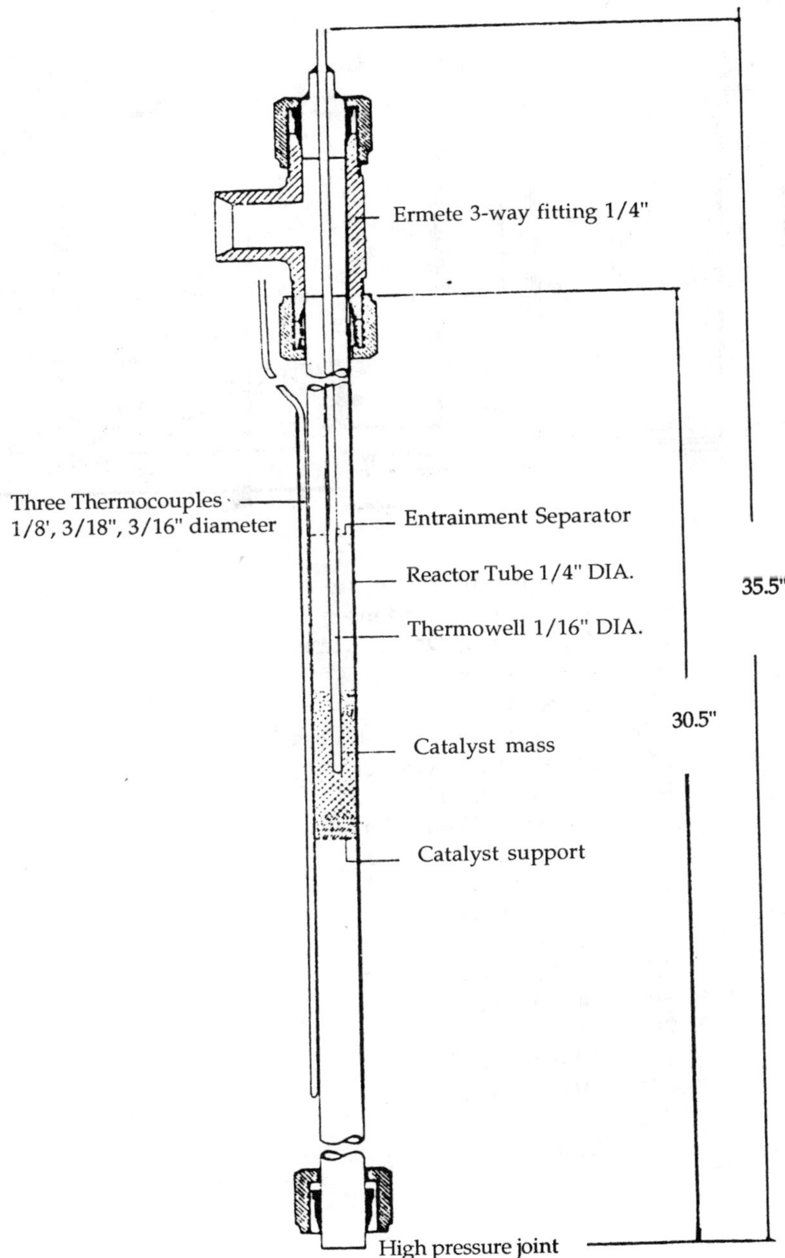

Fig. 3. Reactor No. 2

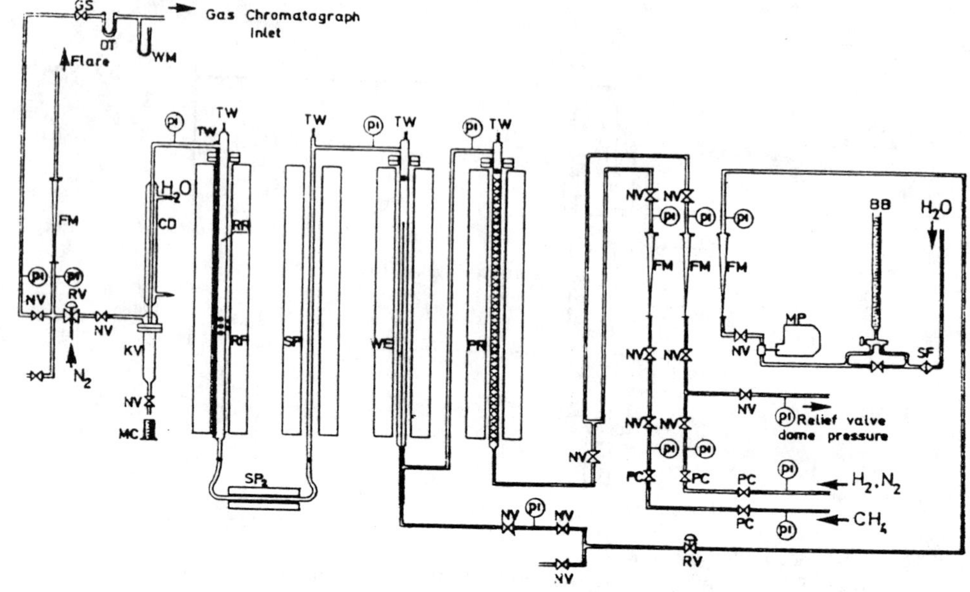

BB Uniform-bore Burette
CD Co-current Condenser
DT Glass $CaCl_2$ Drying Tubes
FM Flowmeter
GS Glass 3-way Valve
KN Pressure Knock-out Vessel
MC Measuring Cylinder
MP Micro-pump
PC Pressure Controller
PR Feedstock Preheater
RF Reactor Furnace
RV Relief Valve
RR Reactor
SF Sintered Glass Filter
SP Superheater No. 1
SP2 Superheater No. 2
TW Thermowell
WE Water Evaporator
WM Water Manometer
• Temperature Control Points

Fig. 4. Schematic Diagram of the Reforming Apparatus

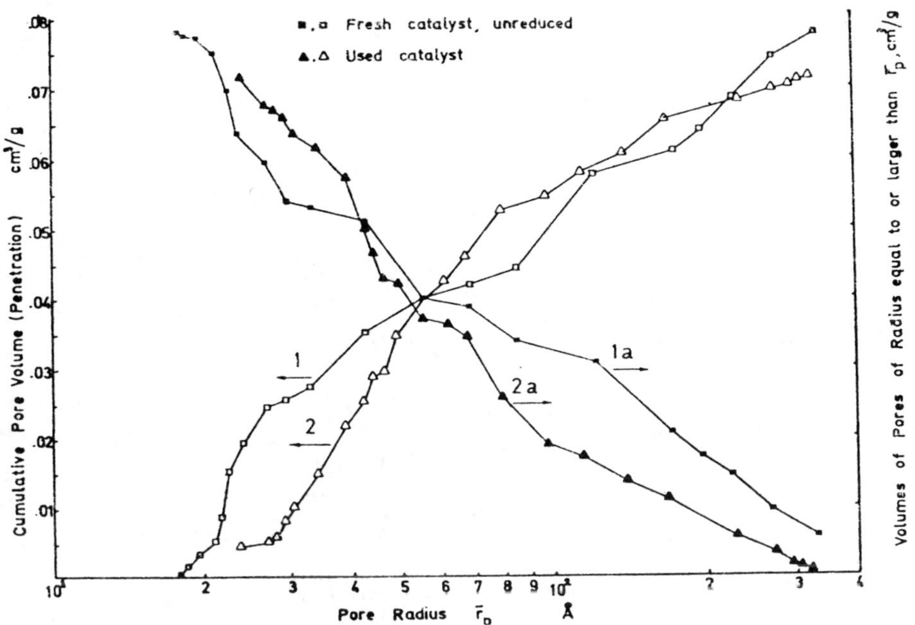

Fig. 5. Pore volumes of used and unused catalyst.

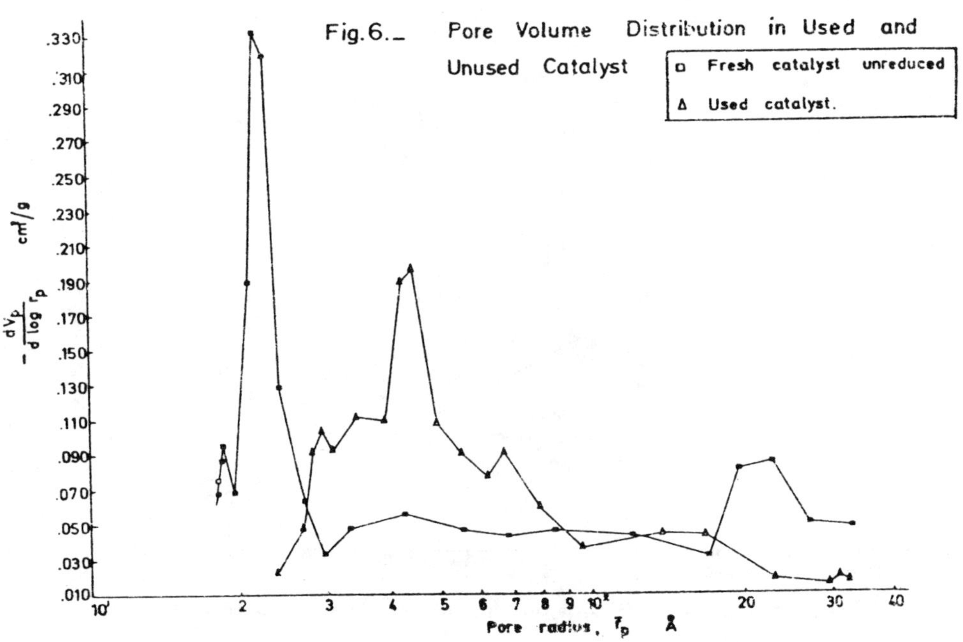

Fig. 6. Pore volume distribution in used and unused catalyst.

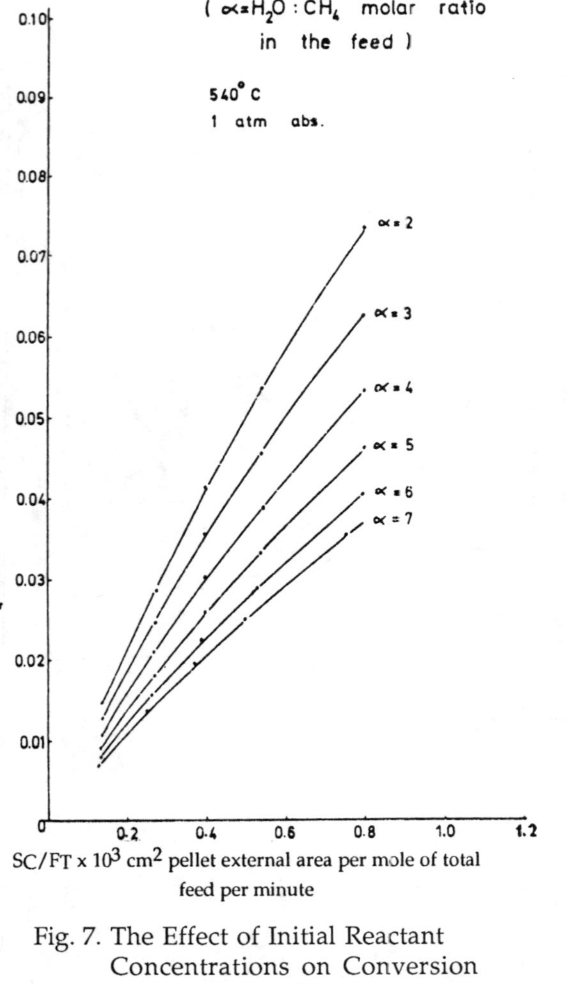

Fig. 7. The Effect of Initial Reactant Concentrations on Conversion

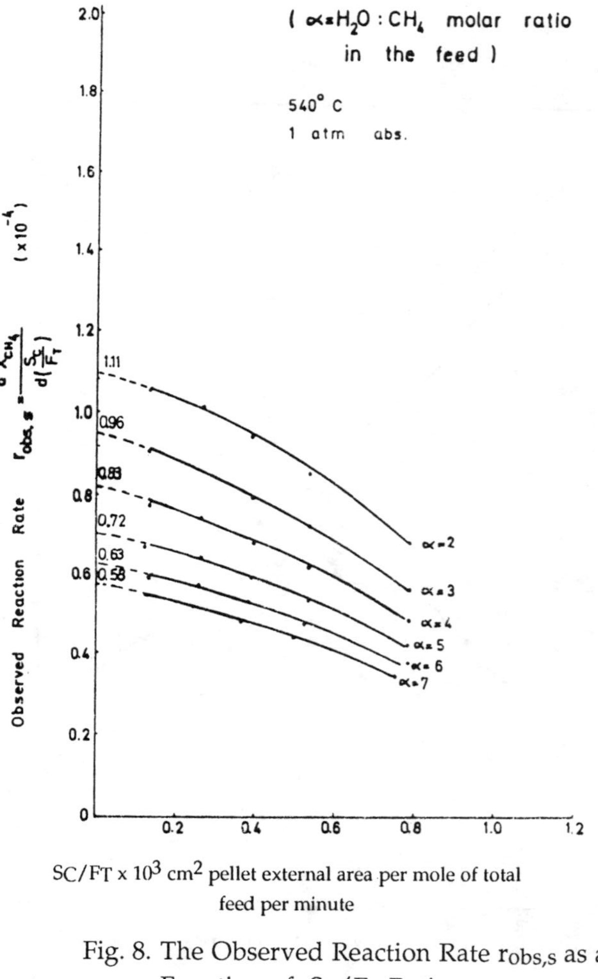

Fig. 8. The Observed Reaction Rate $r_{obs,s}$ as a Function of SC/FT Ratio

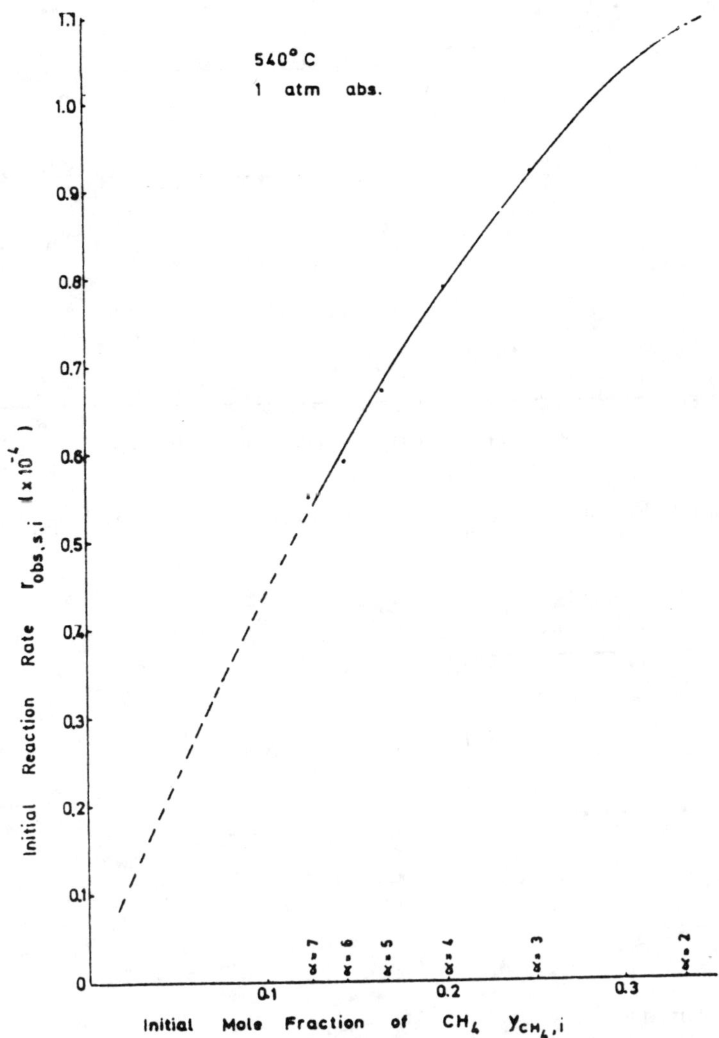

Fig. 9. The Initial Reaction Rate $r_{obs, s, i}$ as a Function of Initial Mole Fraction of CH_4 $y_{CH_4, i}$

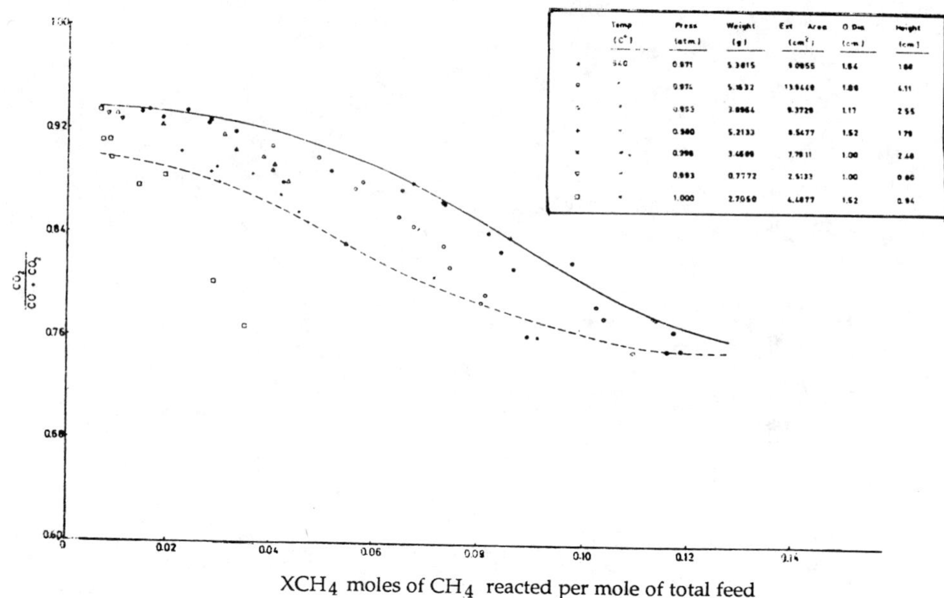

Fig. 10. The Ration $\frac{CO_2}{CO + CO_2}$ as a Function of Conversion X_{CH4} ($\alpha = 3.0$)

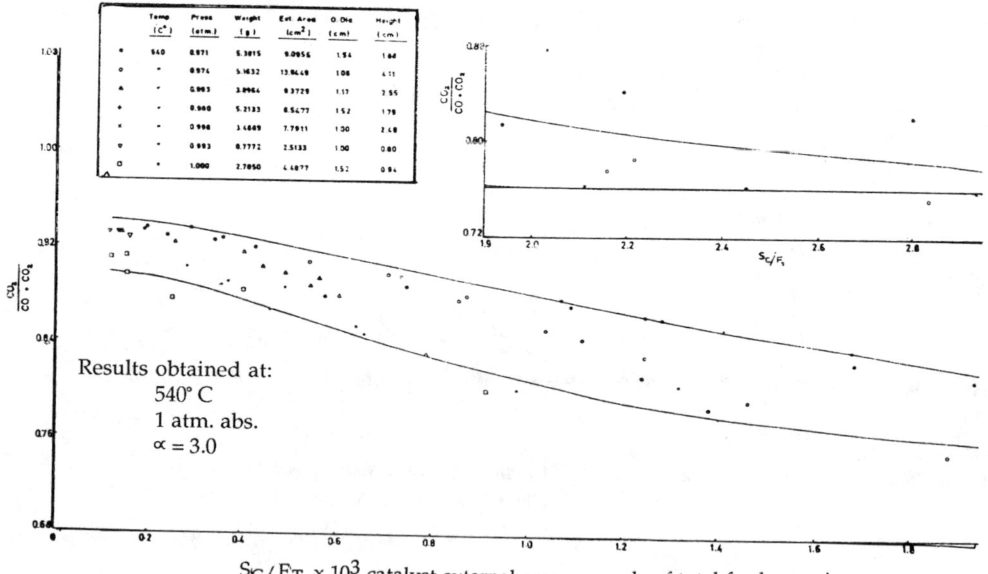

Fig. 11. The Ratio $\frac{CO_2}{CO + CO_2}$ as a Function of the Ratio S_C/F_T

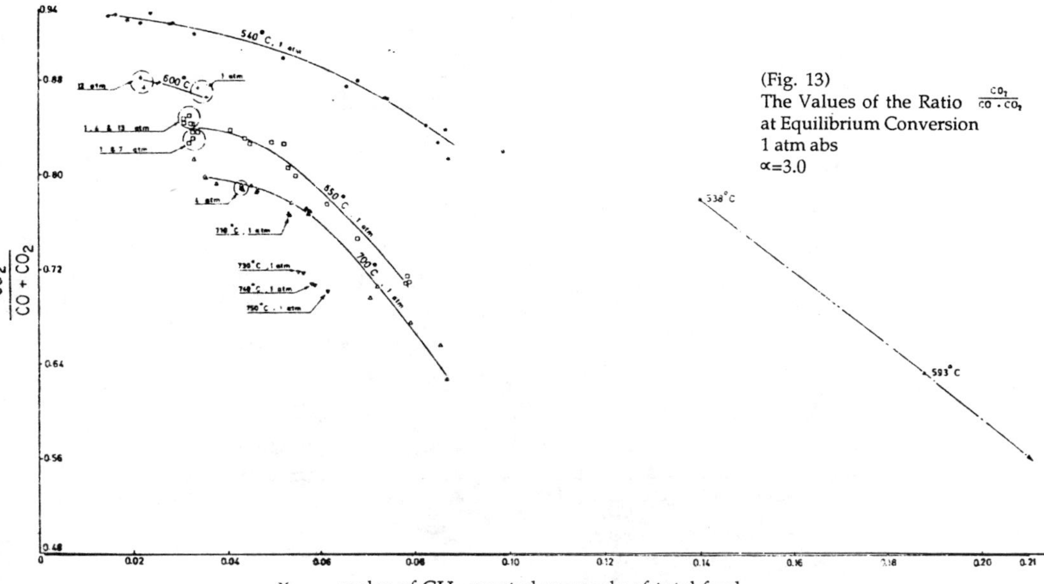

X_{CH4} moles of CH_4 reacted per mole of total feed

Fig. 12. The Ratio $\frac{CO_2}{CO + CO_2}$ as a Function of the Conversion X_{CH4} at Differnet Temperature and Pressure Levels (∝ = 3.0)

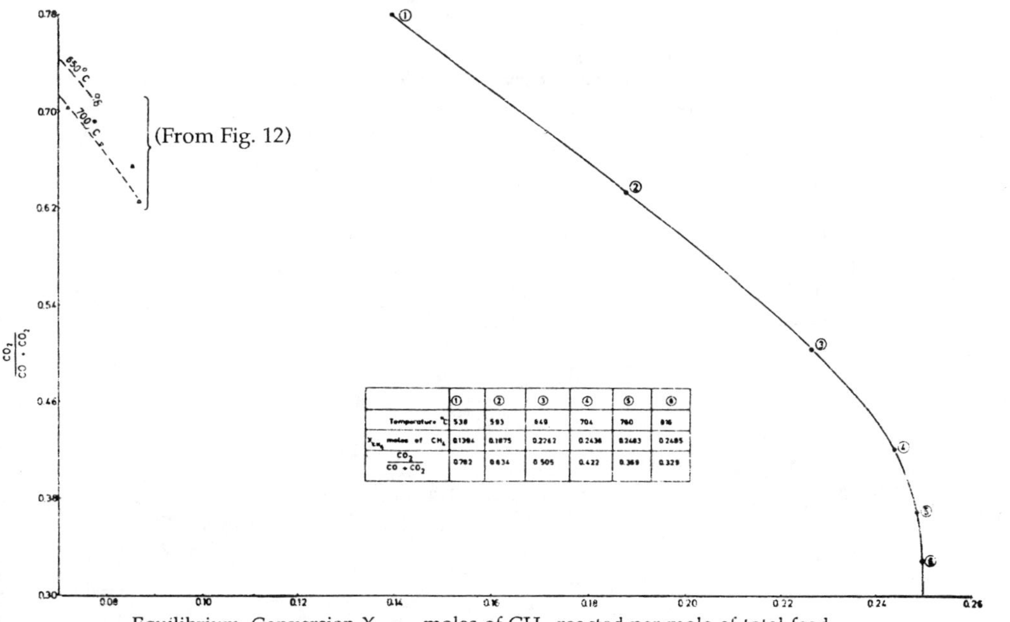

	①	②	③	④	⑤	⑥
Temperature °C	538	593	649	704	760	816
X_{CH_4} moles of CH_4	0.1394	0.1875	0.2262	0.2436	0.2483	0.2485
$\frac{CO_2}{CO + CO_2}$	0.782	0.634	0.505	0.422	0.369	0.329

Equilibrium Conversion X_{CH4} moles of CH_4 reacted per mole of total feed

Fig. 13. The Ratio $\frac{CO_2}{CO + CO_2}$ as a Function of Equilibrium Conversion X_{CH4} at Different Temperatures 1 atm abs. ∝=3.0

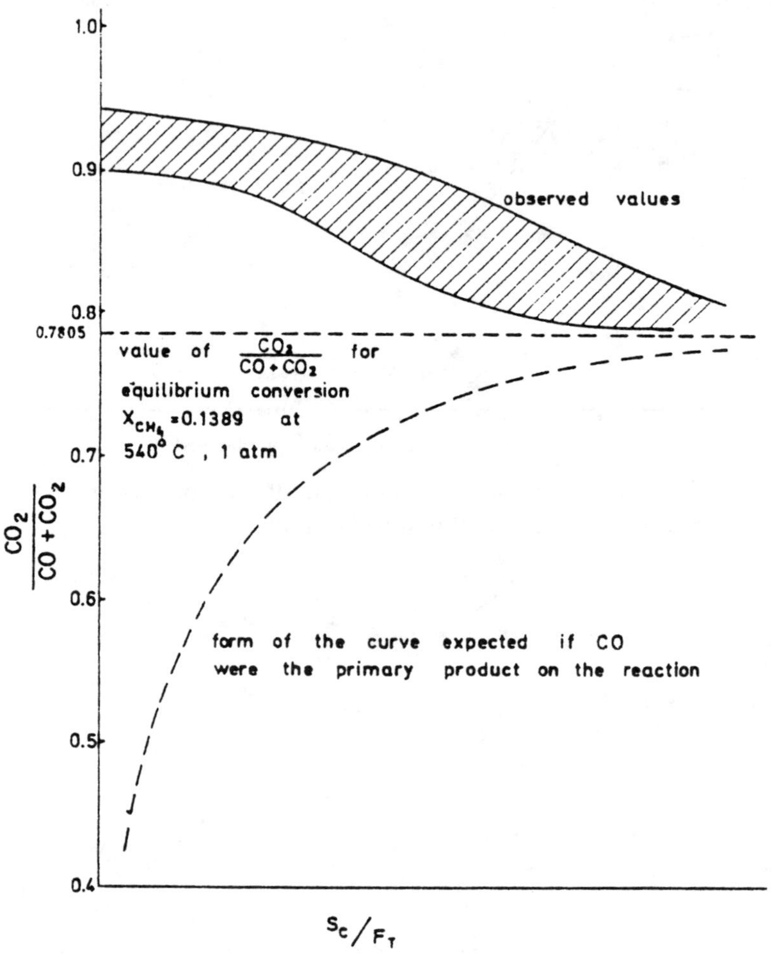

Fig. 14. Possible Variations of $\frac{CO_2}{CO + CO_2}$ Curve as a Function of the Ratio S_C/F_T (for CO_2 or Co as the initial reaction products)

LIQUEFACTION OF BEYPAZARI LIGNITE BY MOLTEN SALT CATALYSIS

Y. Yürum, J. Özkisacik and S. Bektas
Hacettepe University, Department of Chemistry, Beytepe, Ankara 06532, Turkey

Abstract

Liquefaction of Beypazarı lignite in tetralin using $NiCl_2$-KCl-LiCl (14:36:50 molar percentages) as catalyst was investigated. Effects of the catalyst/lignite ratio and temperature were determined in experiments done at 275°C, 300°C and 360°C. Liquid products were separated into oils, asphaltenes and asphaltols by a solvent extraction method. Yield of liquefaction increased with temperature in all experiments, the highest yield was observed in experiments performed at the eutectic temperature of the catalyst mixture. The highest yields of oils were 20% and 30% with a catalyst/coal ratio of 0.5 at 275°C and 300°C, respectively. The activity of the catalyst increased in experiments in which the catalyst was molten. The yield of asphaltenes were not affected with increases in the catalyst/coal ratio in the experiments done at 275°C or 300°C in which the catalyst mixtures were in solid state. Asphaltene yields decreased from 25% to less than 5% with increasing values of catalyst/coal ratio and the asphaltol yields remained constant at 10% between catalyst/coal ratios of 0.25 and 1.00 and suddenly increased to 30% and 40% for catalyst/coal ratios of 1.50 and 2.00, respectively, at 360°C.

1. INTRODUCTION

The use of molten metal halides for hydrocracking of coal and coal extract have been extensively investigated [1-4]. Most of that work was done with $ZnCl_2$ as the catalyst using bituminous coal extract as the feed.

On a batch scale, work has been done using not only extract as feedstock but also coal directly [2, 5]. Bituminous and subbituminous coals were investigated and they were found excellent feedstocks for zinc chloride hydrocracking. Hydrocracking of coal directly in molten zinc chloride on a continuous basis was demonstrated using $ZnCl_2$/coal ratios varying from 0.83 to 1.83 [6]. Molten $SnCl_2$ has been used as hydrocracking catalyst by some workers [7-9]. Investigations of hydroliquefaction [10-15] using molten salts of $ZnCl_2$ and/or $SnCl_2$ and/or alkali metal chlorides or transition metal chlorides ($MoCl_5$, CuCl or $CrCl_3$) showed the combined molten catalysts to be more effective than pure $ZnCl_2$ or $SnCl_2$. It has been shown that mixed halides $AlCl_3$ + $MoCl_3$ and $AlCl_3$ + $NiCl_2$ are effective catalyst combinations in coal conversion to low molecular weight hydrocarbons [16].

$NiCl_2$-LiCl-KCl ternary molten salts were examined as catalysts for hydroliquefaction of subbituminous Wandoan coal at 400°C. Higher conversion and higher hexane soluble yield was apparent with molten catalyst than the corresponding solid catalyst [17].

The objective of the present study is to investigate the effect of $NiCl_2$-LiCl-KCl (14:50:36, mol%; eutectic temperature: 360°C) in hydroliquefaction of a Turkish lignite at temperatures below and at the eutectic temperature of the catalyst mixture. Effects of the catalyst/lignite ratio and temperature were determined in experiments done at 275°C, 300°C and 360°C.

2. EXPERIMENTAL

Beypazarı lignite was used in this study. The elemental analysis of Beypazarı lignite is given in Table I. The lignite sample was ground under a nitrogen atmosphere to -65 mesh ASTM and stored under nitrogen.

The liquefaction experiments of coal was carried out using microreactors of 15 ml capacity. Beypazarı lignite was liquefied by using molten and solid salt mixtures of $NiCl_2$-KCl-LiCl (14:36:50 molar percentages). Effects of the catalyst/lignite ratio and temperature were determined in experiments done at 275°C, 300°C and 360°C. Catalyst/coal ratios = 0, 0.25, 0.50, 1.00, 1.50 and 2.00 were used. In all of the experiments 1g of coal was used. Prior to the experiments 1g of coal and required amount of catalyst mixture was mixed together with 100ml of methanol and stirred for 1 hour. Then the methanol was evaporated under vacuum and the catalyst absorbed coal was charged with 5ml of tetralin into the autoclave. The microreactors were flushed with nitrogen to remove any oxygen present and sealed. In all of the experiments the vapor density at supercritical conditions was 0.3 g/ml. The experimental temperatures of 275°C, 300°C or 360°C were used in different sets of experiments. The temperatures were adjusted by a thermocouple immersed into the reaction mixture and an on-off temperature controller. The heating rate was 10°C/min and the maximum experimental temperature was held constant for 60 minutes. At the end of an experiment the microreactor was cooled to room temperature. The solution in the microreactor was separated from the residue by filtration. The residue was extracted with 500ml of tetrahydrofuran in a soxhlet extractor. Tetrahydrofuran in the solution was separated under vacuum and the extract obtained was added to the filtrate of supercritical experiment. The residue was first washed with hot water to wash the catalyst mixture absorbed in the coal until $AgNO_3$ test in the washings gave negative result for chloride ions. Then the residue was dried at 100°C in a nitrogen atmosphere and weighed to determine the yield of experiment. Percent yield was calculated by the following equation,

$$\% \text{ Yield} = (W_c - W_r)100/W_c, \text{ where,}$$

W_c: weight of dry coal charged to the autoclave
W_r: weight of the residue after supercritical and soxhlet extraction.

Liquid products were separated into oils, asphaltenes and asphaltols by a solvent extraction method given elsewhere [18].

The eutectic temperature of the catalyst $NiCl_2$-KCl-LiCl (14:36:50 mol%) was measured by using a Mettler TA3000 differential scanning calorimeter. The temperature was raised to 500 °C by 5°C/min, in a nitrogen atmosphere. The eutectic temperature was measured as 360°C.

3. RESULTS AND DISCUSSION

Yield of Liquefaction

Figure 1 gives change of yield of liquefaction with catalyst/coal ratio and temperature. Yield of liquefaction increased with temperature in all experiments, the highest yield was observed in experiments performed at the eutectic temperature of the catalyst mixture. The catalyst as it was determined by DSC experiments was molten at this temperature. This indicated that the same catalyst mixture in all catalyst/coal ratios was more efficient in molten phase than it was used as a solid mixture.

As the catalyst/coal ratio was increased from 0 to 0.5 the yield increased from 44% to 59% in experiments done at 275°C, from 49% to 63% in experiments done at 300°C and from 63% to 72% in experiments done at 360°C. Higher catalyst/coal ratios caused a decrease in the yield of experiments. It seems that the most efficent catalyst/coal ratio to obtain the highest yield was 0.5.

Oils

Figure 2 gives the change in oil yield with the catalyst/coal ratio and temperature. Experiments performed with a solid catalyst mixture at 275°C and 300°C produced similar oil yields in all catalyst/coal ratios. The highest yields of oils were 20% and 30% with a catalyst/coal ratio of 0.5 at 275°C and 300°C, respectively. In experiments done at 360°C in which the catalyst mixture was molten the highest yields changed between 54-55% with catalyst/coal ratios of 0.5 and 1.0. It seemed that the activity of the catalyst increased in experiments in which the catalyst was molten. The activity was not changed significantly at temperatures below the eutectic temperature of the catalyst mixture.

Asphaltenes

Change of asphaltene yields with temperature and catalyst/coal ratio are presented in Figure 3. Experiments performed in the absence of catalysts produced increasing amounts of asphaltenes as the temperature was raised from 275°C to 375°C. This indicated that the production of asphaltenes was directly dependent on the temperature of experiment while there was not any catalyst used.

The most interesting feature of the data in Figure 3 is the sharp decrease of asphaltene yields from 25% to less than 5% with increasing values of catalyst/coal ratio at 360°C. This effect was most pronounced in experiments with catalyst/coal ratios higher than 0.5. At these experimental conditions it seems that the asphaltenes degrade into products of smaller molecular weight such as oils and some volatile compounds. This observation is supported by our findings in the oils yields. At the very same conditions there was a considerable increase in the yield of oils as it was stated above. Ozawa et al. [19] found that asphaltenes decomposed into smaller molecular weight hydroaromatics in experiments done at the presence of molten catalysts. Increasing the catalyst/coal ratio to higher values than 0.5 did not further decrease the yield of asphaltenes. The yield of asphaltenes were not affected with increases in the catalyst/coal ratio in the experiments done at 275°C or 300°C in which the catalyst mixtures were in solid state; the yields of asphaltenes remained between 15-20% in all experiments done at 275°C and 300°C.

Asphaltols

Asphaltol yields are given in Figure 4. Asphaltol yields of experiments done at 275°C and 300°C were not affected by the change in catalyst/coal ratio. In experiments done at 360°C, the asphaltol yields remained constant at 10% between catalyst/coal ratios of 0.25 and 1.00 and suddenly increased to 30% and 40% for catalyst/coal ratios of 1.50 and 2.00, respectively. This abrupt increase in the asphaltol yield is in accord with oils and asphaltenes yields encountered in experiments done at 360°C; oils and asphaltenes started to diminish at catalyst/coal ratios of 1.50 and 2.00. It seemed that higher catalyst/coal ratios favored asphaltols formation to the production of less complex and smaller molecular weight compounds found in oils or asphaltenes. This situation was probably due to a sharp increase in the radical concentration [20] and the hydrogen atoms supplied by the tetralin to naphthalene conversion was no more sufficient to saturate the radicals produced and thus high molecular weight material was produced. Gas chromatographic analyses done in the present study revealed that tetralin/naphthalene ratio at 360°C was the least with respect to tetralin/naphthalene ratios obtained in experiments at lower temperatures. This finding supported our claim in relation to insufficient hydrogen atom concentration in experiments done at 360°C.

Residue

The percentage of residual material obtained in experiments is given in Figure 5. In all experiments the residual percentage passed through a minimum value at the catalyst/coal ratio of 0.5. This showed that the optimum catalyst/coal ratio was 0.5 in the reaction of Beypazarı lignite with tetralin. As the temperature was raised from 275°C to 360°C at which the catalyst was molten the percentage residue decreased from over

40% to 25%. This result indicated the high effectivity of the molten salt catalyst used in connection to the conversion of the lignite into greater amounts of soluble products and less amounts of residual material.

4. CONCLUSIONS

Liquefaction of Beypazarı lignite using both solid and molten $NiCl_2$-KCl-LiCl as catalysts gave high conversions of more than 75% and high oils yields of 55%. These findings indicated that the catalyst is highly effective and have high oils selectivity for coal liquefaction. Molten catalysts seemed to be more effective than the solid catalysts. The optimum catalyst/coal ratio was 0.5 in experiments which gave the highest conversions.

REFERENCES

[1] Zielke, C.W., Struck, R.T., Evans, J.M., Costanza, C.P. and Gorin, E., Ind. Eng. Chem. Process Des. Dev., 5 , 151 (1966).

[2] Zielke, C.W., Struck, R.T., Evans, J.M., Costanza, C.P. and Gorin,E., Ind. Eng. Chem. Process Des. Dev., 5 , 158 (1966).

[3] Zielke, C.W., Struck, R.T. and Gorin, E., Ind. Eng. Chem., 8 ,552 (1969).

[4] Struck, R.T., Clark, W.E., Dudt, P.J., Rosenhoover, W.A., Zielke, C.W. and Gorin, E., Ind. Eng. Chem., 8 , 546 (1969).

[5] Zielke, C.W., Rosenhoover, W.A. and Gorin, E., Adv. Chem. Ser., No. 151, 153 (1976).

[6] Zielke, W.C., Klunder, E.B., Maskew, J.T. and Struck, R.T., Ind. Eng. Chem. Process Des. Dev., 19 , 85 (1980).

[7] Larsen, J.W. and Earnest, S., Fuel Process Technol., 2 , 123 (1979).

[8] Sato, Y., Imuta, K. and Yamakawa, T., Fuel, 60 , 1159 (1981).

[9] Mobley, D.P and Bell, A.T., Fuel, 58 , 661 (1979).

[10] Nomura, M., Miyake, M., Sakashita, H. and Kikkawa, S., Fuel, 61 , 19 (1982).

[11] Miyake, M., Sakashita, H., Nomura, M., and Kikkawa, S., Fuel, 61 , 125 (1982).

[12] Nomura, M., Kimura, K. and Kikkawa, S., Fuel, 61, 1119 (1982).

[13] Nomura, M., Sakashita, H., Miyake, M., and Kikkawa, S., Fuel, 62 , 73 (1983).

[14] Nomura, M., Yoshida, T. and Morita, Z., Ind. Eng. Chem. Prod. Res. Dev., 23 , 215 (1984).

[15] Nomura, M., Yoshida, T., Philip, P. and Gilbert, T., Fuel, 64 , 108 (1985).

[16] Butler, R. and Snelson, A., Fuel, 59 , 93 (1980).

[17] Song, C., Nomura, M. and Miyake, M., Fuel, 65 ,922 (1986).

[18] Yürüm and M. Levy, Fuel 64, 102 (1985).
[19] Ozawa, S., Ohsaki, H. and Ogino, Y., Fuel Process.Technol., 17 , 187 (1987).
[20] Petrakis, L. and Grandy, D.W., Fuel, 60 , 120 (1981).

TABLE I: Elemental Analysis of Beypazarı Lignite

	%, daf
Carbon	61.2
Hydrogen	5.5
Sulphur (total)	5.3
Nitrogen	1.9
Oxygen (by difference)	26.1
Sulphur types	
Sulphate	0.9
Pyrite	0.3
Organic	4.1

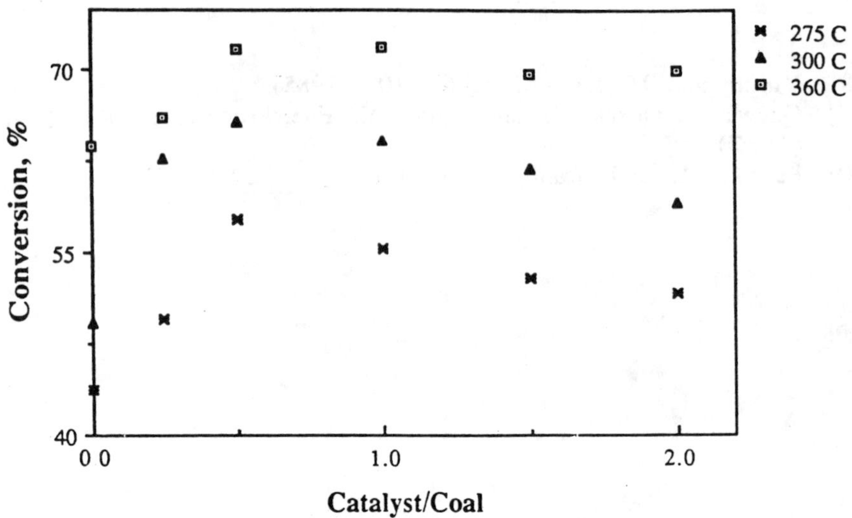

Figure 1. Change of conversion with catalyst/coal ratio.

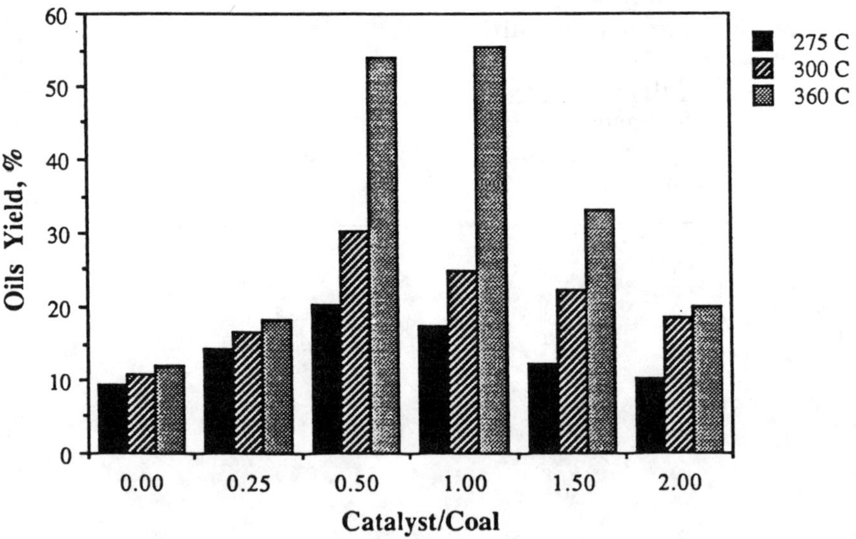

Figure 2. Change of oils yield with catalyst/coal ratio.

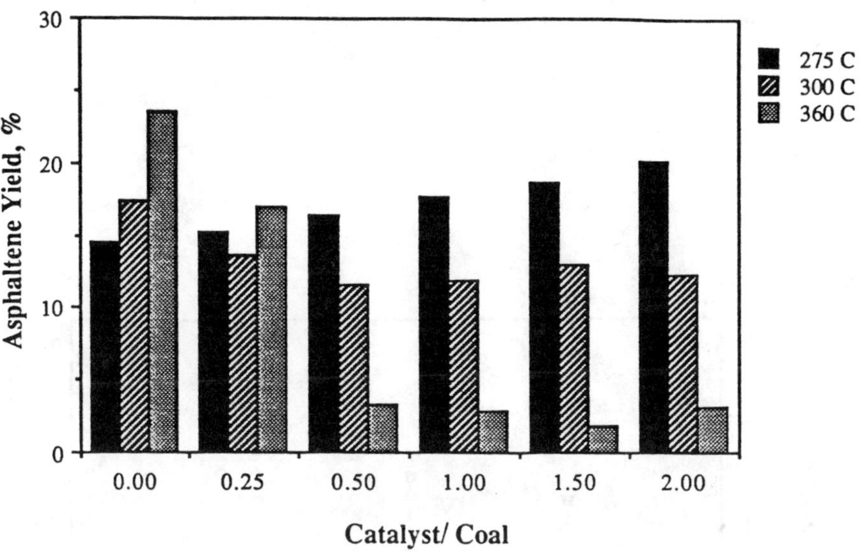

Figure 3. Change of asphaltene yields with catalyst/coal ratio.

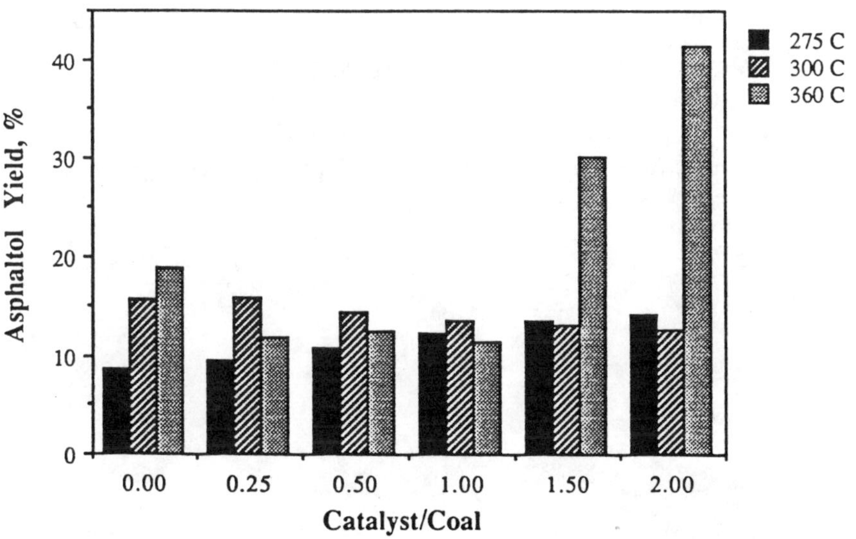

Figure 4. Change of asphaltol yields with catalyst/coal ratio.

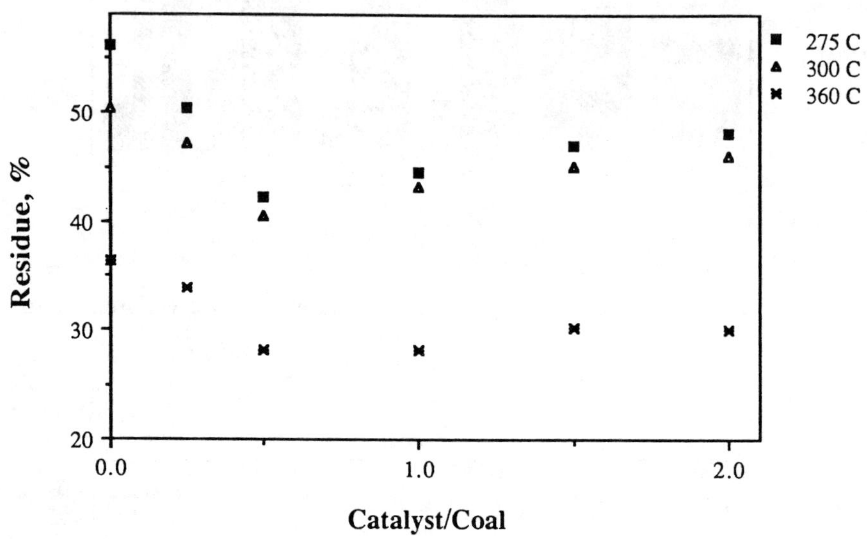

Figure 5. Change of residue with catalyst/coal ratio.

LIQUEFACTION OF SEA-GRASS WITH ACETIC ACID FOR CONVERSION TO CRUDES AT 0.1 MPa INITIAL PRESSURE

F. Taner and A. Olcay
Çukurova University, Arts & Sciences Faculty, Chemistry Department, 01330, Adana/Turkey

Abstract

In this study, aqueous suspensions of the sea-grass (20 % by weight) have been treated with acetic acid (10% by weight of the sea-grass) at 423 K, 473 K, 523 K, and 573 K, to liquefy in a high pressure autoclave for one hour, starting 0.1 MPa initial pressure. After thermochemical treatment, the gas produced was analysed, the reaction mixtures were separated into solid-oil and aqueous phases by filtering. The percentages of the acetone extract and the lignin of the solid-oil phases, the calorific values and also CHNO of the solid oil have been determined. It can be concluded that the solid-oil yield of the waste decreases while liquefying temperature increases. The oil yield (acetone extract) and calorific values of the solid-oil increase, with the increase in thermochemical treatment temperature.

1. INTRODUCTION

The conversion of lignocellulosic materials into liquid fuels has been a subject for a long time. Based on the early studies [1], several investigators have treated lignocellulosic wastes with aqueous alkaline solutions using initial high pressure of CO gas at temperatures around 523 K [2-4]. Some other experiments have been conducted with pressurized hydrogen and transition metal catalyst in aqueous and non aqueous media at about the same temperatures [5,6]. Wood is the main raw material used for conversion processes.

A few studies on the liquefaction of the annual plants and their components, such as carbohydrates and lignin have been carried out by using catalytic hydrogenation [7]. Some experiments on liquefaction of cotton stalk, pulp mill waste, and the sea-grass in aqueous alkaline and acid media have been conducted [8-11]. Since the liquefaction of the lignocellulosic material has been carried out in aqueous medium, the aqueous phases which are rich in organic water-soluble compounds, can be used for some other purposes or has to be treated to prevent the environment from pollution.

2. MATERIAL AND METHODS

Raw material

During the summer months, the Mediterranean coast, which is used as the recreational area, is covered up by the sea-grass that is collected and piled up nearby. About five tons of sea-grass on the coast of 2 km lengh and 10 m width are collected every day. The sea-grass is dragged out by the tide. The sample was taken from the piles made up last summer in December. Since the sample was sandy and salty, it was washed with water and dried in the open atmosphere for further treatment. Cellulose, lignin, ash, and moisture content were determined. Cellulose was determined by digesting with 1% NaOH, 2% Na2SO3 in session and then chlorination, washing, drying and weighing. Lignin content was determined according to the cited method [12]. Lignin and cellulose contents were calculated on maf (moisture and ash free) basis. The results are given in TABLE I.

The sample was washed several times for the removal of the salts present in the sea water. Then the sample was ignated in an oven to determine the ash content of the sample. The ash was analyzed to determine the Mg, Cu, Cr, Fe, Zn and Ni content of the sample by atomic absorbtion spectrophotometer according to Standard Methods [13]. The concentrations of these elements of sea-grass was calculated on a dry basis. The results are shown in TABLE I.

Liquefaction

The aqueous slurry of 100 g dry and ashless sea-grass in 300 mL water was poured into a high pressure autoclave (Cook and Sons Ltd., Cook Vertical Autoclave) and 11 mL acetic acid (99% by weight) in 100 mL water was added to the slurry. The autoclave was sealed and purged with nitrogen gas by pressurezing upto 1 MPa and then venting. The autoclave was heated upto operating temperatures of 423 K, 473 K, 523 K, and 573 K at which the reaction has been completed for one hour, while mixing continuously. Then the autoclave was allowed to cool over night. After cooling, the inside pressure was measured and the gas was analysed with Orsat gas analyser. The rest of the gas was vented and the autoclave was unsealed. The reaction mixture was separated with filtering into solid-oil and aqueous layer. The solid-oil was washed with water and saved for further treatment.

Solid-oil analyses

The amount of soli-oil was determined by drying and weighing. The acetone extract which is defined as oil yield was determined by extracting with acetone in soxhelet apparatus. The lignin content of solid-oil was determined with 72% H2SO4 solution (by weight) [12]. Calorific value was determined with a bomb calorimeter. The elemental composition of the solid-oil was determined with LECO CHNO analyzer. All the results were calculated on maf basis (TABLE II).

3. RESULTS AND DISCUSSION

As described in the text, the sea-grass which is an aquatic plant was treated for the production of liquid fuel and biogas which are considered as convensional energy sources.

Liquefaction products

The results obtained from the liquefaction of the sea-grass had shown that the solid-oil yield of the sea-grass decreases upto 523 K, and then starts to increase while the operating temperature increasing. At low temperature (423 K) the shape of the sea-grass is saved and partial delignification occurs. At 473 K, the delignification occurs mostly and the form of the sea-grass starts to change. It almost looses its appearance at 523 K. Bitumen-like substances form at 573 K.

Percentage of unsoluble amount of solid-oil in 72% sulphuric acid solution is higher (21.5) at 573 K than that of 523 K (19.6). This is not exactly lignin but some kinds of substance which looks like an oily coal and does not dissolve in 72% H_2SO_4 solution. This shows that the low operating temperatures, 423 K and 473 K, which are applied for paper pulp production, cause to delignify the sea-grass. Increament in calorific values of the solid-oil with the increase in operating temperature shows that deoxigenation of the solid occurs and some new compounds which have higher heating value than those of original raw material and the solid-oil obtained at lower temperatures (423 K and 473 K) form. The oil yield also supports these arguments. As the operating temperature increases, the oil yield of the solid-oil also increases. These values are 13.6, 24.2, 38.3, and 61.7 at temperatures of 423 K, 473 K, 523 K, and 573 K, respectively (TABLE II).

The amount of water soluble compound increases upto 573 K. This shows that the raw material, sea-grass, is converted to water soluble compound which produced by degradation of high molecular weight components present in the sea-grass. Therefore, the aqueous layers obtained at around 573 K are rich in organic water soluble compounds which has to be recovered as feedstock or used for some other purposes such as biogas production. According to the components of the gas produced during thermochemical treatment, the volume percentage of the inert at 573 K is higher and more flammable than that of 523 K. This shows that thermochemical degradation of the sea-grass occurs and some low molecular volatile compounds form.

ACKNOWLEDGEMENT

I'd like to thank to Prof. Dr. Gülsen Önengüt to help me in writing and also to my students: G.Ersöz, A. Eratik, and S. Karaaslan to help me in lab. studies.

REFERENCES

1. Appell, H.R.; Fu, Y.C.; Friedman,S.; Yavorsky,P.M.; Wender,I., 1971, "Conversion Organic Wastes to oil, a replenishable energy", US Bureau of Mines, RI 7560.

2. Davis, H.G., 1983,"Direct Liquefaction of Biomass", Final Report and Summary of effort 1977-1983.Lawrence Berkeley Labaratory, LBL-16243.

3. Schlanger, L.L.; Figueroa, C.; Davis, H.G., 1982, "Direct Liquefaction of Biomass Biotech. and Bioeng. Sym.", 10.3.

4. Eager, R.L.; Mathews, J.F.; Papper, J.M., 1982, "Liquefaction of Aspen Poplar Wood", Can. J. Chem. Eng. 60, 289.

5. Fredon, C.; Soyer, N.; Bruneau, C.; Brault, A., 1983, "Chemical Study of the thermal and Catalytic Liquefaction of poplarwood", Proc. 2nd Int. Conf. on Energy from Biomass, Apl. Sci. Publ. Ltd. London New York, 930.

6. Kranitch,W.C.; Weiss,A.H., 1980, "Oil and Gas from cellulose by catalytic Hydrogenation", Chem., J. Chem. Eng. 58.735.

7. Meier,D.; Larimer, D.R.; Faix, O., 1985, "Direct Thermochemical Liquefaction of Plant Biomass using hydrogenation conditions", Proc. 3rd int. Conf. Energy from Biomass, Apl. Sci. Publishers, London-New York, 929.

8. Taner, F., 1986, "Thermochemical Treatment of Cotton Stalk with NaOH", Int. Conf. on Renew. Energy Sources, May 18-23, Madrid.

9. Taner, F.; Boztepe, H.; Kimyonsen, U., 1987, "Liquefaction of Cotton Stalk with Acid", Submitted to Conf. on Energy from Biomass and Wastes", March 16-20, Orlando, Florida.

10. Taner, F.; Boztepe, H.; Kimyonsen, U., 1987, "Thermochemical Treatment of the Solid waste obtained from NaOH Pulp and Paper Factory with 15% Acetic Acid and for Conversion to Crudes", Submitted to 2. Euro. Conf. on Env. Tech., June 22-26, Amsterdam.

11. Taner, F.; Kimyonsen, U.; Eratik, A.; Ersöz, G.; 1989, "Liquefaction of Sea-grass with NaOH and Anaerobic Biodegradation of aqueous layers", Submitted to 5th Euro. Conf. on Biomass for Energy and Industries, Oct.9-13, Lisbon, Portugal.

12. Pfeffer, J.T., 1980, "Biological Conversion of Biomass to Methane, Final Report", Report No. UILU-ENG-80-2009, Dept. of Civil Eng. Univ. Illinois, Urbane, Illinois 61801.

13. American Public Health Association, 1980, Standart Methods for the Examination of Water and Wastewater, 15th edition, Washington,D.C.

TABLE I: SOME PROPERTIES OF THE SEA-GRASS

Ash	:	13.6 (on a dry basis)
Lignin	:	22.3% (maf)
Cellulose	:	47.5% (maf)
Hemicellulose	:	30.2% (by difference)
Calorific value	:	16.9 MJ/kg (maf)
Mg	:	6988 ppm (on a dry basis)
Fe	:	908.8 ppm (on a dry basis)
Zn	:	67.0 ppm (on a dry basis)
Cu	:	Trace
Cr	:	Trace
Ni	:	Trace

TABLE II: SOME PROPERTIES OF THE PRODUCTS FROM THE LIQUEFACTION PROCESSES

Operating temperatures	423 K	473 K	523 K	573 K
Solid-oil%of Sea-grass	58.0	46.0(53)	42.0(40)	41.0
Oil yield of solid-oil	14.0	18.0(17)	15.0(9)	13.0
Higher heating v.(MJ/kg)	19.5	25.2(19.8)	25.2(20.6)	23.0
Lignin % of solid-oil	38.0	32.1	47.5	49.4
Pressures;				
operating (MPa)	1.3	6.8	2.0	7.5
final (MPa)	0.2	0.5	1.8	2.0
Gas composition (volume percent, dry :				
CO_2	13	68 (28)	20 (25)	75
CO	1	4 (1)	2 (1)	2
*inerts	80	28 (71)	86 (75)	25
Elemental composition of the solid-oil (% by weight)				
C	57.58	68.26	70.64	75.70
H	5.65	5.22	6.10	5.78
N	2.57	2.81	3.48	3.59
O(by difference)	34.20	23.71	19.78	14.93
C/H (mol/mol)	0.85	1.09	0.97	1.09

* : inerts are flammable

(..) : results obtained when any chemical was not used

DISPOSABLE CATALYSTS FOR COAL GASIFICATION

J.F. Akyurtlu, A. Akyurtlu, K. Jothimurugesan, and A.A. Adeyiga
Department of Engineering, Hampton University, Hampton, Virginia 23668, U.S.A.

Abstract

The main objective of this investigation is to evaluate the catalytic activity of (K_2SO_4 + $FeSO_4$) mixtures in the steam gasification of coal char and to compare it with that of K_2CO_3. The research is planned so that the effects of chemical and physical properties of char upon gasification rate and product composition will be evaluated quantitatively. Thus, the ultimate analysis and mineral matter content, and the surface area and porosity of parent coal and char samples, are determined in addition to the measurements of gasification rates and product distributions. The gasification rates are normalized with respect to the surface area in order to facilitate the comparison of different catalytic systems.

The (K_2SO_4 + $FeSO_4$) mixtures are attractive as gasification catalysts because they are disposable and of comparable activity to K_2CO_3: Iron is a very active catalyst for gasification and it is readily available as $FeSO_4$. However, a good contact between iron and carbon cannot be maintained and poisoning of iron by sulfur is excessive. So, the (K_2SO_4 + $FeSO_4$) mixtures couple the excellent adhesion and contact properties of potassium with the high catalytic activity of iron.

This catalyst system, also, represents a possible solution to one of the economic problems with catalytic coal gasification, namely using low cost raw materials which do not have to be recovered from gasification residues.

1. INTRODUCTION

Processes for the conversion of coal by steam, carbon dioxide and hydrogen to gaseous fuels or to gaseous chemical feedstocks are the underlying basis for interest in catalytic gasification of carbon. Application of catalysis to coal gasification has the objective of achieving sufficiently high rates of reaction at reduced temperature and shift the product gas distribution to the desired direction [1]. In broad classification there are two types of catalysts:

◊ Transition metal catalysts: which generally catalyze CH_4 forming reactions

◊ Non-transition metal catalysts: which generally catalyze CO and H_2 forming reactions

Transition metal catalysts promote hydrogasification processes which produce methane. They involve different mechanisms than those of steam and carbon dioxide gasification that are promoted primarily by the non-transition metal salts.

The most effective catalysts for the carbon gasification by steam and CO_2 are alkali and alkaline earth metal salts. For alkali metal carbonates on coals and chars, the relative activity decreases with atomic weight: Cs > K > Na > Li [1,2]; this order is reversed for graphite due to the presence of relatively large concentrations of bound hydrogen and other heteroatoms in the nongraphitic substances. A critical requirement for an effective gasification catalyst is the possession of an oxygen-bearing anion.

Type of carbonaceous material that is gasified is an important factor on catalyst activity: the degree of catalytic effect increases with the increasing rank of the coal chars. Thus, the effectiveness of the catalyst is related inversely to the gasification activity of the carbonaceous substance in the absence of added catalyst [3]. This variation in catalyst effectiveness is attributable to differences in calcium ion content, the fraction of volatile matter and the pore size distribution. Results of various investigations suggest that effective gasification catalysts can be used with any type of carbonaceous material, thus suggesting that a single mechanism governs the action of these catalysts [1].

Gasification rate is strongly dependent on the gaseous reactant. The addition of alkali metal carbonate catalyst enhances the reactivity of the oxidizing gases with all chars; presence of product gases like H_2 and CO, inhibit the gasification rate on those catalysts.

Method of adding the catalyst to the carbonaceous material is not critical because upon heating in contact with carbon, the catalyst becomes extremely mobile and distributes itself over the entire char surface [4, 5]. Thus, physical admixing is equally as effective as impregnation or ion exchange. Rate of gasification of chars impregnated with alkali metal carbonates increases with an increasing metal/carbon atom ratio up to a saturation level, typically about 0.1 [6].

Thus, effective coal/char gasification catalysts are ionic salts with oxygen-bearing anions or anions that are converted to oxygen-containing species under gasification conditions. Also, cation of the salt reacts with carbonaceous material to form active intermediates for gasification. The anion of the salt modifies the formation or character of this active species. Majority of investigators favor involvement of a redox cycle in which the catalyst is alternatively reduced by the carbon and oxidized by the gaseous reactant, steam:

$$M_2CO_3 + C \longrightarrow 2M + CO + CO_2 \quad (1)$$

$$2M + 2H_2O \longrightarrow 2MOH + H_2 \quad (2)$$

$$2MOH + CO_2 \longrightarrow M_2CO_3 + H_2O \quad (3)$$

A metal-oxygen-carbon complex analogous to a phenolate salt is hypothesized as a reaction intermediate [1].

2. MIXTURES OF POTASSIUM SULPHATE AND FERROUS SULPHATE SALTS AS STEAM GASIFICATION CATALYSTS

The activity of a coal gasification catalyst depends on three properties [7]:

1. convertibility of the catalyst raw material into the active state;

2. intrinsic activity of the catalyst; and,

3. stability of the catalyst under gasification conditions.

In the steam gasification of coal, potassium is a promising catalyst [8]. The basis of its superior catalytic activity is a high degree of dispersion which results from the liquid/solid interface between K and C and the excellent wettability of the carbon surface. However, relevant precursors of potassium, like K_2CO_3 are expensive. Also, substantial quantities are lost by reaction with the silicates of the mineral matter of coal and the recovery of potassium from the resulting compounds is a problem which has not yet been solved.

Besides potassium and other alkali metals and alkaline earth metals, some transition metals have shown catalytic activity in steam gasification. Iron seems to be the most attractive. The necessary catalyst amounts are in the region of 1% (mass) and suitable catalyst precursors like $FeSO_4$ are readily available. However, a problem with iron arises from the solid/solid interface which may be the reason for bad contact between catalyst and carbon. Also, poisoning of iron by sulphur is excessive. In a recent study, it was attempted to evaluate the performance of a novel catalyst system which couples the excellent adhesion and contact properties of potassium with the remarkable catalytic activity of iron, namely mixtures of K_2SO_4 and $FeSO_4$ [9]. The study was performed on a PVC coke as a model carbon and it was shown that certain mixtures of K_2SO_4 and $FeSO_4$ exhibit similar catalytic activity to the more expensive but the most active K_2CO_3. The reason for the activity of this mixture was found to be an accelerated activation of the thermally and chemically stable K_2SO_4 by the presence of iron compound. This system represents a possible solution to one of the economic problems with catalytic coal gasification, namely using low cost raw materials which do not have to be recovered from gasification residues (disposable catalysts). The above-mentioned study was performed on a model coke. However, it is known that the properties of the parent coal introduce the major factors which control the reactivity of chars in the gasification reactions, namely the concentrations of active carbon sites, ease of accessibility of the reactant gas to the active sites and presence of catalytic inorganic impurities. The present research addresses these points.

3. EXPERIMENTAL WORK

An experimental work is undertaken with the objectives of determining the catalytic activity of (K_2SO_4 + $FeSO_4$) mixtures in steam gasification of chars of coals of different rank; deriving a rate expression; and comparing the performance of (K_2SO_4 + $FeSO_4$) mixtures with that of K_2CO_3. The main variables to be studied are the type of coal; composition of (K_2SO_4 + $FeSO_4$) mixtures; catalyst loading; temperature of gasification; and the composition of gaseous reactants.

In the first phase of the study, the effect of the catalyst composition on the gasification rate of Pittsburgh (PSOC-1451) HVA coal char in the presence of H_2O/H_2 mixtures was investigated. The proximate and ultimate analysis of the coal obtained from Penn State Coal Bank is presented in Table I. To enable the interpretation of gasification data, an experimental program was also undertaken to elucidate the catalyst behavior under gasification conditions. The second phase of the gasification program was designed with the objective of determining the gasification rate of PSOC-1451 coal under steam/N_2 mixtures.

The conditions used for the first and second phases of the gasification experiments are summarized in Table II. The activation method for the catalyst mixture was optimized before performing the second phase gasification experiments. The optimized activation procedure is as follows:

◊ Heat the catalyst-char mixture up to 725°C at 25K/min under 10% H_2 - 90% N_2; then switch to N_2 until 925°C.

◊ Under N_2, lower temperature to 750°C; maintain this temperature for 15 minutes. Then raise the temperature up to the reaction temperature and introduce H_2O-N_2 mixture.

Equipment

The gasification experiments were performed in the setup shown on Figure 1. The main part of the system is a CAHN TG-121 thermogravimetric analyzer.

Charring Procedure

Coal is charred by heating at 25 K/min up to 1200 K under flowing nitrogen and keeping it at that temperature for a soak time of 30 minutes. Preliminary experiments indicated that soak time does not affect the mass loss from the parent coal; all volatile matter of coal is lost at 1200 K; and as heating rate is increased mass loss increases, effect of heating rate being more pronounced at the higher final temperatures. The N_2- and CO_2-surface areas of the chars prepared under different conditions of final temperature and heating rate were measured by using a QUANTASORB Surface Analyzer; the results of those measurements are presented in Table III.

4. RESULTS AND DISCUSSION

A selected sample of the results of the experiments performed to elucidate the catalyst behavior under gasification conditions are shown on Figures 2 - 5. Based on mass measurements, it has been hypothesized that the following reactions take place:

Transformations of $FeSO_4 \cdot 7H_2O$

a) In N_2 and N_2/H_2O:

$$FeSO_4 \cdot 7H_2O \longrightarrow FeSO_4 \cdot 4H_2O + 3H_2O \quad (4)$$

$$FeSO_4 \cdot 4H_2O \longrightarrow FeSO_4 \cdot H_2O + 3H_2O \quad (5)$$

$$FeSO_4 \cdot H_2O \longrightarrow FeSO_4 + H_2O \quad (6)$$

$$FeSO_4 \longrightarrow FeO_x + SO_2 + (1 - \frac{x}{2})\ O_2 \qquad (7)$$

$x \approx 1.26$ in N_2
$x \approx 1.05$ in presence of H_2O

b) In H_2/H_2O:

Reaction 4 - 6

$$FeSO_4 + 4H_2 \longrightarrow FeS + 4H_2O \qquad (8)$$

$$FeS + H_2O \rightleftharpoons FeO + H_2S \qquad (9)$$

$$FeO + H_2 \rightleftharpoons Fe + H_2O \qquad (10)$$

$$FeS + H_2 \rightleftharpoons Fe + H_2S \qquad (11)$$

FeS, FeO and Fe are expected to be in equilibrium.

c) Char/$FeSO_4 \cdot 7H_2O/N_2$:

Reactions 4 - 6

$$FeSO_4 + C \longrightarrow FeO + SO_2 + CO \qquad (12)$$

d) Char/$FeSO_4 \cdot 7H_2O/H_2 - H_2O$:

Reactions 4 - 6 and 8 - 11

$$FeO + C \longrightarrow Fe + CO \qquad (13)$$

$$FeO + CO \longrightarrow Fe + CO_2 \qquad (14)$$

$$FeS + CO \longrightarrow Fe + COS \qquad (15)$$

All iron is expected to be in metallic Fe form under these conditions.

Transformations of K_2SO_4

a) In N_2 and N_2/H_2O:

$$K_2SO_4 \longrightarrow K_2O + SO_2 + \tfrac{1}{2}\ O_2 \text{ (very unfavorable)} \qquad (16)$$

No effect of H_2O

b) In H_2/H_2O:

$$K_2SO_4 + 4H_2 \longrightarrow K_2S + 4H_2O \qquad (17)$$

$$K_2S_4 + H_2O \rightleftharpoons KHS + KOH \qquad (18)$$

$$KHS + H_2O \rightleftharpoons KOH + H_2S \qquad (19)$$

$$2KOH + H_2 \rightleftharpoons 2K + 2H_2O \qquad (20)$$

KHS, KOH and K are expected to be in equilibrium. There is the possibility of vaporization of K and KOH.

c) In N_2/H_2:

Reaction 17

$$K_2S + H_2 \rightleftharpoons 2K + H_2S \qquad (21)$$

K_2S and K are expected to be in equilibrium. There is the possibility of vaporization of K and K_2S.

d) Char/K_2SO_4/N_2:

$$K_2SO_4 + 4C \rightleftharpoons K_2S + 4CO \qquad (22)$$

$$K_2SO_4 + 2C \rightleftharpoons K_2S + 2CO_2 \qquad (23)$$

$$K_2SO_4 + 4CO \rightleftharpoons K_2S + 4CO_2 \qquad (24)$$

$$K_2S + CO \longrightarrow 2K + COS \qquad (25)$$

e) Char/$K_2SO_4/H_2 - H_2O$:

Reaction 17

$$K_2S + 2H_2O \rightleftharpoons 2KOH + H_2S \qquad (26)$$

Reaction 20

$$KOH + \tfrac{1}{2}C \rightleftharpoons K + \tfrac{1}{2}CO_2 + \tfrac{1}{2}H_2 \qquad (27)$$

$$KOH + CO \rightleftharpoons K + CO_2 + \tfrac{1}{2}H_2 \qquad (28)$$

All potassium is expected to be in metallic form under these conditions.

A sample of the thermograms obtained for the first phase of the gasification experiments are presented on Figures 6 and 7. The results of the whole program is shown on Table IV and Figure 8. It may be observed from Figure 8 that there is a maximum in the activity of (K_2SO_4 + $FeSO_4$) mixtures with respect to the catalyst composition. This optimum composition is approximately 10% Fe - 90% K (atomic). This behavior may be due to several factors; namely, a reduction in the melting point of K_2SO_4 in the presence of $H_2 - H_2O$ as K/Fe increases, going through a minimum [9]; prevention of iron agglomeration due to better melt formation and catalysis of K_2SO_4 dissociation by Fe.

The second phase of the gasification experiments were performed according to the procedure summarized in Section 3 and their results are presented on Table IV. The conclusion that may be drawn is that the gasification activity is much higher. This fact may be due to the elimination of H_2 - inhibition since steam - N_2 mixtures were employed during gasification. Another important outcome was that the synergistic interaction between K and Fe have disappeared: Only H_2 was employed during the new activation procedure thus involving only reactions (8) and (11). However, in the presence of $H_2 - H_2O$ mixtures, as in the activation procedure employed in phase one, reactions (8)

- (11) take place. FeO is more readily reduced to Fe than FeS [10].

5. CONCLUSIONS

The main conclusion derived from the experiments performed is that synergistic catalytic effect between potassium and iron occurs in the presence of hydrogen, thus promoting hydrogasification and consequently increasing methane production.

An additional experimental program has been undertaken to elucidate the behavior of catalyst mixtures under different atmospheres. These experimental results will be compared with the results obtained from thermodynamic calculations being done on the catalyst systems. This will be the subject of a subsequent paper.

Acknowledgements

This study has been performed under the U. S. DOE Grant DE-FG 22-87 PC 79934.

REFERENCES

1. Wood, B. J. and Sancier, K. M.: Catal. Rev. - Sci. Eng. 26, 233 (1984).

2. Spiro, C. L.; McKee, D. W.; and Juntgen, H.: Fuel 62, 196 (1983).

3. McKee, D. W.; Spiro, C. L.; Kosky, P. G.; and Lamby, E. J.: Fuel 62, 217 (1983).

4. Coates, D. J.; Evans, J. W.; Cabrera; A. L.; Somorjai, G. A.; and Heinemann, H.: J. Catal. 80, 215 (1983).

5. Marsh, H. and Mochida, I.: Fuel 60, 231 (1981).

6. Spiro, C. L.; McKee, D. W.; Kosky, P. G.; Lamby, E. J.; and Maylotte, D. H.: Fuel 62, 323 (1983).

7. Hüttinger, K. J. and Minges, R.: Fuel 64, 364 (1985).

8. McKee, D. W.: Fuel 62, 170 (1983).

9. Adler, J. and Hüttinger, K. J.: Fuel 63, 1393 (1984).

10. Hüttinger, K. J. and Minges, R.: Fuel 64, 364 (1985).

TABLE I: ANALYSIS OF PITTSBURGH HVA COAL

Proximate (As Received Basis)		Ultimate (On Dry Basis)	
Component	%	Component	%
Moisture	2.54	Ash	13.67
Ash	13.32	Carbon	71.88
Volatile Matter	33.56	Hydrogen	4.67
Fixed Carbon	50.58	Nitrogen	1.36
		Total Sulfur	1.37
		Oxygen (diff)	7.05

TABLE II: OPERATING CONDITIONS

	Phase I	Phase II
Heating Rate:	25 K/min	25 K/min
Temperature:	1123 K for 1 hour	1123 K for 1 hour
Metal to carbon atom ratio:	0.02	0.02
Char particle size:	150-355 μm	53-106 μm
Reactant Gas:	H_2O-H_2 (y_{H_2O} = 0.3)	H_2O-N_2 (y_{H_2O} = 0.3)

TABLE III: SURFACE AREAS OF PITTSBURGH PSOC 1451 CHAR

	N_2 surface area, m^2/g	CO surface area, m^2/g
Raw Coal 1451	1.2	35.2
1451 Char/1073/25K/min	3.9	128.3
1451 Char/1200/25	2.2	31.5
1451 Char/1200/10	3.4	29.8
1451 Char/1200/50	2.2	35.6
1451 Char/1200/200	2.4	32.7

TABLE IV: RESULTS OF GASIFICATION EXPERIMENTS

Nominal Catalyst Composition, atom %K	% of Fixed Carbon Reacted	
	Phase 1	Phase 2
No Catalyst	4.2	41.8
0	6.5	39.5
30	8.2	66.2
50	10.2	84.5
75	12.2	96.2
85	17.1	--
90	23.9	100
95	21.4	--
100	15.6	99.5
K_2CO_3	46.1	

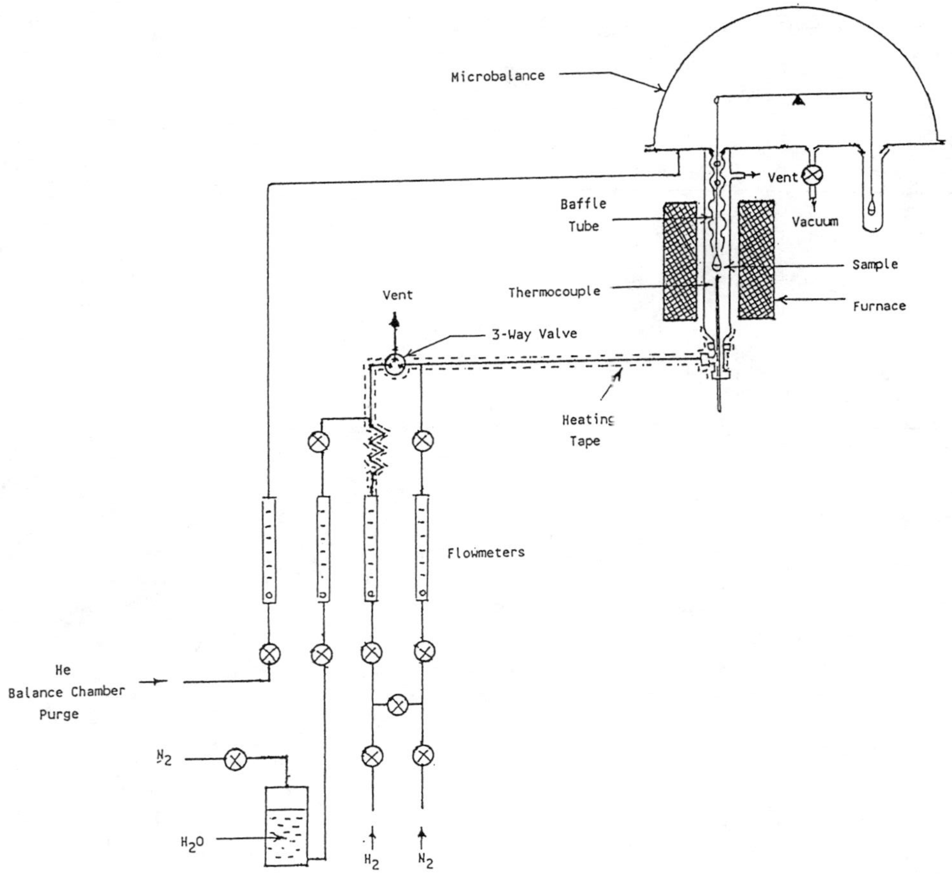

Figure 1. - Experimental Setup

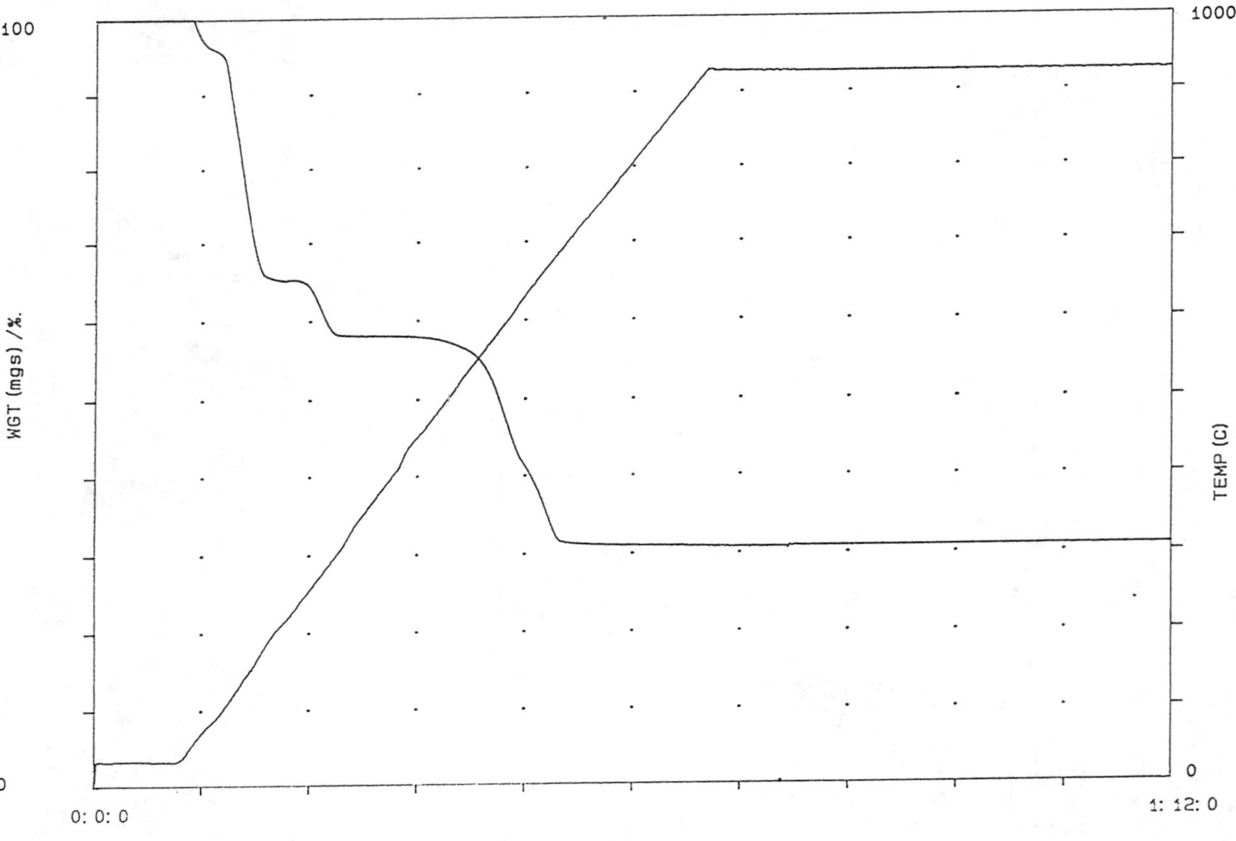

Figure 2. - Thermogram for the Treatment of $FeSO_4 \cdot 7H_2O$ in $H_2O - N_2$

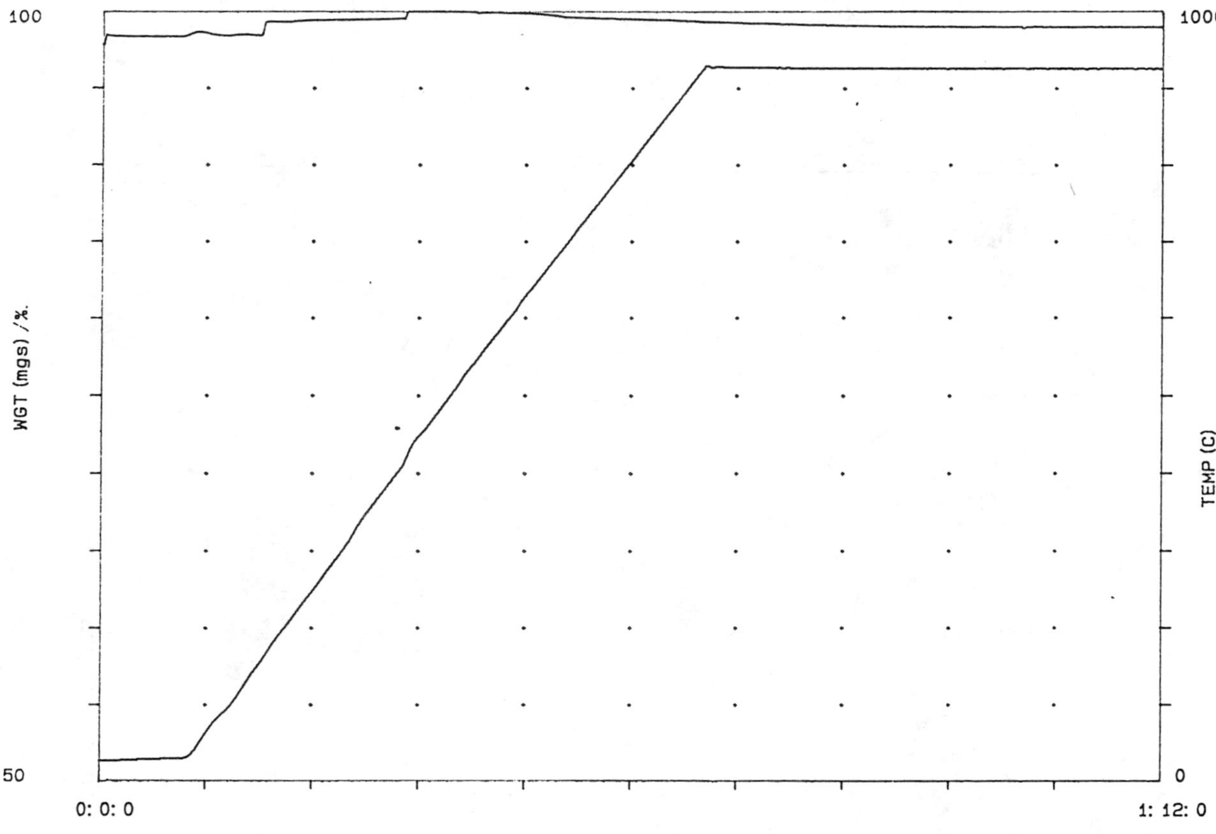

Figure 3. - Thermogram for the Treatment of K_2SO_4 in H_2O - N_2

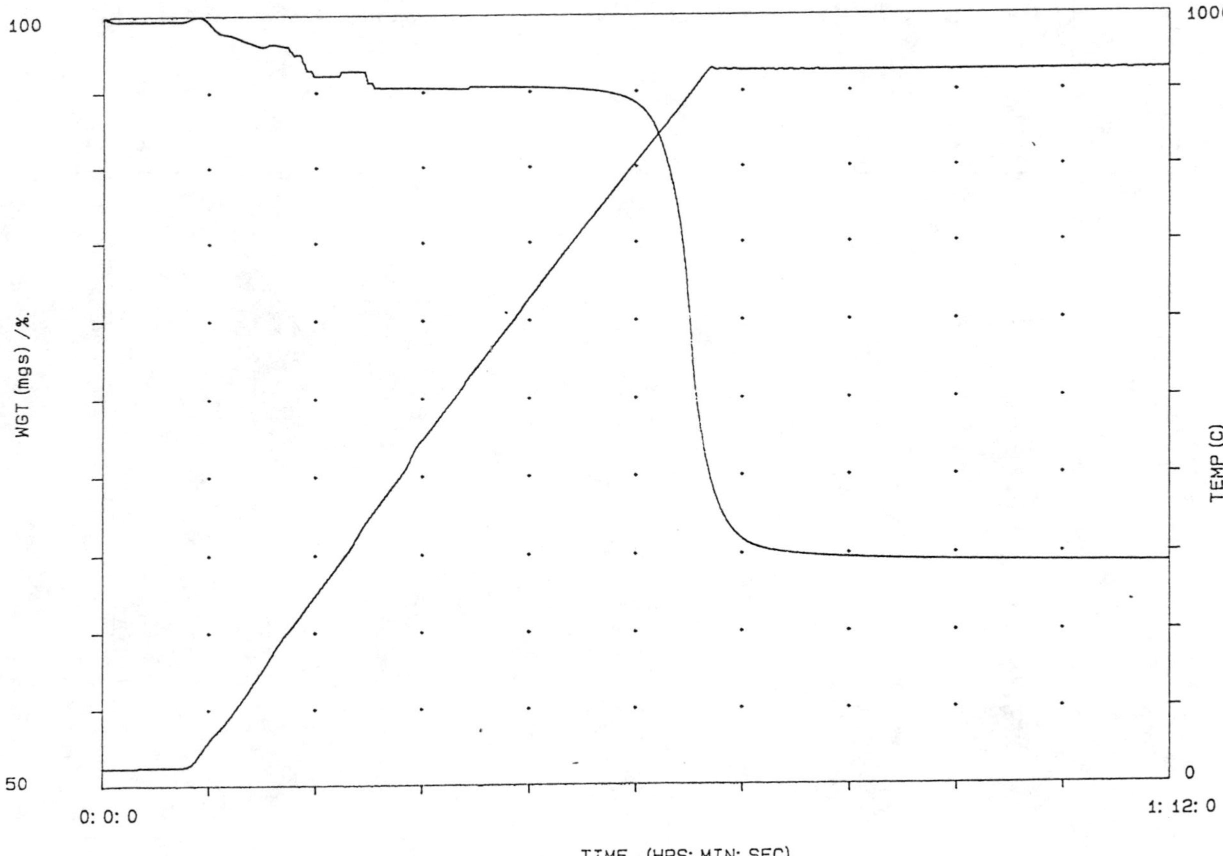

Figure 4. - Thermogram for the Treatment of K_2CO_3 in H_2O - N_2

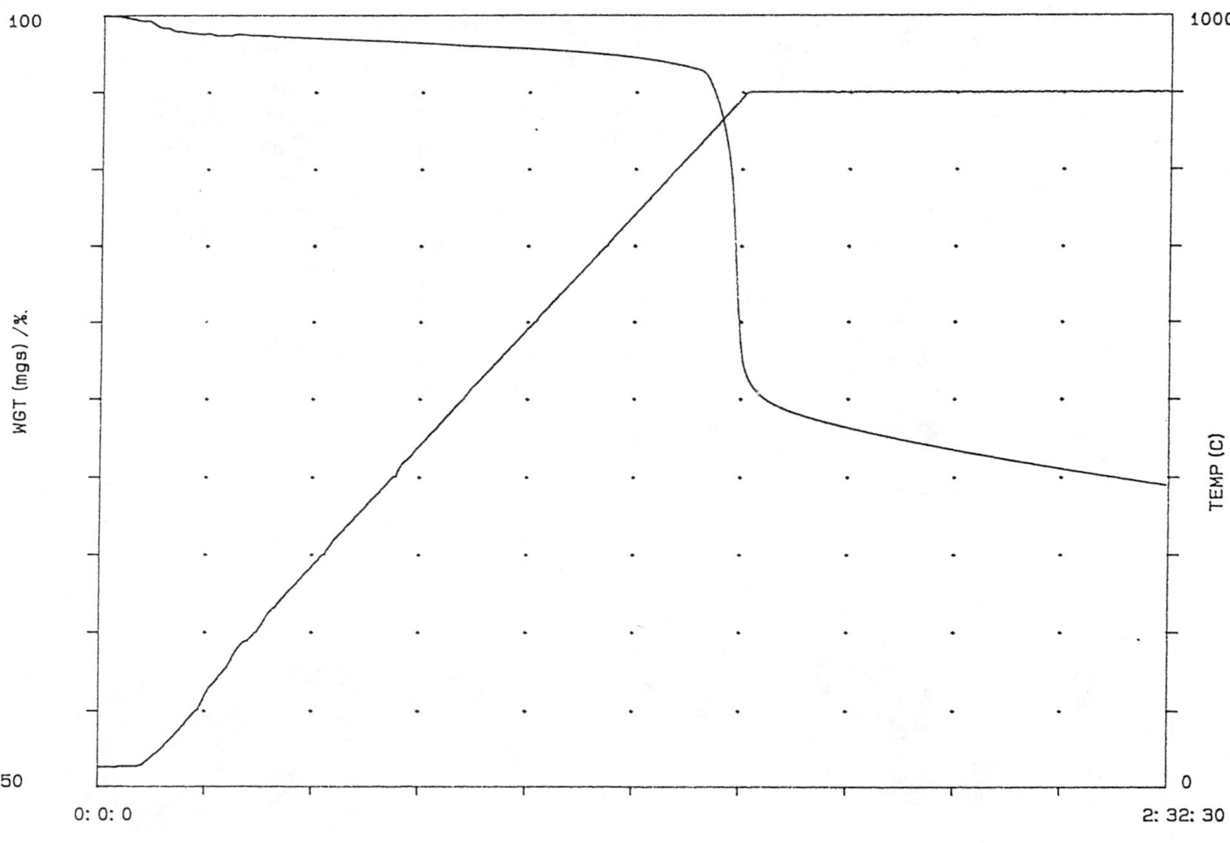

Figure 5. - Thermogram for the Treatment of K_2SO_4 in N_2 in the Presence of PSOC-1451 Char

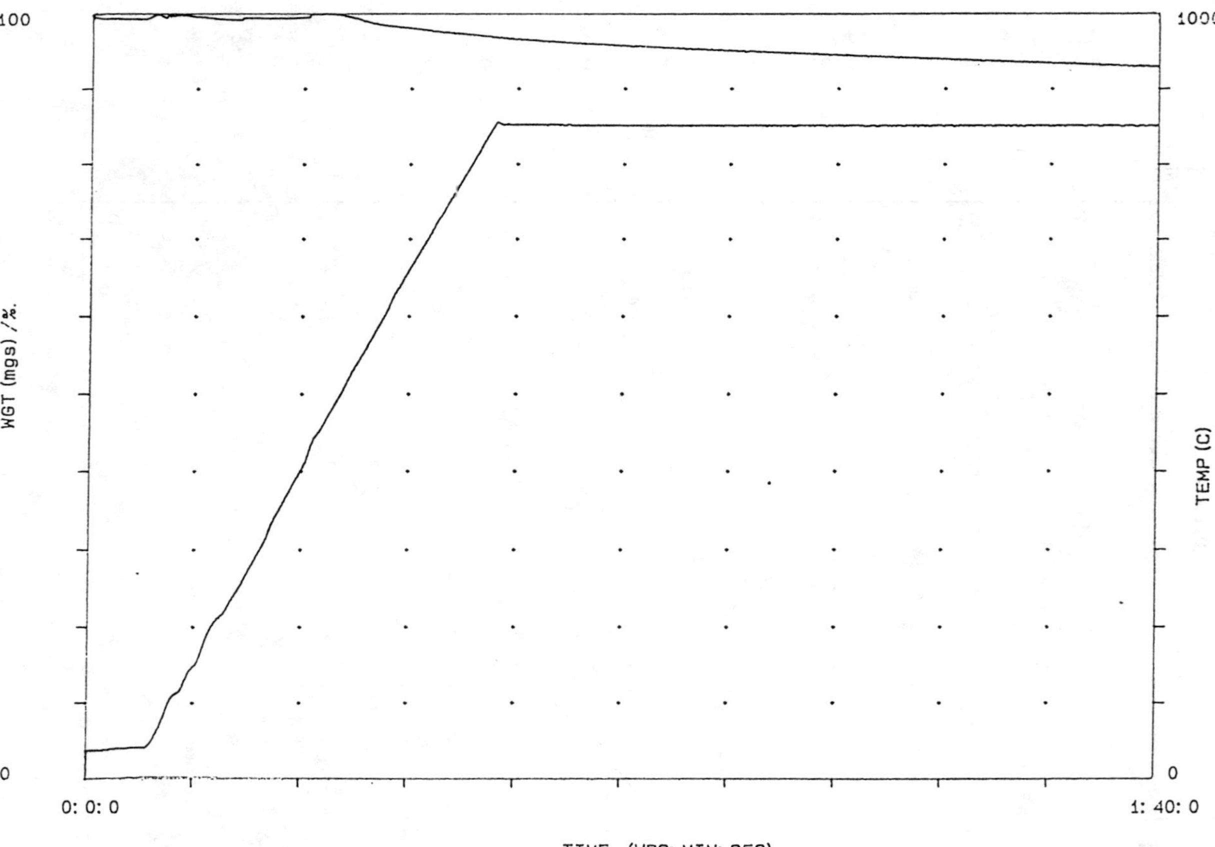

Figure 6. - Thermogram for the Steam Gasification of PSOC-1451 Char in the Presence of H_2

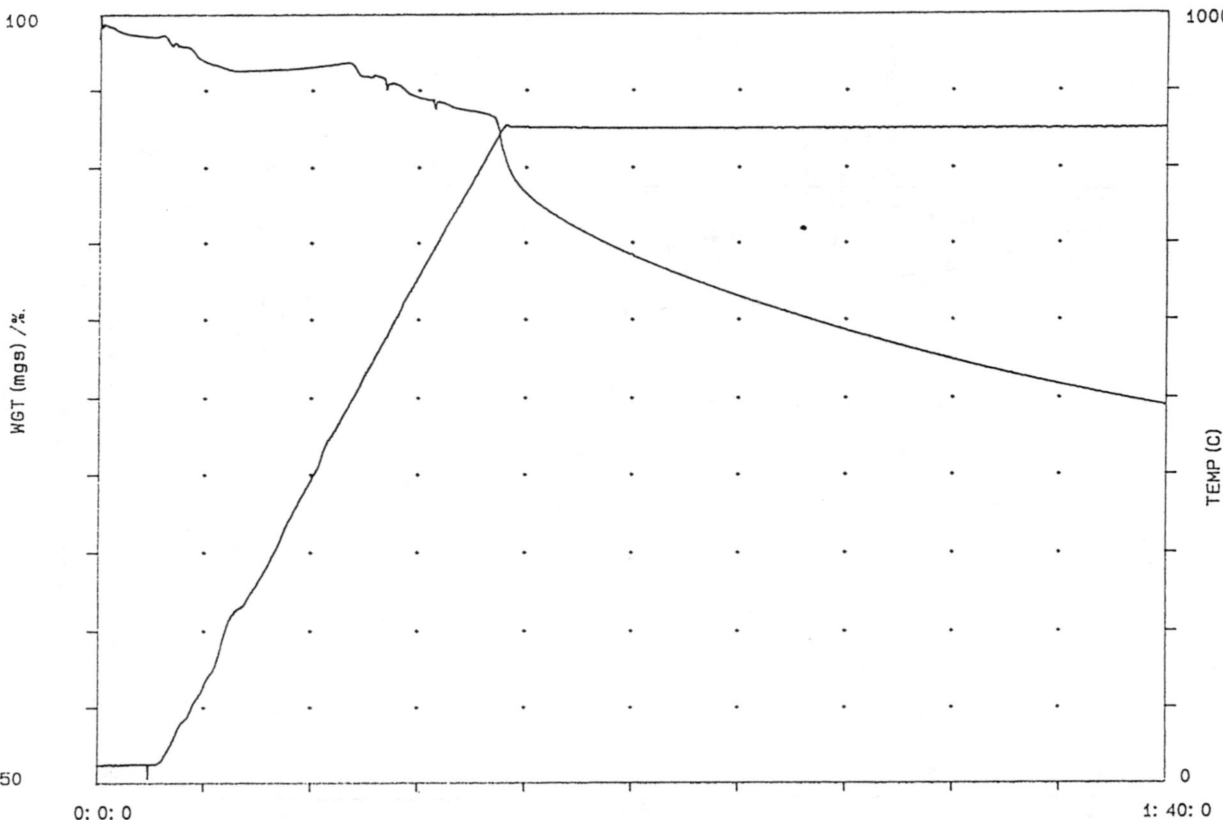

Figure 7. - Thermogram for the Catalytic Steam Gasification of PSOC-1451 Char in the Presence of H_2 with 10% Fe and 90% K (atomic)

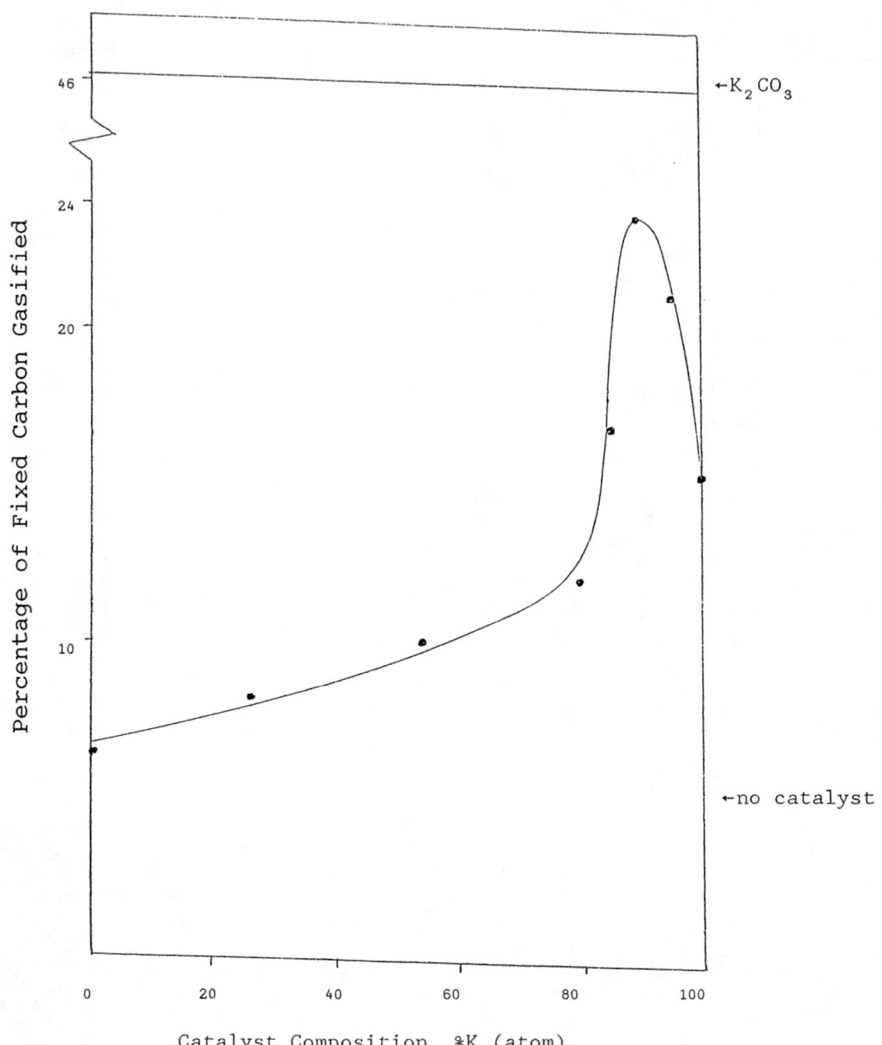

Figure 8. - Extent of Fixed Carbon Gasification as a Function of Catalyst Composition (Based on Table IV)

LOW-TEMPERATURE CARBONIZATION OF TURKISH LIGNITES

T. Durusoy, G. Güruz*
Department of Chemical Engineering,
Hacettepe University, Beytepe, Ankara, Turkey
* Department of Chemical Engineering,
Middle East Technical University, Ankara, Turkey

Abstract

Carbonization experiments have been performed on samples of Seyitömer, Tunçbilek, Elbistan and Kangal lignites in the temperature range of 200-600°C in a fixed bed reactor system. The lignite samples used were identified in terms of ultimate, proximate, petrographic, rational, sulfur forms and ash composition analysis. Variation of semi-coke, tar and gaseous product yields with carbonization temperature has been studied and decrease in semi-coke and increase in tar and gaseous product yields with increasing temperature are observed for all samples. Reduction in total sulfur content of the lignites after carbonization is also experimentally determined. Gaseous product has been analyzed by Gas-Chromatography and H_2S and H_2O contents were determined gravimetrically.

In order to study the effect of C, H and O contents of original lignites on the yield and the distribution of the products, a linear regression analysis has been carried out and various relations have been obtained.

1. INTRODUCTION

When solid fuels are heated in an oxygen free atmosphere, they pyrolize into smaller molecules which may be classified as tar, gaseous products and a residue called semi-coke. This process can be classified as low-temperature carbonization if the temperature is in the range of 300-650°C [1].

The distribution of products obtained by low-temperature carbonization depend on coal composition and the carbonization temperature. Selvig and Ode [2] have shown that the semi-coke yield tends to decrease and the liquid and gaseous product yields tend to increase with increasing volatile matter content for a carbonization temperature of 500°C. They also indicated that no direct correlation existed with the total carbon content of the coal. For a temperature range of 350-900°C, Dicker et al [3] have shown that the semi-coke yield tends to decrease and liquid and gas yields tends to increase with temperature. A study has been made by Schafer [4] of the evolution of water, carbon dioxide and carbon monoxide during the pyrolysis of Australian Yallourn brown coal and on the way in which the evolution is influenced by the exchange of the carboxyl groups in the coal with magnesium and barium cations. A range of Canadian coals were subjected to variable heating rate conditions in a variety of atmospheres by Stange and Sears [5]. Heating rate was found to have little effect on total weight loss of the coal, but a dramatic effect on the actual composition of products. Arendt and Heek [6] pyrolysed five German hard coals with a range of volatile matter contents using two different apparatuses, differing mainly in the heating rates. Pressure and gas atmosphere were also varied and the product gas composition was determined by quantitatively analysing for H_2, CH_4, C_2H_4, C_2H_6, CO, CO_2 and H_2O. Under an inert gas atmosphere, high heating rates resulted in slightly higher yields of liquid products. Hydropyrolysis of Godavari (Indian) coal has been performed by Sharma, et.al.[7] at 650°C using $3K_{min}^{-1}$ heating rate in the presence of a mixture of steam and hydrogen in different ratios at 66 bar pressure. It was indicated that the yields of tar products were enchanced in the presence of steam. In order to investigate the effect of precarbonization temperature and the reduction of the oxygen content on the liquefaction of lignites Quchi K. et al. [8] carried out a study with Soya Koishi and North Dakota lignites at 350, 400 and 420°C. With the aim of sulfur transformation and removal during carbonization different investigations have been carried out [9,10] using coals with a range of properties. For pyrolysis under nitrogen sulfur removal up to 59% at 740°C [9] and up to 62 in the temperature range of 700-800°C [10] were obtained.

The aim of the present work is to investigate the effect of lignite properties on the product distribution at different carbonization temperatures and is a part of the study [11] carried out with some selected Turkish lignites.

2. EXPERIMENTAL

Seyitömer, Tunçbilek, Elbistan and Kangal lignites were selected for this study The lignite samples were identified in terms of ultimate, proximate, petrographic, rational, sulfur forms and ash composition analyses. The results are shown in Table I for the four lignites.

Carbon and hydrogen contents were determined by "Heraeus Combustion Apparatus, Type Standard for Microanalytical Determination" and nitrogen by Kjeldahl instrument. Total sulfur was determined by the Eschka method according to ASTM D3177, sulphate and pyritic sulfur forms according to ASTM D2492 and organic sulfur was determined from the difference. Combustible sulfur was calculated as the difference between total sulfur and ash sulfur, determined by the Eschka Method. "Leitz MPV 2 Orthoplan Microscopefotometer" microscope was used for the petrographic analysis of the lignite samples. Rational analysis is carried out in order to determine the bituminous, humic acid, ligneous matter and cellulose matter contents of the lignite samples. Coal samples is extracted with benzol:alcohol (1:1) mixture and bituminous fraction is calculated from the weight of the extract. Humic acid fraction is found by NaOH extraction, followed by ether extraction to obtain the ligneous matter content. Cellulose matter content was found by difference [12] .

The lignite sample was ground in a ball mill and sized to -65 mesh size.

The apparatus used is shown in Figure 1. The furnace made from pressure-rated SS 304 type stainless steel tubing, 3.5 cm. in diameter and 30 cm. in length was used for the carbonization experiments. The furnace was heated from the outside with an electric heater, the temperature was measured using an iron-constantan thermocouple and controlled with Siemens 3TA21 Type controller.

Tar formed is collected in a vessel. $CaCl_2$ trap and a washing bottle containing $Cd(CH_3COO)_2$ solution were used in order to remove the water and the hydrogen sulfide present in the outlet gas stream. Gas product was collected in a cylindrical vessel for future analysis. Hydrogen, carbon dioxide, carbon monoxide, methane, ethane, propane and butane contents were determined using a gas chromatograph with the specification given below.

Gas Chromatograph	:	Hewlett-Packard, Model 5890A
Column	:	Poropak Q
Detector Type	:	Thermal Conductivity Detector
Detector Temperature	:	80-150°C
Column Temperature	:	40-120°C
Enjection Temperature	:	65-120°C
Carrier Gas	:	Nitrogen (30 ml/min)

In order to investigate the effect of temperature on product distribution,experiments were done at nine different temperatures in the range of 200 to 600°C.

3. RESULTS AND DISCUSSIONS

For each set of experiment semi-coke and tar yield values were calculated with respect to the weight of the original lignite fed to the system and the gas yield was obtained by difference. Table II summarizes the semi-coke, tar and gas yield values for four lignite samples.

As expected [3] a decrease in semi-coke yield and an increase in tar and gas yields were observed for all lignite samples. For the carbonization temperatures studied, the highest semi-coke yield values were obtained from Tunçbilek lignite and the lowest from Seyitömer lignite. This can be considered as an indication that the original volatile matter content being a more effective parameter than the fixed carbon in determining semi-coke yield values. A reverse trend was observed in the gas and tar yield values for the four lignite samples studied.

The results of the proximate analysis of the semi-cokes obtained at different temperatures are shown in Table III. Volatile matter content of semi-cokes decreased while the ash content increased continuously as a function of carbonization temperature. Fixed carbon values were higher for the semi-cokes compared to lignites for temperatures above 350°-400°C, but the increase was not a continuous function of temperature.

The decrease in volatile matter content was calculated taking into account the semi-coke yield values and are shown Figure 2 as a function of temperature. The trend observed indicates that carbonization starts above 250°C and the difference between different lignite samples based on the original volatile matter content is more pronounced for temperatures above 400°C.

The reduction in the total sulfur content was also determined and are shown in Figure 3. Sulfur transformation reactions starts above 300°C except for Tunçbilek lignite, which has a lower combustible sulfur content. Sulfur removal up to 60 % was observed with the lignite sample having the highest total sulfur content which is mainly in organic form.

Hydrogen, carbon dioxide, carbon monoxide, methane, ethane, propane, butane, hydrogen sulfide and water yield values were determined as a function of temperature. Some of the results obtained for the main constituents are summarized in Figures 4 to 7 for Seyitömer, Tunçbilek, Elbistan and Kangal lignites, respectively. The highest carbon dioxide content was obtained with Seyitömer lignite compared to others which has the highest carbon content based on ultimate analysis. Following a parallel trend to that of sulfur removal values in semi-cokes, highest hydrogen sulfide content was observed with Elbistan lignite. In general; hydrogen, carbon dioxide and methane contents increased with increasing temperature. The increase in the hydrogen sulfide content is less pronounced compared to other constituents except for the case of the lignite having the highest sulfur content.

The lignite samples used in the experiments were thoroughly identified as shown in Table I. As an initial step to generalize the carbonization behaviour of Turkish lignites the yield values were correlated through a linear regression analysis. Carbonization temperature and the carbon, hydrogen and oxygen contents of the original lignites were chosen as the important variables after a detailed analyses [11]. For the lignite samples investigated the following relations are suggested for the yields of semi-coke, tar and gaseous products respectively,

$$X_{semi\text{-}coke} = 3828\ T^{-0.63}\ C^{0.35}\ H^{-0.53}\ O^{-0.16}$$

$$X_{tar} = 1.1 \times 10^{-22}\ T^{8.10}\ C^{-1.50}\ H^{-2.50}\ O^{0.66}$$

$$X_{gas} = 8.3 \times 10^{-5}\ T^{2.84}\ C^{-4.02}\ H^{5.02}\ O^{0.79}$$

REFERENCES

1. Wilson, Jr.P.J., Clendenin, J.D., Chemistry of Coal Utilization, H.H. Lowry (Ed.) Suppl.Vol., John Wiley and Sons, New York, 395-440, 1961.

2. Selvig, W.A., Ode, W.H., "Low-temperature Carbonization Assays of North American Coals", Bul. 571, Bureou of Mines, U.S. Government Printing Office, Washington, 1957.

3. Dicker, P.H., Woods, B., Martin, T.G., Gaines, A.F., Griffiths, D., Report No: 17.9, National Coal Board, Chelthenham, 1960.

4. Schafer, H., Fuel, 59, 295-301, 1980.

5. Stangeby, P.C., Sears, P.L., Fuel, 50, 131-135, 1981.

6. Arendt, P., Heek, K.H., Fuel, 60, 779-787, 1981.

7. Sharma, K.D., Sulimma, A., Heek, K.H., Fuel, 65, 1571-1574, 1986.

8. Quchi, K., Shiraishi, K., Itoh, H., Makabe, M., Fuel, 60, 471-478, 1981.

9. Maa, P.S., Lewis, R.C., Hamrin, Jr. C.E., Fuel, 54, 62, 1975.

10. Gürüz, G., Fuel Processing Technology, 5, 183-201, 1982.

11. Durusoy, T., Ph.D. Thesis, Hacettepe University, Ankara, Turkey, 1989.

12. Kreulen, J.W., Sechs Abhandlungen-Über Braunkohlen/Lignite. Vol.6, 72-78, Freiberger Forschungsheft Nr. A/244, Academie-Verlag, Berlin, 1962.

TABLE I. ANALYSES OF LIGNITE SAMPLES

	Seyitömer	Tunçbilek	Elbistan	Kangal
Proximate Analysis (%)				
Moisture	12.2	4.7	28.4	18.9
Volatile Matter	44.3	31.4	42.0	37.4
Ash	14.0	34.1	22.6	36.1
Fixed Carbon	29.5	29.8	7.0	7.6
Ultimate Analysis (mf%)				
Carbon	56.1	45.6	39.7	35.4
Hydrogen	4.2	3.2	3.0	2.8
Sulfur (combustible)	0.9	0.3	2.0	0.1
Ash	15.9	35.8	31.5	44.5
Nitrogen	1.7	1.5	1.1	0.8
Oxygen	21.2	13.6	22.7	16.4
Petrographic Analysis (%org. b)				
Huminite	90.7	93.0	89.2	72.2
Liptinite	8.9	6.6	10.0	17.7
Inertinite	0.4	0.4	0.8	10.1
Rational Analysis (%)				
Bituminous	4.1	4.4	5.7	4.3
Humic Acids	25.3	7.2	57.2	21.1
Ligneous matter contents	70.1	86.9	35.1	74.6
Cellulose matter contents	0.5	1.5	2.0	0.0
Analysis of Sulfur Forms (mf%)				
Pyritic	0.5	0.5	0.1	0.2
Sulphate	0.2	0.1	0.1	0.3
Organic	1.0	0.6	2.8	0.1
Total	1.7	1.2	3.0	0.6
Combustible	0.8	0.9	1.0	0.5
Ash Sulfur				
Analysis of ash composition (%)				
SiO_2	38.7	46.5	18.0	74.8
Al_2O_3	15.8	21.7	10.4	12.0
Fe_2O_3	15.1	13.8	6.5	4.1
CaO	11.8	5.4	47.5	4.2
MgO	8.1	7.7	2.7	0.3
SO_3	10.5	4.8	14.6	4.6
Na^+	0.1	0.0	0.1	0.0
K^+	0.0	0.0	0.1	0.0

TABLE II. RESULTS OF YIELD VALUES

(% mf)

T, °C	SEYİTÖMER			TUNÇBİLEK			ELBİSTAN			KANGAL		
	S-Coke	Tar	Gas	S-Coke	Tar	Gas	S-Coke	Tar	Gas	S-Coke	Tar	Gas
200	94.5	0.4	5.1	98.1	0.0	1.9	96.9	0.2	2.9	97.6	0.1	2.3
250	91.6	0.3	8.1	98.2	0.0	1.8	95.4	0.1	4.5	95.5	0.1	4.4
300	89.8	1.4	8.8	97.5	0.4	2.1	91.9	3.0	5.1	91.2	2.9	5.9
350	84.5	7.4	8.1	93.9	3.5	2.6	89.4	4.1	6.5	88.6	3.9	7.5
400	76.5	10.9	12.6	90.5	3.2	6.3	78.9	10.7	10.4	83.7	7.4	8.9
450	69.9	16.4	13.7	85.0	6.9	8.1	73.8	12.9	13.3	80.5	10.1	9.4
500	66.4	18.3	15.3	81.4	11.3	7.3	71.3	13.7	15.0	75.5	13.2	11.3
550	65.2	16.7	18.1	78.4	11.5	10.1	67.2	15.2	17.6	76.0	11.1	12.9
600	58.5	22.7	18.8	75.7	13.5	10.8	64.6	16.6	18.8	73.9	13.1	13.0

TABLE III. PROXIMATE ANALYSES RESULTS AND SEMI-COKE YIELD VALUES

	SEYİTÖMER				TUNÇBİLEK				EBBİSTAN				KANGAL			
	V.M.	Ash	F.C.	S-Coke Yield	V.M.	Ash	F.C.	S-Coke Yield	V.M.	Ash	F.C.	S-Coke Yield	V.M.	Ash	F.C.	S-Coke Yield
Linyit	50.5	15.9	33.6	-	32.9	35.8	31.3	-	58.7	31.5	9.8	-	46.1	44.5	9.4	-
200°C	52.9	16.1	31.0	94.5	33.2	36.2	30.6	98.1	60.0	32.2	7.8	96.9	46.8	45.2	8.0	97.6
250°C	54.4	15.9	29.7	91.6	33.1	36.3	30.6	98.2	60.6	33.0	6.4	95.4	47.7	45.5	6.8	95.5
300°C	51.2	16.5	32.3	89.8	30.8	36.3	32.9	97.5	58.1	34.2	7.7	91.9	46.1	47.3	6.6	91.2
350°C	49.0	17.6	33.4	84.5	29.0	36.0	35.0	93.9	54.6	35.1	10.3	89.4	42.7	49.3	8.0	88.6
400°C	46.8	18.9	34.3	76.5	25.3	37.5	37.2	90.5	52.7	36.6	10.7	78.9	39.2	51.6	9.2	83.7
450°C	40.2	21.6	38.2	69.9	23.9	39.0	37.1	85.0	43.2	37.4	19.4	73.8	32.5	53.3	14.2	80.5
500°C	33.5	22.0	44.5	66.4	20.5	42.2	37.3	81.4	33.3	39.4	27.3	71.3	28.0	55.4	16.6	75.5
550°C	24.0	22.1	53.9	65.2	17.2	43.0	39.8	78.4	26.0	42.0	32.0	67.2	20.9	55.9	23.2	76.0
600°C	19.9	23.7	56.4	58.5	13.9	45.4	40.7	75.7	19.9	43.3	36.8	64.6	16.7	58.1	25.2	73.9

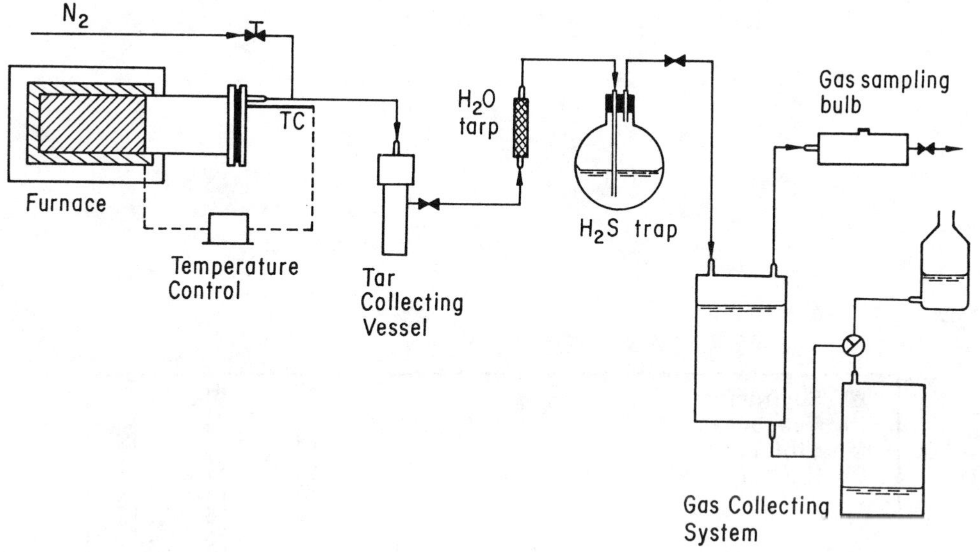

Fig.1.-Experimental Set-up.

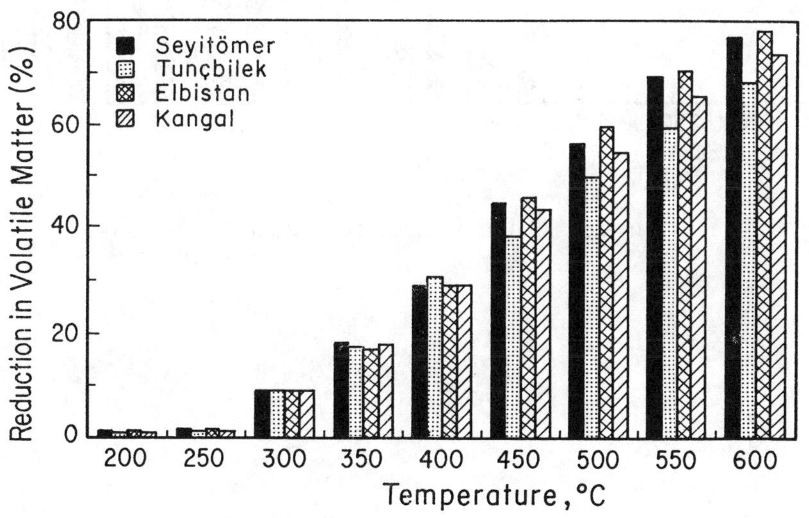

Fig.2.-Reduction in Volatile Matter with Carbonization Temperature.

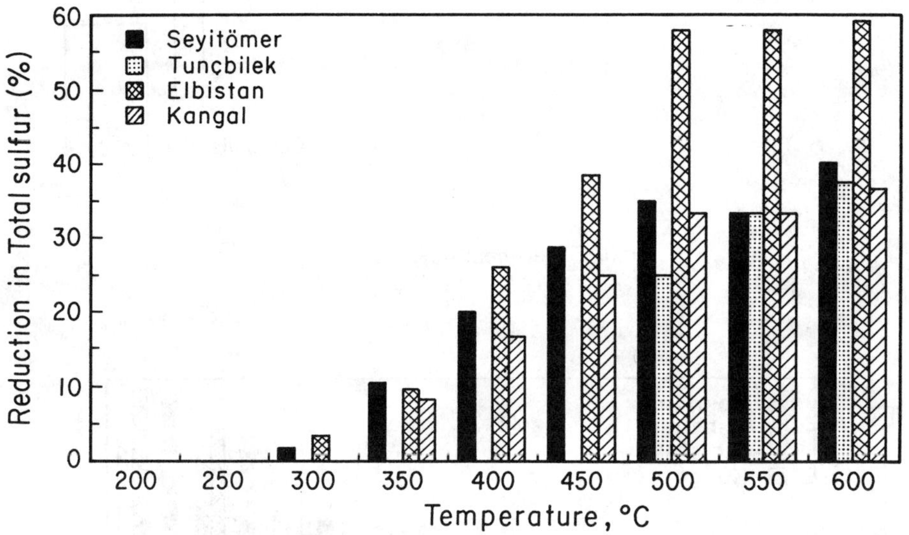

Fig.3.-Reduction in Total Sulfur with Carbonization Temperature.

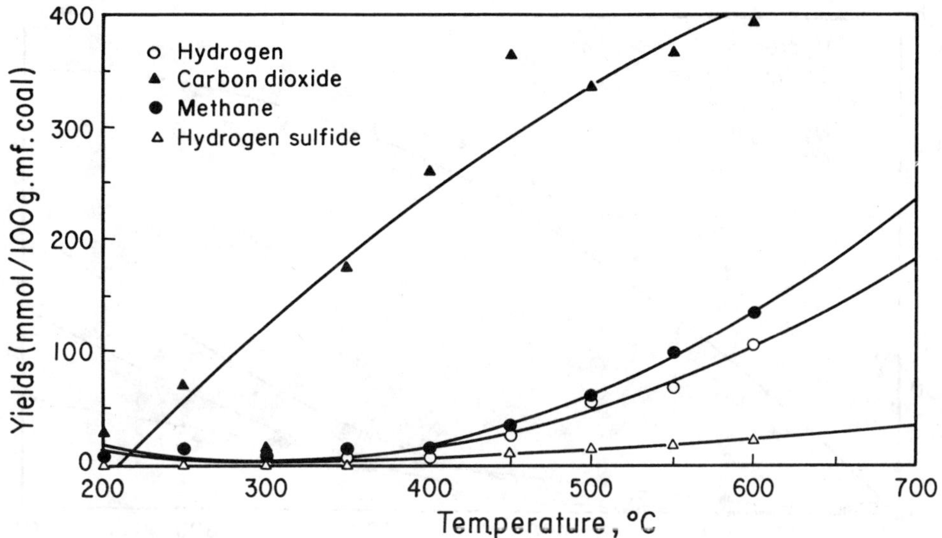

Fig.4.-H_2, CO_2, CH_4 and H_2S Contents for Seyitömer Lignite

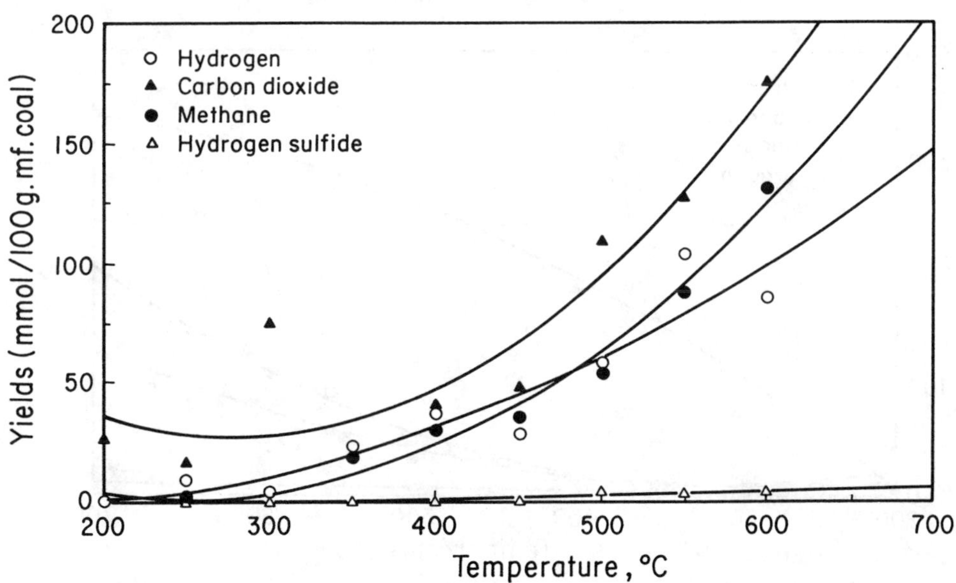

Fig.5.-H_2, CO_2, CH_4 and H_2S Contents for Tunçbilek Lignite

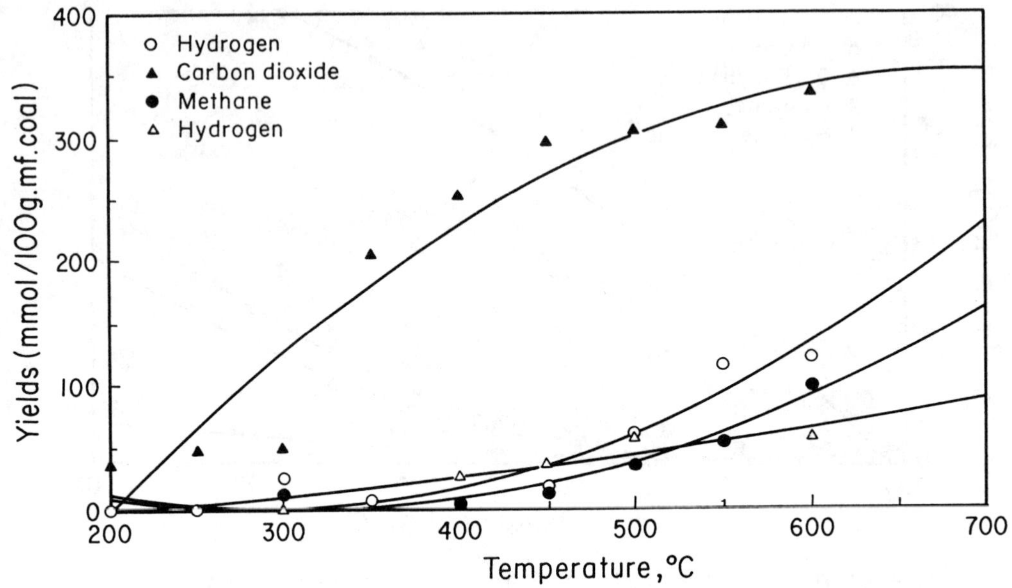

Fig.6.-H_2, CO_2, CH_4 and H_2S Contents for Elbistan Lignite

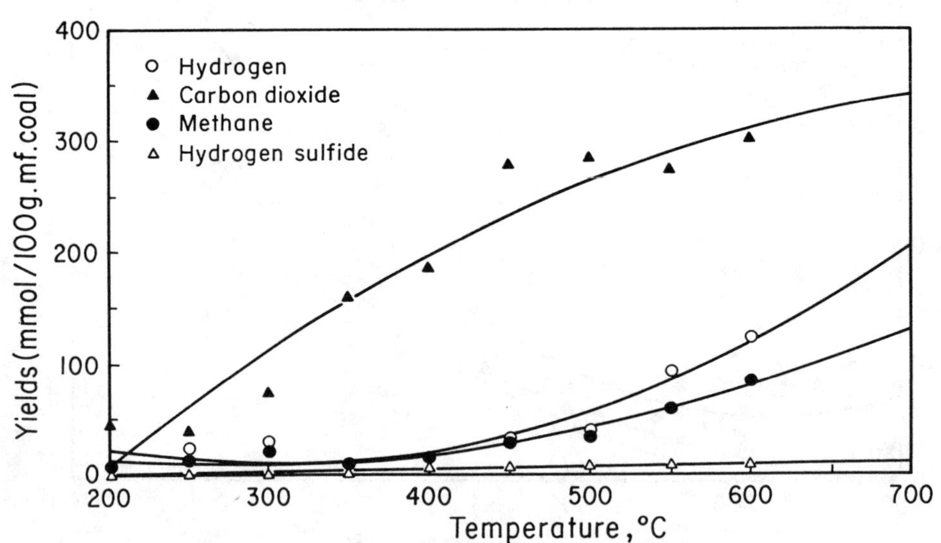

Fig.7.-H_2, CO_2, CH_4 and H_2S Contents for Kangal Lignite

THE CHARACTERIZATION OF COAL DERIVED LIQUIDS OBTAINED UNDER VARIOUS LIQUEFACTION CONDITIONS

G. Gürüz, N. Baç, and E. Inanç
Department of Chemical Engineering
Middle East Technical University, Ankara, 06531 Turkey

Abstract

Coal derived liquids from Mengen (Turkey) Lignite were separated into oils, asphaltenes and preasphaltenes. The oils were further separated by column chromatography while the asphaltenes were separated into basic and acid/neutral fractions. The preasphaltenes were divided into two fractions as carbene (CS_2 Solubles) and carboid (CS_2 Insolubles). The effects of reaction temperature, hydrogen pressure, solvent to coal ratio and catalyst loading on subfractions of oils, asphaltenes and preasphaltenes are investigated and a positive trend was observed with respect to the yield of hexane eluted oil.

The existence of the maximum in the desired product of asphaltenes with respect to increasing values of process variables can be attributed to the formation of aliphatics in oil fraction due to hydrogenation reactions.

In terms of preasphaltene subfractions, yield of carbene and carboid fractions decrease continuously with catalyst loading, solvent to coal ratio and pressure. Carbene yield defined as highly aromatic molecules increases with temperature since the formation of aromatics is favored at high temperatures.

The results of H-NMR and IR Spectroscopies indicate significant observations in terms of the reaction mechanism and the structure of coal.

1. INTRODUCTION

Separation of coal liquids into a number of well defined fractions based on the solubility characteristics in organic solvents is necessary for the characterization of the liquid products. Coal derived liquids are usually separated into three fractions as oils, asphaltenes and preasphaltenes.

Oils are defined as the fraction which is soluble in hexane and are the desired products of coal liquefaction processes. The molecular weight is in the range of 200-300 and consist of ethers, thioethers and non-basic N_2 compounds in decreasing order. Asphaltenes are soluble in toluene and are predominantly monofunctional compounds in the molecular weight range of 300-700. Preasphaltenes are soluble in tetrahydrofuran and contain polyfunctional compounds with MW ranging between 400-2000.

Product identification of these products is possible by fractionating these into subfractions. The analyses of these, provide information about the chemical changes occurring during the processing and also allow the effect of process variables to be moniroted. Fractionation of oils and asphaltenes can be made by liquid column chromatography (or elution chromatography).

Schwager and Yen [1] examined the characterization of coal liquids taken from three direct liquefaction processes (Synthoil, SRC, COED) by separating these into five fractions. They reported that under hydrogenolysis conditions, the conversion of coal to oil proceeded through asphaltene as an intermediate product. Schweighardt et al [2] studied the chromatographic and NMR analyses of SYNTHOIL hydrodesulfurization of coal liquid products. Separation of coal derived asphaltenes from five major processes in United States (Synthoil, H-Coal, COED, SRC and CATALYTIC in SRC) were studied by a simple, rapid chromatographic procedure on silicagel [3] as an alternative to the acid-base separation proposed by Stenberg et al [4]. Barthle et al [5] studied the chemical structure of two extracts prepared by supercritical extraction of low-rank coals with toluene, with and without hydrogen gas. Identification of the extracts have been carried out by using solvent and chromatographic fractionation followed by ultimate analysis. Structural analyses of neutral, acidic and basic heteroatom compounds in products from coal hydrogenation were studied by Yokoyama et al [6]. Kershaw and Kelly [7] separated the flash pyrolysis tars obtained from three different Australian coals into oils, asphaltenes and preasphaltenes. Marsh et al [8] applied an alumina adsorption together with chromatographic method for separating aromatic species in coal derived oils by using the modified method of Hirsh [9]. Hayamizu et al [10] examined the characterization of asphaltenes and preasphaltenes obtained under various coal liquefaction conditions. They concluded that the trend of increasing aromaticity of the solid products with increasing reaction temperature is general for coal liquefaction reactions. Mulligan et al [11] used a chromatographic system where the separation is based mainly on a combination of size exclusion and adsorption processes. The basis of the fractionation procedures is to use organic solvents of increasing polarity such as n-pentane, benzene and tetrahydrofuran to give fractions of increasing molar mass and heteroatom content.

In this study coal derived liquids obtained from hydrogenation of Mengen lignite [12,13] were separated by liquid column chromatography and were studied through ^{1}H-NMR and IR analyses for further identification.

2. EXPERIMENTAL

The liquefaction characteristics of Mengen lignite has been investigated in the presence of cobalt-molybdenium on alumina catalyst in a 1 lt batch autoclave system with anthracene oil used as solvent [12]. The experiments were carried out in the range of 15-60 atm for initial hydrogen pressure, 360-440°C for reaction temperature, 1-5 for solvent to coal ratio and 0-20 % of coal for catalyst loading which were chosen as process variables. The results of total conversion, liquid yield and liquid product distribution determined as preasphaltenes, asphaltenes and oils as a function of process variables is summarized in Table I.

Coal derived liquids in terms of oils (hexane solubles), asphaltenes (toluene solubles) and preasphaltenes (tetrahydrofuran solubles) were further separated into subfractions by solvent fractionation scheme based on using organic solvents of increasing polarity [14]. Fractionation scheme is summarized in Figure 1.

Oils were divided into subfractions by a silicagel (70-230 Mesh) column using hexane, toluene and methanol successively. Solvents were removed in a rotary evaporator under vacuum. The asphaltenes were separated into basic and acid/neutral fractions by passing HCl gas through toluene solutions [1]. Acidic components remaining in solution were further separated into three major fractions using benzene, diethylether and methanol as solvents by elution chromatography [1]. Methanol, a more polar solvent, was used as the third solvent to increase the percent recovery and the selection was in agreement with the discussion of Marsh et al [8]. Preasphaltenes were divided into two fractions, as carbenes, high molecular weight and highly aromatic solids, and carboids, polymerized coke like materials [1].

Subfractions of oil, asphaltenes and preasphaltenes obtained under a hydrogen pressure of 35 atm., at a temperature of 400 ^{o}C, with a solvent to coal ratio of 2 and using 10 % catalyst were analyzed by ^{1}H-NMR and IR Spectroscopies. ^{1}H-N.M.R. spectroscopy with 60 MHz operating frequency and CS_2 as solvent was used for this purpose. Infrared spectra of each liquid subfraction in KBr disk was run on a Hitachi Infrared Spectrophotometer for a wavenumber range of 400 to 4000 cm^{-1}.

3. RESULTS AND DISCUSSION

The effect of reaction temperature, hydrogen pressure, solvent to coal ratio and catalyst loading on subfractions of oils, asphaltenes and preasphaltenes were investigated for seventeen experiments and the results of the analyses are represented in Table II and in Figures 2 to 4.

The effect of process variables on oil subfractions is given in Figure 2. As the temperature is increased (Figure 2.a), the aliphatic fraction (hexane eluate) increases and the presence of a maximum in the aromatic fraction (toluene eluate) can be explained by the formation of small aliphatic and monoaromatic compounds due to the breakage of aromatic and substitution groups [10]. A slight increase in the methanol eluted fraction with increasing temperature is though to be due to thermal cracking of asphaltenes into oils. The increase in the yield of hexane eluate after a certain pressure value (Figure 2.b) can be explained by the formation of small molecules, stabilization by hydrogen gas and yielding more aliphatic hydrocarbons. The results of ^{1}H-N.M.R. and IR analyses indicate that the toluene eluted fraction contains ring joining alkenes and the presence of a maximum with respect to pressure is due to formation of n-alkenes and monoaromatics and the stabilization by hydrogen gas. Methanol eluted fraction of oil generally consists of hydroxyl groups and cyclic molecules containing heteroatoms. The slight variation in the yield of this minor fraction with pressure is probably due to the heteroatom removal reactions. As the solvent to coal ratio is increased (Figure 2.c) the yield of the major subfraction of oil, hexane eluate, increases. By increasing the solvent to coal ratio, the dissolution of the liquid products and solubility of molecular hydrogen are increased. Since the solvent is used as a bridge between the coal and the hydrogen gas, increase in the solvent to coal ratio increases the hydrogenation reactions. The yield of the two major subfraction of oil indicate a minimum and a maximum respectively around a catalyst loading of 10 % (Figure 2.d).

The effect of process variables on asphaltene subfraction is shown in Figure 3. The increase in the basic solid fraction of asphaltene (Figure 3.a) can be explained by the formation of small cyclic molecules containing nitrogen groups from the thermal cracking of coal. The behaviour in the benzene eluted fraction is in agreement with the proposed mechanisms of coal dissolution [3] and the asphaltenes being the intermediate product. Yield of the basic fraction of asphaltenes (Figure 3.b) gives a minimum at a certain pressure value which can be explained by the nitrogenation reactions occuring at low pressures. The yield of the two major subfractions of asphaltenes indicate a minimum and a maximum respectively around a solvent to coal ratio of 3-4 (Figure 3.c). The trend in the basic fraction of asphaltene with respect to catalyst loading can be explained by the positive effect of catalyst on dehydrogenation reactions (Figure 3.d).

The effect of process variables on preasphaltene subfractions is summarized in Figure 4. The increase in the yield of carbene fraction with temperature (Figure 4.a) can be explained by the favorable formation of aromatics at higher temperatures. The positive effect of pressure on liquefaction reaction can be observed by a decrease on the yield of carbene fraction (Figure 4.b). The positive effect of the solvent on the coal derived liquids can be seen by the decrease in the yield of carbene and carboid fractions (Figure 4.c). Yield of carbene and carboid fractions (Figure 4.d) decrease continuously with catalyst loading indicating the positive effect of the presence of a catalyst on product distribution.

The distribution of hydrogen for the aromatic fractions of all eluates for Experiment No: 17 is given in Table III. The percent of H_β and H_γ type hydrogen atoms is generally higher in oil fractions since oil consists of more aliphatics and lesser amount of aromatics. H_F , H_α and H_n type of hydrogen atoms observed in oil fractions represent aromatic ring substitution and ring joining methylene structures. It is observed that three fractions of asphaltene and carbene fraction of preasphaltenes contain small organic molecules which result from thermal breakage of coal structure.

Functional group analyses of the same experiments is also carried out by IR Spectroscopy and the results are summarized in Table IV. The results indicate that in oil fractions, the most intense bands are the alkane groups $-CH_3$, CH_2 near 1450 cm^{-1} and the aromatic substitution groups $-CH_3$, $-CH_2$ near 2900 cm^{-1} The number of benzene ring substitution groups increased going from hexane to methanol fractions. The infrared spectra of the asphaltenes indicate the presence of an increased content of polar groups; phenolic hydroxyl groups near 3400 cm^{-1} and carboxyl groups near 1700 cm^{-1}. All fractions contain ether group near 1100 cm^{-1}, and also contain 1,3,5 - benzene ring substitution near 675-865 cm^{-1}. Oxygen is present as ethers, carboxylic acids and free phenolic oxygen.

The above results gave some information interms of the chemical structure of Mengen Lignite and the similarity with the one proposed by Wender [15]. Also an important information can be gained interms of the reaction mechanism of liquefaction reaction. Preasphaltenes are thought to be first product in coal liquefaction so that they must contain polyfunctional groups coming from the structure of the coal. Results also show that they have lesser functional groups than asphaltenes indicating the conversion of preasphaltenes to asphaltenes. Asphaltenes contain 1,3,5- benzene ring substitution, however, oils contain meta-, ortho di- and 1,3,5- benzene ring substitution. Depending upon the result mentioned, asphaltenes are found out as an intermediate product of coal liquefaction producing oils. Based on above discussion the reaction mechanism for the liquefaction of the reaction mechanism for the liquefaction of Mengen Lignite under the range of the investigated conditions can be proposed as

Coal → Preasphaltenes → Asphaltenes → Oils.

4. NOMENCLATURE

Asp.	:	Asphaltene
Cat, %	:	Catalyst loading (% Coal)
C_a	:	Asphaltene Yield
C_g	:	Gas Yield
C_l	:	Liquid Yield
C_o	:	Oil Yield
C_p	:	Preasphaltene Yield
C_t	:	Total Conversion
daf	:	Dry Ash Free
^{1}H-N.M.R	:	Proton Nuclear Magnetic Resonance
H_{ar}	:	Aromatic Hydrogen
H_{OH}	:	Phenolic Hydrogen
H_F	:	Ring Joining Methylene (Ar-CH_2-Ar)
H_n	:	CH_2 and CHβ to an aromatic ring (including tetralin and indan structures)
H_α	:	CH_3, CH_2 and CHα to an aromatic ring
H_β	:	β-CH_3, CH_2 and CH γ or further from an aromatic ring.
H_γ	:	CH_3 γ or further from aromatic ring
IR	:	Infrared
S/C	:	Solvent to Coal ratio
T	:	Temperature (^oC)
δ	:	Chemical Shift (ppm

REFERENCES

1. Schwager, I. and Yen, T.F., Prepr. Am.Chem.Soc., Div. Fuel Chem., 21(5), 199 (1976).

2. Schweighardt, F.K., Retcofsky, H.L. and Friedel, R.A., FUEL, Vol. 55, 313 (Oct. 1976).

3. Schwager, I. and Yen, T.F., FUEL, Vol. 58, 219 (March 1979).

4. Stenberg, H.W., Raymond, R. and Schweighardt, F.K., Science, Vol. 188, 49 (1975).

5. Barthle, K.D., Ladner, W.R., Martin, T.G., Snape, C.E. and Williams, D.F., FUEL, Vol. 58, 413 (June 1979).

6. Yokoyama, S., Uchino, H., Katoh, T., Sanada, Y. and Yoshida, T., FUEL, Vol. 60, 254 (March 1981).

7. Kershaw, J.R. and Kelly, B.A., Fuel Processing Technology, Vol.7, 145 (1983).

8. Marsh, M.K., Smith, C.A., Snape, C.E. and Stokes, B.J., J.Chromatogr., Vol. 283, 173 (1984).

9. Hirsh, D.E., Hopkins, R.L., Coleman, H.J., Cotton, F.O. and Thompson, C.J., Anal.Chem., Vol.44, No.6, 917 (May 1972).

10. Hayamizu, K., Ohshima, S., Shimada, K., Suzuki, M. and Hayashi, S., Fuel Processing Technology, Vol.11, 47 (1985).

11. Mulligan, M.J., Thomas, K.M. and Tytko, A.P., FUEL, Vol.66, 1472 (Nov. 1987).

12. Erdem, S., M.S. Thesis in Chem.Eng., Middle East Technical University, Ankara (1987).

13. Baç, N., Erdem, S., Gürüz, G., İnanç, E. and Orbey, H., Fuel Processing Technology, (submitted)

14. İnanç, E., M.S. Thesis in Chem.Eng., Middle East Technical University, Ankara (1989).

15. Wender, I., ACS Div. Fuel Chem. Reprints, 20(4), 16 (1975).

TABLE I. LIQUID FRACTIONS OBTAINED FROM MENGEN LIGNITE, ERDEM (1987)

Exp. No.	H_2 Press (atm)	Temp. (^oC)	S to C Ratio	Cat. (%C)	Total Conver. Ct.	Liquid Yield CL	Gas Yield Cg	Preasp. Yield Cp	Asp. Yield Ca	Oil Yield Co
1	15	400	2.0	10.0	45.23	27.44	17.79	9.39	11.13	6.92
2	25	400	2.0	10.0	48.25	30.12	18.13	10.09	12.92	7.11
3	45	400	2.0	10.0	55.21	33.19	22.02	6.45	14.98	11.76
4	60	400	2.0	10.0	55.74	36.37	19.37	8.92	14.94	12.51
5	35	360	2.0	10.0	39.65	18.92	20.73	10.87	4.84	3.21
6	35	380	2.0	10.0	41.73	23.51	18.22	10.36	5.71	7.44
7	35	420	2.0	10.0	55.27	37.04	18.23	9.87	17.56	9.61
8	35	440	2.0	10.0	61.82	37.74	24.08	9.21	18.81	9.72
9	35	400	5 00	10.0	64.41	39.19	25.22	4.76	17.66	16.77
10	35	400	4.00	10.0	55.63	34.52	21.11	7.64	14.27	12.61
11	35	400	1.25	10.0	44.44	29.86	14.58	13.84	9.48	6.54
12	35	400	1.00	10.0	42.77	28.73	14.04	17.01	7.41	4.31
13	35	400	2.0	0.0	48.24	28.75	19.49	15.16	8.82	4.77
14	35	400	2.0	5.0	48.71	30.63	18.08	14.53	9.91	6.19
15	35	400	2.0	15.0	55.74	37.21	18.53	8.48	15.81	12.92
16	35	400	2.0	20.0	58.89	42.01	16.88	6.96	17.52	17.53
17	35	400	2.0	10.0	52.62	32.11	20.51	9.01	13.01	10.09

TABLE II. EFFECT OF PROCESS VARIABLES ON SUBFRACTIONS (DAF BASIS ON COAL)

		Oils			Asphaltenes				Preasphaltenes	
					Basic S.		Acidic Liquid			
EXP	T(°C)	% Hexane Eluate	% Toluene Eluate	% Methanol Eluate	% Basic Solid	% Benzene Eluate	% D.E.Ether Eluate	% Methanol Eluate	% Carbene	% Carboid
5	360	2.14	0.75	0.32	1.23	2.57	0.57	0.47	4.47	6.4
6	380	5.75	0.93	0.76	1.22	3.22	1.02	0.25	4.51	5.85
17	400	4.285	5.39	0.415	2.12	7.92	2.67	0.30	6.75	3.61
7	420	2.64	4.87	2.10	6.33	4.96	4.18	2.09	7.63	2.24
8	440	7.09	1.74	0.89	14.24	2.56	1.395	0.615	7.84	1.37
	P(atm)									
1	15	5.10	1.15	0.67	4.14	4.20	1.55	0.97	8.66	0.73
2	25	4.25	1.57	1.29	3.33	6.58	2.23	0.78	8.97	1.12
17	35	4.285	5.39	0.415	2.12	7.92	2.67	0.30	6.75	3.61
3	45	5.42	4.83	1.51	1.73	10.32	1.59	1.34	4.69	1.76
4	60	10.23	1.75	0.53	6.68	6.60	0.98	0.68	6.265	2.655
	S/C									
12	1.	1.34	1.85	1.12	2.05	3.365	1.396	0.599	13.51	3.50
11	1.25	4.00	1.89	0.65	2.11	5.11	1.89	0.37	10.38	3.46
17	2.	4.285	5.39	0.415	2.12	7.92	2.67	0.30	6.75	3.61
10	4.	9.04	2.01	1.56	1.69	9.72	2.34	0.52	5.36	2.28
9	5.	14.48	2.04	0.25	5.26	8.98	2.24	1.18	2.94	1.82
	%Cat									
13	0	-	-	-	2.32	3.66	1.43	1.41	12.10	3.06
14	5	5.5	0.585	0.105	2.14	4.98	2.66	0.13	11.56	2.97
17	10	4.285	5.39	0.415	2.12	7.92	2.67	0.30	6.75	3.61
15	15	6.02	4.62	2.28	4.38	7.00	2.66	1.77	6.29	2.19
16	20	-	-	-	8.61	5.24	2.14	1.53	4.44	2.52

TABLE III. THE DISTRIBUTION OF HYDROGEN BASED ON ^{1}H-N.M.R. SPECTROSCOPY RESULTS (EXPERIMENT NO: 17)

	Chemical Shift ppm	Percent Hydrogen of Total						
		Oils			Asphaltenes			Preasphaltenes
		Hexane Eluate	Toluene Eluate	Methanol Eluate	Benzene Eluate	Diethylether Eluate	Methanol Eluate	CS_2 Solubles
H_{ar}	6.0-9.0	67.44	65.22	48.15	8.16[1]	-	-	-
H_{OH}	5.0-6.0	-	-	-	4.08	-	23.53	-
H_F	3.3-4.5	0.775	-	-	6.12	-	26.47	-
	3.0-3.3[2]	2.325	-	-	-	-	-	-
H_α	2.0-3.3	-	19.565	17.04	65.31	-	50.00	-
	2.0-3.0	18.605	-	-	-	19.355	-	27.78
H_n	1.6-2.0	0.775	3.26	16.30	12.45	48.390	-	44.44
H_β	1.0-1.6	7.75	11.955	16.30	4.08	29.03	-	25.00
H_γ	0.5-1.0	2.33	-	2.21	-	3.225	-	2.78

(1) $\delta < 8.1$ ppm Sterically hindered aromatic hydrogen

(2) Methylenes of acenapthenes and indenes

TABLE IV. RELATIVE INTENSITIES OF SUBFRACTIONS BASED ON Ar-C=C (EXPERIMENT NO: 17)

Wave Number (cm^{-1})		Oils			Asphaltenes			Preasp.
		Hexane Eluate	Toluene Eluate	Methanol Eluate	Benzene Eluate	D.E.Ether Eluate	Methanol Eluate	Carbene
-OH	3400	1.12	0.92	1.00	1.03	0.92	5.69	1.174
$-CH_3$, $-CH_2$	2900	2.06	1.44	1.06	1.64	1.24	5.54	4.83
C=0 (Carboxylic acid)	1700	-	-	-	1.46	-	2.69	-
Ar-C=C	1600	1.00	1.00	1.00	1.00	1.00	1.00	1.00
C-C-0 stretch	1470	-	-	-	-	-	3.31	-
$-CH_3$, $-CH_2$ (alkenes)	1450	1.47	1.40	0.94	1.54	0.92	-	4.35
$C-CH_3$ (in plane bending)	1380	-	-	-	-	-	-	3.30
>C=0	1360	-	-	-	1.23	-	-	-
$4(-CH_2)$ in trans conformation	1330	-	-	-	-	-	3.62	-
Saturated tertiary, alcohols highly C-0 symetrical secondary	1200-1125	1.18	-	-	-	-	4.00	1.26
C-O-C (ether)	1100	-	-	-	1.26	1.16	3.38	
α-unsaturated secondary, Alicylic secondary C-0 (5-or 6-membered ring) saturated primary	1060	-	-	-	-	-	5.85	-
Highly -unsaturated C-0 tertiary	920	-	-	-	-	-	4.69	-
1,3,5-(benzene ring substitution)	865-810 730-675	-	-	0.65	1.23	0.56	2.46	-
Ortho,di-(benzene ring Substitution)	760-750	-	1.60	-	-	-	-	-
Meta-(benzene ring substitution) out of plane vibrations	770-780 710-690	2.47	-	-	-	-	-	1.13

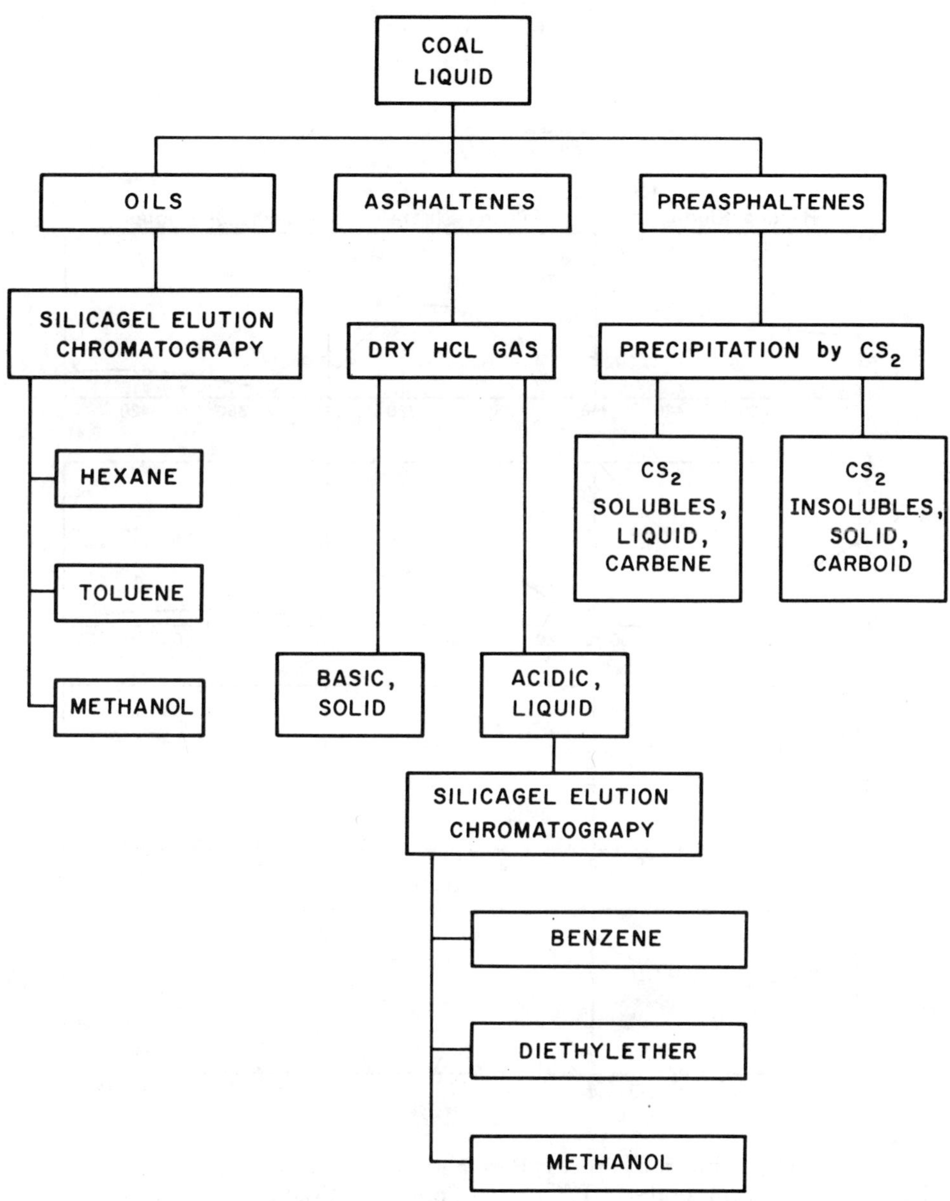

Fig.1. -Fractionation Scheme of Coal Derived Liquids Obtained from Mengen Lignite

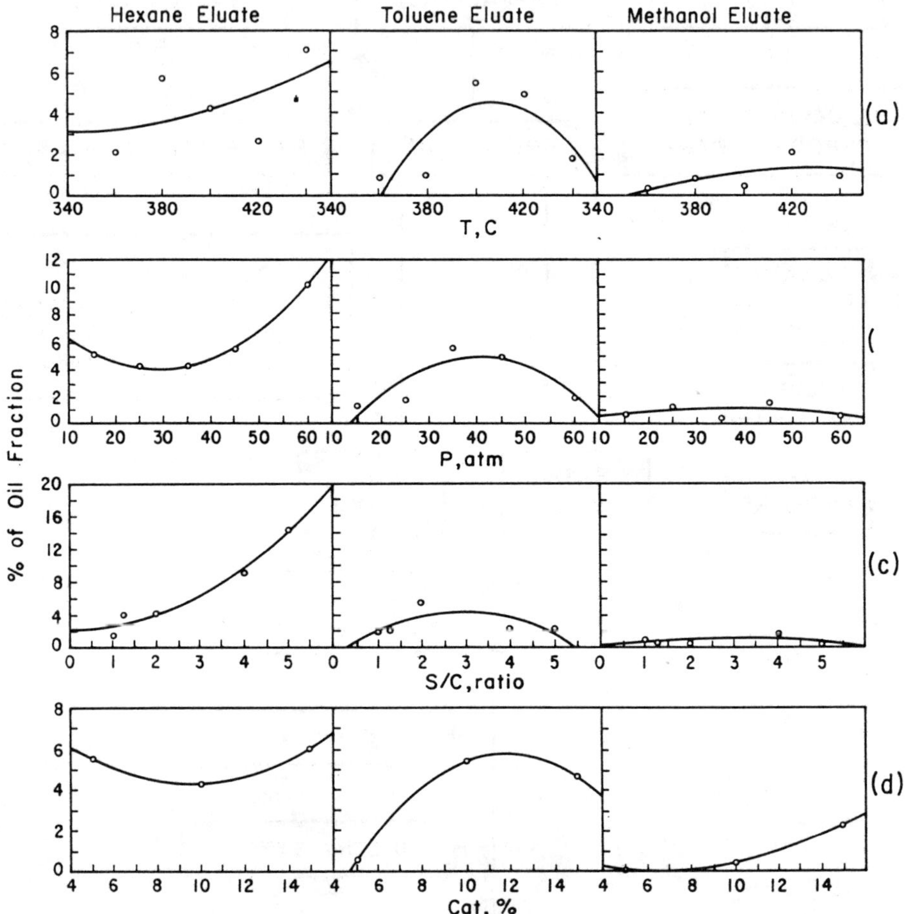

Fig. 2.-Effect of Process Parameters on Subfractions of Oils.
(a) Temperature, T(°C) (c) Solvent to Coal Ratio, S/C
(b) Pressure, P(atm) (d) Catalyst Loading, (% coal)

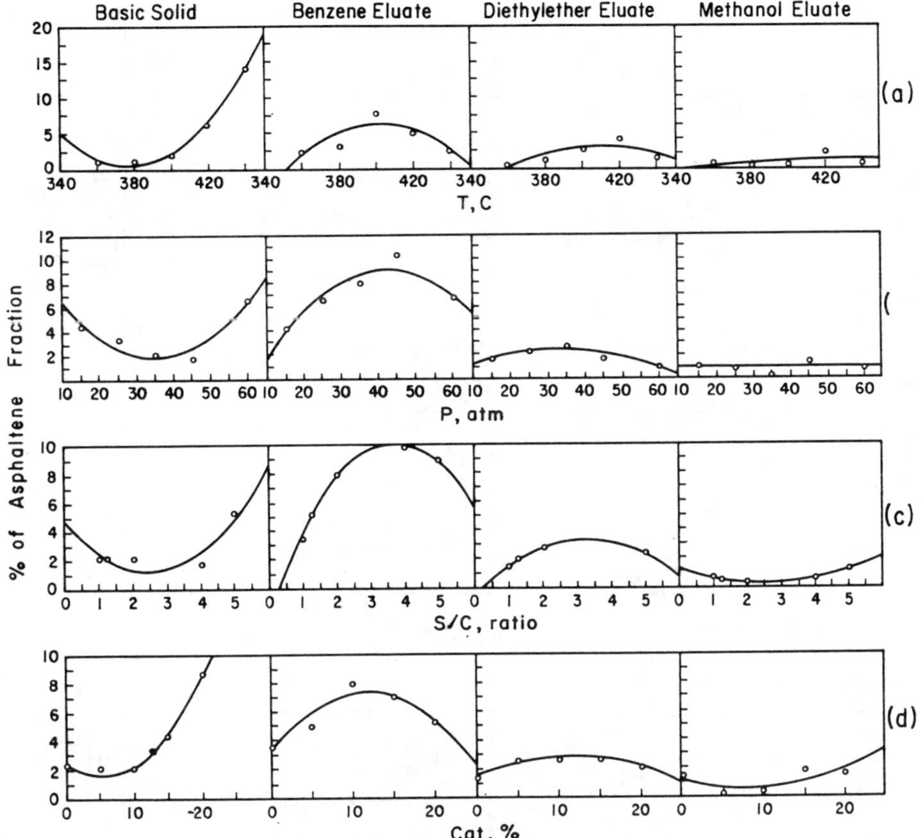

Fig. 3.-Effect of Process Parameters on Subfractions of Asphaltenes
(a) Temperature, T(°C)
(b) Pressure, P (atm)
(c) Solvent to Coal Ratio, S/C
(d) Catalyst Loading, (% coal)

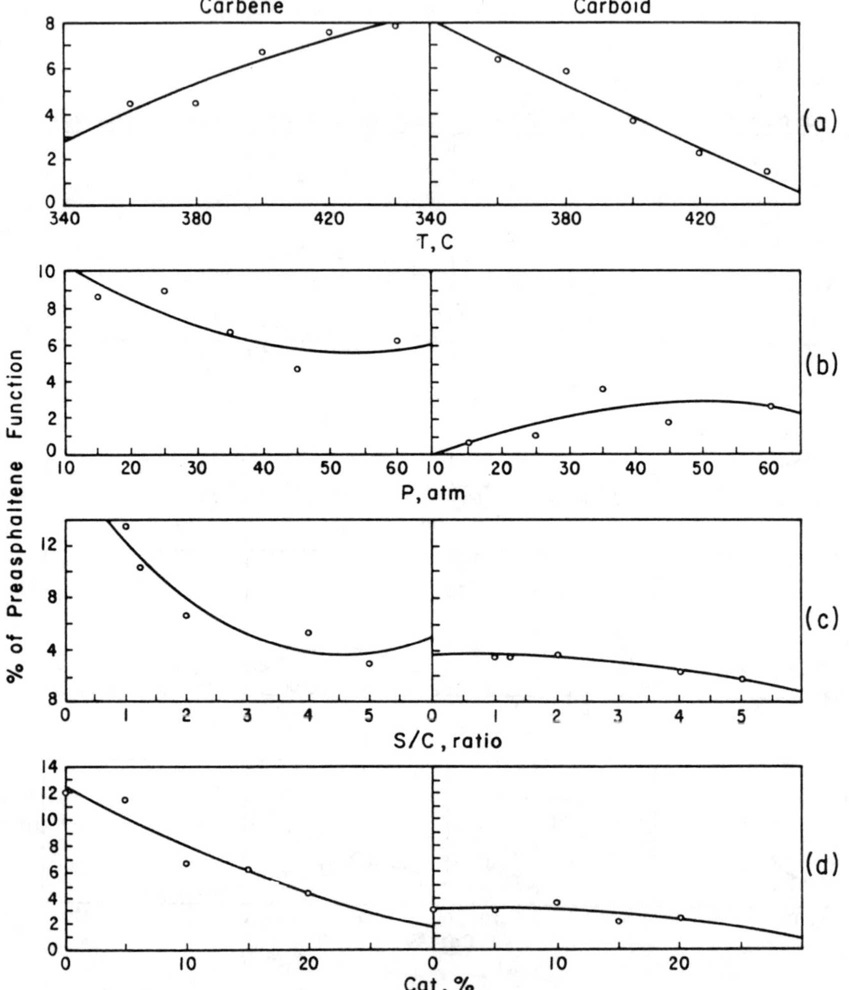

Fig. 4.- Effect of Process Parameters on Subfractions of Preasphaltenes
(a) Temperature, T(°C)
(b) Pressure, P(atm)
(c) Solvent to Coal Ratio, S/C
(d) Catalyst Loading, (% coal)

INFLUENCE OF HIGH ABSORBED IRRADIATION DOSES ON CONVERSION OF CO2-H2S MIXTURES

V.R. Rustamov, L.T. Bugayenko, V.K. Kerimov, Sh. N. Ali-zade, and Kh. Ya. Nasirova
Sector of Radiation Research, Academy of Sciences of Azerbaijan SSR,
Narimanov Avenue, 3I"A", Baku 370I43, USSR

Abstract

Advanced radiolytic methods of obtaining hydrogen are generally associated with the carbon dioxide decomposition. The effect of component ratios and reaction temperature on the radiolysis of CO_2-H_2S (5-50% vol) mixtures has been studied. Hydrogen sulphide processing is attractive both due to increasing resourses (up to I0-20% in natural gas of some large gasfields) and feasibility of simultaneous hydrogen and sulphur production. The total efficiency of converting irradiation energy into chemical one is 25 to 40% depending on the hydrogen sulphide concentration temperature in the mixture.

I. INTRODUCTION

In papers on the radiolysis of CO_2-H_2S mixtures [I, 2] it has been shown that sulphur destruction is performed both due to direct irradiation and the radiolysis products CO_2. The radiolysis mechynism included the following basic reactions:

$$CO_2 \rightsquigarrow CO + O \quad (I)$$

$$O + H_2S \xrightarrow{K_2} OH + HS \quad (2)$$

$$OH + H_2S \longrightarrow H_2O + HS \quad (3)$$

$$2\,HS \longrightarrow H_2S + I/8\,S_8 \quad (4)$$

$$H_2S \longrightarrow H_2 + I/8\,S_8 \quad (5)$$

High yield of end products in this system allows ($\Sigma G \geqslant I0$) to consider the radiolytic destruction of sulphur to be one of the methods of purification of the hydrogen-sulphide-containing residues of natural gas with simultaneous sulphur and synthesis gas ($CO + H_2$) production. The present paper covers investigations of the kinetics of end products accumulation, of their influence on routes of kinetic curves at high absorbed irradiation doses. This allows to see the degree of conversion of the initial mixture in a single step with no separation of conversion products from the initial components.

2. EXPERIMENTAL METHODS

For the purpose of investigation CO_2 (75 vol %)-H_2S(25 vol%) mixture with high yield of conversion products as shown in [I] has been employed.

The irradiation of mixture under investigation has been conducted under the static conditions in ampoules made of quartz glass in ^{60}Co γ-plant. The irradiation dose rate defined by methane and ethylene dosimeters amounted to 30kGy/hour. Pressure of the mixture under investigation is 0.I MPa, irradiation temperature is 3I5 K. Technique of ampoules' cleaning and preparation and of products analysis (CO, H_2, COS) is described in [I].

The process of CO_2-H_2S mixture radiolysis in the quartz-glass ampoules covered in [I]. In the first case H_2, CO, H_2O and sulphur were formed. In the second case, it was also observed the formation of COS which was not found at the radiolysis in quartz ampoules within the whole dose interval under investigation.

3. RESULTS AND DISCUSSION

Fig. I gives the dependence of CO, H_2 and sulphur (the last is taken as a sum of two first products according to the given chain of reactions) concentration on the absorbed dose. In gaseous phase dose rate variation resulted from sulphur precipitation is taken into consideration. As seen, the accumulation curves are practically linear at the section of 5 MGy and the average radiation-chemical yields at this section are G(CO)=3.3, G(H_2)= 2,I, G(S)=G(-H_2S)= G(CO) - G(H_2)=5.4. With higher doses the products yields decrease. It is also seen that CO yield decreases more than H yield.

In the above process hydrogen is formed only in direct irradiation (throgh the intermediate formation of hydrogen atom or directly). Therefore, as given in Fig. 2, quantity of hydrogen is proportional to the dose absorbed in hydrogen sulphide and defined with regard to decrease of hydrogen sulphide concentration. However, the observed yield of direct effect is G(H_2)= II.0 $\pm$ 0.8. This value is significantly higher than the content of G(H_2) in pure H_2S which is 7.5 $\pm$ 0.5 at dose rates about those we used and agrees with values obtained at high dose rates when G(H_2)= II-I3 [3]. Presence of the second component probably makes it easier to realize the "high-dose-rate" reactions of H_2 formation at low dose rate, such as reactions with S atoms [4] But yet, the proportionality of dose absorbed in one component is insufficient for unambigous statement that this process is conducted only with direct effect. Thus, within the experiment accuracy the linear dependence given in Fig.2 can be obtained when it is assumed that the process of H_2 formation is realized by two ways: due to the direct effect with yield of 7.5 and due to the energy transfer from excited CO_2 molecules:

$$CO_2^* + H_2S \longrightarrow CO_2 + H_2S^* \qquad (6)$$

$$H_2S^* \longrightarrow H_2 + I/8\ S_8 \qquad (7)$$

Taking this assumption, $G(CO_2^*)$ can be calculated using the data in Fig.2 by the following expression:

$$G(CO_2^*) = \frac{N(H_2)\ \frac{11.0 - 7.5}{7.5}}{D_{CO_2}} \cdot 10^2 ,$$

where D_{CO_2} is a dose absorbed by CO_2.

So, we obtain $G(CO_2^*) = 0.86 \pm 0.11$ for this mixture, which conforms to $G(CO_2^*) = 1.10 \pm 0.12$ for pure CO_2. Naturally, that the target excited state of CO_2 differs from the state obtained by CO and O atoms at the radiolysis of CO_2. There is no information on the excited state in the literature.

The CO accumulation curve with 5% accuracy is described by the second order parabola:

$$[CO] = (-0.00082 + 0.0359D - 0.00136D^2) \cdot 10^{19} \text{ molec/cm}^3.$$

Initial CO formation for this mixture is 3.59 which is 4.55 when taken for pure CO_2 being in good agreement with data cited in [3] . This value shows that in this mixture CO formation is the effect of direct irradiation of CO_2.

It is not easy to explain the quick decrease of CO yield within the dose range higher than 5 MGy. In this system oxygen, O atoms and OH radicals may act as oxidizers. Oxygen is not accumulated in the system. Rate constants for OH radicals and CO and H_2S are $1.5 \cdot 10^{-13}$ and $4.8 \cdot 10^{-12}$ $cm^3 molec^{-1} c^{-1}$ [5] , i.e. hydrogen sulphide catches OH 30-times quicker than carbon oxide does. So this reaction may be ignored at the observed ratio of concentrations of competitive molecules. Rate constants for CO and H_2S oxidation by oxygen ions and O atoms at 0.1 MPa are $3.5 \cdot 10^{-16}$ [6] and $2.2 \cdot 10^{-14}$ [5] $cm^3 molec^{-1} s^{-1}$. And in this case CO yield can not decrease so sufficiently as it is observed. The conducted kinetic analysis shows that in the system under investigation the dependence of CO yield on dose can not be explained by competition for a natural oxidizing particle save for negative oxygen ions (O_2^-, O^-) resulted from electron capture.

Due to the fact that CO are accumulated in large amounts in the system it is necessary to take into account its distruction by direct irradiation. Keeping in mind the most effective competition for atomic hydrogen using reactions (2) and (8) [6].

$$CO + CO_4^- \xrightarrow{k_8} CO_3^- + CO_2 \qquad K_8 \simeq 3 \cdot 10^{-15} cm^3/mol \cdot s \quad (8)$$

$$CO_3^- + CO_2 \xrightarrow{k_9} 2CO_2 + \bar{e} \quad (9)$$

we obtained the expression for direct irradiation effect on CO by method of stationary concentrations:

$$G_{-CO} = \left[G^{o}(CO) - G(CO) - \frac{G^{o}(CO)}{1 + \frac{k_2 \cdot [H_2S]}{k_8 \cdot [CO] \cdot M}} \right] \cdot \frac{\dot{D}_{CO_2}}{\dot{D}_{CO}}$$

where $\dot{D}_{CO}$ is rate of dose absorbed in CO, $\dot{D}_{CO_2}$ is rate of dose absorbed in CO_2. The obtained data give the value of CO destruction yield with direct irradiation effect $G_{-CO} = 25 \pm 5$. This value is significantly higher than yield of direct irradiation effect on pure CO -8.5 [3] . Either mechynism of radiolysis of $CO_2 - H_2S - CO - H_2 - H_2O$ system includes not mentioned reactions of the process of CO conversion in given systems with direct irradiation effect differs from the appropriate process in pure CO. To understand this we must conduct further investigations.

Conclusion

The degree of $CO_2 + H_2S$ initial system processing with sulphur and synthesis-gas output at the dose of 5 MGy is 16% of initial compositions in useful products. So, the adequate efficient conversion of ionizing irradiation energy takes place with simultaneous solution of problems of ecology for hydrogen-sulphide-containing nytural gases.

REFERENCES

1. Rustamov V.R. et al., "Khimiya vysokikh energii", 1987, v.21, No 5, p. 434.
2. Rustamov V.R. et al., Hydrogen energy progress VII, Moscow, USSR, 25-29 September, Pergamon Press, 1988, No 2,pp.943-954.
3. Willis C., Boyd A.W., Int.J.Radiat.Phys.Chem., 1976, v.8, No 1, p. 97.
4. Rustymov V.R. et al., "Khimiya vysokikh energii", 1989, v. 23, No 4, p.300.
5. Bauich D.L. et al., J.Phys.Chem.Ref.Data, 1984, v.13, No 4, p, 1259.
6. Norfolk D.J., Skinner R.F., Williams W.J.:- Hydrocarbon chemistry in irradiated CO_2-CO_2-CH_4-H_2O-H_2 mixtures.- Radiat.Phys. and Chem., 1983, v. 21, No 3, pp. 307-319.

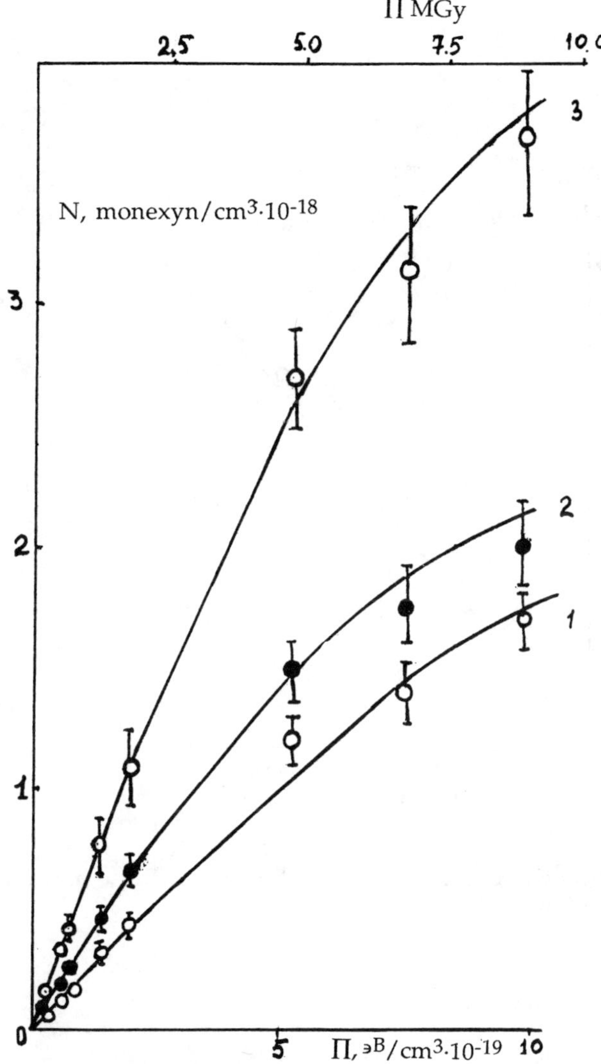

Fig. 1. The dependence of hydrogen (1), carbon oxide (2), sulphur (3) concentrations on dose.

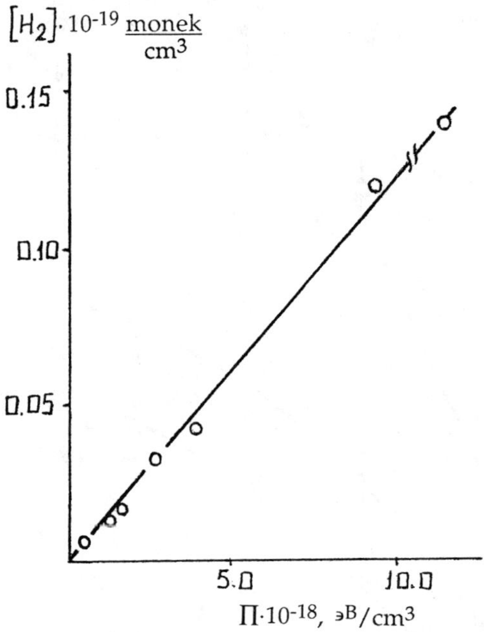

Fig. 2 Hydrogen concentration as a function of dose absorbed in hydrogen sulphide

THERMAL-RADIATION DECOMPOSITION OF CARBON DIOXIDE HYDROGEN SULPHIDE MIXTURES WITH SMALL DOSES OF OXYGEN ADDITIVES

V.R. Rustamov, L.T. Bugayenko, V.K. Kerimov, Kh.B. Gezalov, Sh. N. Ali-zade, and Kh. Ya. Nasirova
Sector of Radiation Research, Academy of Sciences of Azerbaijan SSR,
Narimanov Avenue, 3I"A", Baku 370I43, USSR

Abstract

Thermal-radiation decomposition of CO_2-H_2S mixtures with small doses of oxygen additives has been studied. It has been shown that the process is conducted in the presence of straight unbranched chains. Carbonyl sulphide and sulphur ($G \geq IO^3$) are identified as basic products. The efficient activation energy was 75 $\pm$ IO kJ/mol.

I. INTRODUCTION

Sulphur production from the waste gases is an important task both from economical and ecological point of view. Nowdays for this purpose it is widely used Claus process, which can be chemically interpreted as follows [I] :

$$2H_2S + SO_2 \rightsquigarrow \frac{3}{n} Sn + 2H_2O + IOO \text{ kJ/mol}$$

A significant disadvantage of the process is production of one product (sulphur) whereas the loose-bonded hydrogen in H_2S is transformed into water, i.e. it is ignored as energy resource.

Recently, plasmochemical [2] and radiation-stimulated processes [3] of decomposition of hydrogen sulphide and mixtures based on it have been suggested. They help to obtain hydrogen (carbon monoxide) and sulphur simultaneously thereby increasing the economic indices of the process. In radiation-chemical processing of CO_2-H_2S mixtures ("sour" components of natural gas) the radiolysis products yield amounts to IO ÷ I2 molec/IOO eV and the process itself can be conducted up to deep (20%) conversion at one step [4] .

2. EXPERIMENTAL METHODS

The present paper covers the thermal-radiation decomposition of CO_2-H_2S mixtures with small doses of oxygen additives (3-5 vol %) as we have no data on effect of high temperatures on the radiolysis of CO_2-H_2S mixtures and there is a possible chain formation of products in the presence of small doses of oxygen additives. For this purpose we studied the effect of temperatures from 50 to 450°C on radiolysis products yields. The mixtures were exposed to Co^{60} γ-ray irradiation under static conditions in ampoules made of quartz glass. The absorbed dose rate was evaluated by hydrogen formation from methane: D=IOFp/s. In calculation of product yields we took into account the electron densities ratio. of CH_4 and system under investigation. Total pressure of gas mixture in all experiments was O.I MPa. After production (H_2S) and purification the initial gases of various ratio were delivered to ampoules (V = 20 cm^3) through the calibration unit. Temperature was controlled with accuracy of ± IO°C. Thermal-radiolysis products were analysed on the following chromatographs: hydrogen and carbon monoxide (Gasokhrom 3IOI), carbonyl Sulphide (IXM -8MD).

Kinetic diagram of the process of CO_2-H_2S mixtures thermal-radiolysis without oxygen is written as follows:

$$CO_2 \rightsquigarrow CO, O \quad (I)$$

$$H_2S \rightsquigarrow H, SH \quad (2)$$

$$SH + CO_2 \xrightarrow{T} COS + OH - I.5 \text{ eV} \quad (3)$$

$$OH + H_2S \longrightarrow H_2O + SH \quad (4)$$

From reactions (I-4) it follows that the limiting step of such chain process would be reaction (3) which is endothermal by I.5 eV. To endure this a small dose of oxygen (3-5 vol %) was added to CO_2-H_2S mixture. In this case formation of products is too possible as the secondary reactions of product formation were either quick exothermal reactions or endothermal reactions by 0.4 eV (7).

$$SH + O_2 \dashrightarrow SO + OH \qquad +I\ eV \qquad (5)$$

$$SO + CO_2 \dashrightarrow SO_2 + CO \qquad +0.I8\ eV \qquad (6)$$

$$CO + SH \dashrightarrow COS + H \qquad -0.42\ eV \qquad (7)$$

Thus, the total process of thermal-radiation decomposition of CO_2-H_2S mixtures in the presence of O_2 may be:

$$CO_2 + 3H_2S + O_2 \xrightarrow{T} COS + \frac{2}{n} Sn + 3H_2O \qquad (8)$$

Besides, generation of active particles (H, O, SH) and formation of hydrogen, carbon monoxide and sulphur arise from direct irradiation.

3. RESULTS AND DISCUSSIONS

Fig.I shows, for example, the kinetic curves of accumulation of products (H_2,CO,COS) at thermal radiolysis of (430°C) CO_2 (62%) - H_2S (35%) - O_2 (3%) mixture. As seen, concentration of hydrogen and carbon monoxide are proportional to the absorbed irradiation dose. Curve of carbonyl sulphide accumulation tends to saturation at absorbed doses $D \sim 40$ kGy. The initial radiation-chemical yields of gaseous products were G_H= 6.4, G_{CO}= 24.0, G_{COS}= 400. Similar results were also obtained for CO_2-H_2S-O_2 (75:20:5) mixtures at the same temperature of irradiation but in this case the carbonyl sulphide yield is unsignificantly lower G_{COS}= 296. At more lower temperatures of irradiation of CO_2-H_2S-O_2 (62:35:3) mixtures yields of all products are lower. The most significant decrease is for carbonyl sulphide. Data on products yields are given in Table I.

Using the data on products yields of CO_2-H_2S radiolysis mixture to 200°C [3] and results of this work, we can obtain the temperature dependence of radiolysis gaseous products yield given in Fig.2. As seen, change from the low-temperature mode (T 300°C) is abrupt. The intermediate product inducing sulphur hydrogen oxidation in sulphur monoxide (SO) which is probably oxidized up to sulphur dioxide (SO_2) with formation of carbon monoxide. The above facts allow to suppose that the process proceeds in the presence of straight unbranched chains. Sulphur in this process is obtained through SO_2 and hydrogen sulphide reaction and its yield is probably equal to the yield of carbonyl sulphide (G_S= 400).

$$SO_2 + 2H_2S \longrightarrow \frac{3}{n} Sn + 2H_2O$$

As supposed, carbonyl sulphide is obtained through reaction (7) which is endothermal by 0.42 eV (42 kJ/mol). The obtained value of the effective activation energy of carbonyl sulphide formation is higher.

In the temperature range of 300-430°C the Arrhenius dependence is satisfied when activation energy is E_{ef}= 75± IO kJ/mol (Fig.3) which differs from the value 60 kJ/mol obtained elsewhere [5]. It should be noted that value of effective energy of activation was obtained in paper [5] from hydrogen formation. The kinetic curve of carbonyl sulphide accumulation (Fig.I) shows that along with this product formation one must take into account the following reactions of its consumption:

$$COS + O_2 \xrightarrow{T} CO + SO_2$$

$$2COS \xrightarrow{T} CO_2 + CS_2$$

$$COS + H_2O \xrightarrow{T} CO_2 + H_2S$$

Thus, thermal radiolysis of CO_2-H_2S-O_2 mixtures leads to formation of carbonyl sulphide and sulphur with total yield of $G \simeq IO^3$. Usage of carbonyle sulphide as half-product for formation of hydrogen was studied in paper [6]. It has been found that 350-400°C carbonyl sulphide is oxidized by oxygen to formytion of sulphur gas (SO_2) and carbon monoxide. The process conversion is 92%. At the subsequent stage carbon monoxide is converted into hydrogen.

Conclusion

In conclusion, it should be noted that the oxidation processes of radiolysis of "sour" components of natural gas (CO_2,

H_2S) allow to obtain simultaneously two products (COS,Sn), as distinct from Claus processes (Sn), and high-radiation-chemical yields demonstrate the possible application of thermal-radiation process for the energy-bearing agents and silphur production.

REFERENCES

I. Matros Yu.Sh., Zagoruiko A.N., Reports of the USSR Academy of Sciences, I987, v. 294, No 6, p. I424.
2. Nester S.A., Rusanov V.D., Fridman A.A., Khimiya vysokikh energii, I988, v. 22, No 5, p.46I.
3. Rustamov V.R., Kerimov V.K. et al., Khimiya vysokikh energii, I987, v. 2I, No 5, p. 434.
4. Rustamov V.R., Bugayenko L.T. et al., Khimiya vysokikh energii, I989, v. 23, No 4, p.300.
5. Kurbanov M.A., Rustamov V.R., Mamedov Kh.F., Khimiya vysokikh energii, I988, v. 22, No 3, p. 2I8.
6. United States Patent No 4432960, Thermochemical Method of H_2 production from H_2S, published 84.02.2I, v.I039, No 3.

TABLE I: RADIATION-CHEMICAL YIELDS OF THERMAL-RADIOLYSIS PRODUCTS FOR CO_2-H_2S-O_2 (62:35:3) MIXTURE

T, °C	G, molec/IOO eV				ΣG
	H_2	CO	COS	Sn	
300	4.0	2.4	I6.0	38.4	60.8
370	I4.8	I6.4	I20.0	252.2	383.4
430	I6.4	24.0	400.0	830.4	~I200.0

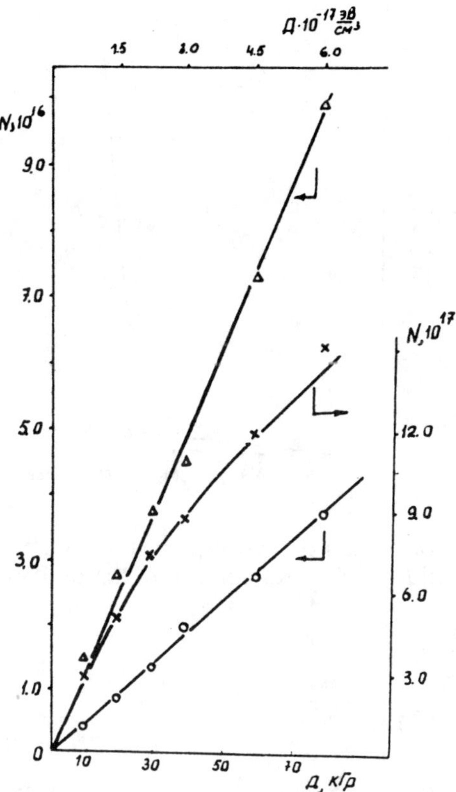

Fig. 1. Curves of accumulation of hydrogen (o), carbon monoxide (Δ), carbonyl sulphide (x) plotted against the absorbed irradiation dose T=430°C, D = LOGg/s

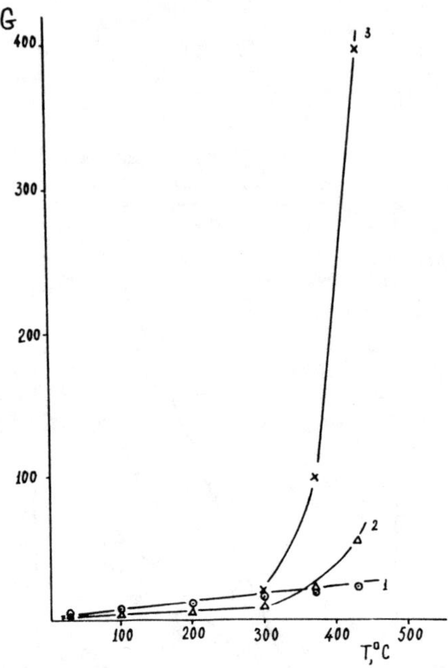

Fig. 2. Radiation yields of hydrogen (1), carbon monoxide (2), and carbonyl sulphide (3) as a function of radiation temperature.

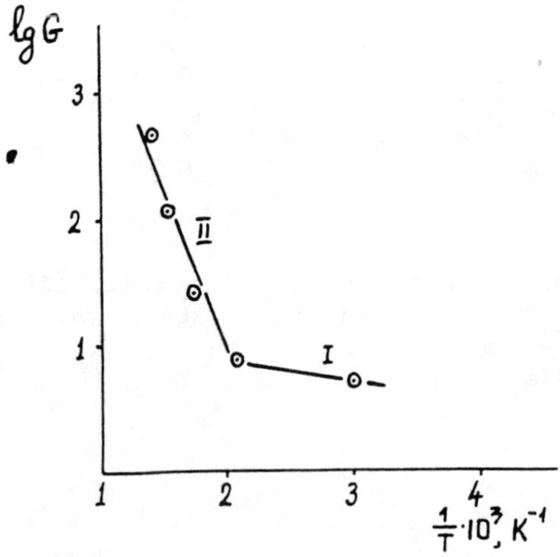

Fig. 3. Products radiation yield logarithm ($lg\sum G_{pr.}$) as a function of inversion temperature.

SUBJECT INDEX